CAMBRIDGE STUDIES IN
ADVANCED MATHEMATICS 27



# ALGEBRAIC NUMBER THEORY

*Already published*

1 W.M.L. Holcombe *Algebraic automata theory*

2 K. Petersen *Ergodic theory*

3 P.T. Johnstone *Stone spaces*

4 W.H. Schikhof *Ultrametric calculus*

5 J.-P. Kahane *Some random series of functions, 2nd edition*

6 H. Cohn *Introduction to the construction of class fields*

7 J. Lambek & P.J. Scott *Introduction to higher-order categorical logic*

8 H. Matsumura *Commutative ring theory*

9 C.B. Thomas *Characteristic classes and the cohomology of finite groups*

10 M. Aschbacher *Finite group theory*

11 J.L. Alperin *Local representation theory*

12 P. Koosis *The logarithmic integral I*

13 A. Pietsch *Eigenvalues and s-numbers*

14 S.J. Patterson *An introduction to the theory of the Riemann zeta-function*

15 H.J. Baues *Algebraic homotopy*

16 V.S. Varadarajan *Introduction to harmonic analysis on semisimple Lie groups*

17 W. Dicks & M. Dunwoody *Groups acting on graphs*

18 L.J. Corwin & F.P. Greenleaf *Representations of nilpotent Lie groups and their applications*

19 R. Fritsch & R. Piccinini *Cellular structures in topology*

20 H Klingen *Introductory lectures on Siegel modular forms*

21 P. Koosis *The logarithmic integral II*

22 M.J. Collins *Representations and characters of finite groups*

24 H. Kunita *Stochastic flows and stochastic differential equations*

25 P. Wojtaszczyk *Banach spaces for analysts*

26 J.E. Gilbert & M.A.M. Murray *Clifford algebras and Dirac operators in harmonic analysis*

28 K. Goebel & W.A. Kirk *Topics in metric fixed point theory*

29 J.F. Humphreys *Reflection groups and Coxeter groups*

30 D.J. Benson *Representations and cohomology I*

31 D.J. Benson *Representations and cohomology II*

33 C. Soulé, D. Abramovich, J.-F. Burnol & J. Kramer *Lectures on Arakelov Geometry*

# ALGEBRAIC NUMBER THEORY

A. Fröhlich, FRS
*Imperial College, London and*
*Robinson College, Cambridge*

M. J. Taylor
*UMIST, Manchester*

Published by the Press Syndicate of the University of Cambridge
The Pitt Building, Trumpington Street, Cambridge CB2 1RP
40 West 20th Street, New York 10011-4211, USA
10 Stamford Road, Oakleigh, Melbourne 3166, Australia


First published 1991
First paperback edition 1993, 1994

Printed in Great Britain at the
Athenaeum Press Ltd, Newcastle upon Tyne

*Library of Congress cataloguing in publication data available*
*British Library cataloguing in publication data available*

ISBN 0 521 36664 X hardback
ISBN 0 521 43834 9 paperback

To our long suffering wives,
Ruth and Sharon

# Contents

# Preface

There are many attractive and instructive topics which can, and should, be included in an introductory, but moderately ambitious, text book on algebraic number theory. But – as if by a conspiracy of silence – they are usually either omitted altogether, or, at best, are treated inadequately in the existing array of texts available. One of our aims in writing this book has been to try to break free from this standard mould, and to fill these gaps. As instances we mention cubic and biquadratic fields, Gaussian periods, Brauer relations, module theory over a Dedekind domain, an algebraic number theoretic treatment of binary quadratic forms, tame ramification and the two-classgroup of a quadratic field.

Conceptually the book breaks fairly neatly into two parts: the first four chapters and the final chapter are, for the most part, of a theoretical nature, though we always take care to fix abstract ideas by means of worked examples; the remaining three chapters are devoted to giving a detailed study of various arithmetic objects in situations of particular interest.

Throughout the text we have laid great stress on worked examples; it is a depressing fact that many number theorists have never acquired sufficient technique to perform number theoretic calculations in anything but a quadratic field. Again, this is, to some extent, the fault of the existing literature, where scant emphasis is placed on calculations.

On the whole we have opted for schematic exposition, rather than attempting an evolutionary or historical development of the subject matter. Thus, for instance, Diophantine equations now become an application of the theory; whereas, of course, historically they were the principal motivation for the development of the theory.

This book has its origins in various lecture courses given by the authors at Cambridge Part III and M.Sc. level; however, in its later sections, we go beyond these courses. What we require is roughly the knowledge of a third year undergraduate. More precisely, somewhere along the line the reader will be supposed to have some familiarity with elementary point set topology, with elementary Galois theory, and with basic module theory, including tensor products and the structure theorem for finitely generated modules over a principal ideal domain.

In its undiluted form, the book is best suited to a two semester course at Masters level. We therefore now describe how it may be used for a more minimal, one semester course. Needless to say, in general the prospective lecturer should concentrate on the main theorems and omit much of the other surrounding material.

In all cases we would suggest that Chapter 1 be used as background material; it is only intended to serve as a ready source of reference for a number of elementary algebraic facts which may be new to the reader. By contrast, (II, §1) up to and including Theorem 5 and then (II.1.31, 32) are fundamental; they contain the genesis of a whole host of ideas which are basic to the subject. The notions of valuation and absolute value in (II, §2) are used repeatedly throughout; and from (II, §3) one needs the basic definitions together with Theorems 10 and 11; however, the final section on module theory for Dedekind domains can, for the most part, be omitted without serious disadvantage.

In Chapter III we consider the behaviour of many of the concepts introduced in the previous chapter when they are extended from a given number field to an extension field. Again much of the material can be omitted for the purposes of a short course. In §1, Theorem 18 can certainly be omitted. In §2 the minimalist may omit the general definition of a discriminant, and get by with the absolute discriminant of (II, §1); on the other hand, Theorem 22 (Dedekind's theorem on the ramification of divisors of the discriminant) and Theorem 23 (Kummer's criterion for the decomposition of prime ideals in an extension) are quite important. All of §4 can be omitted, and the only results of §3 which are essential are those that concern cyclotomic extensions.

Chapter IV concerns the application of convex body theory to the study of classgroups and units, i.e. the topic usually referred to as Minkowski methods. This is the most standard chapter in the book; it is also the briefest and some form of it will be required.

Having dealt with the basic theory, the reader is then in the agreeable position of being faced with a kind of Smørgasbord of choices of applications and further developments. We therefore list some of the options available. Chapter V begins with the important and traditional theory of quadratic fields. This would seem to us to be an essential ingredient of any course in algebraic number theory. The chapter then concludes with sections on biquadratic and cubic number fields. Chapter VI is devoted to the theory of cyclotomic fields. In our opinion, this is the most elegant chapter of the book. In Chapter VII we consider various kinds of Diophantine equations: Fermat's Last Theorem, quadratic forms and finally various cubic results. The main text then concludes with sections on Dedekind zeta-functions and $L$-functions. Since these are the most powerful methods in the whole book, it would be regrettable if too much of this chapter were omitted.

As additional aids to the reader we also include an Appendix on the character theory of Abelian groups, and a wide range of exercises.

The authors wish to express their thanks to Robin Chapman for his detailed reading of the initial manuscript of the book: his comments and suggestions have been of great help. Thanks also go to David Burns for his mathematical reading of the proof script, and to Jean Cougnard for giving us access to his tables of biquadratic number fields. In addition we wish to thank Bryan Birch for permission to use many of the exercises from his Oxford question sheets.

Last – but by no means least – it is a pleasure to acknowledge the help and co-operation that David Tranah has provided, on behalf of Cambridge University Press, in the production of this text.

# Notation

$\mathbb{N}$, $\mathbb{Z}$, $\mathbb{Q}$, $\mathbb{R}$, $\mathbb{C}$ denote the natural numbers, the integers, the rational numbers, the real numbers, the complex numbers respectively.

For two sets $A$ and $B$, $A \subset B$ means $A$ is contained in $B$, and $A \subsetneq B$ denotes strict containment. The number of elements in a finite set $A$ will be denoted by $|A|$ or by $\mathrm{card}(A)$.

The term ring is always to be understood as meaning a commutative ring with a multiplicative identity; $S$ will be called a subring of a ring $R$ if, in addition to the usual requirements, their multiplicative identities coincide. The group of units of a ring $R$, i.e. the group of multiplicatively invertible elements in $R$, will be denoted by $R^*$. Ideals of $R$ will also be referred to as $R$-ideals. Our use of the term prime is slightly non-standard and therefore deserves special mention: unless stated to the contrary, we always use this term to mean non-zero prime ideal.

The sequence $A \xrightarrow{f} B \xrightarrow{g} C$ is aid to be exact if $\mathrm{im}(f) = \ker(g)$, where $A$, $B$, $C$ are modules over a given ring $R$, and $f$, $g$ are $R$-homomorphisms. Thus, in particular $0 \to A \xrightarrow{f} B$ is exact at $A$ if $f$ is injective; $B \xrightarrow{g} C \to 0$ is exact at $C$ if $g$ is surjective. With the same notation the triangle

$$\begin{array}{ccc} & B & \\ {\scriptstyle f}\nearrow & & \searrow{\scriptstyle g} \\ A & \xrightarrow{h} & C \end{array}$$

is said to *commute* if $g \circ f = h$. Similar terminology applies to commutative squares etc.

Given two groups $H \subset G$, of finite index, we write $[G\colon H]$ for that index. The degree of a finite extension of fields $L/K$ is denoted by $(L\colon K)$; if the extension is Galois, we write $\mathrm{Gal}(L/K)$ for the Galois group of $L/K$. Finally, we remark that in general we use the abbreviation iff for "if, and only if".

# Introduction

The purpose of this section is to give an overview of the aims of algebraic number theory and to provide motivation for the study of the subject. We shall be concerned with generalisations of the integral domain $\mathbb{Z}$ of ordinary integers which are called rings of algebraic integers: an algebraic integer is a root of a monic polynomial in $\mathbb{Z}[X]$. Many of the definitions and results of ordinary number theory have natural extensions in algebraic number theory, and, in fact, are often better understood in this wider context. Frequently the study of a suitable ring of algebraic integers will help in the solution of a problem which initially had been stated entirely in terms of ordinary integers: for instance, questions concerning the integral (or rational) solutions of an equation with integral (or rational) coefficients can frequently be dealt with by the study of a suitable ring of algebraic integers. We shall consider a number of instances of this phenomenon.

For ease of exposition we shall introduce a number of concepts in a rather informal manner; they will, of course, all receive a full and formal definition later on. In the same way all results quoted in this introduction will subsequently be proved in the text.

We begin by considering the classical problem of when the prime number $p$ can be represented as the sum of the squares of two integers. The following was first proved by Lagrange:

*An odd prime number $p$ can be written as $p = x^2 + y^2$ with $x, y \in \mathbb{Z}$ iff $p \equiv 1 \bmod (4)$.*

*Proof.* The necessity of the condition follows from the fact that $x^2+y^2 \not\equiv -1 \bmod (4)$ for all $x, y \in \mathbb{Z}$.

Now assume that $p$ is a prime number with $p \equiv 1 \bmod (4)$. The multiplicative group $\mathbb{F}_p^*$ of the non-zero elements of the field of $p$ elements is cyclic of order $p-1$; it therefore possesses an element of order 4. Since the class of $-1$ is the unique element of order 2 in $\mathbb{F}_p^*$, we deduce that $m^2 \equiv -1 \bmod (p)$ for some integer $m$, that is to say

$$p \mid m^2 + 1. \tag{1}$$

It is at this stage that we go over to the ring $\mathbb{Z}[i]$ of algebraic integers of the form $x + iy$, for $x, y \in \mathbb{Z}$, $i^2 = -1$. These algebraic integers were much used by Gauss, and are therefore called the *Gaussian integers.* We write

$$N: \mathbb{Z}[i] \to \mathbb{Z} \tag{2}$$

for the *norm map*

$$\begin{cases} N(z) = z\bar{z} \\ N(x+iy) = x^2 + y^2. \end{cases} \tag{2a}$$

$N$ provides a Euclidean norm on $\mathbb{Z}[i]$, so that $\mathbb{Z}[i]$ is a principal ideal domain. We note that $z$ is a unit of $\mathbb{Z}[i]$ iff $N(z) = 1$, that is to say for $z = \pm i, \pm 1$. By (1) we know that $p$ divides the product of Gaussian integers $(m+i)(m-i)$; if $p$ divided either of the factors, then by applying complex conjugation it would divide the other and so $p \mid 2i$ which is absurd. Since $\mathbb{Z}[i]$ is a principal ideal domain, the usual arguments of elementary arithmetic apply; we therefore conclude that $p$ is not a prime element in $\mathbb{Z}[i]$ and so we have a factorisation

$$p = (x+iy)(x'+iy')$$

as a product of two Gaussian integers, neither of which is a unit. Taking norms yields the equality

$$p^2 = (x^2+y^2)(x'^2+y'^2)$$
$$x^2 + y^2 \neq 1 \neq x'^2 + y'^2; \quad x, y, x', y' \in \mathbb{Z}$$

Therefore $p = x^2 + y^2$ as we required.

□

This example illustrates the general pattern which we have to follow when attacking Diophantine problems of this type. First, and foremost, we see that we need to work in subrings of some *algebraic number field $K$* (i.e. a finite extension field $K$ of $\mathbb{Q}$). The most important such subring is $\mathfrak{o}_K$, the set of all algebraic integers in $K$: later we shall show that $\mathfrak{o}_K$ is indeed a ring. In the example considered above, the relevant number field is $\mathbb{Q}(i)$, whose ring of algebraic integers is $\mathbb{Z}[i]$.

Next we consider two further similar problems. First we check that if an odd prime number $p$ is of the form

(3) $$p = x^2 - 2y^2$$

with $x, y \subset \mathbb{Z}$; then indeed

(4) $$p \equiv \pm 1 \bmod (8).$$

In fact the converse also obtains. To show this we need to work in $\mathbb{Z}[\sqrt{2}]$, the ring of algebraic integers of the number field $\mathbb{Q}(\sqrt{2})$. We define a norm

(5) $$\begin{cases} N_2 \colon \mathbb{Z}[\sqrt{2}] \to \mathbb{Z} \\ N_2(x + y\sqrt{2}) = x^2 - 2y^2. \end{cases}$$

The proof works as before because $\mathbb{Z}[\sqrt{2}]$ is again a principal ideal domain. There is, however, one important difference from the previous example. One readily verifies that there are at most a finite number of integral solutions $(x, y)$ of the equation $p = x^2 + y^2$: this reflects the fact that the group $U_K$ of units of $\mathfrak{o}_K$ is finite when $K = \mathbb{Q}(i)$, as we have already noted. For $K = \mathbb{Q}(\sqrt{2})$ this is no longer the case. An algebraic integer $u + v\sqrt{2}$, $(u, v \in \mathbb{Z})$ is a unit of $\mathbb{Z}[\sqrt{2}]$ iff

$$N_2(u + v\sqrt{2}) = u^2 - 2v^2 = \pm 1.$$

If, in particular, $N_2(x + y\sqrt{2}) = p$ with $x, y \in \mathbb{Z}$, and if $N_2(u + v\sqrt{2}) = +1$; then clearly

$$p = N_2((x + y\sqrt{2})(u + v\sqrt{2})) = N_2([xu + 2yv] + [xv + yu]\sqrt{2}).$$

The structure of $U_{\mathbb{Q}(\sqrt{2})}$ is therefore quite crucial, and, as a special case of Dirichlet's Unit Theorem, we know that $U_{\mathbb{Q}(\sqrt{2})}$ is the direct product of $<\pm 1>$ and the infinite cyclic group generated by $(1+\sqrt{2})$: in particular note that $N_2(\pm(1 + \sqrt{2})^{2n}) = 1$ for all $n$. Observing that $(1 + \sqrt{2})^{-1} = -(1 - \sqrt{2})$, we obtain a recursive description of $U_{\mathbb{Q}(\sqrt{2})}$, and hence, given one solution $p = x^2 - 2y^2$, we can describe them all. More precisely, for $n \geq 0$ we define

$$x_0 = 1 \quad x_{n+1} = x_n + 2y_n$$

$$y_0 = 1 \quad y_{n+1} = x_n + y_n.$$

Then $U_{\mathbb{Q}(\sqrt{2})}$ is the set of algebraic integers $\{\pm x_n \pm y_n\sqrt{2}\}$.

For our next example we consider the corresponding question of when a prime number $p$ can be written in the form $p = x^2 + 6y^2$ with $x, y \in \mathbb{Z}$. Naturally we try to work in $\mathbb{Z}[\sqrt{-6}]$, the ring of algebraic integers in $\mathbb{Q}(\sqrt{-6})$. However, in $\mathbb{Z}[\sqrt{-6}]$

$$-2 \cdot 3 = -6 = \sqrt{-6} \cdot \sqrt{-6}$$

and, using the norm, we see that $-2$, $3$ and $\sqrt{-6}$ are all irreducible elements of $\mathbb{Z}[\sqrt{-6}]$; thus $\mathbb{Z}[\sqrt{-6}]$ is not a principal ideal domain, since it does not possess unique factorisation. We therefore find that, as it stands, the method of proof employed for the sum of two squares cannot be applied in this situation.

Whilst rings of algebraic integers are not in general principal ideal domains, and so do not possess unique factorisation of elements, they do still retain many important algebraic properties of $\mathbb{Z}$. In particular they possess unique factorisation of non-zero ideals: that is to say, given a number field $K$, every non-zero $\mathfrak{o}_K$-ideal can be written uniquely (up to order) as a product of prime ideals of $\mathfrak{o}_K$. This fact enables us to solve a number of Diophantine problems: later on we shall see quite how it permits us to extend the previous method and thereby solve the $p = x^2 + 6y^2$ problem.

We now turn to a famous classical problem of a different type: the problem known as "Fermat's Last Theorem". This is really a conjecture which asserts that for any natural number $\ell > 2$, if $(x, y, z)$ is a solution of the equation

$$X^\ell + Y^\ell = Z^\ell \tag{6}$$

then

$$xyz = 0.$$

This, as yet unsolved, problem has had a tremendous influence on the development of algebraic number theory. It is not hard to show the conjecture when $\ell = 4$, and to reduce the problem to the case when $\ell$ is an odd prime. Indeed, supposing $\ell$ to be an odd prime, we work in the field

$$\mathbb{Q}[\ell] = \mathbb{Q}(\eta) \qquad \eta = e^{2\pi i/\ell}$$

of $\ell$-th roots of unity. Given a solution $(x, y, z)$ to (6), we obtain a factorisation

$$\prod_{j=0}^{\ell-1} (x + \eta^j y) = z^\ell. \tag{7}$$

Proofs of the impossibility of (6) and (7) will then depend on the number theoretic properties of $\mathbb{Q}[\ell]$. In Chapter VII we shall show that, for certain prime numbers $\ell$, $\ell$ must divide $xyz$ for any solution to (6).

Returning to the problem of representing an odd prime $p$ as a sum of squares $x^2 + y^2$, we can obtain a reformulation in terms of the behaviour of prime ideals. Indeed, if $p \equiv -1 \bmod (4)$, then $p\mathbb{Z}[i]$ is a prime ideal of $\mathbb{Z}[i]$; while if $p \equiv 1 \bmod (4)$, then $p = x^2 + y^2 = (x + iy)(x - iy)$ with neither of $x \pm iy$ a unit, and so in this case $p\mathbb{Z}[i]$ is not a prime ideal of $\mathbb{Z}[i]$. Analogous interpretations also apply to the other examples which we have considered.

Quite generally we should now consider ideals, rather than elements, for the purposes of factorisation. Given an extension $L/K$ of algebraic number fields, we wish to find a *decomposition law* which determines, for prime ideals $\mathfrak{p}$ of $\mathfrak{o}_K$, the factorisation of the extended ideal of $\mathfrak{p}\mathfrak{o}_L$ into prime ideals of $\mathfrak{o}_L$.

This question, and variants of it, lies right at the heart of modern algebraic number theory. Under certain circumstances we can even turn the question right around, and parametrize certain extensions of K in terms of decomposition laws of the prime ideals of $\mathfrak{o}_K$.

In the text we shall describe the decomposition law for all quadratic extensions $\mathbb{Q}(\sqrt{m})/\mathbb{Q}$, as well as for cyclotomic extensions $\mathbb{Q}[\ell]/\mathbb{Q}$. This then leads us to yet another important topic: the comparison of decomposition laws. This technique yields an elegant proof of one of the most beautiful theorems of classical number theory: the law of quadratic reciprocity. For two distinct, odd primes $p$ and $q$, this asserts that

$$\left(\frac{p}{q}\right)\left(\frac{q}{p}\right) = (-1)^{\frac{p-1}{2}\cdot\frac{q-1}{2}}.$$

Here $\left(\frac{a}{p}\right)$ denotes the Legendre symbol defined, for $(a,p)=1$ by

$$\left(\frac{a}{p}\right) = \begin{cases} +1 & \text{if } x^2 \equiv a \bmod (p) \text{ is soluble for } x \in \mathbb{Z}; \\ -1 & \text{otherwise.} \end{cases}$$

This result, together with the above mentioned style of proof, provides a further example of an application of the properties of algebraic integers to obtain a result for ordinary integers. Moreover, there is a far reaching generalisation of the quadratic reciprocity law: for every natural number $\ell$ and for every algebraic number field $K$ containing all the $\ell$-th roots of unity, there is an $\ell$-power reciprocity law. This result, however, lies beyond the scope of this book.

We have already seen how $U_K$, the group of units of $\mathfrak{o}_K$, can play an important role in the study of Diophantine equations. We now introduce a further group which plays a central role in algebraic number theory, and which is again of great importance in solving Diophantine equations. This group should be thought of as measuring the deviation of a ring of algebraic integers from being a principal ideal domain. In order to define it we first need to introduce an equivalence relation on the non-zero $\mathfrak{o}_K$-ideals by declaring that two such ideals $\mathfrak{a}, \mathfrak{b}$ lie in the same class iff there exist non-zero elements $\alpha, \beta$ of $\mathfrak{o}_K$ such that $\alpha\mathfrak{a} = \beta\mathfrak{b}$. One easily verifies that this is indeed an equivalence relation, and we write $c(\mathfrak{a})$ for the class of $\mathfrak{a}$: furthermore, the rule $c(\mathfrak{a})c(\mathfrak{b}) = c(\mathfrak{a}\mathfrak{b})$ gives a well-defined multiplication on the set of classes, and one shows that these classes thereby form an abelian group, which is called the *ideal class group* $C_K$ of $K$. It is this group which helps to solve the $p = x^2 + 6y^2$ problem.

It is clear that $\mathfrak{o}_K$ is a principal ideal domain precisely when $C_K = \{1\}$. We shall see that $C_K$ is always finite, and we shall denote its order, the class number of $K$, by $h_K$.

In practice, the calculation by hand of class groups and unit groups can become rather unwieldy when the field $K$ has large degree over $\mathbb{Q}$. This therefore encourages us to look for alternative means for their calculation whenever possible. In fact, there is a wide variety of *ad hoc* methods for determining, or at least estimating, $h_K$ in particular cases. There is, however, a quite general formula for $h_K$, which is obtained by considering a certain function called the Dedekind $\zeta$-function of K. We recall that the Riemann $\zeta$-function is defined for real $x$ with $x > 1$ by

$$\zeta(x) = \sum_{n=1}^{\infty} \frac{1}{n^x} = \prod_p (1 - p^{-x})^{-1}$$

where the product extends over all prime numbers $p$. For a given number field $K$ and a non-zero $\mathfrak{o}_K$-ideal $\mathfrak{a}$, we write $\mathbf{N}\mathfrak{a}$ for the (finite) index $[\mathfrak{o}_K : \mathfrak{a}]$. Analogously, for $x > 1$, we define the Dedekind $\zeta$-function $\zeta_K(x)$ by

$$\zeta_K(x) = \sum_{\mathfrak{a}} \mathbf{N}\mathfrak{a}^{-x} = \prod_{\mathfrak{p}} (1 - \mathbf{N}\mathfrak{p}^{-x})^{-1}$$

where the sum extends over all non-zero ideals of $\mathfrak{o}_K$, while the product runs over all non-zero prime ideals of $\mathfrak{o}_K$. We note that $\zeta_{\mathbb{Q}}(x)$ is of course the Riemann $\zeta$-function.

It will be shown that $(x-1)\zeta_K(x)$ has a limiting value $h_K j_K$ as $x$ tends to 1 from the right, where the further constant $j_K$ incorporates information of $\mathfrak{o}_K$ and $U_K$. A precise expression for $j_K$ will be given in Chapter VIII; it implies in particular that $j_K \neq 0$.

Yet another generalisation of the Riemann $\zeta$-function is the Dirichlet $L$-function of a residue class character $\phi$, defined by

$$L(x, \phi) = \sum_{n=1}^{\infty} \frac{\phi(n)}{n^x} = \prod_p (1 - \phi(p)p^{-x})^{-1}.$$

A residue class character $\phi$ is a function on $\mathbb{Z}$ with complex values, multiplicative in the variable $n$ and only dependent on the residue class of $n$ modulo a fixed natural number $m$. The precise definition will be given in §2 of Chapter VI. For the particular character $\epsilon$, with $\epsilon(n) = 1$ for all $n$, of course $L(x, \epsilon) = \zeta(x)$. The importance of these $L$-functions is twofold. Firstly the Dedekind $\zeta$-function for $K = \mathbb{Q}[m]$, the field of

$m$-th roots of unity, can be written as a product

$$\zeta_{\mathbb{Q}[m]}(x) = \prod_{\phi} L(x, \phi) \tag{8}$$

where the product extends over certain residue class characters $\phi$, including $\phi = \epsilon$. For the factors in (8) belonging to $\phi \neq \epsilon$, $L(x, \phi)$ is finite at $x = 1$. As a consequence of the limit formula for $(x-1)\zeta_{\mathbb{Q}[m]}(x)$ and the product (8), one derives an equation

$$\prod_{\phi \neq \epsilon} L(1, \phi) = h_K j_K. \tag{9}$$

The factors $L(1, \phi)$ can be explicitly computed and (9) in fact contains a host of arithmetic information, both on class numbers $h_K$, and on such questions as the distribution of quadratic residues modulo a prime $p$.

The other place where $L$-functions play a central role is in the proof of the Dirichlet theorem on primes in arithmetic progressions. This asserts that, for a natural number $m$ and an integer $a$ coprime to $m$, there exist infinitely many prime numbers $p$ with $p \equiv a \bmod (m)$. The key point in the proof is the verification that, for $\phi \neq \epsilon$, $L(1, \phi)$ is non-zero. This follows, however, from (9). Thus once again we have seen how the proof of a theorem on ordinary primes depends on the arithmetic of a certain number field. Looking at it from the point of view of number fields, the Dirichlet theorem can, on the other hand, be interpreted as asserting the existence of infinitely many primes in $\mathbb{Z}$ with given decomposition type in $\mathbb{Q}[m]$. In this form the theorem does in fact generalize to arbitrary number fields $K$; this, however, lies outside the framework of this book.

# I

# Algebraic Foundations

In this chapter we seek to lay down the algebraic foundations which will be needed. In the first section we recall a number of basic results from field theory, and we then briefly consider the theory of finite commutative algebras over a field. In the second section we introduce the notion of integrality and the Noetherian properties for modules and rings.

The reader is advised not to spend too much time on Chapter I, and to move to Chapter II - where the algebraic number theory really begins. Indeed, for the reader who has already encountered these four topics – in some form or other – it is suggested that he start straightaway at Chapter II; and then refer back to Chapter I as is necessary.

## §1 Fields and Algebras

As has been seen in the introduction, many arithmetic problems require us to work with fields. First and foremost among the types of field which we need to consider are algebraic number fields, that is to say extensions of finite degree over the rationals. Reduction techniques lead us to work with finite fields, which turn up as the residue class fields of rings of algebraic integers modulo prime ideals. Finally we shall have to consider various "completions", of which the fields of real or of complex numbers are the most familiar examples.

We shall freely assume basic field theory, including Galois theory, but we include here a treatment of certain special topics, which are important for us, which however are often not dealt with in courses on field theory or which are not done in the way in which we shall need them.

In addition to fields we shall have to consider algebras, more precisely finite commutative algebras over a field, and we shall include in this section everything that we need on this matter. These algebras turn up naturally in two ways, as illustrated by the following two examples.

Firstly let $d$ be a square-free integer, $d \neq 1$, and consider the quadratic extension of $\mathbb{Q}$ given by $\mathbb{Q}(\sqrt{d})$. Then one has to consider the tensor product $\mathbb{Q}(\sqrt{d}) \otimes_{\mathbb{Q}} \mathbb{R} = A$. If we write $\mathbb{Q}(\sqrt{d})$ as $\mathbb{Q}[X]/(X^2 - d)$ (the residue class ring of $\mathbb{Q}[X]$ mod $(X^2 - d)$), then the tensor product is just $A = \mathbb{R}[X]/(X^2 - d)$. This is typical of the kind of algebra which we need to consider. It may turn out to be a field, as in the case $d = -1$, when $A = \mathbb{R}(\sqrt{-1}) = \mathbb{C}$; however, this need not be the case, for instance when $d = 2$, $A = \mathbb{R}[X]/(X^2 - 2)$ which has a zero divisor given by the class of $X - \sqrt{2}$, since $(X - \sqrt{2})(X + \sqrt{2})$ is zero in $A$.

The second situation where we need algebras occurs when we want to extend prime ideals. Recall that $\mathbb{Z}/7\mathbb{Z}$ is a finite field. Let $\mathbb{Z}[\sqrt{d}]$, with $d$ as before, denote the ring of numbers of the form $a + b\sqrt{d}$, $a, b \in \mathbb{Z}$, and consider $B = \mathbb{Z}[\sqrt{d}] \otimes_{\mathbb{Z}} (\mathbb{Z}/7\mathbb{Z})$. Then $B$ is just $\mathbb{Z}[\sqrt{d}]/7\mathbb{Z}[\sqrt{d}]$. This may be a field: $\mathbb{Z}[i]/7\mathbb{Z}[i]$ is an instance of such an occurrence. This will not, however, be the case in general as is shown by the zero divisor represented by $3 + \sqrt{2}$ in $\mathbb{Z}[\sqrt{2}]/7\mathbb{Z}[\sqrt{2}]$ (for $7 = (3 + \sqrt{2})(3 - \sqrt{2})$).

*Polynomials.* Let $F[X]$ denote the integral domain of polynomials in an indeterminate $X$ over a field $F$. The derivative $f'(X) = f'$ of a polynomial $f = f(X) = a_0 + a_1X + \cdots + a_nX^n$, with coefficients $a_j$ in $F$, is defined formally by

$$f' = a_1 + 2a_2X + \cdots + na_nX^{n-1}.$$

One then verifies the usual rules

$$(1.1) \qquad \begin{aligned} (f+g)' &= f' + g' \\ (fg)' &= fg' + f'g \end{aligned}$$

for $f, g$ in $F[X]$. From the very definition of the derivative we deduce

**(1.2)** *If $F$ has characteristic $0$, then $f' = 0$ iff $f$ is a constant. If $F$ has characteristic a prime number $p$, then $f' = 0$ iff $f(X) = g(X^p)$ for some $g \in F[X]$.*

For every non zero polynomial $f \in F[X]$, there exists a field $L$ (e.g. an algebraic closure of $F$ or just a splitting field for $f$ over $F$) so that in $L[X]$

$$f(X) = c\prod_{j=1}^{n}(X - \alpha_j), \quad c \neq 0. \tag{1.3}$$

We say that $f$ is a *separable polynomial* if the roots $\alpha_j$ are all distinct. (Note that this property is independent of the particular field $L$, or of the original field $F$, as long as $f \in F[X]$.)

**(1.4)** *$f$ is separable iff* $\mathrm{HCF}(f, f') = 1$.

Recall that $F[X]$ is a Euclidean Domain, and so is a unique factorisation domain. Thus, given two polynomials $f, g$ in $F[X]$ the $\mathrm{HCF}(f, g)$ is defined and is unchanged if we go over to a bigger coefficient field. (In the sequel we shall often abuse notation and write $(f, g)$ for $\mathrm{HCF}(f, g)$.) We can now assume that $f$ splits as in (1.3). Suppose now that $\alpha_1 = \alpha_2$, i.e. $f = (X - \alpha_1)^2 f_1$; then an application of (1.1) shows that $(X - \alpha_1) \mid f'$; hence $(X - \alpha_1) \mid (f, f')$ and so $(f, f') \neq 1$. Conversely, if $(X - \alpha_1) \mid (f, f')$, then $f = (X - \alpha_1)g$, $f' = (X - \alpha_1)h$, and we also have $f' = (X - \alpha_1)g' + g$. Thus, comparing the two expressions for $f'$, we get $(X - \alpha_1) \mid g$ and so $(X - \alpha_1)^2 \mid f$.

From this last result we deduce that $f$ is separable iff $f'(\alpha_j) \neq 0$ for each $j$, with $f$ as in (1.3). Now suppose $f$ to be monic, that is to say $c = 1$ in (1.3); we then define its discriminant $\mathrm{Disc}(f)$ by

$$\mathrm{Disc}(f) = (-1)^{n(n-1)/2}\prod_{j=1}^{n} f'(\alpha_j) \tag{1.5}$$

where $n$ denotes the degree of $f$. From (1.4) we deduce

**(1.6)** *$f$ is separable iff* $\mathrm{Disc}(f) \neq 0$.

Next we shall derive yet another criterion for separability. Define the polynomial $V(T_1, \ldots, T_n)$ in $n$ indeterminates over $L$, as the determinant

$$V = V(T_1, \ldots, T_n) = \det\begin{pmatrix} 1 & 1 & \ldots & 1 \\ T_1 & T_2 & \ldots & T_n \\ \vdots & \vdots & & \vdots \\ T_1^{n-1} & T_2^{n-1} & \ldots & T_n^{n-1} \end{pmatrix}. \tag{1.7}$$

As each term of the expansion of $V$ has total degree $n(n-1)/2$, the total degree of $V$ in all indeterminates is at most $n(n-1)/2$. On the other hand subtracting the $j$th column in turn from each other column, we see that

$$\prod_{\substack{i=1 \\ i \neq j}}^{n} (T_i - T_j)$$

divides $V$; moreover this is true for each $j$. Now the $T_i - T_j$ are non-associate irreducible elements of the unique factorisation domain $L[T_1, \ldots, T_n]$, for all $1 \le i < j \le n$. We therefore conclude that

$$P = \prod_{1 \le i < j \le n} (T_i - T_j) \mid V.$$

Comparing degrees shows that $V = aP$ for some constant $a$. Comparing coefficients of $T_2 T_3^2 \ldots T_n^{n-1}$, we find that $a = (-1)^{n(n-1)/2}$, and so

$$V = (-1)^{n(n-1)/2} \prod_{1 \le i < j \le n} (T_i - T_j). \tag{1.8}$$

Thus we have shown

**(1.9)** *$f$ is separable iff $V(\alpha_1, \ldots, \alpha_n) \neq 0$.*

The determinant $V(\alpha_1, \ldots, \alpha_n)$ – called the *Vandermonde* determinant – and the discriminant of $f$ are connected by the equation

$$V(\alpha_1, \ldots, \alpha_n)^2 = \operatorname{Disc}(f). \tag{1.10}$$

Indeed, by repeated application of (1.1), we get

$$f'(X) = \sum_{k=1}^{n} \prod_{\substack{i=1 \\ i \neq k}}^{n} (X - \alpha_i)$$

whence

$$f'(\alpha_j) = \prod_{\substack{i=1 \\ i \neq j}}^{n} (\alpha_j - \alpha_i)$$

and so (1.10) follows from (1.5) and (1.8).

The theory of algebraic extensions of fields is particularly nice for base fields $F$ which have the property that any irreducible polynomial in $F[X]$ is separable. Such a field is said to be *perfect*. The majority of fields that we encounter will be perfect. In fact, one knows:

**(1.11)**

($a$) *All fields of characteristic* 0 *are perfect.*

($b$) *All finite fields are perfect.*

*Proof.* Let $f$ be a monic irreducible polynomial in $F[X]$. If $d = (f, f')$ has positive degree, then $d$ must divide $f$ and so $f = d$, since $f$ is irreducible. This therefore implies $f' = 0$. Now apply (1.2). If the characteristic of $F$ is zero, $f$ would be a constant – a contradiction. Thus, in this case, $F$ is indeed perfect.

Next let $F$ be a finite field of characteristic $p$. By (1.2) $f' = 0$ implies that

$$f = b_0 + b_1X^p + b_1X^{2p} + \cdots + b_mX^{mp}$$

with $b_i \in F$. We shall show

**(1.12)** *Let $F$ be a finite field of characteristic $p$. Then the map $c \mapsto c^p$ is an automorphism of $F$ (the so-called Frobenius automorphism).*

First, in order to complete the proof of (1.11.$b$), we now see that with $b_i = c_i^p$ for all $i$,

$$f = (c_0 + c_1X + \cdots + c_mX^m)^p.$$

Thus f is reducible, and there is no inseparable, irreducible polynomial in $F[X]$.

□

*Proof of (1.12)* The map $c \mapsto c^p$ is additive, i.e. is an endomorphism of the additive group of $F$. As such its kernel is zero ($c^p = 0$ implies $c = 0$), and it is therefore injective. As $F$ is a finite set, any injective map $F \to F$ is bijective. Clearly $(c_1c_2)^p = c_1^pc_2^p$, and so $c \mapsto c^p$ is an automorphism of the field $F$.

□

*Example of a non-perfect field.* Let $K$ be a field of characteristic $p > 0$; let $F = K(T)$, the field of rational functions over $K$ in an indeterminate $T$. We shall show that the polynomial $f(X) = X^p - T \in F[X]$ is irreducible. It is certainly inseparable, as $f' = 0$.

Let $t$ be a root of $f(X)$. Then $f(X) = (X - t)^p$ in $E[X]$, $E = F(t)$. The minimal polynomial $m_{F,t}(X)$ of $t$ over $F$ is therefore of the form $(X - t)^r$, where $1 \leq r \leq p$. We have to show that $r = p$. If $r < p$, then $m_{F,t}(0) = \pm t^r \in F$. Since $t^p = T \in F$, it follows that $t \in F$, $t = h(T)/g(T)$ say, for polynomials $h, g$ in $K[T]$. Thus

$$T = \frac{h(T)^p}{g(T)^p} = \frac{h_1(T^p)}{g_1(T^p)}$$

where $h_1, g_1$ are polynomials in $K[T]$. This means that $g_1(T^p)T$ is a polynomial in $T^p$, which is absurd.

*Separable field extensions.* Let $E/F$ be a finite extension of fields, i.e. one of finite degree $(E : F)$. Then one knows that, given an embedding, i.e. an injective homomorphism $\sigma\colon F \hookrightarrow L$, with $L$ algebraically closed, there exist at most $(E : F)$ embeddings $\sigma_i\colon E \hookrightarrow L$ extending $\sigma$ (i.e.

such that $\sigma_i|_F = \sigma$). If there are $(E : F)$ such embeddings $\sigma_i$ for some $\sigma: F \hookrightarrow L$, then we say that $E/F$ is a *separable* extension. (In this case any embedding $\sigma': F \hookrightarrow L'$, $L'$ an algebraically closed field, will have $(E : F)$ extensions to $E$.) A simple counting argument then gives:

**(1.13)** *Let $F \subset E \subset H$ be finite extensions. Then the extensions $H/E$, $E/F$ are separable iff $H/F$ is.*

An element $\alpha$, algebraic over a field $F$, is said to be *separable over $F$* if its minimal polynomial $m_{F,\alpha}(X)$ over $F$ is separable. (We shall always use this notation for minimal polynomials.) As any factor of a separable polynomial is separable, it will suffice to show that $\alpha$ is the root of some separable polynomial in $F[X]$.

Suppose that in some algebraically closed field $L \supset F$

$$m_{F,\alpha}(X) = \prod_{i=1}^{n} (X - \alpha_i).$$

If $m_{F,\alpha}$ is separable, then the given embedding $F \hookrightarrow L$ has $n$ extensions to $F(\alpha)$ induced by $\alpha \mapsto \alpha_i$. Conversely, if $\sigma$ admits $n$ distinct extensions to $\sigma_i$, then setting $\alpha_i = \alpha^{\sigma_i}$ in the above, we see that $m_{F,\alpha}$ is a separable polynomial in $F[X]$. Thus we have shown

**(1.14)** *$\alpha$ is separable over $F$ iff $F(\alpha)/F$ is a separable extension.*

Combining (1.13) and (1.14), it follows that:

**(1.15)** *If $E/F$ is separable, then every element $\beta$ of $E$ is separable over $F$.*

Finally we quote

**(1.16) (Theorem of the primitive element).** *Let $E = F(\alpha, \beta)$ with $\alpha, \beta$ algebraic over $F$, and $\alpha$ separable over $F$. Then there exists an element $\gamma \in E$ such that $E = F(\gamma)$.*

By repeated application of (1.16), and using also (1.14), (1.15), we conclude

**(1.17)** *A finite extension $E/F$ is separable iff $E = F(\alpha)$ with $\alpha$ separable over $F$.*

In the sequel let $E/F$ denote a separable finite extension of degree $n$, and let $L$ denote an algebraically closed field containing $F$. Write $\sigma_i$,

$i = 1, \ldots, n$, for the $n$ embeddings $E \to L$ which extend the inclusion map $F \hookrightarrow L$. If $u_1, \ldots, u_n$ is a basis of $E$ over $F$, we define

(1.18*a*) $$V^*(u_1, \ldots, u_n) = \det(u_j^{\sigma_i})_{j,i}.$$

Then

(1.18*b*) $$V^*(u_1, \ldots, u_n) \neq 0.$$

To prove this first consider a further basis $w_1, \ldots, w_n$ of $E$ over $F$; then we may write

$$w_k = \sum_{j=1}^{n} c_{k,j} u_j \quad k = 1, \ldots, n$$

where $c_{k,j} \in F$ and $\det(c_{k,j}) \neq 0$. More generally

$$w_k^{\sigma_j} = \sum_{j=1}^{n} c_{k,j} u_j^{\sigma_i}.$$

Therefore

(1.19) $$V^*(w_1, \ldots, w_n) = V^*(u_1, \ldots, u_n) \det(c_{k,j})$$

whence we see that in order to prove (1.18.*b*) for an arbitrary basis of $E$ over $F$, it will suffice to prove it for one special such basis. For this we use an element $\alpha$, as in (1.17). Then its minimal polynomial $m_{F,\alpha} = f$, say, has degree $n$, and so $1, \alpha, \ldots \alpha^{n-1}$ is a basis of $E$ over $F$ and

(1.20) $$V^*(1, \alpha, \ldots, \alpha^{n-1}) = V(\alpha^{\sigma_1}, \ldots, \alpha^{\sigma_n})$$

cf. (1.7). By (1.9) we then derive the desired result.

Next we define the *trace*

(1.21.*a*) $$t_{E/F}(\alpha) = \sum_i \alpha^{\sigma_i}$$

and the *norm*

(1.21.*b*) $$N_{E/F}(\alpha) = \prod_i (\alpha^{\sigma_i})$$

and more generally the *characteristic polynomial* of $\alpha$ with respect to $E/F$

(1.21.*c*) $$\prod_i (X - \alpha^{\sigma_i}) = c_{E/F,\alpha}(X).$$

Clearly

(1.22) $$\begin{cases} t_{E/F}(\alpha_1 + \alpha_2) = t_{E/F}(\alpha_1) + t_{E/F}(\alpha_2) \\ t_{E/F}(c\alpha) = c t_{E/F}(\alpha) \\ N_{E/F}(\alpha_1 \alpha_2) = N_{E/F}(\alpha_1) N_{E/F}(\alpha_2) \end{cases}$$

for $\alpha, \alpha_1, \alpha_2 \in E$, $c \in F$. We assert that

(1.23) $$t_{E/F}(\alpha) \in F, \quad N_{E/F}(\alpha) \in F$$

and that more generally all the coefficients of $c_{E/F,\alpha}$ lie in $F$. To see this, for instance in the case of the trace, we use Galois theory and observe that the images $F^{\sigma_i}$, for all $\sigma_i$, lie in a Galois extension $H$ of $F$ of finite degree. Any automorphism $\omega$ of $H$ over $F$ will permute the $\sigma_i$, i.e. for some $j$, $(b^{\sigma_i})^{\omega} = b^{\sigma_j}$ for all $b \in F$. Therefore $t_{E/F}(\alpha)^{\omega} = t_{E/F}(\alpha)$ for all such $\omega$; hence by Galois theory $t_{E/F}(\alpha) \in F$. We shall obtain a further proof of (1.23) below (see (1.27)), which also makes it clear that the maps $t_{E/F}, N_{E/F}: E \to F$ do not depend on the choice of $L$.

We define the *discriminant* $d(u_1, \ldots, u_n)$ of a basis $u_1, \ldots, u_n$ of $E$ over $F$ by

$$d(u_1, \ldots, u_n) = \det\bigl(t_{E/F}(u_i u_j)\bigr). \tag{1.24}$$

By multiplying the matrix $(u_j^{\sigma_i})_{j,i}$ by its transpose, we conclude that

$$d(u_1, \ldots, u_n) = V^*(u_1, \ldots, u_n)^2 \tag{1.25}$$

Thus by (1.10) and (1.20)

$$d(1, \alpha, \ldots, \alpha^{n-1}) = \mathrm{Disc}(f) \tag{1.25.a}$$

where $f(X) = m_{F,\alpha}(X)$. By (1.25) and (1.18.*b*)

$$d(u_1, \ldots, u_n) \neq 0. \tag{1.25.b}$$

It follows that given any $a \in E$, there exists $b \in E$ with $t_{E/F}(ab) \neq 0$. In particular the linear mapping $t_{E/F}: E \to F$ is surjective.

In addition we suppose $F/K$ to be a finite separable extension of degree $m$. Then for $\alpha \in E$ we have the so-called transitivity formulae for trace and norm:

$$\begin{aligned} t_{E/K}(\alpha) &= t_{F/K}(t_{E/F}(\alpha)) \\ N_{E/K}(\alpha) &= N_{F/K}(N_{E/F}(\alpha)). \end{aligned} \tag{1.26}$$

We shall prove the first formula. Let $\{\tau_k\}$ denote the distinct embeddings of $F \to L$ which extend the inclusion map $K \hookrightarrow L$. We write $E'$ for the composite in $L$ of the fields $E^{\sigma_i}$; then the extension $E'/F$ is finite and separable. Let $\overline{\tau_k}$ denote an extension of $\tau_k$ to $E' \to L$. Then the maps $\sigma_i \overline{\tau_k}$, for fixed $k$ and varying $i$, when restricted to $E$ are all distinct and are all extensions of $\tau_k: F \to L$; hence they comprise all such extensions. Therefore the maps $\sigma_i \overline{\tau_k}$ ($i = 1, \ldots, n$; $k = 1, \ldots, m$) are precisely the distinct embeddings $E \to L$ which fix $K$; we therefore have equalities:

$$\begin{aligned} t_{E/K}(\alpha) &= \sum_{i,k} \alpha^{\sigma_i \overline{\tau_k}} = \sum_k (t_{E/F}(\alpha))^{\overline{\tau_k}} \\ &= \sum_k t_{E/F}(\alpha)^{\tau_k} = t_{F/K}(t_{E/F}(\alpha)). \end{aligned}$$

*Algebras.* A finite commutative $F$-algebra is a commutative ring $A$ containing the field $F$ as a subring with the same multiplicative identity;

thus, in particular, $A$ is an $F$-vector space, and the second part of the definition of such an algebra is that its dimension over $F$ be finite.

For future reference note that if $F$ is a field and we are given an isomorphism of fields $F \cong F_1$, then a finite commutative $F_1$-algebra $A$ can be viewed as a finite commutative $F$-algebra via that isomorphism.

Let $A$ denote a finite commutative $F$-algebra. For $a \in A$ the map $b \mapsto ab$ (all $b \in A$) defines a linear transformation $\ell_a$ of $A$ (viewed as an $F$-vector space). Any finite extension field $E$ of $F$ is a finite commutative $F$-algebra. If moreover $E/F$ is separable, then taking $A = E$ we have:

$$(1.27.a) \qquad \begin{cases} \operatorname{trace}(\ell_a) = t_{E/F}(a) \\ \det(\ell_a) = N_{E/F}(a) \\ \det(X1_A - \ell_a) = c_{E/F,a}(X) \end{cases}$$

*Proof.* Let $u_1, \ldots, u_n$ denote an $F$-basis of $E$. Then

$$u_j a = \sum_{k=1}^{n} c_{jk} u_k \quad (j = 1, \ldots, n).$$

More generally for a complete set of embeddings $\sigma_i \colon E \to L$ extending $F \hookrightarrow L$

$$u_j^{\sigma_i} a^{\sigma_i} = \sum c_{jk} u_k^{\sigma_i} .$$

In other words we have a matrix equation

$$(u_j^{\sigma_i})(\operatorname{diag}_i(a^{\sigma_i})) = (c_{jk})(u_k^{\sigma_i}) .$$

By (1.18.$b$) $\det(u_j^{\sigma_i}) \neq 0$. Therefore the trace, the determinant and the characteristic polynomials of $(c_{jk})$ and $(\operatorname{diag}(a^{\sigma_i}))$ coincide. The former matrix represents $\ell_a$ with respect to the given basis and the latter matrix has trace $\sum a^{\sigma_i}$, determinant $\prod a^{\sigma_i}$ and characteristic polynomial $c_{E/F,a}(X)$.

In view of the above, for any finite commutative algebra $A$ we are led to define

$$(1.27.b) \qquad t_{A/F}(a) = \operatorname{trace}(\ell_a) \quad N_{A/F}(a) = \det(\ell_a).$$

In the same way, given an $F$-basis $v_1, \ldots, v_n$ of A, we define the discriminant

$$(1.28) \qquad d(v_1, \ldots, v_n) = \det(t_{A/F}(v_i v_j)).$$

The structure theory of finite commutative algebras has two basic ingredients which go beyond the usual theory of fields. The first of these is the decomposition as a product via the associated idempotents. Recall that an idempotent of a finite commutative algebra $A$ is an element $e \in A$ with the property that $e^2 = e$. There are two obvious idempotents: the multiplicative identity $1_A = 1$; the zero $0_A = 0$. If $e$ is any non-zero idempotent of $A$, then clearly $Ae = \{ae \mid a \in A\}$ is a commutative ring under the operations inherited from $A$. It contains a copy of $F$ under the map $c = c1_A \mapsto ce$, and clearly it is finite dimensional since $\dim_F(Ae) \leq \dim_F(A)$; thus $Ae$ is indeed a finite commutative $F$-algebra.

Suppose that $A_i$ $(i = 1, \ldots r)$ are finite commutative $F$-algebras. Their product $\prod_1^r A_i$ is defined as consisting of sequences $(a_1, \ldots, a_n)$ with $a_i \in A_i$; addition and multiplication are defined componentwise, so that for example

$$(a_1, \ldots, a_r)(b_1, \ldots, b_r) = (a_1 b_1, \ldots, a_r b_r).$$

This is a commutative ring with multiplicative identity $(1_{A_1}, \ldots, 1_{A_r})$, and it contains an isomorphic copy of $F$ via the map $c \mapsto (c, \ldots, c)$. It is therefore a finite dimensional vector space over $F$, and

$$\text{(1.29)} \qquad \dim_F(\prod_1^r A_i) = \sum_{i=1}^r \dim_F(A_i).$$

The element $(a_1, \ldots, a_r)$ is an idempotent whenever each $a_i$ is either 1 or 0.

Suppose now that we have a decomposition of a finite commutative algebra $A$ as a direct product of algebras, as above, i.e. we have an isomorphism

$$\text{(1.30)} \qquad A \cong \prod_1^r A_i.$$

Let $e_j$ be the element of $A$ which is mapped onto $(0, \ldots, 0, 1_{A_j}, 0, \ldots, 0)$ (1 in the $j$th position, 0 otherwise). Then clearly the $e_j$ are all non-zero idempotents, that is to say

$$\text{(1.31.}a\text{)} \qquad e_j^2 = e_j \neq 0.$$

In addition note that

$$\text{(1.31.}b\text{)} \qquad e_i e_j = 0 \quad \text{whenever } i \neq j,$$

$$\text{(1.31.}c\text{)} \qquad 1_A = \sum_{j=1}^n e_j.$$

Any two idempotents satisfying (1.31.$b$) are said to be *orthogonal.* The ordered set $\{e_j\}$ is called the *system of orthogonal idempotents* associated with the product decomposition (1.30). Clearly, we have an isomorphism of finite commutative $F$-algebras

$$\text{(1.31.}d\text{)} \qquad A_i \cong Ae_i$$

and also

$$\text{(1.31.}e\text{)} \qquad \dim_F(A) = \sum_{j=1}^{r} \dim_F(A_j).$$

Conversely, given a system of orthogonal idempotents, as in (1.31), we obtain a homomorphism of $F$-algebras

$$\text{(1.32)} \qquad \begin{gathered} f\colon A \to \prod_{j=1}^{r} Ae_j \\ f(a) = (ae_1, \ldots, ae_r). \end{gathered}$$

$\mathrm{Ker}(f) = (0)$, for if $ae_j = 0$ for all $j$, then $a = \sum ae_j = 0$; hence $f$ is injective. It is also surjective, since $a_1e_1 + \cdots + a_re_r$ $(a_j \in A)$ maps onto $(a_1e_1, \ldots, a_re_r)$. Obviously the system of idempotents associated with (1.32) is the given one $\{e_j\}$. On the other hand, if $\{e_j\}$ is associated with a given direct product decomposition (1.30); then by (1.31.$d$) the decomposition (1.32) is isomorphic componentwise to the original one in (1.30).

Next we suppose we are given a product decomposition

$$A = A_1 \times \cdots \times A_r$$

and we suppose that some $A_j$ is itself a product $A_j = A_{j,1} \times A_{j,2}$. Then we obtain a further decomposition

$$A = A_1 \times \cdots \times A_{j,1} \times A_{j,2} \times \cdots \times A_r.$$

By (1.31.$e$) this process of refining product decompositions of $A$ must terminate, i.e. $A$ will possess a decomposition (1.30) in which each factor $A_i$ is *indecomposable* (cannot be written non-trivially as a product). An idempotent $e$ is said to be *primitive* if it is non-zero and if $e = e' + e''$, $e'$ and $e''$ both idempotents with $e'e'' = 0$, necessarily implies that $e' = 0$ or $e'' = 0$. We can then show

**(1.33)** *In the product (1.30), each $A_i$ is indecomposable iff the associated system of idempotents is one of primitive idempotents.*

To prove (1.33), it clearly suffices to show

**(1.34)** *Let $e$ be a non-zero idempotent of a finite commutative algebra $A$; then $Ae$ is indecomposable iff $e$ is primitive.*

*Proof.* If $e$ is not primitive, then we can write $e = e' + e''$, $e'e'' = 0$, $e'$ and $e''$ both non-zero idempotents. Then, just as in (1.32), we get

$$Ae \cong Ae' \times Ae''$$

Conversely, if $Ae$ is not indecomposable, then the construction leading from (1.30) to (1.31) (with $1_{Ae} = e$) yields a system of orthogonal idempotents $\{e_j\}$, $(j = 1, \ldots, r)$, $r > 1$, of $Ae$. We set $e' = \sum_{j=2}^{r} e_j$ so that $e = e_1 + e'$, $e_1 e' = 0$ and $e$, $e'$ are both non-zero idempotents.

□

Now let $\{e_j\}$, $(j = 1, \ldots, r)$ denote a system of orthogonal primitive idempotents of $A$ associated with a product decomposition

$$A = \prod_{i=1}^{r} A_i$$

with the $A_i$ all indecomposable. We have already seen that such a decomposition must exist; next we consider the matter of uniqueness and show

**(1.35)** *The $e_j$ $(j = 1, \ldots, r)$ are the only primitive idempotents in $A$.*

To achieve this we show

**(1.36)** *If $e'$, $e''$ are distinct primitive idempotents of $A$, then $e'e'' = 0$.*

(1.35) is an immediate consequence of (1.36): for, if $e'$ is a primitive idempotent of $A$ which is not contained in the $\{e_j\}$; then

$$e' = e' \cdot 1 = \sum e' e_j = 0$$

which is a contradiction.

□

*Proof of (1.36).* If $e' = e'e''$ and $e'' = e'e''$, then clearly $e' = e''$. Therefore, on interchanging $e'$, $e''$ if necessary, we may assume that $e' \neq e'e''$, and so

$$e' = e'e'' + e'(1 - e''). \tag{1.37}$$

Since the two right-hand terms are orthogonal idempotents, and since $e'$ is primitive we deduce that $e'e''$ must be zero.

□

Consider now a product decomposition as in (1.30) and fix an $F$-basis $a_{i,k}$ $(k = 1, \ldots, n_i)$ of $A_i$ for each $i$. Let $b_{i,k}$ $(k = 1, \ldots, n_i,\ i = 1, \ldots, r)$ denote the elements of A whose image in $A_i$ is $a_{i,k}$, and whose image is 0 in $A_j$ for $j \neq i$; then $\{b_{i,k}\}$ is an $F$-basis of $A$. If $c \in A$ has image $(c_1, \ldots, c_r)$ in $\prod A_i$ then we obtain

$$\det(X1_A - \ell_c) = \prod_{1}^{r} \det(X1_{A_i} - \ell_{c_i})$$

for the characteristic polynomials; hence, with the definitions of (1.27)

$$t_{A/F}(c) = \sum_i t_{A_i/F}(c_i) \tag{1.37}$$
$$N_{A/F}(c) = \prod N_{A_i/F}(c_i).$$

Finally, in the same way, we obtain the formula

$$d(b_{1,1}, \ldots, b_{r,n_r}) = \prod_i d(a_{i,1}, \ldots, a_{i,n_i}). \tag{1.38}$$

The second basic ingredient in the description of a finite commutative $F$-algebra $A$ is the *radical* of $A$, $\mathrm{Rad}(A)$; this is defined as the subset of $A$ consisting of its *nilpotent* elements, that is

$$\mathrm{Rad}(A) = \{a \in A | a^n = 0, \text{some } n \in \mathbb{N}\}.$$

**(1.39)** *$\mathrm{Rad}(A)$ is an ideal of $A$, and hence is an $F$-subspace. If $\overline{A} = A/\mathrm{Rad}(A)$, then $\mathrm{Rad}(\overline{A}) = (0)$.*

*Proof.* If $a^n = 0 = b^m$ then by the Binomial theorem $(a+b)^{n+m-1} = 0$, and $(ac)^n = a^n c^n = 0$ for any $c \in A$; thus $\mathrm{Rad}(A)$ is an $A$-ideal. It then follows that $\overline{A}$ is a finite commutative $F$-algebra. If $a + \mathrm{Rad}(A) \in \mathrm{Rad}(\overline{A})$, then $a^n \in \mathrm{Rad}(A)$, and so $a$ itself must be nilpotent; this shows $\mathrm{Rad}(\overline{A}) = (0)$.

□

*Example.* If $f \in F[X]$ has degree $n > 0$, then the residue class ring $A = F[X]/(f)$ is a finite commutative $F$-algebra. Clearly it is a commutative ring; the constants form a subring isomorphic to $F$; we note that $\dim_F(A) = n$. Now assume that $f = g^m$ for a separable polynomial $g \in F[X]$. Then $\mathrm{Rad}(A)$ is generated by the class of $g$ mod $(f)$; thus $\mathrm{Rad}(A) = 0$ iff $m = 1$ and so

$$\overline{A} = A/\mathrm{Rad}(A) \cong F[X]/(g).$$

If $e$ is an idempotent of $A$, then its class $\overline{e}$ mod $\mathrm{Rad}(A)$ is an idempotent of $\overline{A}$.

**(1.40)** *The map $e \mapsto \overline{e}$ is a bijection from the set of idempotents of $A$ to those of $\overline{A}$.*

**Corollary.** *$A$ is indecomposable iff $\overline{A}$ is indecomposable.*

*Proof.* First note that if $e$ is a non-zero idempotent of $A$ then $\overline{e} \neq \overline{0}$ in $\overline{A}$: for otherwise $e \in \mathrm{Rad}(A)$; hence $e^n = 0$ for some $n \in \mathbb{N}$; and so

$e = 0$ since $e$ is idempotent. We now prove injectivity. Suppose $\overline{e} = \overline{e}'$ for idempotents $e$, $e'$ in $A$. By the above, if either $e$ or $e'$ is non-zero, then so is the other. So now assume this to be the case. We can then write each idempotent as a sum of primitive idempotents, say $e = \sum c_j$, as before. Thus in $\overline{A}$

$$\overline{0} \neq \overline{e}_1 = \overline{e}_1\overline{e} = \overline{e}_1\overline{e}'.$$

Since $\overline{e}_1\overline{e}' \neq \overline{0}$, it follows that $e_1$ occurs in the expression for $e'$ as a sum of primitive idempotents. This applies to all primitive idempotents in turn, and so $e = e'$.

Next we prove surjectivity. Given $\overline{e}$ an idempotent of $\overline{A}$, we have to find an idempotent of $A$ which maps onto it. Clearly we may suppose $\overline{e}$ to be non-zero. Choose $e_1 \in A$ with $e_1 + \mathrm{Rad}(A) = \overline{e}$, and set $f_1 = 1 - e_1$. As $\overline{(1-e_1)} \cdot \overline{e}_1 = (1-\overline{e})\overline{e} = \overline{0}$, we know that $e_1 f_1 \in \mathrm{Rad}(A)$, and so $(e_1 f_1)^n = 0$ for some $n \in \mathbb{N}$. Put $e_2 = e_1^n$, $f_2 = 1 - e_2$, $f = f_2^n$, $e = 1 - f$. Then $e$ is still in the class of $\overline{e}$; moreover $e = (1 - f_2^n) = (1 - f_2)a$ for some $a \in A$, and so $e = e_2 a = e_1^n a$. Also $f = f_2^n = (1-e_2)^n = (1-e_1^n)^n = f_1^n b$ for some $b \in A$. Thus $ef = abe_1^n f_1^n = 0$; hence $e^2 = e$, and so $e$ is the required idempotent.

□

If $\{e_j\}$ is a system of non-zero orthogonal idempotents, then it is clear that

$$\mathrm{Rad}(A) \cap Ae_j = \mathrm{Rad}(Ae_j)$$

$$\mathrm{Rad}(A) = \bigoplus_j \mathrm{Rad}(Ae_j). \tag{1.41}$$

Assuming that $\{e_j\}$ corresponds to a product decomposition $A \cong \prod A_i$, as in (1.30), then, under the isomorphism (1.31.$d$),

$$\mathrm{Rad}(A_i) \cong \mathrm{Rad}(Ae_i) \tag{1.42.a}$$

and so

$$\mathrm{Rad}(A) \cong \prod_i \mathrm{Rad}(A_i). \tag{1.42.b}$$

We shall subsequently need two results on radicals. The first of these is

**Theorem 1.** *Let $A$ denote a finite commutative $F$-algebra. Then the following two conditions are equivalent:*

($a$) $\mathrm{Rad}(A) = (0)$

($b$) $A \cong \prod A_i$ *with each* $A_i$ *a field.*

*If in addition $F$ is perfect, then these conditions are also equivalent to each of the following:*

(*c*) $d(v_1, \ldots, v_n) \neq 0$ *for some F-basis* $v_1, \ldots, v_n$ *of A.*
(*d*) $d(v_1, \ldots, v_n) \neq 0$ *for all F-bases* $v_1, \ldots, v_n$ *of A.*

*Remark.* By its very definition it follows that $\mathrm{Rad}(A)$ is contained in every maximal $A$-ideal. Conversely, by (1.39) together with the first two parts of Theorem 1, we see that $\mathrm{Rad}(A)$ is the intersection of the maximal ideals given by the kernels of the maps $A \to A_i$ i.e. of all maps from $A$ onto a field. This then shows that $\mathrm{Rad}(A)$ is the intersection of all maximal $A$-ideals.

*Proof.* (*d*) obviously implies (*c*). If $v_1, \ldots, v_n$ and $w_1, \ldots, w_n$ are both $F$-bases of $A$, and if we have $w_i = \sum c_{ij} v_j$, $(c_{ij} \in F)$; then $\det(c_{ij}) \neq 0$ and, by the definition of $d$,

$$d(w_1, \ldots, w_n) = \det(c_{ij})^2 d(v_1, \ldots, v_n).$$

This shows that (*c*) implies (*d*).

Next, assume $F$ to be perfect. The implication (*b*)⇒(*c*) follows from (1.38) and (1.25.*b*). For the implication (*d*)⇒(*a*) (which in fact also holds for non-perfect $F$), we use the following fact.

**(1.43)** $t_{A/F}(a) = 0$ *if a is nilpotent.*

Indeed, if $\mathrm{Rad}(A) \neq (0)$, choose $a \in \mathrm{Rad}(A)$, $a \neq 0$. Extend to a basis $a = a_1, \ldots, a_n$ of $A$. Then $aa_j \in \mathrm{Rad}(A)$ for each $j$, and so by (1.43) each $t_{A/F}(a_1 a_j) = 0$. Thus

$$d(a_1, \ldots, a_n) = \mathrm{Det}(t_{A/F}(a_i a_j)) = 0$$

since the matrix has a row of zeros.

In order to prove (1.43), we suppose that $a^m = 0$, $m > 0$. Thus the minimal polynomial of $\ell_a$ divides $X^m$, and so is of the form $X^s$. Therefore all the eigenvalues of $\ell_a$ are zero, which establishes (1.43).

The implication (*b*)⇒(*a*) follows from (1.42.*a*), (1.42.*b*) and from the obvious fact that the radical of any field is zero. Lastly the implication (*a*)⇒(*b*) is an immediate consequence of the following result

**(1.44)** *An indecomposable finite commutative F-algebra A with* $\mathrm{Rad}(A)$ $= (0)$ *is a field.*

*Proof.* Let $a \in A$, $a \neq 0$. Then

$$a^n A = \{x \in A \mid x = a^n y \text{ some } y \in A\}$$

is an ideal of $A$, hence is an $F$-subspace, and of course $a^{n+1} A \subset a^n A$.

As $\dim_F(A)$ is finite, and since $\dim_F(a^{n+1}A) \le \dim_F(a^n A)$, it must happen that $a^{n+1}A = a^n A$ for some $n$. Therefore $a^{2n}A = a^n A$, and so $a^n = a^{2n}b$ for some $b \in A$. Hence $a^n b = (a^n b)^2$, that is to say $a^n b$ is an idempotent. Since $A$ is indecomposable $a^n b$ must be 0 or 1. If $a^n b = 0$, then $a^n = a^{2n}b = a^n \cdot a^n b = 0$, and so $a = 0$ since $\mathrm{Rad}(A) = 0$. If, however, $a^n b = 1$, then clearly $a$ possesses a multiplicative inverse. This shows that $A$ is therefore a field.

□

Next we have to introduce the procedure of extending the base field of a finite commutative algebra. (Compare the examples mentioned at the beginning of this section.) Even if in the first instance we are interested in algebraic extensions of $\mathbb{Q}$, extension to $\mathbb{R}$ leads us necessarily to finite commutative algebras. The main tool here is the tensor product; however, we shall only need a special case of it here. Let $A$ denote a finite commutative algebra over a field $F$, and let $K$ be any field which contains $F$. Recall that the tensor product $A \otimes_F K$ is, in the first place, the additive group of finite sums $\sum_i a_i \otimes c_i$, $(a_i \in A,\ c_i \in K)$ with relations

$$\begin{aligned} a \otimes (c + c') &= a \otimes c + a \otimes c' \\ (a + a') \otimes c &= a \otimes c + a' \otimes c \\ af \otimes c &= a \otimes fc \end{aligned} \tag{1.45}$$

for $a, a' \in A$, $c, c' \in K$, $f \in F$. We then define a multiplication on $A \otimes_F K$ by the rule

$$(\sum_i a_i \otimes c_i)(\sum_j a'_j \otimes c'_j) = \sum_{ij} a_i a'_j \otimes c_i c'_j. \tag{1.46}$$

This is compatible with the relations (1.45), so that $A \otimes_F K$ is then a commutative ring containing the copy $1 \otimes K = \{1 \otimes c \mid c \in K\}$ of the field $K$. If $\{a_i\}$ is an $F$-basis of $A$, then $\{a_i \otimes 1\}$ is a $K$-basis of $A \otimes_F K$. Thus $A \otimes_F K$ is a finite commutative $K$-algebra and

$$\dim_K(A \otimes_F K) = \dim_F(A). \tag{1.47}$$

The $K$-algebra $A \otimes_F K$ has the following universal mapping property. Given a commutative, not necessarily finite-dimensional, $K$-algebra $B$ and a homomorphism $A \xrightarrow{g} B$ of rings and $F$-vector spaces, then there is a unique homomorphism $A \otimes_F K \xrightarrow{g^*} B$ of rings and of $K$-vector spaces such that the following diagram commutes:

(1.48)

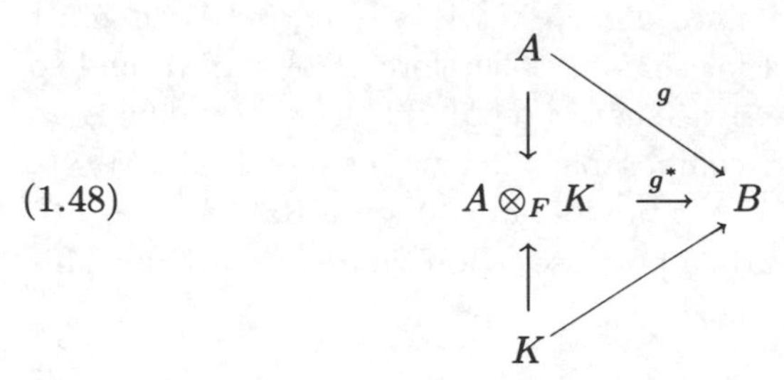


Here $K \to B$ is $c \mapsto c \cdot 1_B$; $K \to A \otimes_F K$ is $c \mapsto 1 \otimes c$; $A \to A \otimes_F K$ is $a \mapsto a \otimes 1$. Note that the diagram

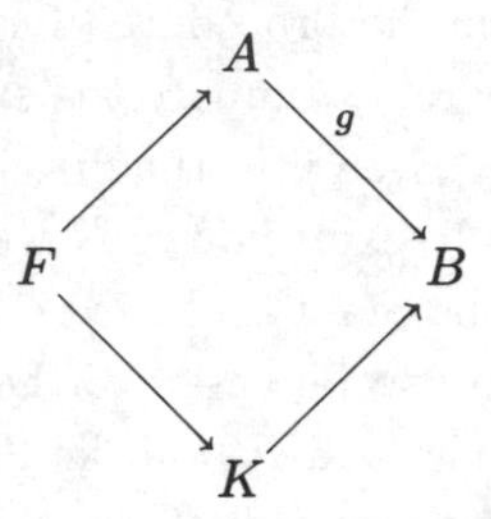


commutes, since $g(f1_A) = fg(1_A) = f \cdot 1_B$ for any $f \in F$.

This mapping property is a special case of that for the general tensor product, but it is also easy to verify directly that $g^*(a \otimes c) = g(a) \cdot c$ and that this extends to arbitrary elements of $A \otimes_F K$.

As an illustration of the tensor product in practice we show

**(1.49)** *Let $E_1$, $E_2$ be finite extension fields of $F$ contained in an algebraic closure $F^c$ of $F$, and let $E_1E_2$ denote their compositum in $F^c$. Then there is a homomorphism $h\colon E_1 \otimes_F E_2 \to E_1E_2$ with $h(x_1 \otimes x_2) = x_1x_2$. This is an isomorphism iff $E_1$ and $E_2$ are linearly disjoint over $F$.*

*Proof.* The existence of $h$ with the given property $h(x_1 \otimes x_2) = x_1x_2$ is a special case of the universal property (1.48). Obviously $h$ is surjective, and so, viewing it as a homomorphism of $E_1$ vector spaces, $h$ will be bijective iff $\dim_{E_1}(E_1 \otimes_F E_2) = (E_1E_2 : E_1)$. But $\dim_{E_1}(E_1 \otimes_F E_2) = (E_2 : F)$ by (1.47); the result then follows since $(E_1E_2 : E_1) = (E_2 : F)$ iff $E_1$ and $E_2$ are linearly disjoint over $F$. □

In the exercises given at the end of this chapter, we describe the decomposition, as a product of fields, of the tensor product of two arbitrary finite Galois extensions of $F$.

The last of the major results for algebras which we require is

**(1.50)** *Let $A$ be a finite separable extension field of $F$. Then $A \otimes_F K$ is a product of finite separable extension fields of $K$.*

*Remark.* This result fails to hold if we omit the word separable in both places; see example (1.52) below.

The proof of (1.50) is based on

**(1.51)** *Let $A = F[X]/fF[X]$ for some non-constant polynomial $f$ in $F[X]$. Then*

$$A \otimes_F K \cong K[X]/fK[X].$$

*Proof.* We use the universal mapping property (1.48) with

$$B = K[X]/fK[X],$$

$g$ the embedding

$$F[X]/fF[X] \hookrightarrow K[X]/fK[X].$$

We then get a map $g^*(\overline{h} \otimes c) = \overline{hc}$, where $\overline{\phantom{x}}$ denotes residue class mod $f$. Clearly $g^*$ is surjective. To prove bijectivity we use the fact that $g^*$ is a linear mapping, and so it suffices to verify equality of dimension. Indeed,

$$\begin{aligned} \dim_K(A \otimes_F K) &= \dim_F(A) = \deg(f) \\ &= \dim_K\big(K[X]/fK[X]\big). \end{aligned}$$

□

(1.52) *Example.* Recall the earlier example of a non-perfect field: let $k$ denote a field of characteristic $p > 0$, let $F = k(T)$ for an algebraic indeterminate $T$ over $k$, and set $A = F[X]/(X^p - T)F[X]$. Then $A$ is a field. Let $K$ denote any field containing $F$ together with an element $t$ such that $t^p = T$. Then

$$A \otimes_F K \cong K[X]/(X - t)^p K[X]$$

and the latter algebra contains the non-zero nilpotent element $\overline{(X - t)}$.

*Proof of* (1.50) By (1.17) we may take $A = F[X]/fF[X]$ where $f$ is an irreducible, separable polynomial in $F[X]$. We have to show that $K[X]/fK[X]$ is a product of separable extension fields of $K$. By the Chinese Remainder Theorem we have the product decomposition

$$K[X]/fK[X] \cong \prod_i K[X]/f_iK[X] \tag{1.53}$$

where the $f_i$ denote the irreducible factors of $f$ in $K[X]$. As $f_i \mid f$, we know that each $f_i$ is separable, and so each factor $K[X]/f_iK[X]$ is a finite separable extension of $K$.

$\square$

In the last example it is instructive to describe concretely the idempotents corresponding to the product decomposition (1.53). This also yields another proof, which makes no direct reference to the Chinese Remainder Theorem.

Let $\alpha = X + fF[X]$, so that $A = F[\alpha]$. We identify $A$ with its image in $K[X]/fK[X]$, so that by (1.51) $A \otimes_F K = K[\alpha]$. Note that $\alpha$ has minimal polynomial $f$ in both $A/F$ and $A \otimes_F K/K$. Let $f_i$ run through the irreducible factors of $f$ in $K[X]$, and set $g_i = \prod_{j \neq i} f_j$; therefore, by the separability of $f$, we know that

$$f_i g_i = f. \tag{1.54}$$

Since the highest common factor of the $g_i$ is 1, we can find polynomials $h_i$, such that

$$\sum_i e_i = 1 \quad \text{where} \quad e_i = g_i h_i. \tag{1.55}$$

If $i \neq k$, then $f \mid g_i g_k$, and so $f \mid e_i e_k$. Hence in $K[\alpha]$, we have

$$e_i(\alpha) e_k(\alpha) = 0 \quad \text{when } i \neq k \tag{1.56}$$

while by (1.55)

$$\sum_i e_i(\alpha) = 1. \tag{1.57}$$

It follows that the $\{e_i(\alpha)\}$ are orthogonal idempotents in $K[\alpha]$. If we had $e_i(\alpha) = 0$; then it would follow that $f_i(X) \mid h_i(X)$. Since $f_i(X) \mid g_j(X)$ when $i \neq j$, we would then have $f_i(X) \mid 1$ – which is absurd. Thus we have shown that the $\{e_i(\alpha)\}$ are all non-zero. Therefore we obtain a product decomposition

$$K[\alpha] \cong \prod_i K[\alpha] e_i(\alpha).$$

Since $f(X) \mid e_i(X) f_i(X)$, we know that $f_i(\alpha) e_i(\alpha) = 0$. So, because $f_i$ is irreducible in $K[X]$, we can deduce that $K[\alpha]e_i(\alpha) \cong K[X]/f_iK[X]$, which is a finite separable extension of $K$.

## §2 Integrality and Noetherian properties

In the course of the introduction we have seen the fundamental importance of rings of algebraic integers. Part of the next chapter will be devoted to a systematic treatment of these rings. In preparation for this we shall deal here, in a more general ring theoretic setting, with two basic properties: integrality and the Noetherian finiteness condition. Many other rings, such as those arising in algebraic geometry, also possess these properties. We shall see that integrality also corresponds to a finiteness property.

Throughout this section $\mathfrak{O}$ will denote a ring extension of a ring $\mathfrak{o}$ where it is always assumed that both rings have the same identity. If $a_1, a_2, \ldots, a_n$ are elements of $\mathfrak{O}$ then $(a_1, \ldots, a_n)\mathfrak{o}$ is the $\mathfrak{o}$-module in $\mathfrak{O}$ spanned by $a_1, \ldots, a_n$, while $\mathfrak{o}[a_1, \ldots, a_n]$ is the subring of $\mathfrak{O}$ generated by $\mathfrak{o}$ and $a_1, \ldots, a_n$. We say that $a \in \mathfrak{O}$ is *integral over* $\mathfrak{o}$ iff there is a monic polynomial $f \in \mathfrak{o}[X]$ such that $f(a) = 0$. Note that in the special case when $\mathfrak{o}$ is a field, then integrality over $\mathfrak{o}$ coincides with the notion of algebraicity over $\mathfrak{o}$.

In practice, this definition of integrality can be quite cumbersome: for instance if $a, b \in \mathfrak{O}$ are both integral over $\mathfrak{o}$, then it could be tedious to have to show that $ab$ is integral over $\mathfrak{o}$ by determining a monic polynomial for which $ab$ is a root. It is therefore often advantageous to have an alternative, equivalent condition:

**(2.1)** *An element $a \in \mathfrak{O}$ is integral over $\mathfrak{o}$ iff the ring $\mathfrak{o}[a]$ is finitely generated as an $\mathfrak{o}$-module.*

*Proof.* Suppose first that $a$ is integral over $\mathfrak{o}$, with $f(a) = 0$, for a monic polynomial $f \in \mathfrak{o}[X]$

$$f = X^n + a_1 X^{n-1} + \cdots + a_n.$$

It will suffice to show that $\mathfrak{o}[a] = (1, a, \ldots, a^{n-1})\mathfrak{o}$. To this end we show inductively that for $m \geq n-1$

$$(1, a, \ldots, a^m)\mathfrak{o} = (1, a, \ldots, a^{n-1})\mathfrak{o}.$$

The case $m = n - 1$ is trivial. More generally, if this result holds for $m - 1$ in place of $m$, then we note that since $a^{m-n} f(a) = 0$

$$a^m = -a_1 a^{m-1} + \cdots - a_n a^{m-n}$$

and so

$$a^m \in (1, a, \ldots, a^{m-1})\mathfrak{o} = (1, a, \ldots, a^{n-1})\mathfrak{o}.$$

Conversely, suppose that $\mathfrak{o}[a]$ is generated over $\mathfrak{o}$ by $f_1(a), \ldots, f_r(a)$ with $f_i(x) \in \mathfrak{o}[X]$. Put $n = \max_i(\deg(f_i))$; then, by hypothesis, we can write $a^{n+1} = \sum a_i f_i(a)$ with $a_i \in \mathfrak{o}$; then $a$ is a root of the monic polynomial $X^{n+1} - \sum a_i f_i(X)$.

□

In certain situations it can be extremely advantageous to replace the role of $\mathfrak{o}[a]$, by any extension ring $\mathfrak{R}$ which is known to be a finitely generated $\mathfrak{o}$-module.

**(2.2)** *An element $a \in \mathfrak{O}$ is integral over $\mathfrak{o}$ iff $a$ lies in some extension ring $\mathfrak{R}$ of $\mathfrak{o}$which is finitely generated as an $\mathfrak{o}$-module.*

*Proof.* If $a$ is integral over $\mathfrak{o}$, then we can take $\mathfrak{R} = \mathfrak{o}[a]$ and use (2.1). Conversely, suppose that $\mathfrak{R}$ is as given above with $\mathfrak{R} = (r_1, \ldots, r_n)\mathfrak{o}$; we can therefore write

$$r_i a = \sum_{j=1}^{n} r_j a_{ji} \quad (a_{ij} \in \mathfrak{o}). \tag{2.2.a}$$

We let $A$ denote the matrix $a1_n - (a_{ji})$, $d = \det(A)$, and let $A^*$ denote the adjugate of $A$, so that $A \cdot A^* = d \cdot 1_n$. Then by (2.2.a) $\mathbf{r}AA^* = 0$ where $\mathbf{r}$ denotes the row vector whose $i$th entry is $r_i$. Thus $r_i d = 0$ for all $i$; as $1 \in \mathfrak{R} = (r_1, \ldots, r_n)\mathfrak{o}$, this implies $d = 0$, and so $a$ is a root of the monic polynomial $\det(X \cdot 1_n - (a_{ji}))$.

□

We call the set of elements in $\mathfrak{O}$ which are integral over $\mathfrak{o}$, the *integral closure* of $\mathfrak{o}$ in $\mathfrak{O}$. Note that, of course, any element $a \in \mathfrak{o}$ is integral over $\mathfrak{o}$, since it trivially satisfies the polynomial $X - a$.

If $\mathfrak{o}$ is an integral domain with field of fractions $K$, then we call $\mathfrak{o}$ *integrally closed* if $\mathfrak{o}$ coincides with its integral closure in $K$.

We now illustrate some of the ideas that we have introduced by briefly considering some particular cases and examples.

First we mention that in (II,§1) it will be shown that any principal ideal domain is integrally closed.

*Example 1.* $\mathfrak{o} = \mathbb{Z}[\sqrt{-3}]$ is not integrally closed, since the non-trivial cube root of unity $\omega = (-1 + \sqrt{-3})/2$ lies in the field of fractions $\mathbb{Q}(\sqrt{-3})$, and is integral over $\mathfrak{o}$ because it satisfies the polynomial $X^2 + X + 1$.

*Example 2.* Let $F$ denote a field and let $\mathfrak{o} = F[T^2, T^3]$ for some indeterminate $T$. Then $\mathfrak{o}$ is not integrally closed since $T = T^3/T^2$ lies in the field of fractions of $\mathfrak{o}$, and satisfies the polynomial $X^2 - T^2 \in \mathfrak{o}[X]$.

We next investigate the transitivity properties of integrality. As a preliminary algebraic result we first show

**(2.3)** *Let $\mathfrak{R} \supset \mathfrak{O} \supset \mathfrak{o}$ denote a tower of ring extensions. If $\mathfrak{R}$ (resp.*

*$\mathfrak{O}$)is a finitely generated module over $\mathfrak{O}$ (resp. $\mathfrak{o}$);then $\mathfrak{R}$ is a finitely generated module over $\mathfrak{o}$.*

*Proof.* If $\mathfrak{R} = \sum_i u_i\mathfrak{O}$, $\mathfrak{O} = \sum_i v_j\mathfrak{o}$, then $\mathfrak{R} = \sum_{i,j} u_i v_j \mathfrak{o}$.

□

**(2.4)**

(*a*) *Let $a_1, \ldots, a_n \in \mathfrak{O}$ all be integral over $\mathfrak{o}$; then $\mathfrak{o}[a_1, \ldots, a_n]$ is a finitely generated $\mathfrak{o}$-module.*

(*b*) *Let $\mathfrak{R} \supset \mathfrak{O} \supset \mathfrak{o}$ be a tower of ring extensions; and suppose that $\mathfrak{O}$ is a finitely generated $\mathfrak{o}$-module. If $r \in \mathfrak{R}$ is integral over $\mathfrak{O}$, then $r$ is integral over $\mathfrak{o}$.*

*Proof.* We first prove (*a*) by induction on $n$. The case $n = 1$ follows from (2.1). Inductively suppose that $\mathfrak{o}[a_1, \ldots, a_{n-1}]$ is a finitely generated module over $\mathfrak{o}$. By (2.1) $\mathfrak{o}[a_n]$ is a finitely generated $\mathfrak{o}$-module, and so $\mathfrak{o}[a_1, \ldots, a_{n-1}, a_n]$ is a finitely generated $\mathfrak{o}[a_1, \ldots, a_{n-1}]$-module; the result now follows from (2.3).

To prove (*b*) note that $\mathfrak{O}[r]$ is a finitely generated $\mathfrak{O}$-module by (2.1), while $\mathfrak{O}$ is a finitely generated $\mathfrak{o}$-module by hypothesis. The result therefore follows from (2.3) and (2.2).

□

**(2.4.*c*) Corollary** *The integral closure of an integral domain $\mathfrak{o}$ is integrally closed.*

*Proof.* Let $\mathfrak{C}$ denote the integral closure of $\mathfrak{o}$, and let $r$ be integral over $\mathfrak{C}$, satisfying the equation $r^n + a_1 r^{n-1} + \cdots + a_n = 0$, with $a_i \in \mathfrak{C}$. The result then follows from (2.4).

□

By (2.4.*a*) we see that if $a, b \in \mathfrak{O}$ are integral in $\mathfrak{o}$, then $\mathfrak{o}[a, b]$ is a finitely generated $\mathfrak{o}$-module; since $ab$, $a + b$, $a - b$ all lie in $\mathfrak{o}[a, b]$, we deduce that these elements themselves are all integral over $\mathfrak{o}$, by (2.2). We have therefore shown

**(2.4.*d*)** *The set of elements in $\mathfrak{O}$ which are integral over $\mathfrak{o}$ form a ring.*

In particular if $\mathfrak{o}$ is an integral domain, then, on taking $\mathfrak{O}$ to be the field of fractions $K$ of $\mathfrak{o}$, we have shown

**(2.5)** *If $\mathfrak{o}$ is an integral domain, then its integral closure in its field of fractions $K$ is a subring of $K$.*

Let $K$ now denote a number field, that is to say a finite extension of the rationals $\mathbb{Q}$. We denote the integral closure of $\mathbb{Z}$ in $K$ by $\mathfrak{o}_K$; thus $\mathfrak{o}_K$ is precisely the set of all algebraic integers in $K$. By (2.5) we know that $\mathfrak{o}_K$ is a ring. The study of the properties of the ring $\mathfrak{o}_K$ is of fundamental importance in algebraic number theory.

Next we give a very useful criterion for determining whether or not a given element is integral over an integral domain $\mathfrak{o}$: needless to say, the main application which we have in mind, is a criterion for determining whether an algebraic number is an algebraic integer or not.

**(2.6)** *Let $\mathfrak{o}$ denote an integrally closed integral domain with field of fractions $K$, and let $E$ denote a finite, separable extension of $K$; then $a \in E$ is integral over $\mathfrak{o}$ iff the characteristic polynomial $c_{E/K,a}(X)$ lies in $\mathfrak{o}[X]$.*

*Proof.* The result is clear if $c_{E/K,a}(X)$ is assumed to have coefficients in $\mathfrak{o}$, since $a$ is a root of $c_{E/K,a}$. Conversely, we now suppose that $a$ is integral over $\mathfrak{o}$, so that $p(a) = 0$ for some monic polynomial $p$ with coefficients in $\mathfrak{o}$. Let $\sigma\colon E \hookrightarrow E^c$ denote a field embedding of $E$, over $K$, into an algebraic closure of $E$. Then $0 = (p(a))^\sigma = p(a^\sigma)$, so that $a^\sigma$ is also integral over $\mathfrak{o}$. Thus if $\{\sigma_i\}$, $i = 1, \ldots, (E : K)$ denotes the full set of embeddings of $E$, over $K$, into $E^c$, then all the symmetric functions in the $\{a^{\sigma_i}\}$ are integral over $\mathfrak{o}$ and lie in $K$; therefore, because $\mathfrak{o}$ is integrally closed, they must all lie in $\mathfrak{o}$. This then shows that $c_{E/K,a}(X) = \prod_i (X - a^{\sigma_i})$ lies in $\mathfrak{o}[X]$.

□

Keeping the above hypotheses, we deduce

**(2.6.$a$) Corollary.** *An element $a \in E$ is integral iff $m_{K,a}(X) \in \mathfrak{o}[X]$.*

*Proof.* If we take $E$ to be $K(a)$ in the above, then $c_{E/K,a} = m_{K,a}$.

□

We have now seen that, in many respects, the notion of integrality is far easier to handle when cast in terms of the finite generation of modules. This is the first of many instances when we need a finiteness condition on the modules we treat. In consequence, it will be particularly advantageous to consider modules, all of whose submodules are automatically finitely generated.

An $\mathfrak{o}$-module $M$ is called a *Noetherian* $\mathfrak{o}$*-module* if all its $\mathfrak{o}$-submodules are finitely generated over $\mathfrak{o}$. We call the ring $\mathfrak{o}$ a *Noetherian ring* if $\mathfrak{o}$ is a Noetherian $\mathfrak{o}$-module, i.e. if all the ideals of $\mathfrak{o}$ are finitely generated over $\mathfrak{o}$.

As illustrative examples, note that any principal ideal domain is automatically a Noetherian ring; also any finite ring is a Noetherian ring, and any finite module is a Noetherian module.

It is frequently useful to interpret the Noetherian property for modules in terms of a chain condition. Let $\{M_i\}$, $i = 1, 2, 3, \ldots$ denote an ascending chain of $\mathfrak{o}$-modules,

$$M_1 \subset M_2 \subset M_3 \cdots \subset M_n \subset M_{n+1} \subset \cdots.$$

We say that the chain stabilizes if there exists an integer $k$ such that $M_j = M_k$ for all $j \geq k$.

**(2.7)** *The following conditions are equivalent:*

(*a*) *$M$ is a Noetherian $\mathfrak{o}$-module;*
(*b*) *every ascending chain of $\mathfrak{o}$-submodules of $M$ stabilizes;*
(*c*) *every non-empty family of $\mathfrak{o}$-submodules of $M$ contains a maximal element (that is to say maximal with respect to the partial ordering induced by inclusion).*

*Proof.* ($a$)$\Rightarrow$($b$): Let $\{M_i\}$ denote a chain of $\mathfrak{o}$-modules and set $\overline{M} = \bigcup_{i=1}^{\infty} M_i$. Then $\overline{M}$ is a submodule of $M$: indeed, if $m, n \in \overline{M}$, then there exists some $k$ such that $m, n \in M_k$, hence $m\lambda + n\mu \in M_k \subset \overline{M}$ for all $\lambda, \mu \in \mathfrak{o}$. By hypothesis we can find $\mathfrak{o}$-generators $m_1, \ldots, m_n$ of $\overline{M}$, and so each $m_i$ belongs to some $M_{\ell_i}$; we then set $\ell = \max_{1 \leq i \leq n}(\ell_i)$; clearly each $m_i \in M_\ell$, so $\overline{M} \subset M_\ell$, therefore $M_\ell = \overline{M}$, and so the chain stabilizes at $\ell$.

($b$)$\Rightarrow$($c$): Suppose that $\mathcal{S}$ is a non-empty family of $\mathfrak{o}$-submodules of $M$ which has no maximal element. Suppose inductively that we have constructed a chain

$$M_1 \subsetneqq M_2 \subsetneqq M_3 \cdots \subsetneqq M_n$$

of submodules in $\mathcal{S}$. By hypothesis $M_n$ is not maximal in $\mathcal{S}$, so there exists $M_{n+1} \in \mathcal{S}$ with $M_{n+1} \supsetneqq M_n$. Proceeding in this fashion, we construct an infinite chain of submodules of $M$ which fails to stabilize.

($c$)$\Rightarrow$($a$): Suppose that $M$ contains a submodule $N$ which is not finitely generated over $\mathfrak{o}$. We let $\mathcal{S}$ denote the set of finitely generated submodules of $N$: $(0) \in \mathcal{S}$ so that $\mathcal{S}$ is non-empty; however, if $P \in \mathcal{S}$

were maximal then, since $N$ is not itself finitely generated, there exists $n \in N$, $n \notin P$, and so $(P, n) \in \mathcal{S}$, contradicting the maximality of $P$.

□

Next we consider the behaviour of the Noetherian property with respect to exact sequences of $\mathfrak{o}$-modules.

**(2.8)** *Let* $0 \to M \xrightarrow{\rho} N \xrightarrow{\pi} P \to 0$ *denote an exact sequence of* $\mathfrak{o}$*-modules; then* $N$ *is a Noetherian* $\mathfrak{o}$*-module iff* $M$ *and* $P$ *are both Noetherian* $\mathfrak{o}$*-modules.*

*Proof.* First suppose that $N$ is Noetherian: then every submodule of $M$ is isomorphic to a submodule of $N$ via $\rho$, and it is therefore finitely generated over $\mathfrak{o}$. If $Q$ is a submodule of $P$, then $\pi^{-1}(Q)$ is a finitely generated $\mathfrak{o}$-submodule of $N$, and hence $Q$ itself must be finitely generated over $\mathfrak{o}$.

Conversely, let $\{N_i\}$, $i = 1, 2, 3\ldots$ denote an ascending chain of $N$-submodules, and suppose that $P$ and $M$ are both Noetherian $\mathfrak{o}$-modules. By (2.7) we can find an integer $n$, such that both chains $\{\pi(N_i)\}$, $\{\rho(M) \cap N_i\}$ stabilize after their $n$th term. We claim that $\{N_i\}$ must stabilize after its $n$th term. Indeed, given $m \geq n$, we are required to show $N_m \subset N_n$. By construction, for $a \in N_m$, we can find $b \in N_n$ such that $\pi(a) = \pi(b)$; so by exactness, $a - b = \rho(c)$ for some $c \in M$. Hence

$$a - b \in \rho(M) \cap N_m = \rho(M) \cap N_n$$

and so $a \in N_n$, as required.

□

We can use (2.8) to construct many new kinds of Noetherian modules; for instance we have

**(2.9)** *Let* $\mathfrak{o}$ *be a Noetherian ring and let* $M$ *be a finitely generated* $\mathfrak{o}$*-module; then* $M$ *is a Noetherian* $\mathfrak{o}$*-module.*

*Proof.* Applying (2.8) repeatedly, we conclude that each free module $\bigoplus_1^n \mathfrak{o}$ is a Noetherian $\mathfrak{o}$-module. Since $M$ is a finitely generated $\mathfrak{o}$-module, it is a homomorphic image of such a free $\mathfrak{o}$-module, and so, by (2.8), $M$ is Noetherian.

□

As an illustration of the power of the Noetherian property, we give an alternative proof of (2.5) when $\mathfrak{o}$ is a Noetherian ring; this proof does not depend on (2.2). So now let $\mathfrak{o}$ denote a Noetherian integral domain with field of fractions $K$, and let $a, b$ be elements of $K$ which are integral over $\mathfrak{o}$; by (2.1), $\mathfrak{o}[a]$ and $\mathfrak{o}[b]$ are finitely generated $\mathfrak{o}$-modules, so that $\mathfrak{o}[a, b]$ is a finitely generated $\mathfrak{o}$-module by (2.3); from (2.9) we deduce that $\mathfrak{o}[a, b]$ is a Noetherian $\mathfrak{o}$-module, and so the $\mathfrak{o}$-submodules $\mathfrak{o}[ab]$, $\mathfrak{o}[a-b]$ are finitely generated $\mathfrak{o}$-modules; thus by (2.1) we conclude that $ab$ and $a-b$ are integral over $\mathfrak{o}$.

Next we consider the question of constructing a useful supply of Noetherian rings. Our first result in this direction is

**(2.10)** *Let $\phi: \mathfrak{o} \to \mathfrak{R}$ denote a surjective homomorphism of rings. If $\mathfrak{o}$ is a Noetherian ring, then $\mathfrak{R}$ is a Noetherian ring.*

*Proof.* We view each $\mathfrak{R}$-ideal as an $\mathfrak{o}$-module via $\phi$. Since $\mathfrak{R}$ is generated over $\mathfrak{o}$ by $1_{\mathfrak{R}}$, by the surjectivity hypothesis we conclude from (2.9) that $\mathfrak{R}$ is a Noetherian $\mathfrak{o}$-module; thus each $\mathfrak{R}$-ideal is a finitely generated $\mathfrak{o}$-module, and hence is a finitely generated $\mathfrak{R}$-module.

□

The most important result for constructing new Noetherian rings is the following result, which is known as Hilbert's Basis Theorem:

**(2.11)** *Let $\mathfrak{o}$ denote a Noetherian ring; then the polynomial ring $\mathfrak{o}[X]$ is also a Noetherian ring.*

Before proving this fundamental result, we firstly deduce three important corollaries.

**Corollary 1.** *If $\mathfrak{o}$ is a Noetherian ring, then the polynomial ring over $\mathfrak{o}$ in any finite number of variables is also a Noetherian ring.*

*Proof.* Suppose inductively that $\mathfrak{o}[X_1, \ldots, X_n]$ is a Noetherian ring; then

$$\mathfrak{o}[X_1, \ldots, X_n, X_{n+1}] = \mathfrak{o}[X_1, \ldots, X_n][X_{n+1}]$$

is a Noetherian ring by (2.11).

□

We shall call $\mathfrak{O}$ a *finitely generated ring extension of* $\mathfrak{o}$, if we can write

$$\mathfrak{O} = \mathfrak{o}[a_1, \ldots, a_n]$$

for some finite set of elements $a_i \in \mathfrak{O}$. The ring $\mathfrak{O}$ is said to be a *finitely generated ring* if $\mathfrak{O}$ is finitely generated over the prime ring $\overline{\mathbb{Z}}$ (i.e. $\overline{\mathbb{Z}}$ is the canonical homomorphic image of $\mathbb{Z}$ in $\mathfrak{O}$). As an immediate consequence of (2.10) and the above corollary, we deduce:

**Corollary 2.** *If $\mathfrak{o}$ is a Noetherian ring and $\mathfrak{O}$ is a finitely generated ring extension of $\mathfrak{o}$, then $\mathfrak{O}$ is a Noetherian ring.*

Since $\overline{\mathbb{Z}}$ is always a principal ideal domain, it is a Noetherian ring and so

**Corollary 3.** *If $\mathfrak{O}$ is a finitely generated ring, then $\mathfrak{O}$ is a Noetherian ring.*

We now prove (2.11). Let $\mathfrak{a}$ denote an $\mathfrak{o}[X]$ ideal: we shall exhibit a finite set of generators for $\mathfrak{a}$ over $\mathfrak{o}[X]$. To this end we let $\mathfrak{b}$ denote the $\mathfrak{o}$-ideal of leading coefficients of elements in $\mathfrak{a}$. Since $\mathfrak{o}$ is Noetherian, we may choose a finite set of generators $\{a_i\}$, $i = 1, \ldots n$ for $\mathfrak{b}$ over $\mathfrak{o}$. We suppose that $a_i$ is the leading coefficient of $f_i \in \mathfrak{a}$: we then set $d_i = \deg(f_i)$; $d = \max_i(d_i)$, and we let $\mathfrak{c}$ denote the $\mathfrak{o}$-module of elements in $\mathfrak{a}$ with degree less than or equal to $d$, together with 0. Since $\mathfrak{c}$ is a submodule of the $\mathfrak{o}$-free module $\sum_{i=0}^{d} X^i \mathfrak{o}$, we deduce that $\mathfrak{c}$ is Noetherian: we let $\{g_j\}$, $j = 1, \ldots, m$, denote a set of generators for $\mathfrak{c}$ over $\mathfrak{o}$. We let $\mathfrak{d}$ denote the $\mathfrak{o}[X]$ ideal generated by all the $f_i$ and $g_j$: we claim $\mathfrak{d} = \mathfrak{a}$. The inclusion $\mathfrak{a} \supset \mathfrak{d}$ is clear. Conversely let $h \in \mathfrak{a}$: we shall show by induction on $m$, the degree of $h$, that $h \in \mathfrak{d}$. If $m \leq d$, the result is immediate. So now we take $m > d$, and we let $a$ denote the leading coefficient of $h$. We write $a = \sum_{i=1}^{n} a_i \lambda_i$ with $\lambda_i \in \mathfrak{o}$. The result then follows from the induction hypothesis, since $h - \sum_i \lambda_i X^{m-d_i} \cdot f_i$ lies in $\mathfrak{a}$ and has degree less than $m$.

□

# II

# Dedekind Domains

## §1 Algebraic Theory

From (I.2.5) we know that the set of algebraic integers in an algebraic number field $K$ forms an integral domain. We shall introduce the ring theoretic notion which describes the basic algebraic properties of such integral domains. After the mathematician who first studied them systematically, they are known as Dedekind domains.

The concept of a principal ideal domain turns out to be too restrictive: as we have already seen in the introduction, the ring of algebraic integers of $\mathbb{Q}(\sqrt{-6})$ possesses non-principal ideals. To be more precise the ring of integers of $\mathbb{Q}(\sqrt{-6})$ is

$$\mathbb{Z}[\sqrt{-6}] = \{a + b\sqrt{-6} \mid a, b \in \mathbb{Z}\}$$

and the subset of numbers of the form $2a + b\sqrt{-6}$ is a $\mathbb{Z}[\sqrt{-6}]$ ideal, which we denote by $I$. We first show directly that $I$ is not principal: if $I = (2x + y\sqrt{-6})$, then $2x + y\sqrt{-6}$, and hence also $2x - y\sqrt{-6}$, divides both 2 and $\sqrt{-6}$. Thus $4x^2 + 6y^2$ divides both 4 and 6 (in $\mathbb{Z}$), and hence also divides 2: this, however, is clearly impossible. Alternatively, we could note that $-6$ possesses two distinct factorisations into irreducible elements of $\mathbb{Z}[\sqrt{-6}]$

$$\sqrt{-6} \cdot \sqrt{-6} = -6 = -2 \cdot 3.$$

So $\mathbb{Z}[\sqrt{-6}]$ is not a unique factorisation domain, and so is certainly not a principal ideal domain.

In a Dedekind domain the unique factorisation into a product of prime elements does not necessarily obtain. The notion of an ideal, however, permits us to establish such a unique factorisation of ideals into prime ideals (see Theorem 2 below). In fact the notion of 'ideal numbers' was first introduced by Kummer precisely to circumvent the problem of non-unique factorisation of elements in rings of integers; the notion of ideal, as we now understand it, was formulated later by Dedekind. Principal ideal domains are instances of Dedekind domains; thus, in particular, $\mathbb{Z}$ is a Dedekind domain. Presently we shall show that if $L$ is a finite separable extension of the field of fractions $K$ of a Dedekind domain $\mathfrak{o}$, then $\mathfrak{O}$, the integral closure of $\mathfrak{o}$ in $L$, is a Dedekind domain. Applying this to $\mathbb{Z}$, we can then deduce that the ring of integers in an algebraic number field $K$ is a Dedekind domain.

**(1.1) Definition.** *An integral domain $\mathfrak{o}$ is a Dedekind domain if*

(*a*) *$\mathfrak{o}$ is a Noetherian ring;*
(*b*) *$\mathfrak{o}$ is integrally closed in its field of fractions $K$;*
(*c*) *all non-zero prime ideals of $\mathfrak{o}$ are maximal ideals.*

Recall that an $\mathfrak{o}$-ideal $\mathfrak{a}$ is said to be maximal if $\mathfrak{a} \neq \mathfrak{o}$ and $\mathfrak{a} \subset \mathfrak{b} \subsetneqq \mathfrak{o}$ for an $\mathfrak{o}$-ideal $\mathfrak{b}$, implies that $\mathfrak{a} = \mathfrak{b}$. As indicated at the start of the book, the term prime ideal is henceforth to be taken as meaning non-zero prime ideal.

The majority of results in this section show that Dedekind domains retain many of the desirable properties of $\mathbb{Z}$.

We begin by showing

**(1.2)** *A principal ideal domain is a Dedekind domain.*

*Proof.* Let $\mathfrak{o}$ denote a principal ideal domain. From (I,§2), we know that $\mathfrak{o}$ is a Noetherian ring.

Now let $(a)$ denote a prime ideal of $\mathfrak{o}$, and let $(b)$ be a maximal ideal which contains $a$. Thus $a = bc$, for some $c \in \mathfrak{o}$. Since $(a)$ is a prime ideal, either $b \in (a)$, and so $(a) = (b)$ as we had to show; or else $c = ad$, $d \in \mathfrak{o}$. In the latter case $a = abd$, so that $b \in \mathfrak{o}^*$ and $(b) = \mathfrak{o}$, which is contrary to hypothesis. Thus the prime ideals are all maximal.

Next suppose that $a/b$ is integral over $\mathfrak{o}$, with $a, b \in \mathfrak{o}$, $b \neq 0$ and $(a, b) = 1$. Then for some $c_0, c_1, \ldots, c_{n+1} \in \mathfrak{o}$ we have

$$(a/b)^n + c_{n-1}(a/b)^{n-1} + \cdots + c_1(a/b) + c_0 = 0$$

and so

$$a^n + bc_{n-1}a^{n-1} + \cdots + b^{n-1}c_1 a + b^n c_0 = 0.$$

Therefore $b \mid a^n$. As $(a,b)=1$, it follows that $b \in \mathfrak{o}^*$, and so $a/b \in \mathfrak{o}$. We have thus shown that $\mathfrak{o}$ is integrally closed in its field of fractions, and hence $\mathfrak{o}$ is a Dedekind domain.

□

We now have to introduce a number of further definitions; these apply to any integral domain $\mathfrak{o}$, whose field of fractions we denote by $K$. Firstly we extend the notion of an $\mathfrak{o}$-ideal to that of a *fractional ideal.* This is an $\mathfrak{o}$-submodule $\mathfrak{b}$ of $K$ of the form

(1.3) $$\mathfrak{b} = c\mathfrak{a} = \{x \in K | x = ca \text{ for some } a \in \mathfrak{a}\}$$

where $c \in K^*$ and $\mathfrak{a}$ is a *non-zero* $\mathfrak{o}$-ideal. Thus a non-zero $\mathfrak{o}$-ideal is the same thing as a fractional $\mathfrak{o}$-ideal which lies in $\mathfrak{o}$. If $\mathfrak{o}$ is a Noetherian ring, then clearly a fractional $\mathfrak{o}$-ideal is finitely generated over $\mathfrak{o}$. Conversely, a non-zero finitely generated $\mathfrak{o}$-submodule $\mathfrak{b}$ of $K$ is of the form $c\mathfrak{a}$, with $\mathfrak{a}$ a non-zero $\mathfrak{o}$-ideal and with $c = b^{-1}$, $b \in \mathfrak{o}$: indeed, we may choose a finite generating set of $\mathfrak{b}$ of the form $\{a_1 b^{-1}, \ldots, a_n b^{-1}\}$ with all $a_j, b \in \mathfrak{o}$; then $\{a_1, \ldots, a_n\}$ is the generating set of an $\mathfrak{o}$-ideal $\mathfrak{a}$.

We define the *product* $\mathfrak{b}_1 \cdot \mathfrak{b}_2$ of fractional ideals of an integral domain $\mathfrak{o}$ to be the additive group of finite sums $\sum x_i y_i$ with $x_i \in \mathfrak{b}_1$, $y_i \in \mathfrak{b}_2$. This is evidently again a fractional $\mathfrak{o}$-ideal: first verify this for $\mathfrak{o}$-ideals, then use (1.3). This multiplication on the set of fractional $\mathfrak{o}$-ideals is associative, commutative and has an identity element $\mathfrak{o}$ since $\mathfrak{b}\mathfrak{o} = \mathfrak{b}$ for all $\mathfrak{b}$. It also preserves inclusions in the sense that

(1.4.*a*) $$\text{if} \quad \mathfrak{a}_1 \subset \mathfrak{a}_2, \quad \text{then} \quad \mathfrak{a}_1\mathfrak{b} \subset \mathfrak{a}_2\mathfrak{b}.$$

In particular note that

(1.4.*b*) $$\text{if} \quad \mathfrak{a} \subset \mathfrak{o}, \text{ i.e. } \mathfrak{a} \text{ is an } \mathfrak{o}\text{-ideal, then } \mathfrak{a}\mathfrak{b} \subset \mathfrak{b}.$$

We extend the notion of an ideal being principal to fractional ideals, in the obvious way: for $a \in K^*$ we write

$$(a) = a\mathfrak{o} = \{x \in K | x = ay \text{ for } y \in \mathfrak{o}\}$$

and any fractional ideal of this form is said to be *principal.* We then clearly have

(1.5) $$(a)(b) = (ab).$$

We can now state the basic theorem on the factorisation of ideals into prime ideals.

**Theorem 2.** *Every non-zero ideal $\mathfrak{a}$ of a Dedekind domain $\mathfrak{o}$ can be written as a product*

$$\mathfrak{a} = \mathfrak{p}_1 \cdots \mathfrak{p}_n$$

*of prime ideals $\mathfrak{p}_i$ of $\mathfrak{o}$; moreover this representation is unique up to the order of the factors.*

Here, and in the sequel, we adhere to the usual convention that the empty product is taken to mean $\mathfrak{o}$.

The above theorem is closely connected with a result concerning the invertibility of fractional ideals. Here we call a fractional $\mathfrak{o}$-ideal $\mathfrak{a}$ *invertible* if there exists a fractional $\mathfrak{o}$-ideal $\mathfrak{b}$ such that $\mathfrak{a}\mathfrak{b} = \mathfrak{o}$.

**Theorem 3.** *Every fractional ideal of a Dedekind domain $\mathfrak{o}$ is invertible.*

In the sequel $\mathfrak{o}$ is still just an integral domain with field of fractions $K$, and any further conditions imposed will always be explicitly stated.

At this stage it is advantageous to introduce an additional operation on fractional $\mathfrak{o}$-ideals. We write

(1.6.*a*) $$\mathfrak{a}^{-1} = \{x \in K \mid x\mathfrak{a} \subset \mathfrak{o}\}.$$

Note that obviously if $a \in K^*$, then

(1.6.*b*) $$(a)^{-1} = (a^{-1})$$

and more generally if $a \in \mathfrak{a}$, then $\mathfrak{a}^{-1} \subset (a^{-1})$. Therefore, if $\mathfrak{o}$ is a Noetherian ring then $\mathfrak{a}^{-1}$ is a finitely generated $\mathfrak{o}$-module, and hence is a fractional $\mathfrak{o}$-ideal. If now $\mathfrak{a}$ is invertible, so that $\mathfrak{a}\mathfrak{b} = \mathfrak{o}$ for some fractional ideal $\mathfrak{b}$, then clearly $\mathfrak{b} \subset \mathfrak{a}^{-1}$; hence $\mathfrak{o} = \mathfrak{a}\mathfrak{b} \subset \mathfrak{a}\mathfrak{a}^{-1} \subset \mathfrak{o}$; thus $\mathfrak{a}\mathfrak{a}^{-1} = \mathfrak{o}$ and so $\mathfrak{b} = \mathfrak{b}\mathfrak{a} \cdot \mathfrak{a}^{-1} = \mathfrak{a}^{-1}$. However, it is important to understand that at this stage we shall only use the notation $\mathfrak{a}^{-1}$ in the sense of (1.6.*a*); that is to say we do not necessarily assume that $\mathfrak{a}$ is invertible.

The proof of both theorems is based on a number of intermediate steps. For any fractional $\mathfrak{o}$-ideal $\mathfrak{a}$ we write

$$R_{\mathfrak{a}} = \{x \in K \mid x\mathfrak{a} \subset \mathfrak{a}\}.$$

Clearly $R_{\mathfrak{a}}$ is an extension ring of $\mathfrak{o}$.

**(1.7)** *If $\mathfrak{o}$ is a Noetherian ring which is integrally closed, then $R_{\mathfrak{a}} = \mathfrak{o}$.*

*Proof.* Let $a_1, \ldots, a_n$ denote a generating set of $\mathfrak{a}$ over $\mathfrak{o}$, and let $b \in R_{\mathfrak{a}}$. Then

$$ba_i = \sum_j c_{ij}a_j \text{ with } c_{ij} \in \mathfrak{o}.$$

Thus $\det(b \cdot 1_n - (c_{ij}))(a_j) = 0$ for each $j$, and so $b$ is a root of the monic polynomial $\det(X \cdot 1_n - (c_{ij}))$, whose coefficients all lie in $\mathfrak{o}$. Hence $b \in \mathfrak{o}$.

□

**(1.8)** *If all the prime ideals of an integral domain $\mathfrak{o}$ are maximal, then an inclusion*

$$\mathfrak{p} \supset \mathfrak{p}_1 \dots \mathfrak{p}_n$$

*where $\mathfrak{p}$ and all the $\mathfrak{p}_j$ are non-zero prime ideals, implies that $\mathfrak{p} = \mathfrak{p}_i$ for some $i$.*

*Proof.* We argue by induction on $r$. If $r = 1$, $\mathfrak{p} \supset \mathfrak{p}_1$, then the fact that $\mathfrak{p}_1$ is maximal implies $\mathfrak{p} = \mathfrak{p}_1$.

Next let $r > 1$. If $\mathfrak{p} \neq \mathfrak{p}_r$, so that $\mathfrak{p} \not\supset \mathfrak{p}_r$, then there exists $c \in \mathfrak{p}_r$, $c \notin \mathfrak{p}$. We choose $b \in \mathfrak{p}_1 \cdots \mathfrak{p}_{r-1}$. Then $bc \in \mathfrak{p}$, and so by primality $b \in \mathfrak{p}$ since $c \notin \mathfrak{p}$; thus $\mathfrak{p} \supset \mathfrak{p}_1 \cdots \mathfrak{p}_{r-1}$.

□

For the moment we call a non-zero ideal $\mathfrak{a}$ of an integral domain *weakly invertible* if we can find $c \in \mathfrak{a}^{-1}$ with $c \notin \mathfrak{o}$. Note that $\mathfrak{o}$ is not weakly invertible. We write $\mathcal{S}$ for the set of weakly invertible $\mathfrak{o}$-ideals, so that $\mathcal{S}$ is partially ordered by inclusion. Next we observe that $\mathcal{S}$ is non-empty if $\mathfrak{o}$ is not a field; for then we can find $a \in \mathfrak{o} \setminus \mathfrak{o}^*$, and it is clear that $(a)$ is weakly invertible: indeed by (1.6.b) $(a)$ is invertible. If $\mathfrak{o}$ is a Noetherian ring, then by (I.2.7) we see that $\mathcal{S}$ possesses maximal elements.

**(1.9)** *Let $\mathfrak{o}$ be a Dedekind domain which is not a field and let $\mathfrak{m}$ denote a non-zero $\mathfrak{o}$-ideal which is maximal in $\mathcal{S}$; then $\mathfrak{m}$ is an invertible prime ideal of $\mathfrak{o}$.*

*Proof.* We first show that $\mathfrak{m}$ is a prime ideal. Let $a \in \mathfrak{o} \setminus \mathfrak{m}$ and suppose that for $b \in \mathfrak{o}$, $ab \in \mathfrak{m}$: we wish to show $b \in \mathfrak{m}$. By hypothesis there exists $c \in \mathfrak{m}^{-1} \setminus \mathfrak{o}$; furthermore by the maximality property of $\mathfrak{m}$, $(\mathfrak{m}+a)c \not\subset \mathfrak{o}$, and so $ac \notin \mathfrak{o}$. Since $ab \in \mathfrak{m}$, we know that $b(ac) = a(bc) \in \mathfrak{o}$; thus $ac \in (b)^{-1}$ and, since $a \in \mathfrak{o}$, we also know that $ac \in \mathfrak{m}^{-1}$; hence $\mathfrak{m} + b\mathfrak{o}$ is weakly invertible, and therefore by maximality $b \in \mathfrak{m}$.

$\mathfrak{m}\mathfrak{m}^{-1} \not\subset \mathfrak{m}$ since by (1.7) $\mathfrak{m}^{-1} \not\subset R_{\mathfrak{m}}$. However, $\mathfrak{m} \subset \mathfrak{m}\mathfrak{m}^{-1} \subset \mathfrak{o}$; so that $\mathfrak{m} \cdot \mathfrak{m}^{-1} = \mathfrak{o}$, since $\mathfrak{m}$ is a maximal ideal.

□

**(1.10)** *Let $\mathfrak{o}$ be a Dedekind domain. A non-zero $\mathfrak{o}$-ideal $\mathfrak{a}$ is invertible* *iff*

$$\mathfrak{a} = \mathfrak{m}_1 \dots \mathfrak{m}_r$$

*where the* $\mathfrak{m}_j$ *are invertible prime ideals of* $\mathfrak{o}$.

As always the empty product is to be interpreted as $\mathfrak{o}$: this is particularly important in giving a sense to many of our results when $\mathfrak{o}$ is a field.

*Proof.* If $\mathfrak{a}$ is a product of invertible prime ideals as in (1.10), then $\mathfrak{a}^{-1} = \mathfrak{m}_1^{-1} \dots \mathfrak{m}_r^{-1}$. Next, if $\mathfrak{a}$ is a proper invertible ideal, then $\mathfrak{a}^{-1} \supsetneqq \mathfrak{o}$ and so $\mathfrak{a}$ is weakly invertible; hence it is contained in a maximal weakly invertible ideal $\mathfrak{m}_1$. So, by (1.9) and (1.4), $\mathfrak{o} \supset \mathfrak{a}\mathfrak{m}_1^{-1} \supset \mathfrak{a}$ and by (1.7) $\mathfrak{a}\mathfrak{m}_1^{-1} \neq \mathfrak{a}$ and $\mathfrak{a}\mathfrak{m}_1^{-1}$ is also invertible. If $\mathfrak{a}\mathfrak{m}_1^{-1} = \mathfrak{o}$, then $\mathfrak{a} = \mathfrak{m}_1$. Otherwise, since $\mathfrak{o}$ is Noetherian, we can repeat this procedure a finite number of times and thereby obtain an equality $\mathfrak{o} = \mathfrak{a}\mathfrak{m}_1^{-1} \dots \mathfrak{m}_r^{-1}$; therefore $\mathfrak{a} = \mathfrak{m}_1 \dots \mathfrak{m}_r$.

□

**(1.11)** *Every prime ideal* $\mathfrak{p}$ *of a Dedekind domain* $\mathfrak{o}$ *is invertible.*

*Proof.* Let $a \in \mathfrak{p}\backslash 0$; then applying (1.10) to the invertible ideal (a), we can write

$$\mathfrak{p} \supset \mathfrak{m}_1 \dots \mathfrak{m}_r$$

where the $\mathfrak{m}_j$ are invertible primes ideals. By (1.8) we conclude that $\mathfrak{p} = \mathfrak{m}_j$ for some $j$.

□

**(1.12)** *Every non-zero ideal* $\mathfrak{a}$ *of a Dedekind domain* $\mathfrak{o}$ *can be written as a product*

$$\mathfrak{a} = \mathfrak{p}_1 \dots \mathfrak{p}_r$$

*with the* $\mathfrak{p}_i$ *prime ideals of* $\mathfrak{o}$.

*Proof.* If $\mathfrak{a} \neq \mathfrak{o}$, then $\mathfrak{a} \subset \mathfrak{p}$ for some prime ideal $\mathfrak{p}$. By (1.11) $\mathfrak{p}$ is invertible, and so $\mathfrak{o} \supset \mathfrak{p}^{-1}\mathfrak{a} \supset \mathfrak{a}$, with $\mathfrak{p}^{-1}\mathfrak{a} \neq \mathfrak{a}$ by (1.7). We then proceed as per the proof of (1.10).

□

**(1.13)** *The representation of* $\mathfrak{a}$ *in* (1.12) *is unique, up to the order of the constituent factors.*

*Proof.* If $\mathfrak{p}_1 \dots \mathfrak{p}_s = \mathfrak{q}_1 \dots \mathfrak{q}_r$ for non-zero prime ideals $\mathfrak{p}_i$, $\mathfrak{q}_j$; then

$\mathfrak{p}_s \supset \mathfrak{q}_1 \dots \mathfrak{q}_r$, and so, by (1.8), renumbering the $\mathfrak{q}_i$ if necessary, we have $\mathfrak{p}_s = \mathfrak{q}_r$. On multiplying the given equation by $\mathfrak{p}_s^{-1} = \mathfrak{q}_r^{-1}$, we get

$$\mathfrak{p}_1 \dots \mathfrak{p}_{s-1} = q_1 \dots \mathfrak{q}_{r-1}$$

and the result follows by induction on $s$.

□

Theorem 2 now follows from (1.12) and (1.13). By (1.10), (1.11) and (1.12) we see that every non-zero $\mathfrak{o}$-ideal is invertible. Now clearly $(\mathfrak{a}c)^{-1} = \mathfrak{a}^{-1}c^{-1}$; so that by (1.3) we conclude that every fractional $\mathfrak{o}$-ideal is invertible: this then establishes Theorem 3.

□

For the remainder of this section $\mathfrak{o}$ will always denote a Dedekind domain with field of fractions $K$. From Theorem 3, we see that the set of fractional $\mathfrak{o}$-ideals, endowed with multiplication, forms an abelian group, which we denote by $I_{\mathfrak{o}}$. The fractional ideal $\mathfrak{a}^{-1}$ introduced in (1.6) is now indeed the inverse of $\mathfrak{a}$ in $I_{\mathfrak{o}}$. If $\mathfrak{a} = \mathfrak{b}(b)^{-1}$ is a fractional $\mathfrak{o}$-ideal, with $\mathfrak{b}$ an $\mathfrak{o}$-ideal, $b \in \mathfrak{o}$, then we may express both $\mathfrak{b}$ and $(b)$ as products of prime ideals. If $(b) = \prod \mathfrak{p}_j$, then of course $(b)^{-1} = \prod \mathfrak{p}_j^{-1}$. We collect together all factors belonging to each prime ideal in the resulting expression; we thereby obtain an equality

(1.14.$a$) $$\mathfrak{a} = \prod_{\mathfrak{p}} \mathfrak{p}^{v_{\mathfrak{p}}(\mathfrak{a})}, \quad v_{\mathfrak{p}}(\mathfrak{a}) \in \mathbb{Z}.$$

This product extends formally over all $\mathfrak{p}$; in effect though, it is only a finite product since the formal exponents $v_{\mathfrak{p}}(\mathfrak{a})$ are zero for all but a finite number of $\mathfrak{p}$. (In the sequel, whenever we write a product $\prod_{\mathfrak{p}} \mathfrak{p}^{v_{\mathfrak{p}}}$ it will be assumed that $v_{\mathfrak{p}} = 0$ p.p., that is to say with only finitely many exceptions.) (1.14.$a$) tells us that the prime ideals of $\mathfrak{o}$ form a generating set of $I_{\mathfrak{o}}$; in fact it is a free generating set in the sense that

$$\prod \mathfrak{p}^{v_{\mathfrak{p}}} = \prod \mathfrak{p}^{u_{\mathfrak{p}}} \Rightarrow v_{\mathfrak{p}} = u_{\mathfrak{p}} \text{ for all } \mathfrak{p}.$$

Indeed, for those $\mathfrak{p}$ such that either $v_{\mathfrak{p}} \neq 0$ or $u_{\mathfrak{p}} \neq 0$, choose $r_{\mathfrak{p}}$ such that $v_{\mathfrak{p}} + r_{\mathfrak{p}} \geq 0$ and $u_{\mathfrak{p}} + r_{\mathfrak{p}} \geq 0$. Then

$$\prod \mathfrak{p}^{v_{\mathfrak{p}}+r_{\mathfrak{p}}} = \prod \mathfrak{p}^{u_{\mathfrak{p}}+r_{\mathfrak{p}}},$$

and, by Theorem 2, $u_{\mathfrak{p}} + r_{\mathfrak{p}} = v_{\mathfrak{p}} + r_{\mathfrak{p}}$; hence $v_{\mathfrak{p}} = u_{\mathfrak{p}}$ for each $\mathfrak{p}$, as required.

The above may be interpreted as saying $\mathfrak{a} \mapsto \bigoplus v_{\mathfrak{p}}(\mathfrak{a})$ induces an isomorphism of groups

(1.14.$b$) $$I_{\mathfrak{o}} \cong \bigoplus_{\mathfrak{p}} \mathbb{Z}.$$

It follows that, given an abelian group $A$ and any set of elements $a(\mathfrak{p}) \in A$, there exists a unique homomorphism

**(1.14.*c*)** $f: I_{\mathfrak{o}} \to A$ *such that* $f(\mathfrak{p}) = a(\mathfrak{p})$ *for all prime ideals* $\mathfrak{p}$.

This result is particularly useful when we are given a map, defined on the non-zero $\mathfrak{o}$-ideals with values in $A$, with the property that $f(\mathfrak{ab}) = f(\mathfrak{a})f(\mathfrak{b})$ for all $\mathfrak{o}$-ideals $\mathfrak{a}, \mathfrak{b}$; for then we can immediately deduce that $f$ extends to a homomorphism

(1.14.*d*)
$$f: I_{\mathfrak{o}} \to A.$$

In the next section we shall see that the functions $v_{\mathfrak{p}}$ defined in $K^*$ by the rule $v_{\mathfrak{p}}(a) = v_{\mathfrak{p}}(a\mathfrak{o})$, lead to a generalisation of the usual notion of absolute value. In fact, they will become a fundamental tool in the development of the whole theory.

## Elementary properties of $v_{\mathfrak{p}}$.

Immediately from the definition we deduce that

(1.15.*a*)
$$v_{\mathfrak{p}}(\mathfrak{ab}) = v_{\mathfrak{p}}(\mathfrak{a}) + v_{\mathfrak{p}}(\mathfrak{b})$$

for fractional $\mathfrak{o}$-ideals $\mathfrak{a}$, $\mathfrak{b}$. Since $v_{\mathfrak{p}}(\mathfrak{o}) = 0$ we conclude that

(1.15.*b*)
$$v_{\mathfrak{p}}(\mathfrak{a}^{-1}) = -v_{\mathfrak{p}}(\mathfrak{a}).$$

From Theorem 2 we deduce that $\mathfrak{a}$ is actually an $\mathfrak{o}$-ideal if, and only if, $v_{\mathfrak{p}}(\mathfrak{a}) \geq 0$ for all $\mathfrak{p}$. We shall say that a fractional ideal $\mathfrak{b}$ divides a fractional ideal $\mathfrak{c}$, written $\mathfrak{b} \mid \mathfrak{c}$, if $\mathfrak{c} = \mathfrak{ba}$ for some $\mathfrak{o}$-ideal $\mathfrak{a}$. We claim that

(1.15.*c*)
$$\begin{aligned} \mathfrak{b} \mid \mathfrak{c} &\iff v_{\mathfrak{p}}(\mathfrak{b}) \leq v_{\mathfrak{p}}(\mathfrak{c}) \text{ for all } \mathfrak{p} \\ &\iff \mathfrak{b} \supset \mathfrak{c}. \end{aligned}$$

The first equivalence follows from what we have just said. By (1.4.*a*) we see that $\mathfrak{b} \mid \mathfrak{c}$ implies that $\mathfrak{b} \supset \mathfrak{c}$. Conversely, if $\mathfrak{b} \supset \mathfrak{c}$ then $\mathfrak{cb}^{-1} \subset \mathfrak{o}$ and so $\mathfrak{b} \mid \mathfrak{b}(\mathfrak{cb}^{-1}) = \mathfrak{c}$.

From (1.15.*c*) we now deduce that

(1.15.*d*)
$$v_{\mathfrak{p}}(\mathfrak{a} \cap \mathfrak{b}) = \sup(v_{\mathfrak{p}}(\mathfrak{a}), v_{\mathfrak{p}}(\mathfrak{b})).$$

The sum $\mathfrak{a} + \mathfrak{b}$, consisting of all sums $a + b$, for $a \in \mathfrak{a}$, $b \in \mathfrak{b}$ is the least fractional ideal, under inclusion, which contains both $\mathfrak{a}$ and $\mathfrak{b}$. Therefore

(1.15.*e*)
$$v_{\mathfrak{p}}(\mathfrak{a} + \mathfrak{b}) = \mathrm{Inf}(v_{\mathfrak{p}}(\mathfrak{a}), v_{\mathfrak{p}}(\mathfrak{b})).$$

From (1.15.*a*,*d*,*e*) we immediately deduce that

(1.15.*f*)
$$v_{\mathfrak{p}}(\mathfrak{a} \cap \mathfrak{b}) + v_{\mathfrak{p}}(\mathfrak{a} + \mathfrak{b}) = v_{\mathfrak{p}}(\mathfrak{ab}).$$

Next we have

**(1.15.$g$)** *Let $\mathfrak{a}$, $\mathfrak{b}$ denote non-zero $\mathfrak{o}$-ideals; then the following statements are equivalent:*

*(i)* $\mathfrak{a} + \mathfrak{b} = \mathfrak{o}$

*(ii)* $\mathfrak{a} \cap \mathfrak{b} = \mathfrak{a}\mathfrak{b}$

*(iii)* $v_{\mathfrak{p}}(\mathfrak{a}) \cdot v_{\mathfrak{p}}(\mathfrak{b}) = 0$ for all $\mathfrak{p}$.

*Proof.* The equivalence of the three conditions follows immediately from (1.15.$f$,$e$).

□

When these conditions hold, we shall say that $\mathfrak{a}$ and $\mathfrak{b}$ are coprime.

For a non-zero $\mathfrak{o}$-ideal $\mathfrak{a}$, we write $\mathfrak{o}/\mathfrak{a}$ for the residue class ring $\mathfrak{o}$ mod $\mathfrak{a}$. We write $x_{\mathfrak{a}}$ for the image in $\mathfrak{o}/\mathfrak{a}$ of $x$ under the quotient group homomorphism $\mathfrak{o} \to \mathfrak{o}/\mathfrak{a}$; thus for $x, y \in \mathfrak{o}$, we have equalities

$$(x + y)_{\mathfrak{a}} = x_{\mathfrak{a}} + y_{\mathfrak{a}}$$

$$(x \cdot y)_{\mathfrak{a}} = x_{\mathfrak{a}} \cdot y_{\mathfrak{a}}.$$

Alternatively we can also adopt the standard congruence notation and write $x \equiv y \bmod \mathfrak{a}$ if $x_{\mathfrak{a}} = y_{\mathfrak{a}}$ (or, equivalently, if $x - y \in \mathfrak{a}$). Then, in the usual way, $x \equiv x_1 \bmod \mathfrak{a}$, $y \equiv y_1 \bmod \mathfrak{a}$ implies congruences

$$x + y \equiv x_1 + y_1 \bmod \mathfrak{a}$$

$$xy \equiv x_1 y_1 \bmod \mathfrak{a}.$$

More generally if $\mathfrak{a}$, $\mathfrak{b}$ denote fractional $\mathfrak{o}$-ideals with $\mathfrak{a} \supset \mathfrak{b}$, then the quotient group $\mathfrak{a}/\mathfrak{b}$ is defined. In this case we need the following result:

**(1.16)** *Let $\mathfrak{p}$ denote a non-zero prime ideal of $\mathfrak{o}$ and let $r \in \mathbb{Z}$; then we have an isomorphism of additive groups*

$$\mathfrak{o}/\mathfrak{p} \cong \mathfrak{p}^r/\mathfrak{p}^{r+1}.$$

*Proof.* Let $a \in \mathfrak{p}^r \setminus \mathfrak{p}^{r+1}$: this is possible since $\mathfrak{p}^r \neq \mathfrak{p}^{r+1}$. The map $x \mapsto xa$ for $x \in \mathfrak{o}$, yields isomorphisms $\mathfrak{o} \cong \mathfrak{o}a$, $\mathfrak{p} \cong \mathfrak{p}a$, and hence induces an isomorphism

$$\mathfrak{o}/\mathfrak{p} \cong \mathfrak{o}a/\mathfrak{p}a = \mathfrak{p}^r\mathfrak{a}/\mathfrak{p}^{r+1}\mathfrak{a}$$

where $(a) = \mathfrak{p}^r\mathfrak{a}$ with $\mathfrak{a}$ an $\mathfrak{o}$-ideal which is coprime to $\mathfrak{p}$. By (1.15.$d$,$e$), we therefore have $\mathfrak{a}\mathfrak{p}^r + \mathfrak{p}^{r+1} = \mathfrak{p}^r$, $\mathfrak{a}\mathfrak{p}^r \cap \mathfrak{p}^{r+1} = \mathfrak{a}\mathfrak{p}^{r+1}$, and so we can now conclude that the map $x \mapsto x + \mathfrak{p}^{r+1}$, for $x \in \mathfrak{a}\mathfrak{p}^r$, induces an isomorphism

$$\mathfrak{p}^r\mathfrak{a}/\mathfrak{p}^{r+1}\mathfrak{a} \cong \mathfrak{p}^r/\mathfrak{p}^{r+1}.$$

□

We now come to an important result which generalises the classical Chinese Remainder Theorem for $\mathbb{Z}$ to all Dedekind domains. A number of alternative forms of this result will be encountered in later sections of this chapter.

**Theorem 4.** *Let $\mathfrak{p}_j$ for $j = 1, 2, \ldots, n$ denote distinct prime ideals of $\mathfrak{o}$, and let $r_j$ for $j = 1, \ldots, n$ denote positive integers. Then the map given by the product of the quotient maps*

$$f: \mathfrak{o} \to \prod_{j=1}^{n} \left[\mathfrak{o}/\mathfrak{p}_j^{r_j}\right]$$

*yields an isomorphism of rings: $\mathfrak{o}/\prod \mathfrak{p}_j^{r_j} \cong \prod(\mathfrak{o}/\mathfrak{p}_j^{r_j})$.*

In terms of congruences this means that given $x_j \in \mathfrak{o}$ for $j = 1, \ldots n$, there exists $x \in \mathfrak{o}$ with $x \equiv x_j \bmod \mathfrak{p}_j^{r_j}$; and, moreover, this uniquely determines the class of $x \bmod \prod_{j=1}^{n} \mathfrak{p}_j^{r_j}$.

*Proof of Theorem 4.* By definition

$$\begin{aligned}\ker f &= \bigcap_{j=1}^{n} \mathfrak{p}_j^{r_j} \\ &= \mathfrak{p}_1^{r_1} \cap \mathfrak{p}_2^{r_2} \cap \ldots \cap \mathfrak{p}_n^{r_n}.\end{aligned}$$

Now $\mathfrak{p}_n^{r_n}$ and $\prod_{j=1}^{n-1} \mathfrak{p}_j^{r_j}$ are coprime, since they satisfy (1.15.$g$ (iii)). Hence by (1.15.$g$. (ii))

$$\left[\prod_{j=1}^{n-1} \mathfrak{p}_j^{r_j}\right] \cap \mathfrak{p}_n^{r_n} = \prod_{j=1}^{n} \mathfrak{p}_j^{r_j}.$$

Thus, applying this argument repeatedly, we conclude that $\bigcap_{j=1}^{n} \mathfrak{p}_j^{r_j} = \prod_{j=1}^{n} \mathfrak{p}_j^{r_j}$. It now remains to show that $f$ is surjective.

Let $x_j$, $1 \le j \le n$, denote prescribed elements of $\mathfrak{o}$. For each $k = 1, \ldots n$, the ideals

$$\mathfrak{p}_k^{r_k} \quad \text{and} \quad \mathfrak{a}_k = \prod_{\substack{j=1 \\ j \neq k}}^{n} \mathfrak{p}_j^{r_j}$$

are coprime, and so we can find $a_k \in \mathfrak{a}_k$, $b_k \in \mathfrak{p}_k^{r_k}$ such that $a_k + b_k = 1$. Thus $x_k a_k \in \mathfrak{p}_j^{r_j}$ for each $j \neq k$; while $x_k b_k = x_k(1 - a_k) \in \mathfrak{p}_k^{r_k}$, so that $x_k \equiv x_k a_k \bmod \mathfrak{p}_k^{r_k}$. In summary, we have shown that $x = \sum_{k=1}^{n} x_k a_k$ has the property that

$$x \equiv x_k \bmod \mathfrak{p}_k^{r_k} \quad \text{for each } k.$$

□

*Example.* Let $\mathfrak{p}_1, \ldots, \mathfrak{p}_n$ denote distinct non-zero prime ideals of $\mathfrak{o}$. We choose as our ideal $\mathfrak{p}_1^2\mathfrak{p}_2\ldots\mathfrak{p}_n$ and we let $x_1 \in \mathfrak{p}_1 \setminus \mathfrak{p}_1^2$, while $x_j = 1$ for all $j > 1$. Applying the theorem, we obtain an element $y_1$ of $\mathfrak{o}$ such that $y_1 \equiv x_1 \bmod \mathfrak{p}_1^2$ and $y_1 \notin \mathfrak{p}_j$ for $j > 1$, i.e. $y_1 \in \mathfrak{p}_1 \setminus \mathfrak{p}_1^2$ and $y_1 \notin \mathfrak{p}_j$ for $j > 1$. In the same way we obtain elements $y_j$ such that $y_j \in \mathfrak{p}_j \setminus \mathfrak{p}_j^2$, $y_j \notin \mathfrak{p}_k$ for $k \neq j$.

Suppose now that in fact $\{\mathfrak{p}_1, \ldots, \mathfrak{p}_n\}$ constitutes the whole set of prime ideals of $\mathfrak{o}$; then $\mathfrak{p}_j = (y_j)$ is a principal ideal, and so we have shown:

**Corollary 1 to Theorem 4.** *A Dedekind domain with only finitely many prime ideals is a principal ideal domain.*

**Corollary 2 to Theorem 4.** *Every fractional $\mathfrak{o}$-ideal can be generated over $\mathfrak{o}$ by two or fewer elements.*

*Proof.* By (1.3) it suffices to prove the result for a non-zero $\mathfrak{o}$-ideal $\mathfrak{a}$. We begin by choosing a non-zero element $a \in \mathfrak{a}$. Thus $(a) \subset \mathfrak{a}$ and we can factorise both ideals

$$\mathfrak{a} = \prod_{i=1}^{l} \mathfrak{p}_i^{n_i}$$

$$(a) = \prod_{i=1}^{l} \mathfrak{p}_i^{m_i}$$

so that $m_i \geq n_i$ for $1 \leq i \leq l$. We choose $b_i \in \mathfrak{p}_i^{n_i} \setminus \mathfrak{p}_i^{n_i+1}$, and use the theorem to find an element $b$ of $\mathfrak{o}$ with the property that $b \equiv b_i \bmod \mathfrak{p}^{n_i+1}$. Since $v_{\mathfrak{p}_i}(b) = v_{\mathfrak{p}_i}(b_i) = v_{\mathfrak{p}_i}(\mathfrak{a})$ for each $i$ we know that $b \in \mathfrak{a}$. On the other hand, if $\mathfrak{q}$ is a prime ideal of $\mathfrak{o}$ which does not lie in the set $\{\mathfrak{p}_1, \ldots \mathfrak{p}_l\}$, then $0 = v_{\mathfrak{q}}(a) = v_{\mathfrak{q}}(\mathfrak{a})$. Thus we have shown that for all primes $\mathfrak{p}$ of $\mathfrak{o}$,

$$\inf(v_{\mathfrak{p}}(a), v_{\mathfrak{p}}(b)) = v_{\mathfrak{p}}(\mathfrak{a}),$$

and so by (1.15.*e*) we conclude that $\mathfrak{a} = (a, b)$.

□

In the last three theorems we have seen that many of the properties of principal ideal domains extend to Dedekind domains. This then motivates the following question: how can we measure to what extent a given Dedekind domain deviates from being a principal ideal domain? There is a group, called the *ideal class group* of $\mathfrak{o}$, which measures precisely this deviation. Recall that by (1.5) the *fractional principal ideals* $(a)$, for $a \in K^*$, form a subgroup $P_{\mathfrak{o}}$ of $I_{\mathfrak{o}}$. We define the ideal classgroup of

$\mathfrak{o}$, which we denote $\mathrm{Cl}(\mathfrak{o})$, to be the quotient group:

$$\mathrm{Cl}(\mathfrak{o}) = I_{\mathfrak{o}}/P_{\mathfrak{o}}. \tag{1.17}$$

The elements of this group are called *ideal classes.* It is immediately clear that

$$\mathrm{Cl}(\mathfrak{o}) = 1 \iff \mathfrak{o} \text{ is a principal ideal domain.}$$

Later we shall see that, if $\mathfrak{o}$ is the ring of algebraic integers of an algebraic number field $K$, then $\mathrm{Cl}(\mathfrak{o})$ is finite. This is the first of the finiteness theorems which depend on the special arithmetic properties of rings of algebraic integers, and which do not extend to Dedekind domains in general.

In dealing with the ideal class group it is useful to be able to restrict oneself to $\mathfrak{o}$-ideals, rather than always having to consider fractional ideals.

**(1.18)** *Every ideal class contains an $\mathfrak{o}$-ideal, and two $\mathfrak{o}$-ideals $\mathfrak{a}_1, \mathfrak{a}_2$ lie in the same class if, and only if, $\mathfrak{a}_1 a_1 = \mathfrak{a}_2 a_2$ for non-zero elements $a_1, a_2$ of $\mathfrak{o}$.*

*Proof.* The first assertion follows from the definition of fractional ideals in (1.3); the second part follows from the fact that $K$ is the field of fractions of $\mathfrak{o}$.

□

**(1.19)** *Given finitely many distinct prime ideals $\mathfrak{p}_i$ of a Dedekind domain $\mathfrak{o}$ and an ideal class $C$, there is an $\mathfrak{o}$-ideal $\mathfrak{a}$ in $C$ which is not divisible by any of the $\mathfrak{p}_i$.*

*Proof.* We begin by choosing an $\mathfrak{o}$-ideal $\mathfrak{b}$ with class $C^{-1}$; we factorise $\mathfrak{b} = \prod \mathfrak{q}_j^{n_j}$, where the $\mathfrak{q}_j$ are distinct prime ideals, extended if necessary to include all the $\mathfrak{p}_i$ by setting the corresponding $n_j = 0$. By Theorem 4 we can find $a$ in $\mathfrak{o}$ such that

$$v_{\mathfrak{q}_j}(a) = v_{\mathfrak{q}_j}(\mathfrak{b})$$

for all $j$. Thus $\mathfrak{a} = a\mathfrak{b}^{-1}$ is an $\mathfrak{o}$-ideal with class $C$, and with the property that $v_{\mathfrak{p}_i}(\mathfrak{a}) = 0$ for each $i$.

□

An immediate consequence of (1.18) is the fact that every ideal class is of the form $\prod C_i^{r_i}$ where the $C_i$ are ideal classes of prime ideals and the $r_i$ are non-negative integers. If $\mathfrak{o}$ is actually the ring of algebraic integers of a number field, then every ideal class contains a prime ideal; unfortunately the proof of this fact lies beyond the scope of this book.

In order to express our definition of the class group in terms of maps, we note that, by (1.5), the map

$$K^* \to I_{\mathfrak{o}}$$
$$a \mapsto (a)$$

is a group homomorphism whose cokernel is $\mathrm{Cl}(\mathfrak{o})$. The kernel

(1.20) $$\mathfrak{o}^* = \ker[K^* \to I_{\mathfrak{o}}]$$

is the unit group of invertible elements of $\mathfrak{o}$, i.e.

$$\mathfrak{o}^* = \{x \in \mathfrak{o}\backslash 0 \mid x^{-1} \in \mathfrak{o}\}.$$

We therefore obtain an exact sequence of groups

(1.20.$a$) $$1 \to \mathfrak{o}^* \to K^* \to I_{\mathfrak{o}} \to \mathrm{Cl}(\mathfrak{o}) \to 1$$

which is important in the description of the structure of $\mathfrak{o}$ and $K$. Note that $\mathbb{Z}^* = \{\pm 1\}$, so that in the rational case the above sequence assumes the particularly simple form

$$1 \to \{\pm 1\} \to \mathbb{Q}^* \to I_{\mathbb{Z}} \to 1.$$

In fact (1.20.$a$) may be viewed as providing a good measure of the complexity of $\mathfrak{o}$. When $\mathfrak{o}$ is a ring of algebraic integers, we adopt the now standard abuse of language and call $\mathfrak{o}^*$ the "group of units of $K$"; we shall denote $\mathfrak{o}^*$ by $U_K$. In Chapter IV we will show that $U_K$ is a finitely generated abelian group, and in fact we shall describe $U_K$ completely, by identifying both its rank and its torsion subgroup.

We now come to the problem of extending Dedekind domains. In the sequel $\mathfrak{o}$ again denotes a Dedekind domain with field of fractions $K$. $L$ is a finite separable extension field of $K$, and $\mathfrak{O}$ denotes the integral closure of $\mathfrak{o}$ in $K$; clearly $\mathfrak{O}$ is an integral domain which contains $\mathfrak{o}$. If $\mathfrak{p}$ is a prime ideal of $\mathfrak{o}$, then we write $\mathfrak{p}\mathfrak{O}$ for the $\mathfrak{O}$-ideal generated by $\mathfrak{p}$. Suppose that $\mathfrak{P}$ is a prime ideal of $\mathfrak{O}$ which contains $\mathfrak{p}\mathfrak{O}$. Since

(1.21) $$\mathfrak{o}/\mathfrak{P} \cap \mathfrak{o} \hookrightarrow \mathfrak{O}/\mathfrak{P}$$

and since a non-trivial subring of an integral domain is an integral domain, it follows that $\mathfrak{P} \cap \mathfrak{o}$ is a prime ideal of $\mathfrak{o}$ which contains $\mathfrak{p}$; moreover, as $\mathfrak{p}$ is maximal, we get $\mathfrak{P} \cap \mathfrak{o} = \mathfrak{p}$. Conversely, this equality implies that $\mathfrak{P} \supset \mathfrak{p}$ and so $\mathfrak{P} \supset \mathfrak{p}\mathfrak{O}$; so we have now shown that

(1.22) $$\mathfrak{P} \cap \mathfrak{o} = \mathfrak{p} \iff \mathfrak{P} \supset \mathfrak{p}\mathfrak{O}.$$

Since $\mathfrak{p}$ is a maximal $\mathfrak{o}$-ideal, $\mathfrak{o}/\mathfrak{p}$ is a field; so by (1.21), we can view $\mathfrak{o}/\mathfrak{p}$ as a subfield of the integral domain $\mathfrak{O}/\mathfrak{P}$. Whenever the equivalent conditions in (1.22) hold, we shall say that $\mathfrak{P}$ lies above $\mathfrak{p}$: we shall give a geometric justification for this terminology after the following theorem:

**Theorem 5.** *Let $L/K$ be a finite separable extension of fields; then*

(*i*) $\mathfrak{O}$ *is a finitely generated* $\mathfrak{o}$*-module which spans* $L$ *over* $K$.

(*ii*) $\mathfrak{O}$ *is a Dedekind domain.*

(*iii*) *Every prime ideal* $\mathfrak{P}$ *of* $\mathfrak{O}$ *lies above a prime ideal* $\mathfrak{p}$ *of* $\mathfrak{o}$*; for every prime ideal* $\mathfrak{p}$ *of* $\mathfrak{o}$ *there exists at least one, and at most a finite number, of prime ideals* $\mathfrak{P}$ *of* $\mathfrak{O}$ *which lie above* $\mathfrak{p}$.

*Remark.* Later we shall obtain considerably more precise results on the number of prime ideals $\mathfrak{P}$ of $\mathfrak{O}$ which lie above a given prime ideal $\mathfrak{p}$ of $\mathfrak{o}$, and on the degree of the field extension $\mathfrak{O}/\mathfrak{P}$ over $\mathfrak{o}/\mathfrak{p}$. (Even at this stage we could easily show that they are both bounded by $(L : K)$.) Let $\mathfrak{a}\mathfrak{O}$ be the fractional $\mathfrak{O}$-ideal generated by the fractional $\mathfrak{o}$-ideal $\mathfrak{a}$, i.e. the set of sums $\sum a_i b_i$, $a_i \in \mathfrak{a}$, $b_i \in \mathfrak{O}$. Then $\mathfrak{a} \mapsto \mathfrak{a}\mathfrak{O}$ is a homomorphism $I_{\mathfrak{o}} \to I_{\mathfrak{O}}$ which is completely determined by the images of the prime ideals in $\mathfrak{o}$.

**Corollary to Theorem 5.** *The ring of algebraic integers of an algebraic number field is a Dedekind domain.*

Before proceeding with the proof of Theorem 5, we first make a few remarks concerning other well-known families of Dedekind domains. If $K$ denotes a field and if $x$ is an algebraic indeterminate, then the polynomial ring $K[x]$ is a Euclidean domain, and so is a principal ideal domain. The field of fractions of $K[x]$ is the field of rational functions $K(x)$; thus if $L$ is a finite, separable extension of $K(x)$, then the integral closure $\mathfrak{O}$ of $K[x]$ in $L$ is a Dedekind domain. The case where $K$ is finite is of particular interest, since then $\mathfrak{O}$ has numerous properties in common with rings of algebraic integers.

Next we consider the special case where $K = \mathbb{C}$. Since $\mathbb{C}$ is algebraically closed, there is a natural bijection between the non-zero prime ideals of $\mathbb{C}[x]$ and the points in the complex plane via $(x - \alpha) \longleftrightarrow \alpha$. In this interpretation the residue class map mod $(x - \alpha)$ is just $g(x) \mapsto g(\alpha)$. Now we let $L = \mathbb{C}(x, y)$ denote a finite extension of $\mathbb{C}(x)$. Then by the above theorem $\mathfrak{O}$, the integral closure of $\mathbb{C}[x]$ in $L$, is a Dedekind domain. Let $f(x, Y)$ denote the minimal polynomial of $y$ over $\mathbb{C}(x)$; for the sake of simplicity, we assume $y$ to be integral over $\mathbb{C}[x]$. If $f$ is non-singular, in the sense that $\frac{\partial f}{\partial x}$ and $\frac{\partial f}{\partial Y}$ never vanish simultaneously at solutions of $f(x, y) = 0$, then the locus of solutions of $f(x, y) = 0$ yields a Riemann surface $R_L$. We write $\pi \colon R_L \to \mathbb{C}$ for the projection map given by reading off the $x$-coordinate of a solution. It can then be shown that, with the notation of the theorem, the prime ideals $\mathfrak{P}$

of $\mathfrak{O}$ correspond to the points of $R_L$ and that the map $\mathfrak{P} \to \mathfrak{P} \cap \mathbb{C}[x]$ corresponds to $\pi$. As an example, if $f(x, y) = y^3 - x(x-1)$, then we can associate the 1-dimensional pictorial representation to $\pi\colon R_L \to \mathbb{C}$

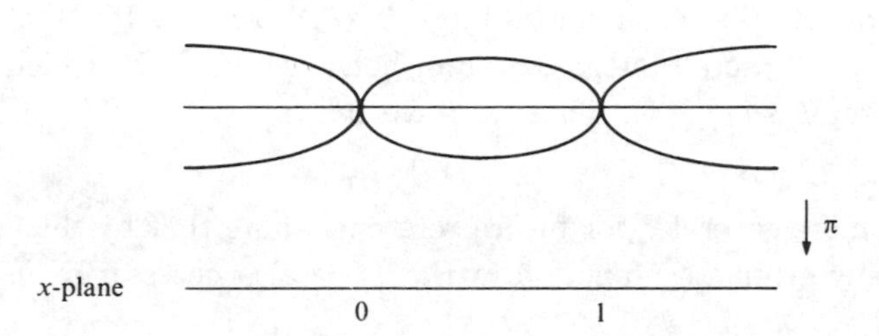


The theory of such function fields lies beyond the scope of this book; we mention it here not only for its intrinsic interest, but also because this is frequently a source of geometric intuition in the solution of arithmetic problems. We shall return to this topic again when we consider elliptic curves in Chapter VI.

*Proof of Theorem 5.* We first show that $\mathfrak{O}$ spans $L$ over $K$; so that, in particular, we will show that $L$ is the field of fractions of $\mathfrak{O}$. Indeed, let $\alpha \in L$; then for some $a_i, b_i \in \mathfrak{o}$ with $b_i \neq 0$ for all $i$, we have

$$\alpha^m + \frac{a_{m-1}}{b_{m-1}}\alpha^{m-1} + \cdots + \frac{a_1}{b_1}\alpha + \frac{a_0}{b_0} = 0.$$

Then writing $b = \prod_{i=0}^{m-1} b_i$, it is clear that

$$(\alpha b)^m + (a_{m-1}\frac{b}{b_{m-1}})(\alpha b)^{m-1} + \cdots + (a_1\frac{b^{m-1}}{b_1})\alpha b + a_0\frac{b^m}{b_0} = 0.$$

Thus $\alpha b$ is a root of a monic polynomial in $\mathfrak{o}[x]$, and so $\alpha b$ must lie in $\mathfrak{O}$.

Next we recall from §1 of Chapter I that the trace form

$$(x, y) \to t_{L/K}(xy)$$

is non-singular (see after (I.1.25.$b$)). For any $\mathfrak{o}$-submodule $X$ of $L$, which spans $L$ over $K$, let

$$X^D = \{x \in L \mid t_{L/K}(xy) \in \mathfrak{o}\ \forall y \in X\}.$$

Note that if $X$ is a free $\mathfrak{o}$-module on a basis $\{x_i \mid 1 \leq i \leq (L:K)\}$, then, by non-singularity there exist unique elements $y_i \in L$ with

$$t_{L/K}(x_i y_j) = \delta_{ij} \quad \text{(the Kronecker delta)}.$$

Thus $\{x_i\}$ and $\{y_j\}$ yield dual $K$-bases of $L$. We assert that $X^D$ is the free $\mathfrak{o}$-module on the $\{y_j\}$. For, clearly the $y_i$ are all linearly independent, and they all lie in $X^D$; moreover, if

$$t_{L/K}(x_i y) = c_i \in \mathfrak{o} \quad \text{for each } i$$

then $t_{L/K}(x_i(y - \sum_j c_j y_j)) = 0$ for each $i$. Therefore, by non-singularity, $y = \sum c_j y_j$, and so $X^D$ is spanned by the $y_j$.

Next let $N$ be a Galois extension of $K$ containing a copy of $L$, and let $\{\sigma\}$ denote the set of embeddings $L \to N$ over $K$. If $x \in \mathfrak{O}$, then of course the $x^\sigma$ are all integral over $\mathfrak{o}$; hence so is $\sum_\sigma x^\sigma$ which from §1 in Chapter 1 is $t_{L/K}(x)$. Thus we have shown

$$t_{L/K}(\mathfrak{O}) \subset \mathfrak{o}. \tag{1.23}$$

Indeed, more generally, for future reference, note that by the same reasoning any symmetric function in the $x^\sigma$ is also necessarily an element of $\mathfrak{o}$.

From (1.23) it follows immediately that

$$\mathfrak{O} \subset \mathfrak{O}^D. \tag{1.23.a}$$

Since $\mathfrak{O}$ spans $L$ over $K$, it contains a basis $\{x_i\}$ of $L$ over $K$. The $\mathfrak{o}$-module $X$ generated by the $x_i$ is then a free $\mathfrak{o}$-module, and

$$X \subset \mathfrak{O} \tag{1.23.b}$$

and hence

$$\mathfrak{O}^D \subset X^D. \tag{1.23.c}$$

By the above we know that $X^D$ is a free $\mathfrak{o}$-module (on $(L : K)$ generators). By (1.23.*a*,*c*) $\mathfrak{O} \subset X^D$; so, because $\mathfrak{o}$ is a Noetherian ring, $\mathfrak{O}$ is finitely generated over $\mathfrak{o}$. Again, since $\mathfrak{o}$ is a Noetherian ring, we deduce that every $\mathfrak{O}$-ideal $\mathfrak{a}$ is also finitely generated over $\mathfrak{o}$; *a fortiori* it is finitely generated over $\mathfrak{O}$, and we have thereby shown that $\mathfrak{O}$ is a Noetherian ring. From (I.2.4.*c*) we know that the integral closure of $\mathfrak{o}$ in $L$ is necessarily integrally closed in $L$; furthermore $\mathfrak{O} \cap K = \mathfrak{o}$ since $\mathfrak{o}$ is integrally closed in $K$.

Next let $\mathfrak{P}$ denote a prime ideal of $\mathfrak{O}$, and let $a \in \mathfrak{P} \setminus 0$. As $a$ is integral over $\mathfrak{o}$, we know that

$$a^n + b_{n-1}a^{n-1} + \cdots + b_1 a + b_0 = 0$$

for some $n$ and some $b_{n-1}, \ldots, b_0 \in \mathfrak{o}$ with $b_0 \neq 0$. But then $b_0 \in \mathfrak{P} \cap \mathfrak{o}$, and so $\mathfrak{p} = \mathfrak{P} \cap \mathfrak{o}$ is non-zero, i.e. in our terminology, is a prime ideal of $\mathfrak{o}$. By (1.22) $\mathfrak{P} \supset \mathfrak{p}\mathfrak{O}$. We shall presently see that this implies that $\mathfrak{P}$ is maximal.

Now consider a prime ideal $\mathfrak{p}$ of $\mathfrak{o}$. Since $\mathfrak{O} \cap K = \mathfrak{o}$, we know that $\mathfrak{p}^{-1} \not\subset \mathfrak{O}$ and so $\mathfrak{O} \supsetneqq \mathfrak{p}\mathfrak{O}$. Since $\mathfrak{O}$ is a finitely generated $\mathfrak{o}$-module, we see that $A = \mathfrak{O}/\mathfrak{p}\mathfrak{O}$ is a finite dimensional commutative $\mathfrak{o}/\mathfrak{p}$-algebra.

So, by Theorem 1, we know that $A$ has finitely many maximal ideals $J_i$, and the inverse image $\mathfrak{P}_i$ of $J_i$ under the quotient map $\mathfrak{O} \to A$ is a maximal ideal of $\mathfrak{O}$. Indeed, for each $i$, $\mathfrak{O}/\mathfrak{P}_i \cong A/J_i$ is a field with finite degree over $\mathfrak{o}/\mathfrak{p}$. (Here we view $\mathfrak{o}/\mathfrak{p}$ as embedded in $A/J_i$.) Moreover, if $\mathfrak{P}$ is a prime ideal of $\mathfrak{O}$ which contains $\mathfrak{p}\mathfrak{O}$, then $\mathfrak{O}/\mathfrak{P}$ is an integral domain and is a finite dimensional algebra over the field $\mathfrak{o}/\mathfrak{p}$; hence by Theorem 1 $\mathfrak{O}/\mathfrak{P}$ is a field; $\mathfrak{P}$ is therefore maximal and so coincides with one of the above $\mathfrak{P}_i$.

□

Using the above theorem in the case when $\mathfrak{o} = \mathbb{Z}$, $K = \mathbb{Q}$, we know that the ring of algebraic integers $\mathfrak{O}$ of a number field $L$ is a finitely generated, torsion free $\mathbb{Z}$-module; hence $\mathfrak{O}$ has a free $\mathbb{Z}$ basis $\omega_1, \ldots, \omega_n$; moreover, as $\mathfrak{O}$ spans $L$ over $\mathbb{Q}$, $\omega_1, \ldots, \omega_n$ is also a basis of $L$ over $\mathbb{Q}$ and so $n = (L : \mathbb{Q})$. In the sequel we shall refer to such a basis $\omega_1, \ldots, \omega_n$ as an *integral basis* of $L$. (The reader is warned not to confuse an integral basis of $L$, with a basis of $L$ which consists of algebraic integers!) By (I.1.25.$b$), we know that the determinant

(1.24)
$$\det(t_{L/\mathbb{Q}}(\omega_i\omega_j)) = \det(\omega_i^\sigma)^2.$$

is a non-zero rational number. (Here $\sigma$ runs through the distinct embeddings of $L$ into $\mathbb{C}$.) The $\omega_i^\sigma$ are all algebraic integers, hence $\det(\omega_i^\sigma)^2$ lies in $\mathbb{Z}$.

If $\mu_1, \ldots, \mu_n$ is another integral basis of $L$, then we have two systems of equations

$$\mu_j = \sum a_{jk}\omega_k \quad a_{j,k} \in \mathbb{Z},$$
$$\omega_j = \sum b_{jk}\mu_k \quad b_{j,k} \in \mathbb{Z}.$$

Thus $(a_{jk})$ and $(b_{jk})$ are mutually inverse matrices whose entries lie in $\mathbb{Z}$; therefore $\det(a_{jk})$ must be $\pm 1$. Squaring we see that

(1.25)
$$\det(t_{L/\mathbb{Q}}(\mu_i\mu_j)) = \det(t_{L/\mathbb{Q}}(\omega_i\omega_j)) = d_L$$

only depends on the number field $L$. This non-zero integer is called the *absolute discriminant* of $L$: it is a key invariant of the field $L$.

Firstly we give two results concerning discriminants

**(1.26) (Stickelberger, Schur).** *If $L$ is an algebraic number field; then either $d_L \equiv 0$ or $d_L \equiv 1 \bmod (4)$.*

*Proof.* We use the formula (1.24) with a chosen integral basis $\{\omega_i\}$ of $L$. From the definition of a determinant, on numbering the embeddings $\sigma_j$

of $L$ into $\mathbb{C}$,

$$\det(\omega_i^{\sigma_j}) = \sum \operatorname{sign}(\pi) \prod_{i=1}^{n} \omega_i^{\sigma_{\pi(i)}} \tag{1.27}$$

where $\pi$ runs through the permutations of $\{1, 2, \ldots n\}$, $n = (L : \mathbb{Q})$. We set

$$A = \sum_{\pi} \prod_{i=1}^{n} \omega_i^{\sigma_{\pi(i)}}$$

$$B = \sum_{\pi \text{odd}} \prod_{i=1}^{n} \omega_i^{\sigma_{\pi(i)}}.$$

Clearly $A, B$ are both algebraic integers, $A \in \mathbb{Q}$, and of course $\det(\omega_i^{\sigma_j}) = A - 2B$. Thus $d_L = A^2 + 4(B^2 - AB)$. Since both $d_L$ and $A^2$ are rationals, we deduce that $B^2 - AB \in \mathbb{Q}$; furthermore because $A$, $B^2 - AB$ are algebraic integers, we conclude that $A, B^2 - AB \in \mathbb{Z}$; therefore $d_L \equiv A^2 \bmod (4)$ and the result follows.

□

In order to determine the sign of $d_L$ we consider the $n = (L : \mathbb{Q})$ embeddings $L \to \mathbb{C}$. Among these some factorise through $L \to \mathbb{R}$. Let $\sigma_1, \ldots, \sigma_s$ denote these real embeddings $\sigma_i: L \to \mathbb{R}$, so that the remaining $n-s$ embeddings are *imaginary* (in the sense that their images are non-real). Given such an imaginary embedding $\sigma_i: L \to \mathbb{C}$ we have a further (distinct) embedding $\sigma_i \cdot \rho$ where $\rho: \mathbb{C} \to \mathbb{C}$ denotes complex conjugation. (Note we write $\sigma_i \cdot \rho$, since such embeddings act from the right.) Thus the number of imaginary embeddings, $n - s$, is an even number, which we denote by $2t$.

$$\operatorname{sign}(d_L) = (-1)^t. \tag{1.28}$$

*Proof.* We again let $\{\omega_i\}$ denote an integral basis of $L$, and, for brevity, we set $\Delta = \det(\omega_i^{\sigma_j})$. Since $\Delta^2 = d_L$, we note that $\Delta^\rho = \Delta$ if $d_L$ is positive, while $\Delta^\rho = -\Delta$ if $d_L$ is negative. However, from the way we have numbered the embeddings, we see that $\rho$ leaves the first $s$ columns invariant, but interchanges in pairs the last $2t$ columns: so that $\Delta^\rho = (-1)^t \Delta$.

□

As a further example of the way Dedekind domains retain desirable properties of $\mathbb{Z}$, we show how the Gauss Lemma generalises:

**Theorem 6.** *If $f(X)$, $g(X)$, $h(X)$ are monic polynomials in $K[X]$,* *with*

$$f(X) = g(X)h(X)$$

*and* $f(X) \in \mathfrak{o}[X]$*; then* $g(X), h(X) \in \mathfrak{o}[X]$.

Note that Theorem 6 provides an alternative proof of the corollary to (I.2.6.$a$) (in the special case of Dedekind domains), and we have

**(1.29)** *The minimal polynomial over* $K$ *of an element* $a$ *which is integral over* $\mathfrak{o}$*, lies in* $\mathfrak{o}[X]$.

To prove Theorem 6 we associate to each non-zero polynomial $f(X)$ in $K[X]$, its so-called content ideal: this is the fractional $\mathfrak{o}$-ideal $\mathfrak{a}_f$ which is generated by the coefficients of $f(X)$. For each prime ideal $\mathfrak{p}$ of $\mathfrak{o}$ we define

$$v_{\mathfrak{p}}(f) = v_{\mathfrak{p}}(\mathfrak{a}_f).$$

We then have

**(1.30)** *For non-zero polynomials* $f_1, f_2$ *in* $K[X]$

$$v_{\mathfrak{p}}(f_1 f_2) = v_{\mathfrak{p}}(f_1) + v_{\mathfrak{p}}(f_2)$$

*for each prime ideal* $\mathfrak{p}$ *of* $\mathfrak{o}$.

In particular we shall call $f$ *primitive* if $v_{\mathfrak{p}}(f) = 0$ for all $\mathfrak{p}$; by the theorem we conclude that the product of two primitive polynomials is primitive.

We first show that (1.30) implies the theorem. Indeed, as $g(X), h(X)$ are both monic, we deduce that $v_{\mathfrak{p}}(g) \leq 0$, $v_{\mathfrak{p}}(h) \leq 0$ for all $\mathfrak{p}$; moreover, since $f$ is monic and in $\mathfrak{o}[X]$, we know that $v_{\mathfrak{p}}(f) = 0$ for all $\mathfrak{p}$. Now from (1.30)

$$v_{\mathfrak{p}}(g) + v_{\mathfrak{p}}(h) = v_{\mathfrak{p}}(f) = 0,$$

hence $v_{\mathfrak{p}}(g) = 0 = v_{\mathfrak{p}}(h)$, and so $g(X)$ and $h(X)$ both lie in $\mathfrak{o}[X]$.

*Proof of* (1.30). By the very definition of $\mathfrak{a}_f$

$$\mathfrak{a}_f = a^{(1)}\mathfrak{o} + \cdots + a^{(k)}\mathfrak{o}$$

where the $a^{(j)}$ range through the non-zero coefficients of $f$. Therefore, by (1.15.$e$),

$$v_{\mathfrak{p}}(f) = \inf_j (v_{\mathfrak{p}}(a^{(j)})).$$

Now let

$$f_1(X) = b_n X^n + \cdots + b_1 X + b_0,$$
$$f_2(X) = c_m X^m + \cdots + c_1 X + c_0.$$

Let $r$ denote the least non-negative integer $r$ such that $v_\mathfrak{p}(b_r) = v_\mathfrak{p}(f_1)$; that is to say

$$\begin{cases} v_\mathfrak{p}(b_r) \leq v_\mathfrak{p}(b_j) & \text{for all } j \text{ with } b_j \neq 0, \\ v_\mathfrak{p}(b_r) < v_\mathfrak{p}(b_j) & \text{for all } j < r \text{ with } b_j \neq 0. \end{cases}$$

We define $c_s$ for $f_2(X)$ in an analogous manner. Thus the coefficient $d_{r+s}$ of $X^{r+s}$ in $f_1 f_2$ is of the form $b_r c_s + d$, where $d$ is a sum of terms each with valuation strictly greater than $v_\mathfrak{p}(b_r c_s)$. Therefore $v_\mathfrak{p}(d_{r+s}) = v_\mathfrak{p}(b_r) + v_\mathfrak{p}(c_s) = v_\mathfrak{p}(f_1) + v_\mathfrak{p}(f_2)$, and so we have shown that

$$v_\mathfrak{p}(f_1 f_2) \leq v_\mathfrak{p}(f_1) + v_\mathfrak{p}(f_2).$$

On the other hand, it is clear that $\mathfrak{a}_{f_1 f_2} \subset \mathfrak{a}_{f_1}\mathfrak{a}_{f_2}$ since every coefficient of $f_1(X)f_2(X)$ is a sum of products $b_j c_k \in \mathfrak{a}_{f_1}\mathfrak{a}_{f_2}$; hence

$$v_\mathfrak{p}(f_1 f_2) \geq v_\mathfrak{p}(f_1) + v_\mathfrak{p}(f_2).$$

□

As an illustration we shall now determine integral bases and discriminants for fields $K$, which are quadratic over $\mathbb{Q}$. The map $m \mapsto \mathbb{Q}(\sqrt{m})$ induces a bijection between square free integers $m$ different from 1 and the quadratic extensions $K/\mathbb{Q}$. In the sequel it is always assumed that $m$ is square-free and $m \neq 1$.

**(1.31)** *An element $\alpha$ in $K = \mathbb{Q}(\sqrt{m})$ is an algebraic integer iff it can be written in the form*

$$(1.31.a) \qquad \begin{cases} \alpha = \dfrac{1}{2}(u + v\sqrt{m}) \quad u, v \in \mathbb{Z} \\ u^2 - mv^2 \equiv 0 \bmod (4). \end{cases}$$

*Proof.* Let $\alpha = x + y\sqrt{m}$ with $x, y \in \mathbb{Q}$. Thus $t_{K/\mathbb{Q}}(\alpha) = 2x$, $N_{K/\mathbb{Q}}(\alpha) = x^2 - my^2$, and from (I.2.6) we already know that $\alpha$ is an algebraic integer iff its characteristic polynomial lies in $\mathbb{Z}[x]$. Therefore $\alpha \in \mathfrak{o}_K$ iff both $2x$ and $x^2 - my^2$ lie in $\mathbb{Z}$. Obviously $2x \in \mathbb{Z}$ precisely when we can write $x = u/2$ for $u \in \mathbb{Z}$; so, in particular, if $\alpha$ can be written as per (1.31.$a$), then $\alpha$ is an algebraic integer. Conversely, if $\alpha \in \mathfrak{o}_K$ then $2x \in \mathbb{Z}$ and so $N_{K/\mathbb{Q}}(2\alpha - 2x) = -4my^2$ must lie in $\mathbb{Z}$. As $m$ is square free we deduce that $y = (1/2)v$ for some $v \in \mathbb{Z}$. The norm condition then yields $(1/4)(u^2 - mv^2) \in \mathbb{Z}$, which shows (1.31.$a$) to be a necessary condition.

□

**(1.32)** *Let $m \equiv 2$ or $m \equiv 3 \bmod (4)$; then $\{1, \sqrt{m}\}$ is an integral basis of $K$, and $d_K = 4m$.*

*Proof.* The congruences $u^2 - 2v^2 \equiv 0 \bmod (4)$ and $u^2 + v^2 \equiv 0 \bmod (4)$ possess only the trivial solution $u \equiv 0 \equiv v \bmod (2)$; thus $\alpha = x + y\sqrt{m} \subset \mathfrak{o}_K$ if, and only if, $x$ and $y$ lie in $\mathbb{Z}$. Therefore $\{1, \sqrt{m}\}$ is a $\mathbb{Z}$-basis of $\mathfrak{o}_K$, and so

$$d_K = \begin{vmatrix} 1 & \sqrt{m} \\ 1 & -\sqrt{m} \end{vmatrix}^2 = (-2\sqrt{m})^2 = 4m.$$

□

**(1.33)** *Let $m \equiv 1 \bmod (4)$; then $\{1, \frac{1+\sqrt{m}}{2}\}$ is an integral basis of $K$ and $d_K = m$. Furthermore $\{1, \frac{m+\sqrt{m}}{2}\}$ is also an integral basis of $K$.*

We adopt the notation of (1.31) for $\alpha \in \mathfrak{o}_K$. The congruence $u^2 - v^2 \equiv 0 \bmod (4)$ is equivalent to $u \equiv v \bmod (2)$; thus we may write $u = v + 2w$ for $w \in \mathbb{Z}$, and so the algebraic integers of $K$ consist of precisely those numbers of the form

$$\alpha = \frac{1}{2}(v + 2w + v\sqrt{m}) = w + v(\frac{1+\sqrt{m}}{2})$$

for $v, w$ in $\mathbb{Z}$. Next we note that

$$d_K = \begin{vmatrix} 1 & \frac{1+\sqrt{m}}{2} \\ 1 & \frac{1-\sqrt{m}}{2} \end{vmatrix}^2 = m.$$

Moreover, since the determinant of the integral transformation

$$1 = 1 + 0\left(\frac{1+\sqrt{m}}{2}\right)$$

$$\frac{m+\sqrt{m}}{2} = \frac{m-1}{2} + 1\left(\frac{1+\sqrt{m}}{2}\right)$$

is 1, we see that $(1, \frac{m+\sqrt{m}}{2})$ is also an integral basis of $K$.

□

We use the above explicit integral bases to establish the following useful fact:

**(1.34)** *If $\alpha \in \mathfrak{o}_K$, then its norm satisfies the congruence*

$$N_{K/\mathbb{Q}}(\alpha) \equiv z^2 \bmod (m)$$

*for some $z \in \mathbb{Z}$.*

*Proof.* For $m \equiv 2$ or $3 \bmod (4)$, this follows immediately from (1.32).

So now we suppose $m \equiv 1 \bmod (4)$; then, by the above, we know that $(1, \frac{m+\sqrt{m}}{2})$ is an integral basis of $K$, and so for $x, y \in \mathbb{Z}$

$$N_{K/\mathbb{Q}}\left(x + y\left(\frac{m+\sqrt{m}}{2}\right)\right) = x^2 + m\left(xy + y^2\left(\frac{m-1}{4}\right)\right)$$
$$\equiv x^2 \bmod (m)$$

since $m \equiv 1 \bmod (4)$.

□

We conclude this section with a special finiteness property for algebraic integers; this result does not hold for arbitrary Dedekind domains.

Given a Dedekind domain $\mathfrak{o}$ and a non-zero $\mathfrak{o}$-ideal $\mathfrak{a}$ we again consider the residue class ring $\mathfrak{o}/\mathfrak{a}$. If this is a finite ring, we call the cardinality of $\mathfrak{o}/\mathfrak{a}$ the *absolute norm* of $\mathfrak{a}$, and we denote it by $\mathbf{N}\mathfrak{a}$; thus, of course, we have an equality of $\mathbb{Z}$-ideals

$$\mathbf{N}\mathfrak{a} \cdot \mathbb{Z} = [\mathfrak{o} : \mathfrak{a}]\mathbb{Z}.$$

**(1.35)** *Suppose that the residue class fields $\mathfrak{o}/\mathfrak{p}$ are finite for all prime ideals $\mathfrak{p}$ of $\mathfrak{o}$. Then the residue class rings $\mathfrak{o}/\mathfrak{a}$ are finite for all non-zero ideals $\mathfrak{a}$ of $\mathfrak{o}$ and*

(1.35.$a$) $$\mathbf{N}(\mathfrak{a}\mathfrak{b}) = \mathbf{N}\mathfrak{a}\mathbf{N}\mathfrak{b}$$

*if $\mathfrak{b}$ denotes a further non-zero $\mathfrak{o}$-ideal; furthermore $\mathbf{N}$ extends to a group homomorphism from $I_\mathfrak{o}$ to $\mathbb{Q}^*$.*

*Proof.* Let $\mathfrak{p}$ denote a prime ideal of $\mathfrak{o}$. If $\mathfrak{o}/\mathfrak{p}$ is finite, then, by (1.16), $\mathfrak{p}^r/\mathfrak{p}^{r+1}$ is also finite, and $[\mathfrak{o} : \mathfrak{p}] = [\mathfrak{p}^r : \mathfrak{p}^{r+1}]$. Therefore $\mathbf{N}(\mathfrak{p}^r) = (\mathbf{N}\mathfrak{p})^r$.

If now $\mathfrak{a} = \prod_i \mathfrak{p}_i^{r_i}$, where the $\mathfrak{p}_i$ denote distinct prime ideals which possess the property that each $\mathfrak{o}/\mathfrak{p}_i$ is finite; then by Theorem 4, $\mathfrak{o}/\mathfrak{a}$ is finite and

$$\mathbf{N}\mathfrak{a} = \prod_i \mathbf{N}(\mathfrak{p}_i^{r_i}),$$

and we have therefore established the formula (1.35.$a$). The fact that $\mathbf{N}$ extends to a group homomorphism now follows at once from (1.14.$d$).

□

If $\mathfrak{o}/\mathfrak{p}$ is finite, then it is a field of characteristic $p$, for some prime number $p$; suppose that its degree over the prime field $\mathbb{F}_p$ is

(1.36) $$(\mathfrak{o}/\mathfrak{p} : \mathbb{F}_p) = f;$$

then it is clear that

(1.36.$a$) $$\mathbf{N}\mathfrak{p} = p^f.$$

Now we consider again a finite extension $L$ of $K$, and we let $\mathfrak{O}$ denote the integral closure of $\mathfrak{o}$ in $L$. From Theorem 5 we know that if $\mathfrak{P}$ is a prime ideal of $\mathfrak{O}$, lying above the prime ideal $\mathfrak{p}$ of $\mathfrak{o}$, then $\mathfrak{O}/\mathfrak{P}$ is a field extension of $\mathfrak{o}/\mathfrak{p}$ of finite degree. We have therefore shown

**(1.37)** *If all the residue class rings of $\mathfrak{o}$ are finite, then the same is true for those of $\mathfrak{O}$. Furthermore, if $\mathfrak{O}/\mathfrak{P}$ is of degree $f_{L/K}(\mathfrak{P})$ over $\mathfrak{o}/\mathfrak{p}$, then*

$$\mathbf{N}\mathfrak{P} = \mathbf{N}\mathfrak{p}^{f_{L/K}(\mathfrak{P})}.$$

In particular from (1.37) we can conclude that the residue class rings of a ring of algebraic integers of a number field are all finite. For algebraic number fields, the norm and the absolute norm are related in the following way:

**(1.38)** *Let $K$ be an algebraic number field. Then for $a \in K^*$*

(1.38.$a$) $$\mathbf{N}(a\mathfrak{o}_K) = |N_{K/\mathbb{Q}}a|,$$

*in particular if $a \in \mathbb{Q}^*$*

(1.38.$b$) $$\mathbf{N}(a\mathbb{Z}) = |a|.$$

*Proof.* Since both sides of (1.38.$a$) are multiplicative, it suffices to consider the case when $a \in \mathfrak{o}_K$. In this case, however, one knows that

$$[\mathfrak{o}_K : a\mathfrak{o}_K] = |det(\ell_a)|$$

where $\ell_a : K \to K$ is given by $\ell_a(x) = ax$. The result then follows from (I.1.27).

□

**(1.39)** *Let $K$ be an algebraic number field with $(K : \mathbb{Q}) = n$ and suppose $\Lambda = \sum_1^n a_i\mathbb{Z}$ is contained in $\mathfrak{o}_K$ with finite index; then*

$$d(a_1, \ldots, a_n) = d_K \cdot [\mathfrak{o}_K : \Lambda]^2.$$

*Proof.* First note that, as in the definition of $d_K$, the value $d(a_1, \ldots, a_n)$ depends only on $\Lambda$, and not on the particular choice of basis. From the theory of modules over a principal ideal domain, we know that we can find a $\mathbb{Z}$-basis $\omega_1, \ldots, \omega_n$ of $\mathfrak{o}_K$ with the property that $\Lambda$ has a basis $\sum b_{ij}\omega_j$ with the matrix $(b_{ij})$ upper triangular with integral entries;

clearly $|\det(b_{ij})| = [\mathfrak{o}_K : \Lambda]$. The result therefore follows from the equality

$$\det(t_{K/\mathbb{Q}}(\sum_k b_{ik}\omega_k \cdot \sum_\ell b_{jl}\omega_\ell)) = \det(b_{ik})^2 d_K.$$

□

## §2 Valuations and absolute values

Let $\mathfrak{o}$ denote a Dedekind domain with field of fractions $K$ and let $\mathfrak{p}$ denote a prime ideal of $\mathfrak{o}$. In the previous section we defined $v_\mathfrak{p}(\mathfrak{a})$, for a fractional $\mathfrak{o}$-ideal $\mathfrak{a}$ to be the exact power of $\mathfrak{p}$ occurring in the factorisation of $\mathfrak{a}$. For $x \in K^*$, we shall write $v_\mathfrak{p}(x)$ for $v_\mathfrak{p}(x\mathfrak{o})$, and so define a map $v_\mathfrak{p}: K^* \to \mathbb{Z}$. We assert that for all $x, y \in K^*$

(2.1.*a*) $$v_\mathfrak{p}(xy) = v_\mathfrak{p}(x) + v_\mathfrak{p}(y),$$

(2.1.*b*) $$v_\mathfrak{p}(x+y) \geq \inf\big(v_\mathfrak{p}(x), v_\mathfrak{p}(y)\big),$$

assuming $x + y \neq 0$ for the moment.Part (a) follows immediately from (1.15.*a*). To prove part (b), note that $(x+y)\mathfrak{o} \subset x\mathfrak{o} + y\mathfrak{o}$ so that by (1.15.*c*,*e*)

$$v_\mathfrak{p}(x+y) \geq v_\mathfrak{p}(x\mathfrak{o} + y\mathfrak{o}) \geq \inf\big(v_\mathfrak{p}(x), v_\mathfrak{p}(y)\big).$$

Since we can always find $x \in \mathfrak{p}$, $x \notin \mathfrak{p}^2$, it follows that $v_\mathfrak{p}$ is a surjective group homomorphism. We can extend $v_\mathfrak{p}$ to all of $K$, in such a way that the properties (2.1.*a*,*b*) still obtain, by setting $v_\mathfrak{p}(0) = \infty$, where the symbol $\infty$ satisfies the rules

$$\infty + \infty = \infty = \infty + r$$

$$\infty > r$$

for all real $r$.

We begin this section by studying axiomatically all maps on $K$ which possess the above properties of $v_\mathfrak{p}$. In the case when $K$ is a number field, we shall see that this provides a striking new way of viewing the prime ideals of the ring of algebraic integers of $K$. It will allow us to introduce absolute values associated with prime ideals quite analogous to the usual real or complex absolute values, and so to bring in topological methods.

We next identify precisely the maps which we wish to study.

**(2.2)** *Denote an arbitrary field by $K$. A surjective map $v: K \to \mathbb{Z} \cup \{\infty\}$ is called a valuation of $K$ if and only if for all $x, y \in K$*

(2.2.*a*) $$v(x) = \infty \quad \text{iff} \quad x = 0;$$

(2.2.*b*) $$v(xy) = v(x) + v(y);$$

(2.2.*c*) $$v(x+y) \geq \inf\big(v(x), v(y)\big).$$

Note for future reference that from (2.2.$b$), it follows that $v(1) = v(-1) = 0$.

Observe that the above conditions imply

**(2.2.d)** *If* $v(x) > v(y)$ *then* $v(x+y) = v(y)$.

Indeed we know $v(x+y) \geq \inf\big(v(x), v(y)\big) = v(y)$. If the inequality were strict, then we would have $v(y) = v(y+x-x) \geq \inf\big(v(y+x), v(x)\big) > v(y)$ which is absurd.

It should be noted that our terminology here is slightly non-standard, in that the above is usually referred to as a discrete valuation.

Given a valuation $v$ of $K$ we define

$$\mathfrak{o}_v = \{x \in K \mid v(x) \geq 0\}.$$

By the above $1 \in \mathfrak{o}_v$; if $x, y \in \mathfrak{o}_v$ then $x+y \in \mathfrak{o}_v$ by (2.2.$c$), while $xy \in \mathfrak{o}_v$ by (2.2.$b$); so that $\mathfrak{o}_v$ is a ring. We call $\mathfrak{o}_v$ the *valuation ring* of $v$ in $K$. Next we set

$$\mathcal{P}_v = \{x \in K \mid v(x) > 0\}.$$

$\mathcal{P}_v$ is readily seen to be a $\mathfrak{o}_v$-ideal, by virtue of (2.2.$b$,$c$): indeed, it is a proper $\mathfrak{o}_v$-ideal since on the one hand $v(1) = 0$, so that $1 \in \mathfrak{o}_v$, $1 \notin \mathcal{P}_v$; while on the other hand $\mathcal{P}_v \neq 0$, since $v$ is surjective so that we can find $x \in K$ with $v(x) = 1$. We shall call $\mathcal{P}_v$ the *valuation ideal* of $v$ in $K$.

Next we choose $x, y \in \mathfrak{o}_v$ with $xy \in \mathcal{P}_v$; then $v(xy) = v(x) + v(y) > 0$; thus, because $v(x) \geq 0$ and $v(y) \geq 0$, we conclude that either $v(x) > 0$ or $v(y) > 0$, i.e. $x \in \mathcal{P}_v$ or $y \in \mathcal{P}_v$. We have therefore shown $\mathcal{P}_v$ to be a prime ideal in $\mathfrak{o}_v$. More strongly we now show

**(2.3)** $\mathfrak{o}_v$ *is a principal ideal domain with unique maximal ideal* $\mathcal{P}_v$, *and* $\mathbb{Z} \cong I_{\mathfrak{o}_v}$ *via* $m \mapsto \mathcal{P}_v^m$.

*Proof.* We note that $x$ is invertible in $\mathfrak{o}_v$ iff $v(x) \geq 0$ and $v(x^{-1}) \geq 0$, i.e. iff $v(x) = 0$; thus we have shown that $\mathfrak{o}_v^* = \mathfrak{o}_v \setminus \mathcal{P}_v$ and so $\mathcal{P}_v$ is indeed the unique maximal ideal of $\mathfrak{o}_v$.

Now let $\mathfrak{a}$ denote an arbitrary non-zero $\mathfrak{o}_v$-ideal. Because the set $\{v(a) \mid a \in \mathfrak{a}\}$ is well-ordered, we can choose $b \in \mathfrak{a}$ with $v(b)$ minimal in this set. We assert that $\mathfrak{a} = b\mathfrak{o}_v$. The inclusion $\mathfrak{a} \supset b\mathfrak{o}_v$ is clear; conversely, if $c \in \mathfrak{a}$, then $v(b) \leq v(c)$ and so $v(b^{-1}c) \geq 0$. Hence $b^{-1}c \in \mathfrak{o}_v$, and

$$c = b \cdot b^{-1}c \in b \cdot \mathfrak{o}_v.$$

Moreover, if $a\mathfrak{O}_v = \mathcal{P}_v$ and $v(b) = n$, we clearly have

$$a^n\mathfrak{O}_v = \mathcal{P}_v^n = b\mathfrak{O}_v = \{c \in K \mid v(c) \geq n\}.$$

□

Given a valuation $v$ on $K$, it follows from (2.3) that $x \in \mathfrak{O}_v$ is a unit iff $v(x) = 0$. Thus the following sequence is exact:

(2.4) $$1 \to \mathfrak{O}_v^* \to K^* \xrightarrow{v} \mathbb{Z} \to 0.$$

This is really a reformulation of the isomorphism in (2.3).

Now write $k_v$ for the residue class field $\mathfrak{O}_v/\mathcal{P}_v$. For $i > 0$, let $U_K^{(i)}$ denote the subgroup of $\mathfrak{O}_v^*$ given by $1 + \mathcal{P}_v^i$. We now show:

**(2.5)**

($a$) $\mathfrak{O}_v^*/U_K^{(1)} \cong k_v^*$

($b$) *For* $i \geq 1$, $U_K^{(i)}/U_K^{(i+1)} \cong \mathcal{P}_v^i/\mathcal{P}_v^{i+1} \cong k_v^+$.

*Proof.* The homomorphism $u \mapsto u \bmod \mathcal{P}_v$ has kernel $U_K^{(1)}$; furthermore any $a \in \mathfrak{O}_v \setminus \mathcal{P}_v$ is a unit and so this map is surjective. This proves ($a$). To show ($b$), consider the map on $U_K^{(i)}$ given by $1 + a \mapsto a \bmod \mathcal{P}_v^{i+1}$. Since for $a, b \in \mathcal{P}_v^i$,

$$(1 + a)(1 + b) - 1 = a + b + ab \equiv a + b \bmod \mathcal{P}_v^{i+1}$$

we see that this is a homomorphism. It is clearly surjective, and its kernel is $1 + \mathcal{P}_v^{i+1}$. Finally the isomorphism $\mathcal{P}_v^i/\mathcal{P}_v^{i+1} \cong k_v^+$ comes from (1.16).

□

Let $\mathfrak{o}$ again denote a Dedekind domain whose field of fractions is $K$. Let $v$ denote a valuation on $K$ with the property that $\mathfrak{o} \subset \mathfrak{O}_v$, that is to say $v$ is non-negative on $\mathfrak{o}$. We set $\mathfrak{p}_v = \mathcal{P}_v \cap \mathfrak{o}$; then $\mathfrak{p}_v$ is a (possibly zero) prime ideal of $\mathfrak{o}$, since $\mathcal{P}_v$ is a prime ideal of $\mathfrak{O}_v$; in fact, $\mathfrak{p}_v$ is non-zero, for otherwise $v$ would be zero on $\mathfrak{o} \setminus 0$, and hence zero on all elements of $K^*$.

Next we wish to compare $v$ and $v_{\mathfrak{p}_v}$. We start by showing that $v(z) = 0$ whenever $v_{\mathfrak{p}_v}(z) = 0$: indeed, we can write such a $z$ as $a'/b'$ with $a', b' \in \mathfrak{o}$, and with $v_{\mathfrak{p}_v}(a') = v_{\mathfrak{p}_v}(b') = \ell$, say; next, on choosing $\pi \in \mathfrak{p}_v^{-\ell} \backslash \mathfrak{p}_v^{-\ell+1}$, and setting $a = \pi a'$, $b = \pi b'$, we have $z = a/b$ with $a, b \in \mathfrak{o} \setminus \mathfrak{p}_v$; thus $a, b \in \mathfrak{O}_v \setminus \mathcal{P}_v$ and so $v(z) = 0$.

Now consider the ideal $\mathfrak{p}_v\mathfrak{O}_v$; by (2.3) it must be of the form $\mathcal{P}_v^e$ for some $e \geq 1$. However if $x \in \mathfrak{o}$, $x \neq 0$, with $v_{\mathfrak{p}_v}(x) = \ell$, then $x\mathfrak{o} = \mathfrak{p}_v^\ell \mathfrak{a}$ with $(\mathfrak{a}, \mathfrak{p}_v) = 1$. By the above work $\mathfrak{a} \cdot \mathfrak{O}_v = \mathfrak{O}_v$; hence $x\mathfrak{O}_v = \mathcal{P}_v^{\ell e}$ and so

$v(x) = ev_{\mathfrak{p}_v}(x)$; thus $v = ev_{\mathfrak{p}_v}$ and in fact $e = 1$, since $v$ and $v_{\mathfrak{p}_v}$ have the same value group – namely $\mathbb{Z}$.

We conclude this sequence of elementary results by showing $\mathfrak{o} + \mathcal{P}_v = \mathfrak{O}_v$. As $\mathfrak{o} \subset \mathfrak{O}_v$, it is clear that $\mathfrak{o} + \mathcal{P}_v \subset \mathfrak{O}_v$. Conversely, for $z \in \mathfrak{O}_v$ we show $z \in \mathfrak{o} + \mathcal{P}_v$. The result is obvious if $v(z) > 0$, so suppose $v(z) = 0$; then, as above, $z = a/b$ with $a, b \in \mathfrak{o} \setminus \mathfrak{p}_v$. Since $\mathfrak{o}/\mathfrak{p}_v$ is a field, we may choose $c \in \mathfrak{o}$ such that $bc \in 1 + \mathfrak{p}_v$. It then follows that

$$z - ac = a(b^{-1} - c) = ab^{-1}(1 - bc) \in \mathfrak{O}_v \mathfrak{p}_v = \mathcal{P}_v.$$

We summarise the above results in

**(2.6)** *If $\mathfrak{o}$ is a Dedekind domain and if $v$ is a valuation of $K$ with $\mathfrak{o} \subset \mathfrak{O}_v$; then $\mathfrak{p}_v$ is a prime ideal of $\mathfrak{o}$; moreover, $v_{\mathfrak{p}_v} = v$, $\mathfrak{p}_v \mathfrak{O}_v = \mathcal{P}_v$, and $\mathfrak{o}/\mathfrak{p}_v \cong \mathfrak{O}_v/\mathcal{P}_v$.*

If, on the other hand, we start with a prime ideal $\mathfrak{p}$ of $\mathfrak{o}$; then note that it is clear from the definitions that

(2.6.*a*)
$$\mathfrak{p}_{v_\mathfrak{p}} = \mathfrak{p}.$$

For the time being, we turn from the general situation to our main concern, namely the case where $\mathfrak{o}$ is a ring of algebraic integers:

**(2.7)** *Let $\mathfrak{o}$ be the ring of algebraic integers of a number field $K$, and let $v$ be a valuation of $K$; then $\mathfrak{o} \subset \mathfrak{O}_v$.*

*Proof.* Since $v(-1) + v(-1) = v(1) = 0$, $v(-1) = 0 = v(1)$, and so by (2.2.*c*) we conclude that $v$ must be non-negative on $\mathbb{Z}$. Since $\mathfrak{o}$ is the integral closure of $\mathbb{Z}$ in $K$, an element $x \in \mathfrak{o}$ will satisy some monic integral polynomial

$$T^n + a_1 T^{n-1} + \cdots + a_n$$

where $a_i \in \mathbb{Z}$. Thus on applying (2.2.c) we conclude that

$$\begin{aligned} nv(x) &\geq \inf_{0 \leq i \leq n-1} \big(v(a_{n-i}) + iv(x)\big) \\ &\geq \inf_{0 \leq i \leq n-1} \big(iv(x)\big) \end{aligned}$$

since the $v(a_{n-1})$ are all non-negative. It therefore follows that $v(x) \geq 0$. □

From (2.6) and (2.7) we deduce

**Theorem 7.** *For the ring of algebraic integers $\mathfrak{o}$ in a number field $K$, the map $\mathfrak{p} \mapsto v_\mathfrak{p}$ sets up a bijection between the prime ideals of $\mathfrak{o}$ and the valuations of $K$.*

It is interesting to note that Theorem 7 fails to hold for the function field $F(X)$, when we replace the ring of algebraic integers by the polynomial ring $F[X]$: let $K = F(X)$ with $F$ a field and $X$ an algebraic indeterminate; then one can define $v_\infty : K \to \mathbb{Z} \cup \{\infty\}$ by setting $v_\infty(f/g) = -\deg(f) + \deg(g)$ for non-zero $f, g \in F[X]$ and $v_\infty(0) = \infty$. It is readily verified that $v_\infty$ is indeed a valuation; however $v_\infty$ is non-positive on $F[X]$, and (2.7) and Theorem 7 both fail to hold in this situation.

We have however

**(2.8)** *Let $K = F(X)$, then $v_\infty$ together with the $v_\mathfrak{p}$ for all prime ideals $\mathfrak{p}$ of $F[X]$, is the full set of valuations of $K$ which are zero on $F^*$. Furthermore, for all $g \in F(X)^*$*

(2.8.*a*) $$v_\infty(g) + \sum_{\mathfrak{p}} f_\mathfrak{p} v_\mathfrak{p}(g) = 0.$$

*where $f_\mathfrak{p} = (\mathfrak{o}/\mathfrak{p} : F)$.*

We remark that of course the prime ideals of $F[X]$ are naturally parameterised by the monic irreducible polynomials in $F[X]$; thus $\mathfrak{o}/\mathfrak{p}$ will always have finite dimension over $F$.

*Proof.* Let $v$ denote a valuation of $K$ which is zero on $F^*$. Firstly suppose that $v(X) \geq 0$, so that $v$ is non-negative on $F[X]$ by (2.2.*c*). Since the ideal group $I_{F[X]} \cong K^*/F^*$ is the free abelian group on the monic irreducible polynomials of $F[X]$, we deduce that $v$ must be positive on some monic irreducible polynomial $f$. If, in addition, $v$ were also positive on a different irreducible polynomial $g$, then we could find $\lambda, \mu \in F[X]$ such that $\lambda f + \mu g = 1$. Thus by (2.2.*c*) we should have $v(1) > 0$, which is absurd. Given a non-zero element $h \in F[X]$, write $h = u f^a \prod {g_i}^{b_i}$, where $u \in F^*$, and the $g_i$ are distinct monic irreducible polynomials which are different from $f$. Then, from the above, we have shown $v(h) = av(f)$. Since $v(K^*) = \mathbb{Z}$, $v(f)$ must be 1, and so $v = v_\mathfrak{p}$ where $\mathfrak{p} = fF[X]$. Next suppose $v(X) = \alpha < 0$. We shall show that for any $h \in F[X]$, $v(h) = \alpha \deg(h)$, by induction on $\deg(h)$. If $\deg(h) = 0$, then $h \in F^*$ and so $v(h) = 0$ by hypothesis: this starts the induction. If $\deg(h) = d > 0$, then we can write $h = aX^d + h_1$ with $a \in F^*$, $\deg(h_1) < d$. So by (2.2.*d*)

$$v(h) = \inf\big(v(X^d), v(h_1)\big) = \alpha d.$$

As $v(K^*) = \mathbb{Z}$, $\alpha$ must be $-1$ and so $v = v_\infty$.

Finally, given $g \in K^*$, we factorise it as $g = u \prod {p_i}^{v_{\mathfrak{p}_i}(g)}$ where $u \in F^*$

and $\mathfrak{p}_i = p_i F[X]$ are the prime ideals of $F[X]$. Reading off degrees gives

$$\deg(g) = \sum_i \deg(p_i) v_{\mathfrak{p}_i}(g)$$

and so (2.8.$a$) now follows, since

$$f_{\mathfrak{p}_i} = (\mathfrak{o}/\mathfrak{p}_i : F) = \deg(p_i).$$

□

In order to describe the arithmetic counterpart to the so-called infinite valuations like $v_\infty$, we need to introduce the notion of an absolute value:

**(2.9)** *Given a field $K$, a function $|.|$ from $K$ to the non-negative reals is called an absolute value if for all $x, y \in K$*

(2.9.$a$) $$|x| = 0 \quad \text{iff} \quad x = 0$$

(2.9.$b$) $$|xy| = |x||y|$$

(2.9.$c$) $$|x + y| \leq |x| + |y|$$

Clearly the function $|x|_0 = 1$ for all $x \in K^*$, $|0|_0 = 0$ is an absolute value. We call this the trivial absolute value. Unless indicated to the contrary, the trivial absolute value is always understood to be excluded.

If more strongly than (2.9.$c$) we have

(2.9.$d$) $$|x + y| \leq \sup(|x|, |y|)$$

then we call $|.|$ an *ultrametric*. Note that if $|.|$ is an ultrametric then $|x| > |y|$ implies $|x + y| = |x|$; for if $|x + y| < |x|$, then we should have $|x| \leq \sup(|x + y|, |-y|) < |x|$, which is absurd.

We call an absolute value $|.|$ on $K$ a *discrete absolute value* if $|K^*|$ is a discrete subgroup of $\mathbb{R}_{>0}$.

**(2.9.$e$)** *If $|.|$ is a discrete absolute value then $|K^*| = \lambda^{\mathbb{Z}}$ for some* $0 < \lambda < 1$. ($\lambda^{\mathbb{Z}}$ is the cyclic subgroup of $\mathbb{R}^*$ generated by $\lambda$).

*Proof.* As $|.|$ is non-trivial, by hypothesis, $|K^*| \cap (0, 1)$ is non-empty. By discreteness we can find a maximal $\lambda \in |K^*| \cap (0, 1)$, and we then choose $x \in K^*$ with $|x| = \lambda$. Given $y \in K^*$, we wish to show $|y| \in \lambda^{\mathbb{Z}}$. The result is clear if $|y| = 1$, and if $|y| > 1$ we can replace $y$ by $y^{-1}$; thus we may assume $|y| < 1$, and so $\lambda^{n+1} \leq |y| < \lambda^n$ for some $n$, i.e. $\lambda < |yx^{-n}| < 1$. Therefore, by the maximality property of $\lambda$, $|yx^{-n}| = \lambda$, and so $|y| = \lambda^{n+1}$.

□

If $K$ is a field and if $|.|$ is an absolute value on $K$, then we call the pair $(K, |.|)$ a *valued field.* A homomorphism of fields $\sigma: K \to L$ is called a *homomorphism of valued fields* between $(K, |.|)$ and $(L, |.|')$ iff $|\sigma(x)|' = |x|$ for all $x \in K$. We illustrate these new concepts with a number of examples, which will be of fundamental importance in the next section.

**(2.10.*a*)** *Let $\sigma: K \hookrightarrow \mathbb{R}$ denote a field embedding; then the function $|x|_\sigma = |x^\sigma|_{\mathbb{R}}$ defines an absolute value, where $|y|_{\mathbb{R}}$ denotes the usual absolute value of a real number $y$, i.e. $|y| > 0$ if $y \neq 0$ and $|y| = \pm y$.*
**(2.10.*b*)** *Let $\sigma: K \hookrightarrow \mathbb{C}$ denote a field embedding; then $|x|_\sigma = |x^\sigma|_{\mathbb{C}}$ defines an absolute value on $K$, where $|y|_{\mathbb{C}}$ denotes the modulus of the complex number $y$. If $\rho$ is a field automorphism of $K$ with the property that $x^{\rho\circ\sigma}$ is the complex conjugate of $x^\sigma$ for all $x \in K$, then $\rho$ induces an automorphism of valued fields on $(K, |.|_\sigma)$. This is of interest in the case when $\sigma$ is imaginary, i.e. takes on some non-real values, i.e. when $\rho \circ \sigma \neq \sigma$.*
**(2.10.*c*)** *Let $K$ denote a field, let $v$ denote a valuation of $K$, and let $\lambda$ denote a real number $0 < \lambda < 1$. Then the function $|x|_v = \lambda^{v(x)}$ defines an ultrametric on $K$ which is also a discrete absolute value.* In fact, at the end of this section, we shall see that all discrete absolute values arise in this manner.

If now $\rho$ is a field automorphism of $K$ such that $v(x^\rho) = v(x)$ for all $x \in K$; then $\rho$ induces an automorphism of valued fields on $(K, |.|_v)$. Here we shall be particularly interested in the case where $K$ is a number field and $v = v_{\mathfrak{p}}$; for then $v(x^\rho) = v(x)$ iff $\mathfrak{p}^\rho = \mathfrak{p}$.

We define an equivalence relation on the set of absolute values of $K$ by saying that two absolute values $|.|$, $|.|'$ are equivalent iff there exists a positive real number $\alpha$ such that $|x|^\alpha = |x|'$ for all $x \in K$. (For positive $\beta$, recall that $\beta^\alpha = \exp(\alpha \log(\beta))$.) Note that if two absolute values are equivalent, and if one is discrete, then so is the other. Moreover if $|.|$ is an ultrametric, then $|.|^\alpha$ is an ultrametric, for any real positive number $\alpha$. Also if $v$ and $v'$ are distinct valuations of $K$ then clearly $|.|_v$ and $|.|_{v'}$ are inequivalent.

The absolute values of an algebraic number field which are equivalent to an absolute value of type (2.10.*a*,*b*) are called *Archimedean absolute values.* Note that if $|.|$ is Archimedean, then in general, for positive $\alpha$, $|.|^\alpha$ will not define an absolute value, e.g. $|.|^2_{\mathbb{C}}$ is not an absolute value on $\mathbb{C}$.

**(2.11)** *Let $|.|_1$, $|.|_2$ be two non-trivial absolute values of $K$. Then $|.|_1$ and $|.|_2$ are equivalent iff we have the inclusion*

(2.11.$a$) $$\{x \in K \mid |x|_1 > 1\} \subset \{x \in K \mid |x_2| > 1\}.$$

*Proof.* If $|.|_1$ and $|.|_2$ are equivalent, then the containment (2.11.$a$) is clear. Conversely, we now suppose (2.11.$a$) to hold, and show $|.|_1$ and $|.|_2$ to be equivalent. By considering the effect of the involution $x \mapsto x^{-1}$, we deduce the containment

(2.11.$b$) $$\{x \in K \mid |x|_1 < 1\} \subset \{x \in K \mid |x|_2 < 1\}.$$

We fix some $x \in K$ with the property $|x|_1 > 1$ and we choose any $y \in K$ with $|y|_1 > 1$; thus we can find a unique positive real number $\alpha$ such that $|y|_1 = {|x|_1}^\alpha$. Let $m, n$ be positive integers such that $\frac{m}{n} > \alpha$. Clearly $|y|_1 = {|x|_1}^\alpha \leq {|x|_1}^{\frac{m}{n}}$, and so $|y^n x^{-m}|_1 < 1$. Hence by (2.11.$b$) $|y^n x^{-m}|_2 < 1$, and therefore $|y|_2 < {|x|_2}^{\frac{m}{n}}$.

In the same way we repeat the argument with $|\frac{m}{n}| < \alpha$, and deduce that $|y|_2 > {|x|_2}^{(\frac{m}{n})}$. In summary approaching $\alpha$ from above and below, through the rationals, we conclude that $|y|_2 = {|x|_2}^\alpha$.

Finally, let $\beta$ be the unique positive real number such that $|x|_1 = {|x|_2}^\beta$; then, by the above work,

$$|y|_1 = {|x|_1}^\alpha = {|x|_2}^{\alpha\beta} = {|y|_2}^\beta.$$

□

We now consider the topological implications of the above result. A valued field $(K, |.|)$ becomes a metric space on putting $d(x, y) = |x - y|$, this, in turn, induces a topology on $K$ which we write as $\mathcal{T}(K, |.|)$.

**(2.12)** *If $K$ has absolute values $|.|_1$ and $|.|_2$, then the topologies*

$$\mathcal{T}(K, |.|_i) \quad i = 1, 2$$

*coincide iff $|.|_1$ and $|.|_2$ are equivalent.*

*Proof.* Recall that the two topologies coincide precisely when $|x_i|_1 \to 0$ iff $|x_i|_2 \to 0$ for $x_i \in K$. Thus if $|.|_1$ and $|.|_2$ are equivalent, then the topologies agree. Conversely, if they are not equivalent, then by (2.11) we can find $x \in K$ with $|x|_2 > 1$, but $|x|_1 \leq 1$; thus $|x^{-n}|_2 \to 0$, but $|x^{-n}|_1$ does not converge to zero.

□

Next we see what happens when several inequivalent absolute values are considered simultaneously:

**Theorem 8 (Weak Approximation).** *Let* $|.|_1, \ldots, |.|_n$, *denote inequivalent absolute values on* $K$. *Given a positive real number* $\epsilon$ *and given* $x_1, \ldots, x_n$ *in* $K$, *then we can find* $y \in K$ *such that for each* $i = 1, \ldots, n$

$$|y - x_i|_i < \epsilon.$$

Before proving this result we consider its relationship to the Chinese Remainder Theorem of the previous section. Let $\mathfrak{o}$ be a Dedekind domain with field of fractions $K$. We let $\mathfrak{p}_1, \ldots, \mathfrak{p}_n$ denote $n$ distinct prime ideals of $\mathfrak{o}$, $x_1, \ldots, x_n$ be elements of $\mathfrak{o}$, and $a$ be a positive integer. Writing $|.|_\mathfrak{p}$ for the absolute value $|.|_{v_\mathfrak{p}}$ of (2.10.$c$), we know, by the Chinese Remainder Theorem, that we can find $z \in \mathfrak{o}$, rather than just in $K$, such that $z \equiv x_i \bmod \mathfrak{p}_i^a$ for each $i$, i.e. $|z - x|_{\mathfrak{p}_i} \leq \lambda^a$.

Prior to proving Theorem 8 we show

**(2.13)** *Given* $n$ *inequivalent absolute values as above, we can find* $a \in K$ *such that*

$$|a|_1 > 1 \qquad |a|_i < 1 \quad \textit{for } i = 2, \ldots, n.$$

*Proof.* We argue by induction on $n$: the case $n = 2$ follows from (2.11). We therefore now consider $n \geq 3$. By induction hypothesis we can find $b \in K$ such that

$$|b|_1 > 1 \qquad |b|_i < 1 \quad \text{for } i = 2, \ldots, n-1.$$

However, since $|.|_1$ and $|.|_n$ are inequivalent, we can find $c \in K$ such that $|c|_1 > 1$, $|c|_n < 1$. For sufficiently large $r$, we then define $a$ by

$$a = \begin{cases} b & \text{if } |b|_n < 1 \\ cb^r & \text{if } |b|_n = 1 \\ \frac{cb^r}{1+b^r} & \text{if } |b|_n > 1. \end{cases}$$

□

*Proof of Theorem 8.* By (2.13) we can choose $a_j \in K$ such that $|a_j|_j > 1$, while $|a_j|_i < 1$ if $i \neq j$. Clearly for large $r$ $|a_j^r(1 + a_j^r)^{-1}|_i$ is very small when $i \neq j$; while $|a_j^r(1+a_j^r)^{-1} - 1|_j$ is also very small. Thus we conclude that, for sufficiently large $r$, the conditions of the theorem are met by

$$\sum_{j=1}^{n}\left(\frac{a_j^r}{1+a_j^r}\right)x_j.$$

□

As an easy application of Theorem 8 we show

**(2.14)** *Let $\sigma_i\colon K \hookrightarrow \mathbb{R}$ denote $m$ distinct field embeddings of an algebraic number field $K$, and let $\epsilon_i$ $(i = 1,\ldots,m)$ each be $\pm 1$. Then there exists $a \in K^*$ with* $\operatorname{sign}(a^{\sigma_i}) = \epsilon_i$.

*Proof.* For $i \neq j$, we can find $b \in K^*$ with $b^{\sigma_i} < b^{\sigma_j}$. Hence there exists a rational number $r$ such that $0 < (b-r)^{\sigma_i} < 1 < (b-r)^{\sigma_j}$. We therefore deduce that the Archimedean absolute values $|.|_i$, corresponding to the $\sigma_i$ (see (2.10.$a$)) are non-equivalent by (2.11). By Theorem 8, we can choose $a \in K^*$ with $a$ close to $\epsilon_i$ under $|.|_i$.

□

We now conclude this section by describing all the equivalence classes of absolute values of $\mathbb{Q}$. As an immediate corollary we obtain the analogue to the sum-formula (2.8.a) which we established for polynomials.

We adopt the language of (2.10) and write $|.|_{\mathbb{R}}$ for the absolute value on $\mathbb{Q}$, deriving from the embedding $\mathbb{Q} \hookrightarrow \mathbb{R}$; and, with $p$ a prime number, we write $|.|_p$ for the so-called *p-adic absolute value* defined by the rule

$$|x|_p = p^{-v_p(x)}$$

for $x \in \mathbb{Q}$, where we abbreviate $v_{p\mathbb{Z}}$ to $v_p$. Thus if

$$x = p^r a/b \quad \text{with } a, b \in \mathbb{Z},\ (ab, p) = 1$$

then

$$|x|_p = p^{-r}.$$

**Theorem 9 (Ostrowski).** *Let $|.|$ denote an absolute value of $\mathbb{Q}$; then $|.|$ is equivalent to exactly one of either $|.|_{\mathbb{R}}$ or $|.|_p$ for some prime number $p$.*

Before proving the theorem we note that the full set of absolute values of $\mathbb{Q}$ are related by the so-called *Product Formula*

**(2.15)** $|x|_{\mathbb{R}} \prod_p |x|_p = 1$ *for each* $x \in \mathbb{Q}^*$.

This is our promised analogue for the rationals of (2.8.a). To prove this result we factorise $x = \pm\prod_p p^{v_p(x)}$; the result then follows since $|x|_{\mathbb{R}} = \prod_p p^{v_p(x)}$, while for each $p$ $|x|_p = p^{-v_p(x)}$.

*Proof of Theorem 9.* Initially suppose that, with $|.|$ a given absolute value, $|n| \leq 1$ for all $n \in \mathbb{Z}$. Then, since $|.|$ is non-trivial, we must have $|p| < 1$ for at least one prime number $p$. If we could find a further prime $q$ with $|q| < 1$; then we could find positive integers $a, b$ such that $|p^a| < \frac{1}{2}$, $|q^b| < \frac{1}{2}$, and integers $\lambda, \mu$ such that $\lambda p^a + \mu q^b = 1$. This then gives

$$1 = |1| = |\lambda p^a + \mu q^b| < \frac{|\lambda| + |\mu|}{2} \leq 1$$

which is absurd. It now follows that if $|p| = |p|_p^\alpha$, then $|x| = |x|_p^\alpha$ for all rationals $x$.

So now assume there is a positive integer $n$ with the property that $|n| > 1$, $|n| = n^\alpha$ say. By (2.9.$c$) we note that

$$|m| \leq m \tag{2.16}$$

for all natural numbers $m$; thus of course $0 \leq \alpha \leq 1$. Given any positive integer $m$ we write $m = \sum_{i=0}^k m_i n^i$ with $0 \leq m_i < n$, $m_k \neq 0$. We then deduce that

$$|m| \leq \sum_{i=0}^k |m_i| n^{\alpha i}$$

$$\leq \sum_{i=0}^k m_i n^{\alpha i}$$

by (2.16)

$$\leq \frac{(n-1)n^{\alpha(k+1)}}{(n^\alpha - 1)}$$

Hence for some constant $c$, independent of $m$, $|m| \leq cm^\alpha$ for all positive integers $m$. Indeed, replacing $m$ by $m^r$, with $r$ large, we see that $|m| \leq c^{(1/r)}|m|_{\mathbb{R}}^\alpha$ for all such $r$; hence we have shown that

$$|m| \leq |m|_{\mathbb{R}}^\alpha \quad \text{for all } m \in \mathbb{Z}. \tag{2.17}$$

With the same notation, we now put $m = n^{k+1} - b$, where $0 \leq b \leq n^{k+1} - n^k$. Then, from (2.17), it follows that

$$|b| \leq b^\alpha \leq (n^{k+1} - n^k)^\alpha$$

and so, by the triangle inequality,

$$\begin{aligned} |m| &\geq |n^{k+1}| - |b| \\ &\geq n^{(k+1)\alpha} - (n^{k+1} - n^k)^\alpha \\ &\geq n^{(k+1)\alpha}\left[1 - (1 - \frac{1}{n})^\alpha\right] \\ &\geq c' n^{(k+1)\alpha} \\ &\geq c' m^\alpha \end{aligned}$$

for a positive constant $c'$ which is independent of $m$. Replacing $m$ by $m^r$ with $r$ large, we deduce that $|m| \geq |m|_{\mathbb{R}}^\alpha$ for all $m \in \mathbb{Z}$; this, in conjunction with (2.17), shows $|.| = |.|_{\mathbb{R}}^\alpha$. The fact that $|.|_{\mathbb{R}}$ and all the $|.|_p$ are inequivalent can be read off from the proof or follows directly by looking at the values $|q|$ for varying primes $q$.

□

Suppose now that $|.|$ is an absolute value on an arbitrary field $K$.

**(2.18)** *If $|.|$ is bounded on $\overline{\mathbb{Z}}$ (the image of $\mathbb{Z}$ in $K$), then $|.|$ is necessarily an ultrametric.*

*Proof.* Let $C$ denote a positive real number with the property that $C \geq |\overline{n}|$ for all $\overline{n} \in \overline{\mathbb{Z}}$. Then, by the Binomial Theorem, we know that for any $x, y \in K$ and for any natural number $m$

$$\begin{aligned} |x+y|^m &\leq \sum_{i=0}^{m} |x|^i |y|^{m-i} \left|\binom{m}{i}\right| \\ &\leq C \sum_{i=0}^{m} |x|^i |y|^{m-i} \\ &\leq C(m+1)\sup(|x|^m, |y|^m). \end{aligned}$$

Hence for all $m > 0$

$$|x+y| \leq (C(m+1))^{(1/m)} \sup(|x|, |y|).$$

The result then follows since $\lim_{m\to\infty} (C(m+1))^{(1/m)} = 1$.

□

If $K$ has positive characteristic, then $\overline{\mathbb{Z}}$ is finite and so $|.|$ is obviously bounded on $\overline{\mathbb{Z}}$. On the other hand if $K$ has characteristic zero and $|K^*|$ is discrete in $\mathbb{R}^*_{>0}$, then so is $|\mathbb{Q}^*|$. Thus the restriction to $\mathbb{Q}$ is either equivalent to the $p$-adic absolute value $|.|_p$ or is the trivial absolute value. In both cases $|.|$ on $\mathbb{Z}$ is bounded by 1.

Applying (2.18) we have now shown

**(2.19)** *If either $|.|$ is a discrete absolute value or if $K$ has positive characterstic, then $|.|$ is an ultrametric.*

In particular it follows that

**(2.20)** *If $|.|$ is a discrete absolute value with value group $\lambda^{\mathbb{Z}}$, where $0 < \lambda < 1$; then the map $v\colon K^* \to \mathbb{Z}$, defined by the equality $|x| = \lambda^{v(x)}$, is a valuation.*

Finally we show

**(2.21)** *Let $|.|$ be a non-trivial absolute value on an algebraic number field $K$. Then the restriction of $|.|$ to $\mathbb{Q}$ is non-trivial.*

*Proof.* There exists $\alpha \in K$ with $|\alpha| > 1$. Raising $\alpha$ to a suitable power, we may suppose $|\alpha| > 2$. Suppose now for contradiction that $|.|$ on $\mathbb{Q}$ is trivial. Of course $\alpha$ satisfies an equation

$$\alpha^n = a_{n-1}\alpha^{n-1} + \cdots + a_1\alpha + a_0 \quad (a_j \in \mathbb{Q}).$$

Therefore

$$|\alpha|^n \leq |\alpha|^{n-1} + \cdots + |\alpha| + 1 = \frac{|\alpha|^n - 1}{|\alpha| - 1}$$

i.e.

$$|\alpha|^{n+1} - |\alpha|^n \leq |\alpha|^n - 1$$

so that $|\alpha|^{n+1} < 2|\alpha|^n$, and hence $|\alpha| < 2$, which is the required contradiction.

□

## §3 Completions

This section is concerned with certain topological developments of the work of the preceding section. A reader who seeks only a basic introduction to algebraic number theory may omit this section, but the main result will be needed for the extension theory of Chapter III.

The absolute values on a field $K$ and their associated topologies can be expected to give great insight into properties of $K$. To make full use of these topological tools, it is desirable that sequences which satisfy Cauchy's convergence criterion should have a limit; thus, given an absolute value $|.|$ on $K$, we wish to embed $K$ in a bigger field, its *completion* with respect to $|.|$, where all Cauchy sequences converge – just in the same way as $\mathbb{Q}$ is embedded in $\mathbb{R}$.

In analysis one studies the behaviour of functions about a point by taking power series expansions. We have already seen in (II,§1) that prime ideals of a number field can be thought of as points. By working in the completion taken with respect to the discrete absolute value associated with a prime ideal $\mathfrak{p}$, we shall be able to expand algebraic numbers by increasing prime powers (the $\mathfrak{p}$-adic expansion); we shall then be able to apply many of the ideas and techniques of analysis to such expansions.

Let $(K, |.|)$ denote a valued field, and, in order to exclude trivialities, we assume that $|.|$ is not the trivial absolute value $|.|_0$, given by $|x|_0 = 1$ for all $x \in K^*$. Suppose that for a given sequence $\{a_n\}_{n\in\mathbb{N}}$ of elements in $K$ there exists an element $l \in K$, so that the limit of real numbers satisfies $\lim_{n\to\infty} |a_n - l| = 0$; we then write

$$\lim_{n\to\infty,|.|} a_n = l \tag{3.1}$$

or just

$$\lim_{n\to\infty} a_n = l$$

when the absolute value $|.|$ is clear from the context.

It is important to be aware that we are concerned with two notions of limit: the limit of real numbers, which are values $|x|$ for $x \in K$, and the limit of sequences in $K$ with respect to $|.|$.

Obviously the limit of a sequence $\{a_n\}$ is unique. A sequence $\{a_n\}$ which satisfies (3.1) is called a *limit sequence* (with respect to $|.|$). We call $\{a_n\}$ a *null sequence* if $\lim_{n\to\infty} a_n = 0$. The sequence $\{a_n\}$ is called a *Cauchy sequence* (with respect to $|.|$) if

$$\lim_{n\to\infty} |a_n - a_{n+r}| = 0$$

uniformly in $r \in \mathbb{N}$, i.e. given $\epsilon > 0$ there must exist $N$ such that $|a_n - a_m| < \epsilon$ for all $m, n \geq N$.

*Note.* If $|.|$ is an ultrametric, then for $\{a_n\}$ to be Cauchy, it suffices that $\lim_{n\to\infty} |a_n - a_{n+1}| = 0$: indeed,

$$|a_n - a_{n+r}| \leq \sup_{0\leq k<r} \Big(|a_{n+k} - a_{n+k+1}|\Big).$$

Returning to the general case again, by the triangle inequality we have

$$|a_n - a_m| \leq |a_n - l| + |a_m - l|,$$

so that every limit sequence is a Cauchy sequence.

The valued field $(K, |.|)$ is said to be *complete* if conversely every Cauchy sequence is a limit sequence, that is to say has a limit $l \in K$. In general this is not always the case, e.g. $(\mathbb{Q}, |.|_{\mathbb{R}})$ is not complete, and $(\mathbb{Q}, |.|_p)$ as in §2 is not complete, see Theorem 9 (consider $p = 5$, $\{a_n = 3^{5^n}\}$) or the example (3.28) below.

It is clear that our definitions depend only on $|.|$ modulo equivalence, i.e. on the topology defined by $|.|$. The same will apply to nearly all definitions, assertions and results which follow; however, we shall refrain from pointing this out at every step.

**Theorem 10.** *Given a valued field $(K, |.|)$, there exists a pair $j, (\overline{K}, \|.\|)$ where $(\overline{K}, \|.\|)$ is a complete valued field and $j: (K, |.|) \to (\overline{K}, \|.\|)$ is a homomorphism of valued fields with the following properties:*

(*a*) $j(K)$ is dense in $\overline{K}$, i.e. every $\epsilon$-neighbourhood $N_\epsilon = \{x \in \overline{K} \mid \|x - b\| < \epsilon\}$ of an element $b$ of $\overline{K}$ contains an element $a$ of $j(K)$.

(*b*) *If $(L, |.|^\times)$ is a complete valued field, and if $k: (K, |.|) \to (L, |.|^\times)$ is a homomorphism of valued fields, then there is a unique homomorphism $k': (\overline{K}, \|.\|) \to (L, |.|^\times)$ of valued fields with $k = k' \circ j$.*

*Either of the above two properties characterises the pair $j, (\overline{K}, \|.\|)$ uniquely to within isomorphism of valued fields.*

**Corollary (3.2).** *If $(K, |.|))$ is complete, then $j: (K, |.|) \cong (\overline{K}, \|.\|)$ is an isomorphism of valued fields.*

In the sequel we refer to $(\overline{K}, \|.\|)$ – strictly speaking with the given $j$ – as the completion of $(K, |.|)$.

*Proof.* Define sum and product of sequences in $K$ by

(3.3.*a*) $$\{a_n\} + \{b_n\} = \{a_n + b_n\}$$

(3.3.*b*) $$\{a_n\} \cdot \{b_n\} = \{a_n b_n\}.$$

Clearly the sequences form a commutative ring $R$ with identity $\{1_n\}$ (i.e. whose terms are all 1).

From now on all Cauchy sequences and limit sequences are always assumed to be taken with respect to $|.|$.

In the usual way one shows

**(3.4)** *The sum and product of limit sequences $\{a_n\}, \{b_n\}$ are limit sequences and*

$$\lim_{n\to\infty} (a_n + b_n) = \lim_{n\to\infty} a_n + \lim_{n\to\infty} b_n$$
$$\lim_{n\to\infty} (a_n \cdot b_n) = \lim_{n\to\infty} a_n \cdot \lim_{n\to\infty} b_n$$

Next we have

**(3.5)** *The sum and product of Cauchy sequences $\{a_n\}, \{b_n\}$ are Cauchy* *sequences; $\{1_n\}$ is a Cauchy sequence; if $\{b_n\}$ is a null-sequence, then $\{a_n\} \cdot \{b_n\}$ is a null sequence.*

The easy proof for the sum is again left to the reader. For the proof of the remainder of (3.5) we first show

**(3.6)** *Let $\{a_n\}$ be a Cauchy sequence. Then there exists a positive real number $k$ such that $|a_n| < k$ for all $n \in \mathbb{N}$.*

*Proof.* First choose $N$ so that $|a_n - a_m| < 1$ for all $n, m \geq N$, and choose

$$k > \max_{1 \leq i \leq N} (1 + |a_i|).$$

Clearly $|a_i| < k$ if $i \leq N$. On the other hand for $n \geq N$

$$|a_n| \leq |a_n - a_N| + |a_N| < k.$$

□

Now let $\{a_n\}, \{b_n\}$ again denote two Cauchy sequences. By (3.6) we can find $k$ such that $|a_i| < k$, $|b_i| < k$ for all $i$. Hence

$$\begin{aligned} |a_n b_n - a_m b_m| &\leq |a_n||b_n - b_m| + |b_m||a_n - a_m| \\ &\leq k(|a_n - a_m| + |b_n - b_m|). \end{aligned}$$

Given $\epsilon > 0$, choose $M \in \mathbb{N}$ such that for all $n, m \geq M$

$$|b_n - b_m|, |a_n - a_m| < \frac{\epsilon}{2k}$$

Then clearly $|a_n b_n - a_m b_m| < \epsilon$ for all $n, m \geq M$ and so $\{a_n\} \cdot \{b_n\}$ is indeed a Cauchy sequence.

If now $\{b_n\}$ is a null sequence, then we may choose $M$ so that $|b_n| < \epsilon/k$ whenever $n \geq M$. Thus $|a_n b_n| < \epsilon$ for $n \geq M$, and so $\{a_n\} \cdot \{b_n\}$ is indeed a null sequence.

**(3.7)** *Let $\{a_n\}$ denote a Cauchy sequence, which is not a null sequence. Then there exists a positive real number $k'$, and $M \in \mathbb{N}$ such that $|a_n| > k'$ for all $n \geq M$.*

*Proof.* Assume for contradiction that this is not so. Then, given $\epsilon > 0$ and $M \in \mathbb{N}$, there exists $l \geq M$ with $|a_l| < \epsilon/2$. If necessary we increase $M$ in order to guarantee that $|a_n - a_m| < \epsilon/2$ for all $n, m \geq M$. Then, for all $n \geq M$, $|a_n| \leq |a_n - a_l| + |a_l| < \epsilon$, which implies that $\{a_n\}$ is a null sequence.

□

**(3.8)** *With $\{a_n\}$, $k'$ and $M$ as in (3.7), let $b_n = a_n^{-1}$ for $n \geq M$, and put $b_n = 1$ for $n < M$; then $\{b_n\}$ is a Cauchy sequence.*

*Proof.* Given $\epsilon > 0$, we choose $n_1 > M$ so that $|a_n - a_m| < \epsilon k'^2$ for all $n, m \geq n_1$; then for all such $n, m$

$$|b_n - b_m| = \frac{|a_n - a_m|}{|a_n a_m|} < \epsilon.$$

□

By (3.5) we know that the Cauchy sequences form a subring $R_0$, with identity $\{1_n\}$, of the ring $R$ of all sequences. By (3.5) we also know that the null sequences form an ideal $\mathfrak{n}$ of $R_0$. It follows from (3.8) that for every Cauchy sequence $\{a_n\} \in R_0 \setminus \mathfrak{n}$, there exists $\{b_n\} \in R_0$ with $\{a_n\}\{b_n\} \equiv \{1_n\} \bmod \mathfrak{n}$; therefore $\mathfrak{n}$ is a maximal $R_0$ ideal and $R_0/\mathfrak{n} = \overline{K}$ is a field.

For an element $\alpha \in \overline{K}$, we say that the Cauchy sequence $\{a_n\}$ represents $\alpha$ if $\alpha = \{a_n\} + \mathfrak{n}$. With every element $a$ of $K$ we associate the constant sequence $\overline{a} = \{a_n\}$, $a_n = a$ for all $n$. The map $a \mapsto \overline{a}$ is a homomorphism $K \to R_0$ of rings. Composing this with $R_0 \to \overline{K}$ gives a homomorphism $j: K \to \overline{K}$ with $j(1_K) = 1_{\overline{K}}$, so that $j$ is immediately seen to be injective.

Next we wish to extend the absolute value $|.|$ from $K$ to $\overline{K}$. By the triangle inequality for $|.|$ we have

$$|a_n| - |a_m| \leq |a_n - a_m|$$
$$|a_m| - |a_n| \leq |a_n - a_m|$$

Writing $|.|_{\mathbb{R}}$ for the ordinary absolute value of $\mathbb{R}$, the above implies

$$||a_n| - |a_m||_{\mathbb{R}} \leq |a_n - a_m|.$$

Therefore, if $\{a_n\}$ is a Cauchy sequence for $|.|$ in $K$, then $\{|a_n|\}$ is a Cauchy sequence for $|.|_{\mathbb{R}}$ in $\mathbb{R}$; hence by the completeness of $\mathbb{R}$ with respect to $|.|_{\mathbb{R}}$, $|a_n|$ has a limit in $\mathbb{R}$. We define

(3.9) $$\|\{a_n\}\| = \lim_{n\to\infty} |a_n|.$$

We have already remarked that if $|.|'$ is an absolute value of $K$ which is equivalent to $|.|$, then the ring $R_0$ and the ideal $\mathfrak{n}$ remain unchanged if we switch from $|.|$ to $|.|'$; thus $\overline{K}$ and $j: K \to \overline{K}$ depend only on the equivalence class of $|.|$. On the other hand of course the absolute value $\|.\|$ of (3.9) depends crucially on the particular choice of absolute value$|.|$; however, if $|.|' = |.|^\alpha$, then clearly

(3.9.a) $$\|.\|' = \|.\|^\alpha.$$

By (3.7)

**(3.10)** *$\|\{a_n\}\| \geq 0$ and $\|\{a_n\}\| = 0$ iff $\{a_n\}$ is a null sequence.*

Next observe that

$$\begin{aligned}
\|\{a_n\} + \{b_n\}\| &= \|\{a_n + b_n\}\| \\
&= \lim_{n\to\infty} |a_n + b_n| \\
&\leq \lim_{n\to\infty} (|a_n| + |b_n|) \qquad (3.11) \\
&= \lim_{n\to\infty} |a_n| + \lim_{n\to\infty} |b_n| \\
&= \|\{a_n\}\| + \|\{b_n\}\|.
\end{aligned}$$

In particular, if $\{b_n\}$ is a null sequence, then by (3.11), $\|\{a_n\}+\{b_n\}\| \leq \|\{a_n\}\|$; in the same way $\|\{a_n\}\| = \|\{a_n + b_n\} - \{b_n\}\| \leq \|\{a_n + b_n\}\|$ and so $\|\{a_n\} + \{b_n\}\| = \|\{a_n\}\|$. Also clearly from (3.9)

$$\|\{a_n\}\{b_n\}\| = \|\{a_n\}\|\|\{b_n\}\|.$$

Summarising we have shown

**(3.12)** *The map $\{a_n\} + \mathfrak{n} \mapsto \|\{a_n\}\|$ is well defined; it is an absolute value on $\overline{K}$ which is denoted by $\|.\|$; moreover, its equivalence class depends only on that of $|.|$.*

If $a \in K$ and if $\overline{a}$ again denotes the corresponding constant sequence, then by (3.9) $\|\overline{a}\| = |a|$, and so $j$ is a homomorphism $(K, |.|) \to (\overline{K}, \|.\|)$ of valued fields.

Next we prove

**(3.13)** *Every element $\alpha$ of $\overline{K}$ is a limit $\alpha = \lim_{n\to\infty,\|.\|} j(b_n)$ $(b_n \in K)$, i.e. $j(K)$ is dense in $\overline{K}$. In fact, if $\alpha$ is represented by the sequence $\{b_n\}$ $(b_n \in K)$, then $\alpha = \lim_{n\to\infty,\|.\|} j(b_n)$.*

*Proof.* Suppose that the Cauchy sequence $\{b_n\}$ represents $x \in \overline{K}$. Then for fixed $m$ and variable $n$ $\{b_n - b_m\}$ represents $x - j(b_m)$. Given $\epsilon > 0$, choose $N$ such that $|b_n - b_r| < \epsilon$ for all $n, r \geq N$; then for all $m \geq N$

$$\|x - j(b_m)\| = \lim_{n\to\infty} |b_n - b_m| \leq \epsilon.$$

□

**(3.14)** *$\overline{K}$ is complete with respect to $\|.\|$.*

*Proof.* Let $\{x_n\}$ be a Cauchy sequence in $\overline{K}$ with respect to $\|.\|$. By (3.13) we know that each $x_n = \lim_{n\to\infty,\|.\|} j(b_{n,m})$ with $b_{n,m} \in K$. For each $n$, fix $m = m(n)$ so that on writing $b(n) = b_{n,m(n)}$, we have

$$\|x_n - j(b(n))\| < \frac{1}{2^n}. \qquad (3.15)$$

Then, by the triangle inequality,

$$\begin{aligned}|b(r)-b(n)| &= \|j(b(r))-j(b(n))\| \\ &< \frac{1}{2^r}+\frac{1}{2^n}+\|x_r-x_n\|.\end{aligned}$$

This implies that $\{b(r)\}$ is a Cauchy sequence with respect to $|.|$. Now let $x$ denote the class of $\{b(r)\}$ in $\overline{K}$. Then by (3.13) $x = \lim_{n\to\infty,\|.\|} j(b(n))$; on the other hand by (3.15) $0 = \lim_{n\to\infty,\|.\|}(x_n - j(b(n)))$, and so we have therefore shown that $x = \lim_{n\to\infty,\|.\|} x_n$.

□

We have now proved all of Theorem 10, except for property $(b)$ and the two uniqueness statements. To establish property $(b)$, let $(L, |.|^\times)$ and $k$ be as given. A Cauchy sequence $\{a_n\}$ in $K$ with respect to $|.|$ determines an element $\alpha$ in $\overline{K}$; as $k$ is a homomorphism of valued fields, the sequence $k\{a_n\} := \{k(a_n)\}$ will be a Cauchy sequence in $L$ with respect to $|.|^\times$. We first show that there is at most one $k' : (\overline{K}, \|.\|) \to (L, |.|^\times)$ with $k' \circ j = k$. Indeed, as $k'$ is a homomorphism of valued fields, $|k'(\alpha) - k' \circ j(a_n)|^\times = \|\alpha - j(a_n)\|$, and so

$$k'(\alpha) = \lim_{|.|^\times} k(a_n). \tag{3.16}$$

To demonstrate the existence of $k'$, suppose that the two Cauchy sequences $\{a_n\}$, $\{a_n'\}$ in $K$ define the same element $\alpha$ of $\overline{K}$; then $\{a_n - a_n'\}$ is a null sequence in $K$ for $|.|$; hence $\{k(a_n)-k(a_n')\}$ is a null sequence in $L$ for $|.|^\times$. We therefore conclude that the rule $k'(\alpha) = \lim_{|.|^\times} k(a_n)$, as in (3.16), defines a unique element of $L$, which depends only on $\alpha$ and not on the choice of the representative Cauchy sequence $\{a_n\}$, i.e. $k'$ is certainly a uniquely defined map $\overline{K} \to L$, with $k' \circ j = k$. Moreover, if $\{a_n\}, \{b_n\}$ are Cauchy sequences representing $\alpha, \beta \in \overline{K}$; then $\{a_n\} \overset{+}{\times} \{b_n\}$ represents $\alpha \overset{+}{\times} \beta$ in $\overline{K}$, and one now sees that $k'(\alpha \overset{+}{\times} \beta) = k'(\alpha) \overset{+}{\times} k'(\beta)$, i.e. $k'$ is a homomorphism.

In order to show that property $(a)$ characterises $(\overline{K}, \|.\|)$ as indicated, suppose that we are given a homomorphism of valued fields $l\colon (K, |.|) \to (K_1, |.|_1)$ with $(K_1, |.|_1)$ complete and with $l(K)$ dense in $K_1$. By part $(b)$ of the theorem there exists a homomorphism of valued fields $k\colon (K, \|.\|) \to (K_1, |.|_1)$ with $k \circ j = l$. We now have to show that $k$ is surjective. Let $y \in K_1$. By the density hypothesis $y = \lim_{|.|_1}(l(a_n))$ with $\{a_n\}$ a Cauchy sequence in $K$. But then $y = k(x)$ where $x = \lim_{\|.\|} j(a_n)$, as required.

Finally, let $(K_1, |.|_1)$ be a complete valued field and let $j_1\colon (K, |.|) \to (K_1, |.|_1)$ be a homomorphism of valued fields, having the analogous property to $(b)$. Then by hypothesis we get homomorphisms of valued fields

$$l: (\overline{K}, \|.\|) \to (K_1, |.|_1)$$
$$l_1: (K_1, |.|_1) \to (\overline{K}, \|.\|)$$

with $l_1 \circ j_1 = j$, $l \circ j = j_1$. Therefore $l_1 \circ l \circ j = j$, and so by the uniqueness of the homomorphism in property $(b)$ $l_1 \circ l = \mathrm{id}_{\overline{K}}$, and similarly $l \circ l_1 = \mathrm{id}_{K_1}$. Thus indeed $l_1$ and $l$ are mutually inverse isomorphisms of valued fields.

□

Let $K$ be an algebraic number field with $|.|$ an absolute value corresponding to an embedding $\sigma: K \to \mathbb{R}$ (see (2.10.$a$)). $(\mathbb{R}, |.|_{\mathbb{R}})$ is complete, and it is already the completion of $\mathbb{Q}$ with respect to the restriction of $|.|$. Therefore, by Theorem 10, $(\mathbb{R}, |.|_{\mathbb{R}})$ is the completion of $(K, |.|)$. Next if $|.|$ is an absolute value of $K$ corresponding to an imaginary embedding $\sigma: K \to \mathbb{C}$, i.e. one which takes some non-real values, then the completion contains $(\mathbb{R}, |.|_{\mathbb{R}})$ properly, i.e. it must be $(\mathbb{C}, |.|_{\mathbb{C}})$.

We now wish to consider in detail the general case of discrete absolute values. From now on in this section we assume that $\mathfrak{o}$ is a Dedekind domain, which is not a field, and let $K$ denote its field of fractions. Let $\mathfrak{p}$ denote a maximal ideal of $\mathfrak{o}$, let $v = v_{\mathfrak{p}}$ be the associated valuation of $K$, and let $|.| = |.|_v$ be the corresponding discrete absolute value as given in (2.10.$c$); that is to say for some real $\lambda$, $0 < \lambda < 1$, we have $|a| = \lambda^{v(a)}$ for all $a \in K$. As previously, let $(\overline{K}, \|.\|)$ denote the completion of $(K, |.|)$.

**Theorem 11.**

($a$) *Under the stated hypotheses, there is a unique valuation $\hat{v}$ of $\overline{K}$ with the property that $\|.\| = |.|_{\hat{v}}$, i.e. such that $\|\alpha\| = \lambda^{\hat{v}(\alpha)}$ for $\alpha \in \overline{K}$. Also $\hat{v}$ restricts to $v$ on $K$ under the map $j$.*

($b$) *Let $\hat{\mathfrak{o}}$ denote the valuation ring of $\hat{v}$, and let $\hat{\mathfrak{p}}$ be the valuation ideal of $\hat{v}$. For $Y \subset \overline{K}$ let $\overline{Y}$ denote the closure of $Y$ in $\overline{K}$ with respect to the topology induced by $\|.\|$. Then, for all $r \in \mathbb{Z}$, $\overline{j(\mathfrak{p}^r)} = \hat{\mathfrak{p}}^r$, and in particular $\overline{j(\mathfrak{o})} = \hat{\mathfrak{o}}$.*

($c$) *The map $j: \mathfrak{o} \to \hat{\mathfrak{o}}$ induces isomorphism $\mathfrak{o}/\mathfrak{p}^r \cong \hat{\mathfrak{o}}/\hat{\mathfrak{p}}^r$.*

*Proof.* The set of non-zero values $\lambda^m$ of $|.|$ is discrete in $\mathbb{R}_{>0}$; hence, by the definition (3.9) of $\|.\|$, this is also the set of non-zero values of $\|.\|$. So we may write $\|\alpha\| = \lambda^{\hat{v}(\alpha)}$ and the fact that $\|.\|$ is discrete implies that $\hat{v}$ is a valuation. This then yields part ($a$).

Suppose $\alpha \notin \hat{\mathfrak{p}}^r$; this then implies that $\|\alpha\| \geq \lambda^{r-1}$. If $\|\alpha - j(a)\| <$

$\lambda^{r-1} - \lambda^r$, then we deduce from the chain of inequalities

$$\|j(a)\| \geq \|\alpha\| - \|\alpha - j(a)\| \geq \lambda^{r-1} - \|\alpha - j(a)\| > \lambda^r$$

that a suitable neighbourhood of $\alpha$ is disjoint from $j(\mathfrak{p}^r)$. Thus indeed $\overline{j(\mathfrak{p}^r)} \subset \hat{\mathfrak{p}}^r$.

The reverse inclusion is a bit more complicated, as $\mathfrak{o}$ need not be the actual valuation ring of $v$. In fact we shall show that

**(3.17)** *If $\alpha \in \hat{\mathfrak{o}}$, then $\alpha$ is represented by a Cauchy sequence $\{a_n\}$ of $K$, with all $a_n \in \mathfrak{o}$.*

Assuming this result for the moment, note that if $\pi \in \mathfrak{p} \setminus \mathfrak{p}^2$, then $\hat{v}(\pi^r) = v(\pi^r) = r$, and so $\hat{\mathfrak{p}}^r = j(\pi^r)\hat{\mathfrak{o}}$. Thus any element of $\hat{\mathfrak{p}}^r$ will be represented by a Cauchy sequence of the form $\{\pi^r a_n\}$ with $a_n \in \mathfrak{o}$. This then shows $\hat{\mathfrak{p}}^r \subset \overline{j(\mathfrak{p}^r)}$.

To prove (3.17) note that by discreteness, an element $\alpha \in \hat{\mathfrak{o}}$ must be represented by a Cauchy sequence $\{a_n\}$, with $v(a_n) \geq 0$ for all sufficiently large $n$ (and of course by changing finitely many $a_n$ we may actually assume that $v(a_n) \geq 0$ for all $n$). Clearly then (3.17) will be an immediate consequence of

**(3.18)** *Given $N \in \mathbb{N}$ and $a \in K$ with $v(a) \geq 0$, there exists $b \in \mathfrak{o}$ with $v(a - b) \geq N$.*

*Proof.* We can write $a = a_1/a_2$, with $a_1, a_2 \in \mathfrak{o}$ and $a_2 \notin \mathfrak{p}$. Thus the class of $a_2$ mod $\mathfrak{p}^N$ is a unit in $\mathfrak{o}/\mathfrak{p}^N$, and so we can find $c \in \mathfrak{o}$ with $ca_2 \equiv 1 \bmod \mathfrak{p}^N$. Now $aca_2 \in \mathfrak{o}$ and $v(a - aca_2) = v(a(1 - ca_2)) \geq N$. □

We now leave the proof of Theorem 11, since part ($c$) will be a trivial consequence of part of the next theorem.

Let $\mathfrak{a}\hat{\mathfrak{o}}$ be the fractional $\hat{\mathfrak{o}}$-ideal generated by the fractional $\mathfrak{o}$-ideal $\mathfrak{a}$. The map $\mathfrak{a} \mapsto \mathfrak{a}\hat{\mathfrak{o}}$ is clearly a homomorphism $I_{\mathfrak{o}} \to I_{\hat{\mathfrak{o}}}$. It is completely described by

**Corollary to Theorem 11.** $\mathfrak{p}\hat{\mathfrak{o}} = \hat{\mathfrak{p}}$ *and if $\mathfrak{q}$ is a prime ideal of $\mathfrak{o}$ other than $\hat{\mathfrak{p}}$, then $\mathfrak{q}\hat{\mathfrak{o}} = \hat{\mathfrak{o}}$.*

*Proof.* $\mathfrak{p}\hat{\mathfrak{o}} \subset \hat{\mathfrak{p}}$ by Theorem 11($b$), and $\mathfrak{p}$ contains an element of $v_{\hat{\mathfrak{p}}}$-value one, by ($a$). Hence $\mathfrak{p}\hat{\mathfrak{o}} = \hat{\mathfrak{p}}$. Again by ($a$) $\mathfrak{q}\hat{\mathfrak{o}} \not\subset \hat{\mathfrak{p}}$, but $\mathfrak{q}\hat{\mathfrak{p}} \subset \hat{\mathfrak{o}}$; hence $\mathfrak{q}\hat{\mathfrak{o}} = \hat{\mathfrak{o}}$. □

For a slightly different treatment, we need a special case of the notion of an *inverse limit.* For a ring $R$, an *inverse system* of $R$-modules $A_n$ and $R$-module homomorphisms $\alpha^n_m : A_n \to A_m$ for $n \geq m$ is such that (i) $\alpha^n_n = \mathrm{id}_{A_n}$; (ii) $\alpha^m_r \circ \alpha^n_m = \alpha^n_r$. The *inverse limit* of such a system, denoted $\varprojlim A_n$ is the $R$-submodule of the direct product $\prod_n A_n$, consisting of those sequences $a = (a(1), a(2), \ldots)$ for which $a(m) = \alpha^n_m(a(n))$ for $n \geq m$. Associated with this we have homomorphisms of $R$-modules

$$\alpha_m : \varprojlim A_n \to A_m$$

called the $m$-th component map, where $\alpha_m(a) = a(m)$. Of course we have the relation $\alpha_r = \alpha^m_r \circ \alpha_m$, for $m \geq r$.

Suppose now we have a second inverse system $\{B_n, \beta^n_m\}$, and that we are given homomorphisms of $R$-modules $\theta_n : B_n \to A_n$ $(n \geq 1)$, so that $\theta_m \circ \beta^n_m = \alpha^n_m \circ \theta_n$. We then derive a homomorphism

$$\varprojlim \theta_n = \theta : \varprojlim B_n \to \varprojlim A_n \tag{3.19}$$

where $\theta(b)(n) = \theta_n(b(n))$. It is readily seen that $\theta$ is an isomorphism if each $\theta_n$ is an isomorphism.

As an illustration of this construction let $B$ be an $R$-module and let $\theta_n : B \to A_n$ be homomorphisms so that for $n \geq m$ the following diagrams commute.

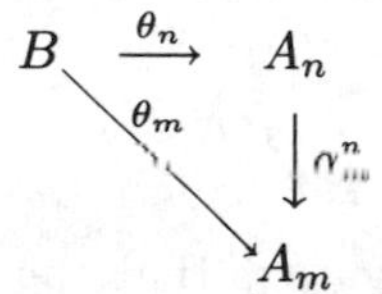


Then we obtain a unique homomorphism $\theta : B \to \varprojlim A_n$ such that the diagram

$$\begin{array}{ccc} B & \xrightarrow{\theta} & \varprojlim A_n \\ & \searrow^{\theta_m} & \downarrow \alpha_m \\ & & A_m \end{array} \tag{3.20}$$

commutes. In fact this property determines $\varprojlim A_n$ uniquely to within isomorphism.

We now apply this construction to the inverse system $\{\mathfrak{o}/\mathfrak{p}^n, \pi^n_m\}$ where $\pi^n_m$ is the residue class map $\mathfrak{o}/\mathfrak{p}^n \to \mathfrak{o}/\mathfrak{p}^m$ $(n \geq m)$. We shall write

$$\begin{aligned} \tilde{\mathfrak{o}} &= \varprojlim \mathfrak{o}/\mathfrak{p}^n \\ \tilde{\mathfrak{p}^r} &= \ker(\pi_r : \tilde{\mathfrak{o}} \to \mathfrak{o}/\mathfrak{p}^r) \qquad (r \geq 0). \end{aligned} \tag{3.21}$$

At present we have 3 distinct symbols in play: $\overline{j(\mathfrak{o})}$, $\hat{\mathfrak{o}}$ and $\tilde{\mathfrak{o}}$. In due course we shall show that they are all essentially the same thing. Thus we are really obtaining three different ways of looking at $\hat{\mathfrak{o}}$: namely as a valuation ring in $\overline{K}$; as the closure of $j(\mathfrak{o})$ in $\overline{K}$; and as the inverse limit of the $\mathfrak{o}/\mathfrak{p}^n$. This latter role gives a new, interesting characterisation of $\overline{K}$ as the field of fractions of $\varprojlim \mathfrak{o}/\mathfrak{p}^n$. This explains the great advantage of inverse systems: for us it is a construction which formalises "coherent systems of congruences" and provides a purely algebraic method for constructing completions.

By the basic property of inverse limits in (3.20),

**(3.22)** *there exists a homomorphism* $l\colon \mathfrak{o} \to \tilde{\mathfrak{o}}$ *so that the diagram*

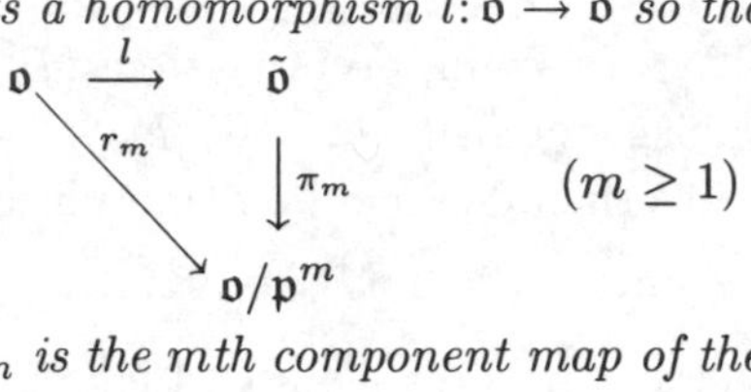


*commutes. Here* $\pi_m$ *is the mth component map of the inverse limit, and* $r_m$ *is the usual residue class map.*

**Theorem 12.**

(*a*) *For each positive integer* $m$, $\pi_m$ *gives rise to an isomorphism* $\tilde{\mathfrak{o}}/\tilde{\mathfrak{p}^m} \cong \mathfrak{o}/\mathfrak{p}^m$.

(*b*) *There is a unique homomorphism* $\hat{l}\colon \hat{\mathfrak{o}} \to \tilde{\mathfrak{o}}$ *such that*

 (*i*) $\hat{l}(j(a)) = l(a)$ *for all* $a \in \mathfrak{o}$

 (*ii*) $\alpha \in \hat{\mathfrak{p}}^r$ *iff* $\hat{l}(\alpha) \in \tilde{\mathfrak{p}^r}$, *for all* $r$.

(*c*) $\hat{l}$ *is an isomorphism* $\hat{\mathfrak{o}} \cong \tilde{\mathfrak{o}}$, *giving rise to isomorphisms* $\hat{\mathfrak{p}}^r \cong \tilde{\mathfrak{p}^r}$.

Observe that Theorem 12 implies that the map $\mathfrak{o}/\mathfrak{p}^r \to \hat{\mathfrak{o}}/\hat{\mathfrak{p}}^r$, induced by $j$, is indeed an isomorphism; thus Theorem 12 implies the still outstanding assertion (*c*) in Theorem 11.

*Proof of Theorem 12.* The maps $r_m$ are surjective; hence by (3.22) so are the maps $\pi_m$. This immediately gives (*a*). To construct $\hat{l}$, let $\{a_n\}$ denote a Cauchy sequence in $K$, with $a_n \in \mathfrak{o}$ for all $n$. This means that given $N \in \mathbb{N}$, there exists $n_0 \in \mathbb{N}$ so that

$$a_n \equiv a_{n_0} \bmod \mathfrak{p}^N \ (\forall n \geq n_0)$$

Let $\tilde{a}(N)$ be the class of $a_{n_0}$ in $\mathfrak{o}/\mathfrak{p}^N$, that is to say it is the "ultimate class of the $\{a_n\}$ mod $\mathfrak{p}^N$". Here we may clearly replace $n_0$ by an integer greater than or equal to $n_0$. Moreover, given a second Cauchy sequence $\{b_n\}$ with $b_n \in \mathfrak{o}$, then, with the obvious notation, $\tilde{a}(N) + \tilde{b}(N) = \widetilde{a+b}(N)$, where $\widetilde{a+b}(N)$ denotes the ultimate class of $\{a_n + b_n\}$ mod $\mathfrak{p}^N$. Also for $N \geq M$, $\tilde{a}(M) = \pi_M^N \tilde{a}(N)$ and so we get a map $\{a_n\} \mapsto \tilde{a} \in \tilde{\mathfrak{o}}$, which defines a homomorphism $R_0(\mathfrak{o}) \to \hat{\mathfrak{o}}$ of additive groups, with $R_0(\mathfrak{o})$ being the ring of Cauchy sequences with entries in $\mathfrak{o}$. Also $\tilde{a}(N) = 0$ for all $N$ iff $\{a_n\}$ is a null sequence; so we get an injective homomorphism

$$R_0(\mathfrak{o})/R_0(\mathfrak{o}) \cap \mathfrak{n} \to \tilde{\mathfrak{o}}.$$

Now by Theorem 11 we know that $\hat{\mathfrak{o}} = \overline{j(\mathfrak{o})}$ is the subset of $\overline{K}$ represented by Cauchy sequences with entries in $\mathfrak{o}$. Therefore the map $\{a_n\} \mapsto \tilde{a}$ gives rise to an injective homomorphism $\hat{l}\colon \hat{\mathfrak{o}} \to \tilde{\mathfrak{o}}$.

Going back to the original construction, observe that if $\{a_n\}$ is a constant sequence $\{a_n = a\}$ for $a \in \mathfrak{o}$, then $\tilde{a}(N)$ is just the residue class $r_N(a)$ of $a$ mod $\mathfrak{p}^N$. This then implies that $\hat{l} \circ j = l$.

In order to prove that $\hat{l}$ is an isomorphism we have to show that it is surjective. Given $\tilde{a} \in \tilde{\mathfrak{o}}$, choose $b_n \in \mathfrak{o}$ such that its class mod $\mathfrak{p}^n$ is $\tilde{a}(n)$. Then $\{b_n\}$ is a Cauchy sequence representing an element $\beta \in \hat{\mathfrak{o}}$ with $\hat{l}(\beta) = \tilde{a}$.

Next we show that for $\alpha \in \hat{\mathfrak{o}}$, $\hat{l}(\alpha) \in \tilde{\mathfrak{p}^r}$ iff $\alpha \in \hat{\mathfrak{p}}^r$; this will then establish property $(b)(ii)$ for $\hat{l}$ and also show that $\hat{l}$ restricted to $\hat{\mathfrak{p}}^r$ is an isomorphism onto $\tilde{\mathfrak{p}}^r$. Suppose that $\alpha$ is represented by the Cauchy sequence $\{a_n\}$ with entries in $\mathfrak{o}$; then $\alpha \in \hat{\mathfrak{p}}^r$ iff for all large enough $n$, $a_n \in \mathfrak{p}^r$, i.e. $\hat{a}(r) = 0$; clearly this occurs iff $\tilde{a} \in \tilde{\mathfrak{p}}^r$.

It now only remains for us to show that properties $(i)$ and $(ii)$ under $(b)$ determine $\hat{l}$ uniquely. We therefore let $l'$ denote a homomorphism $\hat{\mathfrak{o}} \to \tilde{\mathfrak{o}}$ with these given properties. Consider an element $\alpha$ in $\hat{\mathfrak{o}}$ represented by the Cauchy sequence $\{a_n\}$, with entries in $\mathfrak{o}$. As previously, $\hat{l}(\alpha)(N)$ is the ultimate class of $\{a_n\}$ mod $\mathfrak{p}^N$. Now choose $b \in \mathfrak{o}$ with $r_N(b) = \hat{l}(\alpha)(N)$; then, by property $(i)$ $l'(j(b))(N) = l(b)(N) = r_N(b) = \hat{l}(\alpha)(N)$; also, by property $(ii)$, as $\alpha - j(b) \in \hat{\mathfrak{p}}^N$, it follows that $l'(\alpha) - l'(j(b)) \in \hat{\mathfrak{p}}^N$, hence $l'(\alpha)(N) - l'(j(b))(N) = 0$, and so $l'(\alpha)(N) = \hat{l}(\alpha)(N)$. This being true for all $N$, we have $l'(\alpha) = \hat{l}(\alpha)$, and this completes the proof of Theorem 12.

□

Thanks to Theorem 12 we can now drastically simplify our rather elaborate notation: for the remainder of this section we view $K$ as embedded in $\overline{K}$, thus omitting the symbol $j$, and we freely identify $\overline{\mathfrak{o}}$, $\hat{\mathfrak{o}}$ and $\tilde{\mathfrak{o}}$, and likewise $\overline{\mathfrak{p}}$, $\hat{\mathfrak{p}}$ and $\tilde{\mathfrak{p}}$. We shall also write $|.|_{\hat{v}}$ for $\|.\|$.

Note that the last theorem contains a non-trivial statement concerning the special case when the given valued field $(K, |.|)$ is complete (with $|.|$ discrete), and $\mathfrak{o}$ is the valuation ring $\{x \in K \mid |x| \leq 1\}$: namely we now know that $\mathfrak{o} \cong \varprojlim \mathfrak{o}/\mathfrak{p}^n$.

With the notation and hypotheses of the previous theorem we have

**(3.23)** *The homomorphism $\theta: \hat{\mathfrak{o}} \otimes_{\mathfrak{o}} K \to \overline{K}$ induced by $a \otimes x \mapsto ax$ is an isomorphism.*

*Proof.* Choose $\pi \in \mathfrak{o}$ such that $v(\pi) = 1$; then $\overline{K} = \hat{\mathfrak{o}}[\pi^{-1}]$, and so $\theta$ is surjective. Given an arbitrary element $y = \sum_i a_i \otimes x_i$ of $\hat{\mathfrak{o}} \otimes K$, set $n = \min_i(0, v(x_i))$; then $y = (\sum a_i x_i \pi^{-n}) \otimes \pi^n$. Thus $\sum a_i x_i = 0$ implies $y = 0$, i.e $\theta$ is injective.

□

As in ordinary analysis, we say that an infinite series $\sum_{n=N}^{\infty}$ of elements $x_n$ of $\overline{K}$ converges if its sequence $s_m = \sum_{n=N}^{m} x_m$ of partial sums is convergent; in this case we write

$$\lim_{m\to\infty} s_m = \sum_{n=N}^{\infty} x_n.$$

Recall that, because of the ultrametric property, $|x_n| \to 0$ is a sufficient condition for convergence.

Let $R$ now denote a set of representatives in $\hat{\mathfrak{o}}$ for the elements of $k = \hat{\mathfrak{o}}/\hat{\mathfrak{p}}$, where we always assume that the zero of $\hat{\mathfrak{o}}$ represents the zero of $k$. For each $n \in \mathbb{Z}$, let $\pi_n$ be an element of $\overline{K}$ with $\hat{v}(\pi_n) = n$. Any series $\sum_{n=N}^{\infty} r_n \pi_n$ ($r_n \in R$ for each $n$) will converge. If

$$\alpha = \sum_{n=N}^{\infty} r_n \pi_n \tag{3.24}$$

then for all partial sums with sufficiently large $M$

$$\hat{v}\left(\sum_{n=N}^{M} r_n \pi_n\right) = \inf(n \mid r_n \neq 0)$$

whence in the limit

$$\hat{v}(\alpha) = \inf(n \mid r_n \neq 0). \tag{3.24.a}$$

This also makes sense for $\alpha = 0$, since $\hat{v}(0) = \infty$. Moreover

$$\sum_{n=N}^{\infty} r_n \pi_n = \sum_{n=N}^{\infty} r'_n \pi_n \quad (r_n, r'_n \in R)$$

iff

$$\left\{\begin{aligned} \inf(n \mid r_n \neq 0) = \inf(n \mid r'_n \neq 0) = v \text{ say},\\ r_m = r'_m \quad \text{for all } m \geq v. \end{aligned}\right. \tag{3.24.b}$$

We now show

**(3.24.$c$)** *Every element $\alpha$ of $\overline{K}$ can be represented in the form* (3.24).

For the proof we may assume $\alpha \neq 0$, say $\hat{v}(\alpha) = N$. Then $\alpha\pi_N^{-1} \in$

$\hat{\mathfrak{o}} \setminus \hat{\mathfrak{p}}$, hence $\alpha\pi_N^{-1} \equiv r_N \bmod \hat{\mathfrak{p}}$ for some $r_N \in R$, $r_N \neq 0$. Thus $\alpha \equiv r_N\pi_N \bmod \hat{\mathfrak{p}}^{N+1}$. This starts the induction. If now

$$\alpha \equiv \sum_{n=N}^{M} r_n\pi_n \bmod \hat{\mathfrak{p}}^{M+1}$$

then $\left(\alpha - \sum_{n=N}^{M} r_n\pi_n\right)\pi_{M+1}^{-1} \in \hat{\mathfrak{o}}$, hence is congruent to $r_{M+1} \bmod \hat{\mathfrak{p}}$ for some $r_{M+1} \in R$. Therefore

$$\alpha = \sum_{n=N}^{M+1} r_n\pi_n \bmod \hat{\mathfrak{p}}^{M+2}$$

and so $\lim_M \left|\alpha - \sum_{n=N}^{M} r_n\pi_n\right| = 0$.

□

*Example 1.* Let $F$ be a field, let $X$ be an indeterminate over $F$, and let $v$ denote the valuation associated to the prime ideal $(X)$ of $F[X]$. Then $k = F[X]/(X) \cong F$ and we choose $R = F$ as the set of representatives of $k$ in $F[X]$. Thus now $R$ is actually closed under addition and multiplication; in fact it is a subfield of $F(X)$ and so of its completion. By (3.24.c) we see that the completion of $F(X)$ with respect to $|.|_v$ identifies with $F((X))$, the field of finitely tailed Laurent series.

*Example 2.* With the notation of the previous example let $f$ denote a monic irreducible separable polynomial in $F[X]$. Let $|.|_f$ denote the discrete absolute value associated with the prime ideal $(f)$, let $\alpha$ denote a root of $f$, and let $L = F(\alpha)$. We view $L((t))$ as endowed with $|.|_t$, the discrete absolute value associated with the prime ideal $(t)$ of $L[[t]]$. We assert that $(L((t)), |.|_t)$ is the completion of $(F(X), |.|_f)$.

First note that, by example 1, $(L((t)), |.|_t)$ is complete; so by Theorem 10 it suffices to construct a dense homomorphism of valued fields $j\colon (F(X), |.|_f) \to (L((t)), |.|_t)$. We define $j$ by forming the Laurent expansion about $\alpha$ of $h \in F(X)$

$$j(h(X)) = h(t + \alpha) = \sum a_i t^i.$$

To see that $j$ has dense image, we note that on the one hand $j(F[X]) = F[\alpha] \equiv L \bmod (t)$; while on the other hand, because $f$ is separable

$$j(f) = tf'(\alpha) + t^2 a(t)$$

with $f'(\alpha) \neq 0$, for some $a(t) \in L[[t]]$. Thus, given any $b \in L((t))$ we can find elements in $F(X)$ whose $j$ images are arbitrarily close to $b$ with respect to $|.|_t$. Finally note that the above equality implies that for any $h \in F(X)$, $|h|_f = |j(h)|_t$, so that $j$ is a homomorphism of valued fields.

Let $p$ denote a prime number. We next consider the completion of $\mathbb{Q}$ with respect to the metric $|.|_p$ (given prior to Theorem 9). Recall that for $ab^{-1}p^s$ with $a, b \in \mathbb{Z}$, $p \nmid ab$, we have $|ab^{-1}p^s|_p = p^{-s}$. The field which we obtain is denoted by $\mathbb{Q}_p$ and is called the field of $p$-adic numbers or the (rational) $p$-adic field. We write $\mathbb{Z}_p$ for the closure of $\mathbb{Z}$ in $\mathbb{Q}_p$, i.e. for the valuation ring; $\mathbb{Z}_p$ is called the ring of $p$-adic integers. By Theorem 12 and (2.4) we know that $p\mathbb{Z}_p$ is the valuation ideal of $\mathbb{Z}_p$ and that

$$\mathbb{Z}_p/p\mathbb{Z}_p \cong \mathbb{Z}/p\mathbb{Z} \cong \mathbb{F}_p.$$

We can therefore apply (3.24.$c$) to deduce that each element $\alpha \in \mathbb{Q}_p$ can be written in the form $\alpha = \sum_{i \geq N}^{\infty} r_i p^i$ where each $r_i$ belongs to the set $\{0, 1, \ldots, p-1\}$. Thus, very loosely, $p$-adic numbers are "backwards decimals in $p$".

As promised at the start of this section, we provide an instance of how complete fields lend themselves to standard analytic operations. We consider the iterative construction of a root of a polynomial.

**(3.25) (Hensel's Lemma).** *Let $f \in \hat{\mathfrak{o}}[X]$ and suppose that there exists $\alpha_n \in \hat{\mathfrak{o}}$ such that $\hat{v}(f'(\alpha_n)) = c$ and $\hat{v}(f(\alpha_n)) \geq n + c$ where $n > c$. Then there exists $\alpha_{n+1} \in \hat{\mathfrak{o}}$ with $\hat{v}(\alpha_{n+1} - \alpha_n) \geq n$, $\hat{v}(f'(\alpha_{n+1})) = c$ and $\hat{v}(f(\alpha_{n+1})) \geq n+1+c$. Moreover $\alpha = \lim_n \alpha_n \in \hat{\mathfrak{o}}$ has the property that $f(\alpha) = 0$.*

[Here $f'(X)$ denotes the formal derivative of $f(X)$.]

**(3.25.$a$) Corollary.** *Again let $k = \hat{\mathfrak{o}}/\hat{\mathfrak{p}}$, and suppose that the reduced polynomial $\overline{f} \in k[X]$ has a simple root $\overline{\alpha}$ in $k$; then $f$ has a simple root $\alpha$ in $\hat{\mathfrak{o}}$ whose image in $k$ is $\overline{\alpha}$.*

*Proof of Corollary.* Here the iterative procedure of the lemma starts with $\hat{v}(f(\alpha_1)) \geq 1$, $\hat{v}(f'(\alpha_1)) = 0$. Applying the lemma, we obtain a root $\alpha$ of $f$ with $\alpha = \alpha_1 \bmod \hat{\mathfrak{p}}$ and $f'(\alpha) \neq 0 \bmod \hat{\mathfrak{p}}$, i.e. $f'(\alpha) \neq 0$; thus $\alpha$ is indeed a simple root.

*Proof of* (3.25). Again let $\pi$ denote a generator of $\hat{\mathfrak{p}}$ over $\hat{\mathfrak{o}}$. We know that for $\lambda \in \hat{\mathfrak{o}}$ we have

$$f(\alpha_n + \lambda\pi^n) = f(\alpha_n) + \lambda\pi^n f'(\alpha_n) + (\lambda\pi^n)^2 h(\alpha_n, \lambda\pi^n)$$

where $h(X, Y)$ is a polynomial in $X, Y$ over $\hat{\mathfrak{o}}$. Therefore

$$f(\alpha_n + \lambda\pi^n) = f(\alpha_n) + \lambda\pi^n f'(\alpha_n) \bmod \hat{\mathfrak{p}}^{2n}.$$

Hence, if we set $\lambda = -f(\alpha_n)\pi^{-n} f'(\alpha_n)^{-1} \in \hat{\mathfrak{o}}$, and put $\alpha_{n+1} = \alpha_n + \lambda\pi^n$, then clearly $f(\alpha_{n+1}) \in \hat{\mathfrak{p}}^{2n} \subset \hat{\mathfrak{p}}^{n+1+c}$. On the other hand it is immediate that

$$\hat{v}(f'(\alpha_{n+1})) = \hat{v}(f'(\alpha_n)) = c.$$

We now iterate this construction and thereby obtain a Cauchy sequence $\{\alpha_n\}$. We set $\alpha = \lim_{n\to\infty} \alpha_n$. Then $f(\alpha_n)$ is a null sequence and so

$$0 = \lim_n f(\alpha_n) = f(\lim_n \alpha_n) = f(\alpha).$$

□

As an application of Hensel's Lemma we consider the case when $[\hat{\mathfrak{o}} : \hat{\mathfrak{p}}] = q$ is finite; hence $\hat{\mathfrak{o}}/\hat{\mathfrak{p}}$ is a field of prime characteristic $p$. We then denote by $\mu'_{\overline{K}}$ the group of roots of unity in $\overline{K}$ of order prime to $p$. Clearly $\mu'_{\overline{K}} \in \hat{\mathfrak{o}}^*$, and taking residue classes yields a homomorphism

$$\rho \colon \mu'_{\overline{K}} \to (\hat{\mathfrak{o}}/\hat{\mathfrak{p}})^*.$$

**(3.26)** *$\rho$ is an isomorphism.*

*Proof.* Suppose $\mu'_{\overline{K}}$ contains the group $\mu_m$ of primitive $m$-th roots of unity. As $X^m - 1$ is separable over $\hat{\mathfrak{o}}/\hat{\mathfrak{p}}$, $\rho|_{\mu_m}$ is injective. It follows that $\rho$ is injective (hence $\mu'_{\overline{K}}$ is finite). As $X^{q-1} - 1$ is separable, it follows from Hensel's Lemma that $\rho$ is surjective.

□

From (3.26) we know that $\mathbb{Q}_p$ contains $\mu_{p-1}$, the group of $p-1$st roots of unity, and that these are a full set of representatives of the non-zero residue classes of $\mathbb{Z}_p/p\mathbb{Z}_p$. Thus we can also write $\alpha$ uniquely in the form

$$\alpha = \sum_{i \geq N}^{\infty} \zeta_i p^i \tag{3.27}$$

where $\zeta_i \in \mu_{p-1} \cup \{0\}$; this is called the Teichmüller representation of $\alpha$.

We finally apply Hensel's Lemma to quadratic residues in $\mathbb{Q}_p$.

**(3.28)** *Let $\xi \in \mathbb{Z}_p{}^*$; then $\xi \in \mathbb{Q}_p^{*2}$ iff*

$$\begin{cases} \left(\frac{\xi}{p}\right) = 1 & \text{if } p \neq 2 \\ \xi \equiv 1 \bmod 8\mathbb{Z}_2 & \text{if } p = 2. \end{cases}$$

*Proof.* If $y \in \mathbb{Q}_p$ with $y^2 = \xi$, then $2\hat{v}(y) = \hat{v}(\xi) = 0$, and so $y \in \mathbb{Z}_p{}^*$. In particular $y^2 = \xi \bmod p\mathbb{Z}_p$, and so $\xi \bmod p\mathbb{Z}_p$ is certainly a square in $\mathbb{F}_p{}^*$. Moreover, if $p = 2$, we can write $y = 1 + 2\beta \bmod 8\mathbb{Z}_2$ with $\beta \in \mathbb{Z}_2$.

Then $y^2 \equiv 1 + 4(\beta^2 + \beta) \bmod 8\mathbb{Z}_2$. As $\beta^2 + \beta \equiv 0 \bmod 2\mathbb{Z}_2$, it follows that $y^2 \equiv 1 \bmod 8\mathbb{Z}_2$.

Conversely, if $p \neq 2$ and if $y^2 = \xi \bmod p\mathbb{Z}_p$, then by (3.25.$a$) $X^2 - \xi$ has a solution in $\mathbb{Z}_p$, and the result is shown. Finally if $p = 2$ and if $\xi \equiv 1 \bmod 8\mathbb{Z}_2$, then we can apply (3.25) to the polynomial $X^2 - \xi$ (with $n > c = 1$).

□

We shall now review some of the deeper topological implications of the work of this section, presupposing here some familiarity with topological theory beyond what has been needed hitherto.

Earlier we defined inverse limits. We shall first of all show how to endow the inverse limit $A = \varprojlim A_n$ of an inverse system with a topology. It will suffice for our purposes to do this assuming the modules $A_n$ to have the discrete topology – although this can be done in greater generality. A basic neighbourhood of a point $a \in A$ consists then of all $x \in A$ with $\alpha_n(x) = \alpha_n(a)$ for some fixed $n \in \mathbb{N}$, where $\alpha_n \colon A \to A_n$ is the component map.

Now let $\overline{K}, |.|_{\hat{v}}$ be a complete valued field with respect to a discrete absolute value coming from a valuation $\hat{v}$. We had previously identified (algebraically) the valuation ring $\hat{\mathfrak{o}}$ of $\overline{K}$ and the inverse limit $\tilde{\mathfrak{o}} = \varprojlim \hat{\mathfrak{o}}/\hat{\mathfrak{p}}^n$. We now observe that the two topologies in question coincide, i.e. the open subsets of the metric space $(\overline{K}, |.|_{\hat{v}})$ are the same as the open sets obtained from the inverse limit topology. This assertion is just a reinterpretation of Theorem 12 in topological terms.

The topology on $(\overline{K}, |.|_{\hat{v}})$ is Hausdorff, being induced by a metric; this is of course true for any valued field. The fact that $|.|_v$ is discrete yields a stronger result. We note that from the very definition of $\hat{\mathfrak{p}}^r$

$$\begin{aligned}\hat{\mathfrak{p}}^r &= \{x \in \overline{K} \mid \hat{v}(x) > r - 1\}\\ &= \{x \in \overline{K} \mid \hat{v}(x) \geq r\}\end{aligned}$$

and so $\{\hat{\mathfrak{p}}^r \mid r \in \mathbb{N}\}$ is a basis of open neighbourhoods of 0; moreover they are all closed sets; thus $\overline{K}$ is a totally disconnected topological space.

**(3.29)** *If $k = \hat{\mathfrak{o}}/\hat{\mathfrak{p}}$ is finite, then each set $\hat{\mathfrak{p}}^r$ is compact.*

*Proof.* With the previous notation $\hat{\mathfrak{p}}^r = \pi^r \hat{\mathfrak{o}}$ and so it suffices to prove the result for $\hat{\mathfrak{o}}$. Suppose for contradiction that $\mathfrak{C}$ is an open cover of $\hat{\mathfrak{o}}$ which does not contain a finite subcover of $\hat{\mathfrak{o}}$.

Inductively, we claim that, for given $j$, we can find a residue class

$y_j + \hat{\mathfrak{p}}^{j+1}$ which has no finite subcover by $\mathfrak{C}$ and such that $y_j + \hat{\mathfrak{p}}^j = y_{j-1} + \hat{\mathfrak{p}}^j$. So suppose that $y_1, \ldots, y_{j-1}$ have been constructed. Since $\hat{\mathfrak{p}}^j / \hat{\mathfrak{p}}^{j+1} \cong k$ is finite, there are only a finite number of classes $z + \hat{\mathfrak{p}}^{j+1}$ which map to $y_{j-1} + \hat{\mathfrak{p}}^j$ under $\hat{\mathfrak{o}}/\hat{\mathfrak{p}}^{j+1} \to \hat{\mathfrak{o}}/\hat{\mathfrak{p}}^j$. Since $y_{j-1} + \hat{\mathfrak{p}}^j$ has no finite subcover by $\mathfrak{C}$, it follows that at least one of the classes $z + \hat{\mathfrak{p}}^{j+1}$ has the same property. Choose one and write it as $y_j + \hat{\mathfrak{p}}^{j+1}$. This then establishes the inductive step.

To prove the result let $y = \lim_{j\to\infty} y_j$. Then $y$ lies in some open set $U$ of the cover $\mathfrak{C}$; therefore $y_j + \hat{\mathfrak{p}}^{j+1}$ is contained in $U$ for large $j$, and thus contradicts the fact that $y_j + \hat{\mathfrak{p}}^{j+1}$ has no finite subcover of $\mathfrak{C}$.

□

**Corollary.** $\mathbb{Z}_p$ *is compact.*

*Remark.* One can use Tychonoff's theorem to give a quicker proof of (3.29). The hypothesis implies that the discrete sets of $\hat{\mathfrak{o}}/\hat{\mathfrak{p}}^n$ are all finite, and hence compact; therefore their product $\prod_n \hat{\mathfrak{o}}/\hat{\mathfrak{p}}^n$ is compact. However, the inverse limit $\tilde{\mathfrak{o}} = \hat{\mathfrak{o}}$ is a closed subset of the product and is therefore compact. The same reasoning applies to $\hat{\mathfrak{p}}^m$.

Note that the converse of (3.29) holds:

**(3.30)** *If* $\hat{\mathfrak{o}}$ *is compact, then* $k$ *is finite.*

*Proof.* As an image of $\hat{\mathfrak{o}}$ under the continuous surjection $\hat{\mathfrak{o}} \to \hat{\mathfrak{o}}/\hat{\mathfrak{p}}$, $k$ is necessarily compact. As $\hat{\mathfrak{p}}$ is open in $\hat{\mathfrak{o}}$, $\hat{\mathfrak{o}}/\hat{\mathfrak{p}}$ must be discrete. The result follows since any discrete compact set is finite.

□

## §4 Module theory over a Dedekind domain

Our aim here is to develop the theory of finitely generated modules over a Dedekind domain. We shall see that a great part of the theory for principal ideal domains carries over, but, at a crucial place, the ideal classes will have a crucial role to play. One important application will be to the definition of the ideal norm.

This section relies heavily on the three previous ones. The unambitious can omit it without placing himself at too great a disadvantage in the sequel, since the subsequent chapters are largely independent of the results here.

We first recall a number of well-known definitions and results. Here, to begin with, $\mathfrak{o}$ is just a Noetherian integral domain with field of fractions $K$. A torsion element $x$ in an $\mathfrak{o}$-module $M$ is an element with the property that $xa = 0$ for some non-zero element $a \in \mathfrak{o}$. For a finitely generated $\mathfrak{o}$-module $M$, the torsion elements of $M$ form a finitely generated submodule $tM$, and the elements

$$\mathrm{an}(M) = \{a \in \mathfrak{o} \mid tM \cdot a = 0\}$$

form an ideal of $\mathfrak{o}$, and $tM \cdot \mathrm{an}(M) = 0$. We say that $M$ is torsion free if $tM = (0)$, i.e. if 0 is the only torsion element in $M$. Clearly $M/tM$ is always torsion free. We call $M$ a *torsion module* if $M = tM$.

An $\mathfrak{o}$-module $F$ is said to be free (on a set $u_1, \ldots, u_r$ of generators) iff every element $x \in F$ has a representation $x = \sum u_j a_j$ with uniquely determined $a_j \in \mathfrak{o}$. Then, with the above notation, we have:

**(4.1)** *Let $M$ denote a non-zero, finitely generated $\mathfrak{o}$-module. Then the following conditions are equivalent:*

(*a*) *$M$ is torsion free over $\mathfrak{o}$;*

(*b*) *$M$ is isomorphic to a submodule of a free $\mathfrak{o}$-module (of finite rank).*

(*c*) *$M$ is isomorphic to an $\mathfrak{o}$-submodule of a finite dimensional $K$-vector space $V$.*

(*d*) *The homomorphism $M \to M \otimes_{\mathfrak{o}} K$ induced by $m \mapsto m \otimes 1$ is injective.*

We simultaneously show that for $M$ satisfying the conditions of (4.1)

**(4.1.*a*)** *$M$ contains a free $\mathfrak{o}$-module $F$ such that $M/F$ is a torsion $\mathfrak{o}$-module.*

*Proof.* (*d*)⇒(*c*). As $M$ is finitely generated, so is $M \otimes_{\mathfrak{o}} K$ (over $K$), hence this is a finite dimensional $K$-vector space.

(*c*)⇒(*b*). Let $\{v_j \mid j = 1, \ldots, n\}$ denote a $K$ basis of $V$, and let $w_1, \ldots, w_m$ denote a generating set of $M$ over $\mathfrak{o}$. Then there exists $b,\ a_{ij} \in \mathfrak{o}$, with $b \neq 0$, such that (viewing $M$ as embedded in $V$)

$$w_i b = \sum v_j a_{ij}.$$

Thus $M$ is contained in the free $\mathfrak{o}$-module on $\{v_j b^{-1}\}$.

(*b*)⇒(*a*) is obvious. So now we show that (*a*)⇒(*d*). We call elements $x_1, \ldots, x_m \in M$ linearly independent if $\sum x_i a_i = 0$ ($a_i \in \mathfrak{o}$), implies that all the $a_i$ are 0. Since $M$ is non-zero and torsion free, $M$ certainly contains such a linearly independent set with $m \geq 1$. Any finite generating set of $M$ obviously contains some maximal linearly independent set. Let $u_1, \ldots, u_d$ denote such a maximal linearly independent set. Given non-zero $w \in M$, there exists non-zero $c_w \in \mathfrak{o}$ and $b_{w,j}$ $(j = 1, \dots, d)$ in $\mathfrak{o}$ so that

$$w \cdot c_w = \sum_j u_j b_{w,j}.$$

By linear independence, the element $a_{w,j} = b_{w,j}/c_w \in K$ is independent of choices, and $a_{w_1+w_2,j} = a_{w_1,j} + a_{w_2,j}$, $a_{w\lambda,j} = a_{w,j}\lambda$ for $\lambda \in \mathfrak{o}$. Hence if $V$ is a $K$-vector space with basis $v_1, \dots, v_d$; then $f\colon M \to V$, $f(w) = \sum v_j a_{w,j}$ defines an injection $M \hookrightarrow V$. By the universal property of the tensor product, there exists a map $f'\colon M \otimes_{\mathfrak{o}} K \to V$ such that

$$\begin{array}{ccc} M & \xrightarrow{i} & M \otimes_{\mathfrak{o}} K \\ & {\scriptstyle f}\searrow & \downarrow{\scriptstyle f'} \\ & & V \end{array} \tag{4.1.b}$$

commutes. As $f$ is injective, $i$ must be injective.

□

We keep the notation of the proof $(a) \Rightarrow (d)$ and prove (4.1.$a$). First note that $M$ contains the free module $F$ on $u_1, \dots, u_d$. If $w_1, \dots, w_m$ is a generating set of $M$ over $\mathfrak{o}$ then

$$M \cdot \prod_i c_{w_i} \subset F$$

and so $M/F$ is indeed a torsion module.

□

Still keeping the above notation we show that for any module $M$ as in (4.1)

**(4.1.$c$)** $\dim_K(M \otimes_{\mathfrak{o}} K) = d$, *and so every maximal linearly independent set in $M$ has the same cardinality.*

In the sequel we call $d$ the $\mathfrak{o}$-rank of $M$; we denote this important invariant by $\mathrm{rk}_{\mathfrak{o}}(M)$.

*Proof.* By the construction of $V$ in the proof $(a) \Rightarrow (d)$ we know that $\dim_K(V) = d$. Since it is clear that the map $f'$ of (4.1.$b$) is surjective, it is now sufficient to prove that $f'$ is injective. Any element $x$ of $M \otimes_{\mathfrak{o}} K$ is of the form $y \otimes a$, for $a \in K$, $y \in M$. So, if $f'(y \otimes a) = 0$ with $a \neq 0$, then $f(y) = f'(y \otimes 1) = 0$, and therefore $y = 0$, since $f$ is injective.

□

We recall from the theory of principal ideal domains that if $\mathfrak{o}$ is such a domain, then the conditions in (4.1) are equivalent with $M$ being free. If, in addition, $\mathfrak{o}$ has precisely one prime ideal, then an $\mathfrak{o}$ -torsion module $M$ is isomorphic to a direct sum

**(4.2)** $$M \cong \bigoplus_{i=0}^{t} \mathfrak{o}/\mathfrak{p}^{n_i} \quad (n_1 \geq n_2 \geq \cdots \geq n_t)$$

where the $n_i$ are uniquely determined.

Henceforth in this section we always suppose $\mathfrak{o}$ to be a Dedekind domain, which, to avoid trivialities, is not a field. For the time being we fix a prime ideal $\mathfrak{p}$ of $\mathfrak{o}$. Let $M$ be an $\mathfrak{o}$ torsion module, with $\mathrm{an}(M) = \mathfrak{b}\mathfrak{p}^r$ where $\mathfrak{b}$ and $\mathfrak{p}$ are coprime $\mathfrak{o}$-ideals. By the Chinese Remainder Theorem (Theorem 4), there exist elements $a, b \in \mathfrak{o}$ with $b \equiv 0 \bmod \mathfrak{b}$, $b \equiv 1 \bmod \mathfrak{p}^r$, $a \equiv 0 \bmod \mathfrak{p}^r$, $a \equiv 1 \bmod \mathfrak{b}$; hence

$$M = Ma \oplus Mb$$

and $\mathfrak{b} = \mathrm{an}(Ma)$, $\mathfrak{p}^r = \mathrm{an}(Mb)$. We write $M^{(\mathfrak{p})} = Ma$, $M_{(\mathfrak{p})} = Mb$, and so

(4.3) $$M = M_{(\mathfrak{p})} + M^{(\mathfrak{p})}.$$

$M_{(\mathfrak{p})}$ consists of all elements in $M$ which are annihilated by some power of $\mathfrak{p}$, while $M^{(\mathfrak{p})}$ consists of all those elements annihilated by some $c \in \mathfrak{o} \setminus \mathfrak{p}$. The decomposition (4.3) is therefore unique.

**(4.4)** *For any finitely generated* $\mathfrak{o}$*-module* $M$

$$(tM)^{(\mathfrak{p})} = \bigcap_n M\mathfrak{p}^n.$$

*Proof.* First suppose that $M$ is torsion free; so by (4.1) $M$ is contained in some free $\mathfrak{o}$-module $F$. Then $\bigcap_n M\mathfrak{p}^n \subset \bigcap_n F\mathfrak{p}^n$, which is clearly zero. Thus in general, $\bigcap_n M\mathfrak{p}^n \subset tM$. However, since $M$ is finitely generated $tM_{(\mathfrak{p})}\mathfrak{p}^r = 0$ for some sufficiently large $r$, and so $\bigcap_n M\mathfrak{p}^n \subset tM^{(\mathfrak{p})}$. Since $\mathfrak{p}$ is coprime to $\mathrm{an}(tM^{(\mathfrak{p})})$, it follows that $tM^{(\mathfrak{p})}\mathfrak{p} = tM^{(\mathfrak{p})}$, and so $tM^{(\mathfrak{p})} \subset \bigcap_n tM^{(\mathfrak{p})}\mathfrak{p}^n$.

□

This result quickly gives us a special form of Nakayama's Lemma:

**(4.4.a)** *Suppose that the Dedekind domain* $\mathfrak{o}$ *has a unique prime ideal* $\mathfrak{o}$, *and let* $M \subset N$ *be finitely generated* $\mathfrak{o}$*-modules such that* $M + N\mathfrak{p} = N$*; then* $M = N$.

*Proof.* Let $Q = N/M$, then by hypothesis $Q\mathfrak{p} = M + N\mathfrak{p}/M = Q$, and so $Q = \bigcap_n Q\mathfrak{p}^n$; hence by (4.4) $Q = tQ^{(\mathfrak{p})}$. Given $q \in Q$, we know that $qa = 0$ for some $a \in \mathfrak{o} \setminus \mathfrak{p}$. However, since $\mathfrak{p}$ is the unique prime ideal of $\mathfrak{o}$, $a$ must be a unit, and so $q = 0$. This shows $Q = 0$, and so $M = N$.

□

Again let $\mathfrak{p}$ denote a prime ideal of $\mathfrak{o}$, and let $|.|_{\mathfrak{p}}$ denote an associated discrete absolute value. We let $\hat{\mathfrak{o}}$ be the valuation ring and $\hat{\mathfrak{p}}$ be the valuation ideal in the completion $\overline{K}$ of $K$ with respect to $|.|_{\mathfrak{p}}$. Our first aim is to achieve a generalisation of Theorem 12 of §3. As previously $M$ denotes a finitely generated $\mathfrak{o}$-module. The map $x \otimes a \mapsto xa \bmod M\mathfrak{p}^n$ induces a homomorphism

$$M \otimes_{\mathfrak{o}} (\mathfrak{o}/\mathfrak{p}^n) \to M/M\mathfrak{p}^n.$$

In fact we shall show that

**(4.5)** $$M \otimes_{\mathfrak{o}} (\mathfrak{o}/\mathfrak{p}^n) \cong M/M\mathfrak{p}^n.$$

*Proof.* The exact sequence

$$0 \to \mathfrak{p}^n \to \mathfrak{o} \to \mathfrak{o}/\mathfrak{p}^n \to 0$$

gives rise to an exact sequence

$$M \otimes_{\mathfrak{o}} \mathfrak{p}^n \to M \otimes_{\mathfrak{o}} \mathfrak{o} \to M \otimes_{\mathfrak{o}} (\mathfrak{o}/\mathfrak{p}^n) \to 0.$$

But the image of $M \otimes_{\mathfrak{o}} \mathfrak{p}^n$ in $M = M \otimes_{\mathfrak{o}} \mathfrak{o}$ is just $M\mathfrak{p}^n$, which proves (4.5).

□

Now let $\tilde{M} = \varprojlim M/M\mathfrak{p}^n$. By Theorem 12 we have the projection $\hat{\mathfrak{o}} \to \mathfrak{o}/\mathfrak{p}^n$, and so by (4.5) we get homomorphisms

$$M \otimes_{\mathfrak{o}} \hat{\mathfrak{o}} \to M/M\mathfrak{p}^n$$

which are compatible with the maps $M/M\mathfrak{p}^{n+k} \to M/M\mathfrak{p}^n$. Hence, by the universal property of inverse limits (3.20), we obtain a homomorphism

(4.6) $$M \otimes_{\mathfrak{o}} \hat{\mathfrak{o}} \to \tilde{M}.$$

**(4.6.a)** *This map is an isomorphism.*

*Proof.* We shall show that if $N$ is a finitely generated $\hat{\mathfrak{o}}$-module, then the map $N \to \varprojlim \left(N \otimes_{\hat{\mathfrak{o}}} (\hat{\mathfrak{o}}/\hat{\mathfrak{p}}^n)\right)$ is an isomorphism. Applying this result to $N = M \otimes_{\mathfrak{o}} \hat{\mathfrak{o}}$, and noting that

$$M \otimes_{\mathfrak{o}} \hat{\mathfrak{o}} \otimes_{\hat{\mathfrak{o}}} \hat{\mathfrak{o}}/\hat{\mathfrak{p}}^n = M \otimes_{\mathfrak{o}} \hat{\mathfrak{o}}/\hat{\mathfrak{p}}^n = M \otimes_{\mathfrak{o}} \mathfrak{o}/\mathfrak{p}^n = M/M\mathfrak{p}^n$$

we deduce that

$$M \otimes_{\mathfrak{o}} \hat{\mathfrak{o}} = \varprojlim M/M\mathfrak{p}^n = \tilde{M},$$

which establishes (4.6.*a*). In order to prove the assertion on $\hat{\mathfrak{o}}$-modules, we observe that since $\hat{\mathfrak{o}}$ is a principal ideal domain, any such module $N$ is a direct sum of copies of $\hat{\mathfrak{o}}$ and of modules $\hat{\mathfrak{o}}/\hat{\mathfrak{p}}^m$. The result for both these modules is obvious. The general result then follows from the easily seen fact that direct sums commute with the formation of inverse limits.

□

Next we show

**(4.7)** *If $\phi: N \to M$ is an injective homomorphism of finitely generated $\mathfrak{o}$-modules, then the induced homomorphism $\phi': N \otimes_{\mathfrak{o}} \hat{\mathfrak{o}} \to M \otimes_{\mathfrak{o}} \hat{\mathfrak{o}}$ is also injective.*

Combining this with the known properties of the tensor product, we see that if

$$0 \to N \to M \to L \to 0$$

is an exact sequence of finitely generated $\mathfrak{o}$-modules, then

$$0 \to N \otimes_{\mathfrak{o}} \hat{\mathfrak{o}} \to M \otimes_{\mathfrak{o}} \hat{\mathfrak{o}} \to L \otimes_{\mathfrak{o}} \hat{\mathfrak{o}} \to 0 \qquad (4.7.a)$$

is exact.

*Proof of* (4.7). We use (4.6.*a*) and consider $\tilde{\phi}: \tilde{N} \to \tilde{M}$, viewing $N$ as a submodule of $M$. Let $a \in \ker \tilde{\phi}$. We look at the $n$-th component $a(n) \in N/N\mathfrak{p}^n$, and show that it is zero. Indeed, for every $k \geq 0$, $a(n+k) \in \ker\left(N/N\mathfrak{p}^{n+k} \to M/M\mathfrak{p}^{n+k}\right)$, i.e. $a(n+k) \in N \cap M\mathfrak{p}^{n+k}/N\mathfrak{p}^{n+k}$. As $a(n)$ coincides with $a(n+k)$ mod $N\mathfrak{p}^n$, we deduce that

$$a(n) \in \left(N \cap M\mathfrak{p}^{n+k} + N\mathfrak{p}^n\right)/N\mathfrak{p}^n.$$

This is true for all $k$, that is to say

$$a(n) \in \left[\left(\bigcap_k M\mathfrak{p}^{n+k}\right) \cap N + N\mathfrak{p}^n\right] \Big/ N\mathfrak{p}^n.$$

By (4.4)

$$a(n) \in \left[\left(tM^{(\mathfrak{p})} \cap N\right) + N\mathfrak{p}^n\right]/N\mathfrak{p}^n \cong \frac{tM^{(\mathfrak{p})} \cap N}{tM^{(\mathfrak{p})} \cap N\mathfrak{p}^n}.$$

Since the module on the right is zero, it follows that $a(n) = 0$.

□

**(4.8) Corollary.** *If $M$ is also $\mathfrak{o}$-torsion free, then $M \otimes_{\mathfrak{o}} \hat{\mathfrak{o}}$ is $\hat{\mathfrak{o}}$-torsion free.*

*Proof.* By (4.1) $M$ can be embedded in a free $\mathfrak{o}$-module $F$. Then $F \otimes_{\mathfrak{o}} \hat{\mathfrak{o}}$ is free over $\hat{\mathfrak{o}}$, and therefore is $\hat{\mathfrak{o}}$-torsion free. The result then follows from (4.7), which tells us that $M \otimes_{\mathfrak{o}} \hat{\mathfrak{o}}$ is embedded in $F \otimes_{\mathfrak{o}} \hat{\mathfrak{o}}$.

□

Trivially, if $M$ is a torsion $\mathfrak{o}$-module, then $M \otimes_{\mathfrak{o}} \hat{\mathfrak{o}}$ is a torsion $\hat{\mathfrak{o}}$-module. More precisely for such an $M$ we have

**(4.9)** $$M \otimes_{\mathfrak{o}} \hat{\mathfrak{o}} = M_{(\mathfrak{p})} \otimes_{\mathfrak{o}} \hat{\mathfrak{o}} \cong M_{(\mathfrak{p})}$$

and

$$M^{(\mathfrak{p})} \otimes_{\mathfrak{o}} \hat{\mathfrak{o}} = 0.$$

*Proof.* By (4.4) it follows that $M^{(\mathfrak{p})} = \ker(M \to \tilde{M})$ and so the result follows from (4.3) and (4.6.$a$). Alternatively, an element $x \in M^{(\mathfrak{p})}$ is annihilated by some $a \in \mathfrak{o} \setminus \mathfrak{p}$. But this means that $a$ is a unit in $\hat{\mathfrak{o}}$, i.e. $a^{-1} \in \hat{\mathfrak{o}}$, and so $x \otimes 1 \in M^{(\mathfrak{p})} \otimes_{\mathfrak{o}} \hat{\mathfrak{o}}$ is zero. This in turn implies that $M \otimes_{\mathfrak{o}} \hat{\mathfrak{o}} = M_{(\mathfrak{p})} \otimes_{\mathfrak{o}} \hat{\mathfrak{o}}$. However, for all large enough $n$, $M_{(\mathfrak{p})} \otimes_{\mathfrak{o}} \mathfrak{o}/\mathfrak{p}^n \cong M_{(\mathfrak{p})}$, so that in the limit $M_{(\mathfrak{p})} \otimes_{\mathfrak{o}} \hat{\mathfrak{o}} = M_{(\mathfrak{p})}$.

□

From now on we consider simultaneously all prime ideals $\mathfrak{p}$ of $\mathfrak{o}$. We deviate from our previous notation a little and write $K_{\mathfrak{p}}$ for the completion of $K$ "at $\mathfrak{p}$", i.e. with respect to $|.|_{\mathfrak{p}}$, and $\mathfrak{o}_{\mathfrak{p}}$, $\hat{\mathfrak{p}}$ for the valuation ring, and valuation ideal of $K_{\mathfrak{p}}$, respectively. For any $\mathfrak{o}$-module $M$ we write $M_{\mathfrak{p}}$ for $M \otimes_{\mathfrak{o}} \mathfrak{o}_{\mathfrak{p}}$, and $V_{\mathfrak{p}} = V \otimes_K K_{\mathfrak{p}}$ for any $K$-vector space $V$. There is no contradiction between these two notations as the map $v \otimes_{\mathfrak{o}} \lambda \mapsto v \otimes_K \lambda$ $(v \in V,\ \lambda \in \mathfrak{o}_{\mathfrak{p}})$ is an isomorphism.

First note that by repeated application of (4.3) (or by direct appeal to the Chinese Remainder Theorem) we have the direct sum decomposition of a finitely generated $\mathfrak{o}$-torsion module $M$ as

**(4.10)** $$M = \bigoplus M_{(\mathfrak{p})}$$

*where the direct sum runs over all $\mathfrak{p}$, with only finitely many $M_{(\mathfrak{p})}$ non-zero.*

From (4.2), (4.9), using the fact that $\hat{\mathfrak{o}}/\hat{\mathfrak{p}}^n \cong \mathfrak{o}/\mathfrak{p}^n$, for each prime ideal $\mathfrak{p}$ we have a direct sum decomposition

(4.11) $$M_{(\mathfrak{p})} = \bigoplus_{1}^{t_{\mathfrak{p}}} \mathfrak{o}/\mathfrak{p}^{n_i} \quad (n_1 \geq n_2 \geq \cdots \geq n_{t_{\mathfrak{p}}}).$$

From now on, until further notice, we look at finitely generated, torsion free $\mathfrak{o}$-modules $M$. Applying (4.1), we may view $M$ as contained in a finite dimensional $K$-vector space $V$, with $M$ containing a basis of $V$. By (4.1.$c$) the dimension of $V$ is uniquely determined by $M$, and $\mathrm{rk}_{\mathfrak{o}}(M) = \dim_K(V)$. More precisely, from (4.1), there exist free $\mathfrak{o}$-modules $F'$, $F''$ both spanning $V$, with $F' \subset M \subset F''$, on bases say $\{v_i'\}$, $\{v_i''\}$ of $V$. We call such a module $M$ an *$\mathfrak{o}$-lattice in $V$*.

For each prime ideal $\mathfrak{p}$ of $\mathfrak{o}$, we have free $\mathfrak{o}_{\mathfrak{p}}$-modules $F_{\mathfrak{p}}'$, $F_{\mathfrak{p}}''$ and bases $\{v_i' \otimes 1\}$, $\{v_i'' \otimes 1\}$ of $V_{\mathfrak{p}}$, and so by (4.7) $F_{\mathfrak{p}}' \subset M_{\mathfrak{p}} \subset F_{\mathfrak{p}}''$. We thus view $M_{\mathfrak{p}}$ as an $\mathfrak{o}_{\mathfrak{p}}$-module embedded in $V_{\mathfrak{p}}$ and spanning $V_{\mathfrak{p}}$, i.e. $M_{\mathfrak{p}}$ is an $\mathfrak{o}_{\mathfrak{p}}$-lattice in $V_{\mathfrak{p}}$.

**(4.12)** *Let $L$, $M$ both be $\mathfrak{o}$-lattices in $V$; then for almost all $\mathfrak{p}$ $L_{\mathfrak{p}} = M_{\mathfrak{p}}$.*

*Proof.* Choose free $\mathfrak{o}$-lattices $F'$, $F''$ in $V$ with $F' \supset M$, $L \supset F''$. Let $l$ be an automorphism of the $K$-vector space $V$, mapping a basis of $F'$ onto one of $F''$. For almost all $\mathfrak{p}$ the matrix representing $l$ has $\mathfrak{p}$-integral entries and a $\mathfrak{p}$-unit determinant. For all such $\mathfrak{p}$, $F_{\mathfrak{p}}' = F_{\mathfrak{p}}''$; hence $L_{\mathfrak{p}} \supset M_{\mathfrak{p}}$. Similarly, for almost all $\mathfrak{p}$, $M_{\mathfrak{p}} \supset L_{\mathfrak{p}}$. This then establishes the result.

□

Let again $L$, $M$ be $\mathfrak{o}$-lattices in $V$. Torsion free finitely generated modules over a principal ideal domain are free. It follows that, for each prime ideal $\mathfrak{p}$, $L_{\mathfrak{p}}$ and $M_{\mathfrak{p}}$ are free $\mathfrak{o}_{\mathfrak{p}}$-modules of the same rank. Therefore there exists an automorphism $l_{\mathfrak{p}}$ of $V_{\mathfrak{p}}$ with $l_{\mathfrak{p}}(L_{\mathfrak{p}}) = M_{\mathfrak{p}}$. This automorphism is unique modulo $\mathrm{Aut}_{\mathfrak{o}_{\mathfrak{p}}}(M_{\mathfrak{p}})$; hence its determinant is unique modulo $\mathfrak{o}_{\mathfrak{p}}^*$, and so the ideal

$$[L_{\mathfrak{p}} : M_{\mathfrak{p}}] = \mathfrak{o}_{\mathfrak{p}} \det(l_{\mathfrak{p}}) \tag{4.13}$$

is a uniquely defined fractional ideal of $\mathfrak{o}_{\mathfrak{p}}$. By (4.12) we know that $[L_{\mathfrak{p}} : M_{\mathfrak{p}}] = \mathfrak{o}_{\mathfrak{p}}$ for almost all $\mathfrak{p}$. Therefore there is a unique fractional ideal $[L : M]$ such that

$$[L : M]_{\mathfrak{p}} = [L_{\mathfrak{p}} : M_{\mathfrak{p}}]. \tag{4.14}$$

(As explained previously, here we write $\mathfrak{a}_{\mathfrak{p}}$ for $\mathfrak{a} \otimes_{\mathfrak{o}} \mathfrak{o}_{\mathfrak{p}}$, for a given fractional $\mathfrak{o}$-ideal $\mathfrak{a}$.)

If, at any point, there is any possibility of confusion concerning the Dedekind domain in question, we shall write $[L : M]_{\mathfrak{o}}$ in place of $[L : M]$.

Recall that an $\mathfrak{o}$-module $M$ is said to be *projective* if there exists an $\mathfrak{o}$-module $N$ such that $M \oplus N$ is free. If $M$ is projective and if $0 \to L \to N \to M \to 0$ is an exact sequence of $\mathfrak{o}$-modules; then it is a standard fact that $N$ is isomorphic to $M \oplus L$. Also, given homomorphisms of $\mathfrak{o}$-modules $M \xrightarrow{f} T$, $S \xrightarrow{g} T$, with $g$ surjective; then, provided $M$ is projective, there exists an $\mathfrak{o}$-module homomorphism $M \xrightarrow{h} S$ such that $g \circ h = f$. In fact, in the exercises, it is shown that the above give two alternative characterisations of the notion of projectivity. Recall also that direct sums and direct summands of projective modules are projective.

**Theorem 13.**

(*a*) *Every fractional* $\mathfrak{o}$*-ideal* $\mathfrak{a}$ *is projective over* $\mathfrak{o}$.

(*b*) *The non-zero, torsion free, finitely generated* $\mathfrak{o}$*-modules* $M$ *are precisely those satisfying*

$$M \cong F \oplus \mathfrak{a}$$

*with* $F$ *free (possibly* $(0)$*), and* $\mathfrak{a}$ *a fractional* $\mathfrak{o}$*-ideal. In particular they are projective.*

(*c*) *Let* $M$ *be an* $\mathfrak{o}$*-lattice in the* $K$*-vector space* $V$. *Choose a free* $\mathfrak{o}$*-lattice* $F$ *in* $V$*, and define the ideal class* $c(M) \in \mathrm{Cl}(\mathfrak{o})$ *(the so-called Steinitz invariant) as the class of the ideal* $[F : M]$*;* $c(M)$ *is independent of the choice of* $F$.

(*d*)

$$c(M_1 \oplus M_2) = c(M_1)c(M_2)$$

$$c(\mathfrak{a}) = \mathfrak{a} \cdot P_{\mathfrak{o}} \text{ (the usual ideal class of } \mathfrak{a})$$

*and these two properties determine* $c$.

(*e*) $M \cong L \iff \mathrm{rk}_{\mathfrak{o}}(M) = \mathrm{rk}_{\mathfrak{o}}(L)$ *and* $c(M) = c(L)$.

*Proof.* Without loss of generality we may assume that $\mathfrak{o}$ is not a principal ideal domain. First note that the fractional ideals $\mathfrak{a}$ and $a \cdot \mathfrak{a}$, for $a \in K^*$, are isomorphic under the map $x \mapsto a \cdot x$ $(x \in \mathfrak{a})$; in other words two fractional ideals that have the same ideal class are isomorphic as $\mathfrak{o}$-modules. Now let $\mathfrak{a}$ and $\mathfrak{b}$ be coprime $\mathfrak{o}$-ideals. Then the map $\mathfrak{a} \oplus \mathfrak{b} \to \mathfrak{a} + \mathfrak{b} = \mathfrak{o}$ given by $a \oplus b \mapsto a - b$ is surjective and has kernel $\mathfrak{a} \cap \mathfrak{b} = \mathfrak{ab}$ (see 1.15(*g*)). We therefore have an exact sequence of $\mathfrak{o}$-modules

$$(4.15) \qquad 0 \to \mathfrak{ab} \to \mathfrak{a} \oplus \mathfrak{b} \to \mathfrak{o} \to 0.$$

Since $\mathfrak{o}$ is free, it is projective, and so

$$(4.15.a) \qquad \mathfrak{a} \oplus \mathfrak{b} \cong \mathfrak{ab} \oplus \mathfrak{o}.$$

By (1.19), we can choose such an ideal $\mathfrak{b}$ in the inverse of the class of $\mathfrak{a}$; thus for some $a \in K^*$, $\mathfrak{ab} = a\mathfrak{o} \cong \mathfrak{o}$; hence $\mathfrak{a} \oplus \mathfrak{b} \cong \mathfrak{o} \oplus \mathfrak{o}$, which shows that $\mathfrak{a}$ is indeed projective. This establishes (*a*).

To prove (*b*) we argue by induction on $\dim_K(V)$, where $M$ is an $\mathfrak{o}$-lattice in $V$. If $\dim_K(V) = 1$, then $V \cong K$, and $M \cong \mathfrak{a}$ for some fractional ideal $\mathfrak{a}$, as required. Assume now that $\dim_K(V) = r > 1$. Project $V$ onto a one dimensional vector space which we identify with $K$; this will map $M$ onto a fractional $\mathfrak{o}$-ideal $\mathfrak{a}$; by (*a*) we know that $\mathfrak{a}$ is projective, and without loss of generality we may assume $\mathfrak{a} \subset \mathfrak{o}$. Thus $M \cong M_1 \oplus \mathfrak{a}$, where $M_1$ is an $\mathfrak{o}$-lattice in an $r-1$ dimensional $K$-vector space $V_1$. However, by the induction hypothesis, $M_1 \cong F \oplus \mathfrak{b}$, where $F$ is free (possibly zero) and where the ideal $\mathfrak{b}$ can be chosen so that $\mathfrak{a} + \mathfrak{b} = \mathfrak{o}$. By (4.15), $M \cong F \oplus \mathfrak{o} \oplus \mathfrak{ab} = F' \oplus \mathfrak{ab}$, as required.

(*c*) To establish the independence of $c(M)$ from the choice of $F$, we need only observe that if $F$, $F'$ are two free $\mathfrak{o}$-modules spanning $V$, then $[F' : M] = [F' : F][F : M]$, and $[F' : F] = \det(l)\mathfrak{o}$ is principal, where $l \colon F' \cong F$. By definition $[\mathfrak{o} : \mathfrak{a}] = \mathfrak{a}$; therefore $c(\mathfrak{a}) = \mathfrak{a} \cdot P_{\mathfrak{o}}$. Let $M_i$ denote an $\mathfrak{o}$-lattice in $V$ and let $F_i$ be a free $\mathfrak{o}$-module spanning $V$; then clearly $[F_1 \oplus F_2 : M_1 \oplus M_2] = [F_1 : M_1][F_2 : M_2]$. This shows that $c(M_1 \oplus M_2) = c(M_1)c(M_2)$. The uniqueness assertion now follows from (*b*).

For (*e*), first suppose that we are given an isomorphism $\alpha \colon M \cong L$. Extending to an isomorphism $M \otimes_{\mathfrak{o}} K \cong L \otimes_{\mathfrak{o}} K$, and then mapping a free $\mathfrak{o}$-module $F$ spanning $M \otimes_{\mathfrak{o}} K$ onto a free $\mathfrak{o}$-module $F'$ spanning $L \otimes_{\mathfrak{o}} K$, it follows that $[F : M] = [F' : L]$, and hence $c(M) = c(L)$; this also shows that $\mathrm{rk}_{\mathfrak{o}}(M) = \mathrm{rk}_{\mathfrak{o}}(L)$.

As explained previously, if $\mathfrak{a}$ and $\mathfrak{b}$ are fractional $\mathfrak{o}$-ideals with $c(\mathfrak{a}) = c(\mathfrak{b})$; then $\mathfrak{a} \cong \mathfrak{b}$ as $\mathfrak{o}$-modules. Next let $M \cong F \oplus \mathfrak{a}$, $L \cong F_1 \oplus \mathfrak{b}$. If $\mathrm{rk}_{\mathfrak{o}}(M) = \mathrm{rk}_{\mathfrak{o}}(L)$, then $\mathrm{rk}_{\mathfrak{o}}(F) = \mathrm{rk}_{\mathfrak{o}}(F_1)$, and so $F \cong F_1$. Also $c(M) = c(L)$ implies $c(\mathfrak{a}) = c(\mathfrak{b})$; hence $\mathfrak{a} \cong \mathfrak{b}$, and so indeed $M \cong L$.

□

Next we come to a very basic and important invariant of finitely generated torsion $\mathfrak{o}$-modules. Let $T$ be such a module. Then $T \cong L/M$ where $L$, $M$ are torsion free: for instance we may take $L$ to be a free $\mathfrak{o}$-module, with a surjection $\alpha \colon L \to T$, and set $M = \ker \alpha$. Viewing $M$ as embedded in $L$, $M$ and $L$ are both $\mathfrak{o}$-lattices in the same $K$-vector space, and so $[L : M]$ is defined. We define

$$\mathrm{ord}_{\mathfrak{o}}(T) = [L : M].$$

**Theorem 14.**

(*a*) $\mathrm{ord}_{\mathfrak{o}}(T)$ *does not depend on the choice of* $L$ *and* $M$.

(*b*) *For any exact sequence* $0 \to T_1 \to T \to T_2 \to 0$ *of finitely generated torsion* $\mathfrak{o}$*-modules*

(*i*) $\mathrm{ord}_{\mathfrak{o}}(T) = \mathrm{ord}_{\mathfrak{o}}(T_1)\mathrm{ord}_{\mathfrak{o}}(T_2)$.

*Also for any* $\mathfrak{o}$*-ideal* $\mathfrak{a}$

(*ii*) $\mathrm{ord}_{\mathfrak{o}}(\mathfrak{o}/\mathfrak{a}) = \mathfrak{a}$.

*In particular for any prime ideal* $\mathfrak{p}$

(*iii*) $\mathrm{ord}_{\mathfrak{o}}(\mathfrak{o}/\mathfrak{p}) = \mathfrak{p}$.

*Properties* (*i*) *and* (*iii*) *determine* $\mathrm{ord}_{\mathfrak{o}}$ *uniquely.*

(*c*) $(\mathrm{ord}_{\mathfrak{o}}(T))_{\mathfrak{p}} = \mathrm{ord}_{\mathfrak{o}_{\mathfrak{p}}}(T_{\mathfrak{p}})$.

*Proof.* Suppose we can prove the theorem with $\mathfrak{o}$ replaced by $\mathfrak{o}_{\mathfrak{p}}$ for any prime ideal $\mathfrak{p}$. Then, by (4.7.*a*), for $T \cong L/M$, $T \cong L'/M'$, we have $T_{\mathfrak{p}} \cong L_{\mathfrak{p}}/M_{\mathfrak{p}}$, $T_{\mathfrak{p}} \cong L'_{\mathfrak{p}}/M'_{\mathfrak{p}}$, and so

$$[L:M]_{\mathfrak{p}} = [L_{\mathfrak{p}}:M_{\mathfrak{p}}] = [L'_{\mathfrak{p}}:M'_{\mathfrak{p}}] = [L':M']_{\mathfrak{p}}$$

for all $\mathfrak{p}$. Thus both (*a*) and (*b*) will follow from the result for all $\mathfrak{o}_{\mathfrak{p}}$; while by the above we also see that (*c*) is immediate. We may therefore assume for the purposes of the proof, that $\mathfrak{o}$ is a principal ideal domain, so that in particular all finitely generated torsion free $\mathfrak{o}$-modules are now free.

For $i = 1, 2$, let $\beta_i \colon L_i \to T$ be a surjection from a free $\mathfrak{o}$-module $L_i$, spanning a $K$-vector space $V_i$, onto an $\mathfrak{o}$-torsion module $T$. Put $M_i = \ker \beta_i$ and let $l_i$ denote an automorphism of $V_i$ with $l_i(L_i) = M_i$ ($i = 1, 2$). We now wish to show that $\det(l_1)\mathfrak{o} = \det(l_2)\mathfrak{o}$.

To this end we define a homomorphism $\beta \colon L_1 \oplus L_2 \to T$ by $\beta(x_1 \oplus x_2) = \beta_1(x_1) + \beta_2(x_2)$. So $\beta$ is surjective, and we put $M = \ker \beta$. We let $l$ be a $K$-linear automorphism of $V_1 \oplus V_2$ with $l(L_1 \oplus L_2) = M$. It will then suffice to show that

$$\det(l)\mathfrak{o} = \det(l_i)\mathfrak{o} \quad \text{for } i = 1, 2.$$

As $L_2$ is free, there is a homomorphism $\theta \colon L_2 \to L_1$ so that $\beta_1\theta(x_2) = \beta_2(x_2)$, and we use the same symbol for the extension to a $K$-linear map $V_2 \to V_1$.

Now let $\xi$ be the automorphism of $V_1 \oplus V_2$ with

$$\xi(x_1 + x_2) = (x_1 - \theta(x_2)) \oplus x_2.$$

Note that $\det(\xi) = 1$, and for $x_1 \in M_1$, $x_2 \in L_2$ we have

$$\beta((x_1 - \theta(x_2)) \oplus x_2) = \beta_1(x_1) - \beta_1\theta(x_2) + \beta_2(x_2) = \beta_1(x_1) = 0.$$

Thus $\xi(M_1 \oplus L_2) \subset M$. Let $y_1 \oplus y_2 \in M$, with $y_i \in L_i$. As $-\theta(y_2) \oplus y_2 \in M$, it follows that $y_1 + \theta(y_2) \oplus 0 \in M$, i.e. $y_1 + \theta(y_2) \in M_1$. Therefore $y_1 \oplus y_2 \in \xi(M_1 \oplus L_2)$, and so $\xi$ maps $M_1 \oplus L_2$ onto $M$. Also of course $l_1 \oplus \mathrm{id}_{L_2}$ maps $L_1 \oplus L_2$ onto $M_1 \oplus L_2$. Hence $\xi \circ (l_1 \oplus \mathrm{id}_{L_2})$ maps $L_1 \oplus L_2$ onto $M$, as does $l$. Since $\det(\xi \circ (l_1 \oplus \mathrm{id}_{L_2})) = \det(l_1)$, it follows that $\det(l)\mathfrak{o} = \det(l_1)\mathfrak{o}$. In just the same way it follows that $\det(l)\mathfrak{o} = \det(l_2)\mathfrak{o}$.

Property $(ii)$ in $(b)$ now follows immediately from the definition. For property $(i)$, let $L$ be a free $\mathfrak{o}$-module, $L \to T$ a surjection, with kernel $M$. We write $M_1$ for the kernel of the composite map $L \to T \to T_2$ so that

$$\begin{aligned} T_1 &\cong \ker(T \to T_2) \\ &\cong \ker(L/M \to L/M_1) \\ &\cong M_1/M. \end{aligned}$$

Hence $\operatorname{ord}_{\mathfrak{o}}(T) = [L : M] = [L : M_1][M_1 : M] = \operatorname{ord}_{\mathfrak{o}}(T_2)\operatorname{ord}_{\mathfrak{o}}(T_1)$.

Finally, the fact that $\operatorname{ord}_{\mathfrak{o}}$ is determined by property $(i)$, together with its values on the $\mathfrak{o}/\mathfrak{p}$, follows from (4.10), (4.11) using the fact that the $\mathfrak{o}/\mathfrak{p}$ are the only simple $\mathfrak{o}$-modules (up to isomorphism). (Recall: an $\mathfrak{o}$-module is said to be simple if it has no proper submodules.)

□

As a special case of the above, we note

**(4.16)** *If $\mathfrak{o} = \mathbb{Z}$ and $L \supset N$ are $\mathbb{Z}$-lattices in a common vector space $V$, then $[L : N]_{\mathbb{Z}}$ is just the $\mathbb{Z}$-ideal generated by the (ordinary) group index $[L : N]$; and so, for a finitely generated $\mathbb{Z}$-torsion module $T$, $\operatorname{ord}_{\mathfrak{o}}(T)$ is the ideal generated by the group order of $T$.*

We now apply the above work to field extensions. As before, $\mathfrak{o}$ is a Dedekind domain with field of fractions $K$.

**Theorem 15.** *Let $L$ denote a finite, separable extension of $K$, and let $\mathfrak{o}_L$ denote the integral closure of $\mathfrak{o}$ in $L$.*

$(a)$ *If $\mathfrak{A}$ is an ideal of $\mathfrak{o}_L$, then $\mathfrak{o}_L/\mathfrak{A}$ is a finitely generated $\mathfrak{o}$-torsion module.*

*The (relative) ideal norm of $\mathfrak{A}$ with respect to $L/K$ is defined to be*

$$\mathcal{N}_{L/K}(\mathfrak{A}) = \operatorname{ord}_{\mathfrak{o}}(\mathfrak{o}_L/\mathfrak{A}).$$

$(b)$
$$\mathcal{N}_{L/K}(\mathfrak{A}\mathfrak{B}) = \mathcal{N}_{L/K}(\mathfrak{A})\mathcal{N}_{L/K}(\mathfrak{B})$$

*if $\mathfrak{B}$ is a further $\mathfrak{o}_L$ ideal. $\mathcal{N}_{L/K}$ extends to a homomorphism $I_{\mathfrak{o}_L} \to I_{\mathfrak{o}}$.*

$(c)$ *If $a \in \mathfrak{o}_L$, $a \neq 0$, then*

$$\mathcal{N}_{L/K}(a\mathfrak{o}_L) = N_{L/K}(a)\mathfrak{o}.$$

$(d)$ *If the residue class rings of $\mathfrak{o}$ are all finite, then by* (II.1.37) *the same is true for $\mathfrak{o}_L$, and in this case*

$$\mathbf{N}(\mathcal{N}_{L/K}(\mathfrak{A})) = \mathbf{N}(\mathfrak{A}).$$

*Also if $K = \mathbb{Q}$, then $\mathcal{N}_{L/\mathbb{Q}}(\mathfrak{A}) = \mathbf{N}\mathfrak{A} \cdot \mathbb{Z}$.*

(e) *If $\mathfrak{P}$ is a prime ideal of $\mathfrak{o}_L$, with $\mathfrak{p} = \mathfrak{P} \cap \mathfrak{o}$; then $\mathfrak{o}_L/\mathfrak{P}$ is a finite field extension of $\mathfrak{o}/\mathfrak{p}$ of degree $f = f_{L/K}(\mathfrak{P})$ and $\mathcal{N}_{L/K}\mathfrak{P} = \mathfrak{p}^f$.*

*Properties (b) and (e) determine $\mathcal{N}_{L/K}$ uniquely.*

*Furthermore, if $E$ is a finite separable extension of $L$ and if $\mathfrak{A}$ is now an ideal of $\mathfrak{o}_E$, then*

$$\mathcal{N}_{E/K}(\mathfrak{A}) = \mathcal{N}_{L/K}(\mathcal{N}_{E/L}(\mathfrak{A})).$$

*Proof.* First observe that by Theorem 5, $\mathfrak{o}_L$ is a finitely generated $\mathfrak{o}$-module, and so $\mathfrak{o}_L/\mathfrak{A}$ must be finitely generated over $\mathfrak{o}$. Now let $\alpha$ denote a non-zero element of $\mathfrak{A}$. If $\alpha^n + a_{n-1}\alpha^{n-1} + \cdots + a_0$ is the minimal equation of $\alpha$ over $K$, then the $a_i$ lie in $\mathfrak{o}$, and so $a_0 = -(\alpha^n + \cdots + a_1\alpha) \in \mathfrak{A}$, with $a_0 \neq 0$. Thus $(\mathfrak{o}_L/\mathfrak{A}) \cdot a_0 = 0$ and so $\mathfrak{o}_L/\mathfrak{A}$ is indeed a torsion $\mathfrak{o}$-module.

Factorising $\mathfrak{A} = \prod_i \mathfrak{P}_i^{n_i}$, for distinct prime ideals $\mathfrak{P}_i$ of $\mathfrak{o}_L$, we know by the Chinese Remainder Theorem that

$$\mathfrak{o}_L/\mathfrak{A} = \bigoplus_i \mathfrak{o}_L/\mathfrak{P}_i^{n_i}.$$

Therefore by Theorem 14(*b*)(*ii*) $\mathcal{N}_{L/K}(\mathfrak{A}) = \prod_i \mathcal{N}_{L/K}(\mathfrak{P}_i^{n_i})$. But by (II.1.16)

$$\mathfrak{P}_i^r/\mathfrak{P}_i^{r+1} \cong \mathfrak{o}_L/\mathfrak{P}_i$$

and therefore

$$\mathcal{N}_{L/K}(\mathfrak{A}) = \prod_i \mathcal{N}_{L/K}(\mathfrak{P}_i)^{n_i}.$$

This implies (*b*) and shows that $\mathcal{N}_{L/K}$ is determined by its values on prime ideals. Let $\mathfrak{P}$ again be a prime ideal of $\mathfrak{o}_L$. The embedding $\mathfrak{o} \hookrightarrow \mathfrak{o}_L$ yields an embedding $\mathfrak{o}/\mathfrak{p} \to \mathfrak{o}_L/\mathfrak{P}$. Since $\mathfrak{o}_L$ is finitely generated over $\mathfrak{o}$ and since each element of $\mathfrak{o}_L$ is integral over $\mathfrak{o}$, it follows that $\mathfrak{o}_L/\mathfrak{P}$ has finite degree, $f_{L/K}(\mathfrak{P})$ say, over $\mathfrak{o}/\mathfrak{p}$: we call $f_{L/K}(\mathfrak{P})$ the *residue class degree of* $\mathfrak{P}$ in $L/K$. Thus, as an $\mathfrak{o}$-module, $\mathfrak{o}_L/\mathfrak{P}$ is the direct sum of $f_{L/K}(\mathfrak{P})$ copies of $\mathfrak{o}/\mathfrak{p}$. This gives (*e*).

Moreover, if $\mathfrak{P}_1$ is a prime ideal in $\mathfrak{o}_E$ with $\mathfrak{P}_1 \cap \mathfrak{o}_L = \mathfrak{P}$, then by the tower formula for extension degrees

$$f_{E/K}(\mathfrak{P}_1) = f_{E/L}(\mathfrak{P}_1) f_{L/K}(\mathfrak{P}).$$

It therefore follows that $\mathcal{N}_{E/K}(\mathfrak{A}) = \mathcal{N}_{L/K}(\mathcal{N}_{E/L}(\mathfrak{A}))$ for all prime ideals of $\mathfrak{o}_E$, and hence for all non-zero ideals of $\mathfrak{o}_E$.

From (I.1.27) $N_{L/K}(a)$, for $a \in L^*$, is the determinant of the $K$-linear automorphism given by multiplication by $a$; this just means evaluating $[F : Fa]$ for any free $\mathfrak{o}$-lattice $F$ in $L$. Since $\mathfrak{o}_{L,\mathfrak{p}}$ is a free $\mathfrak{o}_\mathfrak{p}$-lattice in

$L_\mathfrak{p}$, for each prime ideal $\mathfrak{p}$ of $\mathfrak{o}$ we know that

$$[F : Fa]_\mathfrak{p} = [\mathfrak{o}_{L,\mathfrak{p}} : \mathfrak{o}_{L,\mathfrak{p}}a]_{\mathfrak{o}_\mathfrak{p}} = [\mathfrak{o}_L : \mathfrak{o}_L a]_\mathfrak{p}$$

and so $N_{L/K}(a) \cdot \mathfrak{o} = [\mathfrak{o}_L : \mathfrak{o}_L a]_\mathfrak{o} = \mathcal{N}_{L/K}(a\mathfrak{o}_L)$.

Finally assume that the residue class fields of $\mathfrak{o}$ are finite. Again it suffices to prove the stated formula for prime ideals $\mathfrak{P}$ of $\mathfrak{o}_L$. Let $q = \mathbf{N}\mathfrak{p}$. Then $\mathbf{N}\mathfrak{p}^f = q^f = \mathbf{N}\mathfrak{P}$ for $f = f_{L/K}(\mathfrak{P})$. The result for $K = \mathbb{Q}$, then follows from (4.16) and (1.38).

□

We conclude this section by briefly considering the behaviour of an $\mathfrak{o}$-lattice $M$ in a $K$-vector space $V$ when we extend $K$ to a finite separable extension $L$ with $\mathfrak{O}$ the integral closure of $\mathfrak{o}$. The map

$$v \otimes_\mathfrak{o} \lambda \mapsto v \otimes_K \lambda \quad v \in M,\ \lambda \in \mathfrak{O}$$

is a homomorphism

$$t_M : M \otimes_\mathfrak{o} \mathfrak{O} \to V \otimes_K L$$

of $\mathfrak{O}$-modules. If $F$ and $G$ are finitely generated free $\mathfrak{o}$-modules spanning $V$ with $F \subset M \subset G$, then $\operatorname{im} t_F \subset \operatorname{im} t_M \subset \operatorname{im} t_G$. Thus $t_M$ is injective, $M \otimes_\mathfrak{o} \mathfrak{O} \cong \operatorname{im} t_M$ for all $\mathfrak{o}$-lattices in the vector space $V \otimes_K L$ over $L$. We may now identify $M \otimes_\mathfrak{o} \mathfrak{O}$ with $\operatorname{im} t_M$ for all $\mathfrak{o}$-lattices $M$ in $V$. If $N$ is a further such $\mathfrak{o}$-lattice then we assert that:

(4.17) $$[M : N]_\mathfrak{o}\mathfrak{O} = [M \otimes_\mathfrak{o} \mathfrak{O} : N \otimes_\mathfrak{o} \mathfrak{O}]_\mathfrak{O}.$$

To prove this observe that by the same reasoning as above we may, for all $\mathfrak{o}$-lattices $M$ in $V$, view $M_\mathfrak{p} \otimes_{\mathfrak{o}_\mathfrak{p}} \mathfrak{O}_\mathfrak{P}$ as a lattice in $V_\mathfrak{p} \otimes_{K_\mathfrak{p}} L_\mathfrak{P}$, $(M \otimes_\mathfrak{o} \mathfrak{O})_\mathfrak{P}$ as a lattice in $(V \otimes_K L)_\mathfrak{P}$, where $\mathfrak{P}$ is a prime ideal of $\mathfrak{O}$ above $\mathfrak{p}$. We have linear maps of $L_\mathfrak{P}$-vector spaces

$$V_\mathfrak{p} \otimes_{K_\mathfrak{p}} L_\mathfrak{P} = V \otimes_K K_\mathfrak{p} \otimes_{K_\mathfrak{p}} L_\mathfrak{P} \to V \otimes_K L_\mathfrak{P}$$
$$v \otimes c \otimes \lambda \mapsto v \otimes c\lambda \quad (v \in V, c \in K_\mathfrak{p}, \lambda \in L_\mathfrak{P})$$

and

$$(V \otimes_K L)_\mathfrak{P} = V \otimes_K L \otimes_L L_\mathfrak{P} \to V \otimes_K L_\mathfrak{P}$$
$$v \otimes b \otimes \lambda \mapsto v \otimes b\lambda \quad (v \in V, b \in L, \lambda \in L_\mathfrak{P})$$

Both maps are clearly surjective and the $L_\mathfrak{P}$-dimensions are the same; hence they are both isomorphisms.

By composing the first of these isomorphisms with the inverse of the second, we obtain an isomorphism

$$s : V_\mathfrak{p} \otimes_{K_\mathfrak{p}} L_\mathfrak{P} \cong (V \otimes_K L)_\mathfrak{P}.$$

By checking the individual steps, we verify that

$$s(M_\mathfrak{p} \otimes_{\mathfrak{o}_\mathfrak{p}} \mathfrak{O}_\mathfrak{P}) = (M \otimes_\mathfrak{o} \mathfrak{O})_\mathfrak{P}$$

for all $M$; therefore

$$(4.17.a)\quad [M_{\mathfrak{p}} \otimes_{\mathfrak{o}_{\mathfrak{p}}} \mathfrak{O}_{\mathfrak{P}} : N_{\mathfrak{p}} \otimes_{\mathfrak{o}_{\mathfrak{p}}} \mathfrak{O}_{\mathfrak{P}}]_{\mathfrak{O}_{\mathfrak{P}}} = [(M \otimes_{\mathfrak{o}} \mathfrak{O})_{\mathfrak{P}} : (N \otimes_{\mathfrak{o}} \mathfrak{O})_{\mathfrak{P}}]_{\mathfrak{O}_{\mathfrak{P}}}$$

($M, N$ $\mathfrak{o}$-lattices in $V$). Next, as $M_{\mathfrak{p}}$, $N_{\mathfrak{p}}$ are free over $\mathfrak{o}_{\mathfrak{p}}$, we verify that

$$(4.17.b)\qquad [M_{\mathfrak{p}} \otimes_{\mathfrak{o}_{\mathfrak{p}}} \mathfrak{O}_{\mathfrak{P}} : N_{\mathfrak{p}} \otimes_{\mathfrak{o}_{\mathfrak{p}}} \mathfrak{O}_{\mathfrak{P}}]_{\mathfrak{O}_{\mathfrak{P}}} = [M_{\mathfrak{p}} : N_{\mathfrak{p}}]_{\mathfrak{o}_{\mathfrak{p}}} \cdot \mathfrak{O}_{\mathfrak{P}}$$

and the right-hand side is by definition equal to $[M : N]_{\mathfrak{p}} \cdot \mathfrak{O}_{\mathfrak{P}}$. But for any fractional $\mathfrak{o}$-ideal $\mathfrak{a}$, we have $\mathfrak{a}_{\mathfrak{p}}\mathfrak{O}_{\mathfrak{P}} = (\mathfrak{a}\mathfrak{O})_{\mathfrak{P}}$ (see Corollary to Theorem 11 and the remark after Theorem 5). Thus finally by (4.17.$a$), (4.17.$b$) and the definition of $[\ :\ ]$ over $\mathfrak{O}$

$$([M : N]_{\mathfrak{o}}\mathfrak{O})_{\mathfrak{P}} = [M \otimes_{\mathfrak{o}} \mathfrak{O} : N \otimes_{\mathfrak{o}} \mathfrak{O}]_{\mathfrak{P}}.$$

This, however, holds for all $\mathfrak{P}$ and thus establishes (4.17).

# III

## Extensions

In this chapter we shall consider the behaviour under field extension, of many of the concepts that we have introduced. The central instance of this theme will be a detailed analysis of the way prime ideals factorise. We shall see that this problem, say for algebraic number fields, is intimately related to the study of the arithmetic of finite extensions of fields which are complete with respect to a discrete absolute value.

### §1 Decomposition and ramification

In this section we use the work of §2, §3 of Chapter II and of §1 of Chapter I to determine the basic properties of absolute values, and hence also of prime ideals, under extension. This therefore provides the foundation for the whole of this chapter.

It will now be convenient not always to be restricted to using variants of the symbol $|.|$ for absolute values; thus we shall also often denote absolute values by the letters $u$ and $w$.

We begin by considering extensions of an absolute value $u$ for a complete valued field $(K, u)$ which has the following property:

**(P).** *For any finite separable extension $E/K$, there is an absolute value $w: E \to \mathbb{R}$ such that $w|_K = u$.*

We shall not give a general proof of this, but we check that this property does hold for the two kinds of complete valued field with which we shall have to deal. If $K$ is $\mathbb{R}$ or $\mathbb{C}$, endowed with the ordinary absolute value (to within equivalence), then (P) is trivial. On the other hand, if $u$ is a discrete absolute value, then by (II.2.20) it corresponds to a unique valuation $v$ on $K$ (see II.2.10.c). We write $\mathfrak{o}$ for the valuation ring of $v$, and $\mathfrak{p}$ for its valuation ideal. So, in terms of the discrete absolute value,

$$\mathfrak{o} = \{x \in K \mid u(x) \leq 1\}.$$

It is important to remember here that a discrete absolute value determines a unique valuation; but that a valuation determines an equivalence class of discrete absolute values. To see that property (P) holds for such $(K, u)$, observe that we may take $w$ to be a suitable power of any absolute value associated with any prime ideal $\tilde{\mathfrak{p}}$ of the integral closure of $\mathfrak{o}$ in $L$. For $\tilde{\mathfrak{p}} \cap \mathfrak{o}$ will be the valuation ideal $\mathfrak{p}$ and so, for $x \in K$, we have $w(x) < 1$ iff $u(x) < 1$.

We now consider the matter of uniqueness for such an extension $u$:

**Theorem 16.** *Let $(K, u)$ be a complete valued field which has property (P), and let $E$ denote a finite separable extension of $K$. Then the extension $w$ of $u$ to $E$ is unique, and $(E, w)$ is complete. Furthermore, for $a \in E$*

(1.1) $$w(a)^{(E:K)} = u(N_{E/K}(a))$$

*and, if $u$ is a discrete absolute value, then so is $w$.*

Let $V$ denote a $K$-vector space and let $u$ denote an absolute value of $K$. Recall that a map $\|.\|: V \to \mathbb{R}_{>0}$ is called a *$u$-norm* if for $x \in V$

**(1.2)** (*a*) *$\|x\| = 0$ iff $x = 0$;*

(*b*) *$\|x\lambda\| = \|x\| \cdot u(\lambda)$ for $\lambda \in K$;*

(*c*) *For $y \in V$ $\|x + y\| \leq \|x\| + \|y\|$.*

Two $u$-norms $\|.\|$ and $\|.\|'$ are said to be equivalent iff there exist positive real numbers $c$, $d$ such that for all $x \in V$

$$c\|x\| \leq \|x\|' \leq d\|x\|.$$

Clearly equivalent norms induce equivalent topologies on $V$. Conversely, suppose that the two norms induce equivalent topologies: if there is no such real number $c$, then we can find a sequence of non-zero $x_n$ in $V$, with $\|x_n\|' \leq \|x_n\|/n$; and multiplying $x_n$ by a suitable scalar in $K$, we can suppose $1 \leq \|x_n\| \leq b$, with $b$ independent of $n$. This however implies that $x_n$ converges to zero with respect to the topology induced by $\|.\|'$ but not with respect to the topology induced by $\|.\|$. This is contrary to hypothesis, and so some such constant $c$ must exist. Interchanging the roles of $\|.\|$ and $\|.\|'$ gives $d$.

**(1.3)** *Let $V$ denote a finite-dimensional $K$-vector space with a u-norm $\|.\|$. Let $x_1, \ldots, x_n$ denote a $K$-basis of $V$; then the map*

$$\|\sum x_i\lambda_i\|' = \sum u(\lambda_i) \quad (\lambda_i \in K)$$

*is also a u-norm on $V$; the two norms $\|.\|$ and $\|.\|'$ are equivalent.*

It is entirely routine to show that $\|.\|'$ is a $u$-norm. By the triangle inequality, for $x = \sum x_j\lambda_j$ we have

$$\|x\| \leq \sum \|x_j\|u(\lambda_j) \leq \left(\sum \|x_j\|\right)\left(\sum u(\lambda_j)\right).$$

It therefore now remains to show that if $\|x\|$ is small, then so is $\|x\|'$. Before proving this fact we first show:

**(1.3.*a*)** *Let $U$ denote a closed linear subspace of $V$ (with respect to the $\|.\|$-topology). Then*

$$\|x + U\|_0 = \inf_{z\in U} \|x + z\|$$

*defines a norm on the quotient space $V/U$.*

*Proof.* As $U$ is closed, it follows that $\|x + U\|_0 = 0$ iff $x \in U$, which establishes property $(a)$. Since $U$ is linear

$$\begin{aligned}\|x\lambda + U\|_0 &= \inf_{z\in U} \|x\lambda + z\lambda\| \\ &= \inf_{z\in U} \|x + z\| \cdot u(\lambda) \\ &= \|x + U\|_0 \cdot u(\lambda)\end{aligned}$$

which gives property $(b)$. Lastly, for property $(c)$, note that for all $z, z' \in U$

$$\|x + z + x' + z'\| \leq \|x + z\| + \|x' + z'\|$$

and so

$$\inf_{z''\in U} \|x + x' + z''\| \leq \inf_{z\in U} \|x + z\| + \inf_{z'\in U} \|x' + z'\|.$$

□

To prove (1.3), first note that the result holds trivially if $V$ is one dimensional, for then

$$\|x\|\|x_1\|' = \|x\|'\|x_1\|$$

for any $x \in V$.

Next observe that in general $x{\cdot}K$ is closed in $V$ for any non-zero $x \in V$: for $x\lambda_i$ converges with respect to $\|.\|$ iff $\lambda_i$ converges with respect to $u$.

We complete our proof of (1.3) by arguing by induction on $\dim_K(V) \geq 2$. Let $\{x_i\}$ be a sequence in $V$ which converges to zero with respect to

$\|.\|$, and write $x_i = \sum z_j \lambda_{i,j}$ ($\lambda_{i,j} \in K$) with respect to a given basis$\{z_j\}$ of $V$; we wish to show that the $u(\lambda_{i,j}) \to 0$ for each $j$ as $i \to \infty$. Applying the induction hypothesis to $V/z_1K$, we conclude, by (1.3.$a$), that $u(\lambda_{i,j}) \to 0$ for each $j > 1$. Repeating with $z_n$ in place of $z_1$, we deduce $u(\lambda_{i,j}) \to 0$ for each $j < n$. This then gives the result.

□

We now prove the theorem. Let $w, w'$ denote extensions of $u$; then $w$ and $w'$ may be viewed as defining $u$-norms on $E$. Therefore, by (1.3), they are equivalent norms, and so, by (II.2.12), $w' = w^c$ for some positive real number. However, since $w'|_K = u = w|_K$ it follows immediately that $c = 1$, and so $w' = w$.

Since $K$ is complete with respect to $u$, it follows that, on choosing a $K$-basis of $E$, then $E$ will be complete with respect to the norm $\|.\|'$ of (1.3). However, by (1.3), $w$ is equivalent (as a norm) to $\|.\|'$, and so $(E, w)$ will be complete.

If $\sigma$ denotes a Galois automorphism of $E/K$, then $w \circ \sigma$ is also an absolute value of $E$ which extends $u$; thus, by uniqueness, it follows that $w \circ \sigma = w$. Now choose $E' \supset E$, with $E'/K$ finite and Galois. Of course for $a \in E$, $N_{E/K}(a) = \prod a^\sigma$, where $\sigma$ runs through a full set of embeddings over $K$ of $E$ into $E'$. Thus, if $w'$ denotes an extension of $w$ to $E'$, then by the above

$$(1.4) \qquad w(a)^{(E:K)} = \prod_\sigma w'(a^\sigma) = w'(N_{E/K}(a)) = u(N_{E/K}(a)).$$

Finally, if $u$ is discrete, then by (1.4) so is $w$.

□

**Corollary.** *If $u$ is a discrete absolute value and if $\mathfrak{o}$ denotes the valuation ring associated with $u$, then $\mathfrak{o}_E$, the integral closure of $\mathfrak{o}$ in $E$, has exactly one non-zero prime ideal.*

*Proof.* If $\mathfrak{o}_E$ had two distinct prime ideals $\mathfrak{P}$, $\mathfrak{P}'$ above the unique prime ideal of $\mathfrak{o}$, then $|.|_\mathfrak{P}$, $|.|_{\mathfrak{P}'}$ would give rise to two inequivalent absolute value extensions of $u$.

□

Next we digress to give an interesting alternative proof of the above theorem, in the case where $u$ is a discrete absolute value.

With the above notation, it suffices to show that $\mathfrak{P}$ is unique (to give uniqueness of extension) and to show that $\mathfrak{o}_E = \varprojlim \mathfrak{o}_E/\mathfrak{P}^n$ (which will imply that $\mathfrak{o}_E$, and hence $E$ is complete). Indeed, if there were two primes above the prime ideal $\mathfrak{p}$ of $\mathfrak{o}$, then the $\mathfrak{o}/\mathfrak{p}$-algebra

$$\mathfrak{o}_E/\mathfrak{p}\mathfrak{o}_E/\mathrm{Rad}(\mathfrak{o}_E/\mathfrak{p}\mathfrak{o}_E)$$

has a non-trivial idempotent. By (I.1.40) this lifts to a non-trivial idempotent $e$, say, of $\mathfrak{o}_E/\mathfrak{p}\mathfrak{o}_E$. Since $\mathfrak{o}$ is a principal ideal domain, by Theorem 5($a$) we know that $\mathfrak{o}_E$ is a free $\mathfrak{o}$-module of rank $(E:K)$; hence by Theorem 12, $\mathfrak{o}_E = \varprojlim \mathfrak{o}_E/\mathfrak{p}^n\mathfrak{o}_E$.

Choose an element $e_1 \in \mathfrak{o}_E$ whose image in $\mathfrak{o}_E/\mathfrak{p}\mathfrak{o}_E$ is $e$; hence $e_1 \not\equiv 0 \bmod \mathfrak{p}\mathfrak{o}_E$ and $e_1^2 \equiv e_1 \bmod \mathfrak{p}\mathfrak{o}_E$. We now construct a sequence $e_n$ of elements of $\mathfrak{o}_E$ with $e_{n+1} \equiv e_n \bmod \mathfrak{p}^n\mathfrak{o}_E$ (so that $e_n \equiv e_1 \bmod \mathfrak{p}\mathfrak{o}_E$) and $e_{n+1}^2 \equiv e_{n+1} \bmod \mathfrak{p}^{n+1}\mathfrak{o}_E$. Thus the limit $\lim_{n\to\infty} e_n = e'$ exists and $e'^2 = e'$ with $e' \equiv e_1 \bmod \mathfrak{p}\mathfrak{o}_E$ (so that $e' \neq 0$, $e' \neq 1$). The proof of the inductive step follows on setting $e_{n+1} = 1-(1-e_n^2)^2$, and on using the argument given in the latter part of the proof of (I.1.40).

Lastly we note that since $\mathfrak{p}\mathfrak{o}_E = \mathfrak{P}^m$ for some $m$, clearly

$$\varprojlim \mathfrak{o}_E/\mathfrak{p}^n\mathfrak{o}_E = \varprojlim \mathfrak{o}_E/\mathfrak{P}^n,$$

which shows $\mathfrak{o}_E = \varprojlim \mathfrak{o}_E/\mathfrak{P}^n$.

Henceforth we relax the condition that $(K,u)$ necessarily be complete, and we let $(K_u, \overline{u})$ denote a completion of $(K,u)$ which has property (P). Throughout $K$ will be viewed as a subfield of $K_u$. Let $L$, $M$ denote fields which contain $K$. As in (I.1.48) we have embeddings $K_u \hookrightarrow K_u \otimes_K L$, $L \hookrightarrow K_u \otimes_K L$. By the universal property for tensor products (loc. cit.), it follows that, given $K$-homomorphisms $K_u \xrightarrow{\pi} M$, $L \xrightarrow{\theta} M$, there is a unique $K$-homomorphism $K_u \otimes_K L \xrightarrow{\pi\otimes\theta} M$ such that the following diagram commutes.

(1.5)

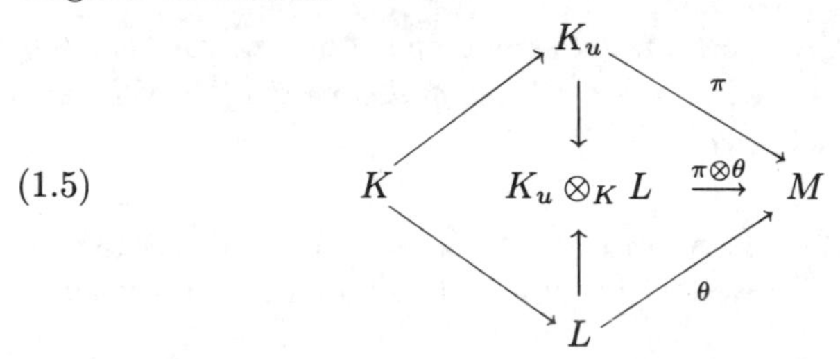


Here, a $K$-homomorphism $\phi\colon K' \to K''$ of fields, both containing $K$, is one which restricts to the identity map on $K$. More generally a $K$-homomorphism $\phi\colon A \to B$ of algebras over $K$ is one which satisfies $\phi(ca) = c\phi(a)$ for all $a \in A$, all $c \in K$. Conversely, given any $K$-homomorphism $K_u \otimes_K L \xrightarrow{\psi} M$, on writing $\pi$ for the map $K_u \to K_u \otimes_K L \xrightarrow{\psi} M$, and $\theta$ for the map $L \to K_u \otimes_K L \xrightarrow{\psi} M$, we see that

$\psi$ can be uniquely written as $\pi \otimes \theta$. Furthermore, if $M \xrightarrow{f} M'$ is a $K$-homomorphism of fields, then $f \circ (\pi \otimes \theta) = f \circ \pi \otimes f \circ \theta$.

Suppose now that $L/K$ is a finite separable extension; then by (I.1.50) the $K_u$-algebra $K_u \otimes_K L$ is a product of finitely many finite separable extensions $L_i$ of $K_u$

$$K_u \otimes_K L = \prod_{i \in I} L_i \tag{1.6}$$

Suppose that, with the above notation, $\pi \otimes \theta$ is surjective; then, via $\pi$, $M/K_u$ will be a finite separable extension; hence, by Theorem 16, there is a unique extension $\overline{u}_{\pi,\theta}$ of $\overline{u}$ to $M$. Restricting along $\theta$, $\overline{u}_{\pi,\theta}$ induces an absolute value $u_{\pi,\theta}$ of $L$. The embeddings

$$(K, u) \hookrightarrow (L, u_{\pi,\theta})$$
$$(L, u_{\pi,\theta}) \hookrightarrow (M, \overline{u}_{\pi,\theta})$$

are both homomorphisms of valued fields. Indeed, by Theorem 16 $(M, \overline{u}_{\pi,\theta})$ is complete, and, since $\pi \otimes \theta$ is surjective and $K$ is dense in $K_u$, $\theta(L)$ must be dense in $M$; thus in fact $(M, \overline{u}_{\pi,\theta})$ is a completion of $(L, u_{\pi,\theta})$. If $M \xrightarrow{f} M'$ is a $K$-isomorphism of fields, then by the uniqueness part of Theorem 16, we know that $\overline{u}_{\pi,\theta} = \overline{u}_{f\circ\pi, f\circ\theta} \circ f$, and so in fact $f$ must be an isomorphism of valued fields $(M, \overline{u}_{\pi,\theta}) \cong (M', \overline{u}_{f\circ\pi, f\circ\theta})$. Therefore by restriction to $L$, it follows that $u_{\pi,\theta} = u_{f\circ\pi, f\circ\theta}$.

We now return to the algebra decomposition (1.6) and apply the preceding discussion to each of the factors $L_i$. These distinct field factors $L_i$ are classified, up to $K_u$-isomorphism, by the primitive idempotents $\{\eta_i\}$, where $(K_u \otimes_K L)\eta_i = L_i$. Thus, in this way, each idempotent $\eta_i$ gives rise to an extension of the form $u_{\pi,\theta}$ of $L$. (Here the map $\pi \otimes \theta$ corresponds to multiplication by $\eta_i$). For brevity we write $u_i$, $\overline{u}_i$ in place of $u_{\pi,\theta}$, $\overline{u}_{\pi,\theta}$.

**Theorem 17.** *The map $\eta_i \overset{b}{\mapsto} u_i$ is a bijection between the primitive idempotents of the $K_u$-algebra $K_u \otimes_K L$ and the absolute values of $L$ which extend $u$. Furthermore the field $K_u \otimes_K L \cdot \eta_i$ is the completion of $L$ with respect to $u_i$, under the embedding*

$$L \to K_u \otimes_K L \to K_u \otimes_K L \cdot \eta_i.$$

*Proof.* We have already seen that $(L, u_i) \hookrightarrow (L_i, \overline{u}_i)$ is a completion. To show that $b$ is surjective, let $w$ denote an extension of $u$ to $L$ and let $\theta' \colon (L, w) \hookrightarrow (L_w, \overline{w})$ denote the completion. By the universal property of completions (see Theorem 10), there exists a homomorphism of valued fields $\pi' \colon (K_u, \overline{u}) \to (L_w, \overline{w})$ which makes the following diagram commute

$$\begin{array}{ccc} (K, u) & \xrightarrow{j} & (K_u, \bar{u}) \\ \downarrow & & \downarrow \pi' \\ (L, w) & \xrightarrow{\theta'} & (L_w, \bar{w}) \end{array} .$$

We derive a commutative diagram

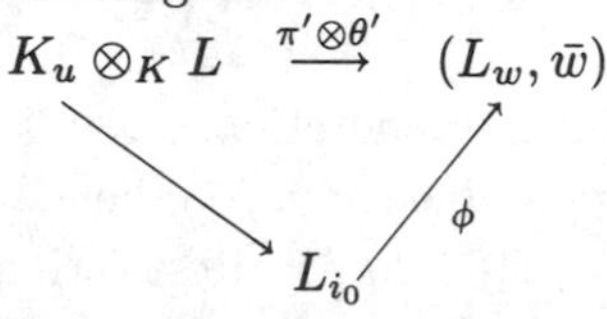


The top row comes from the universal property of tensor products. As $L_w$ is a field, $\pi' \otimes \theta'$ will factorise through a map $x \mapsto x \cdot \eta_{i_0}$ for one of the idempotents $\eta_i$; thus the diagram is indeed commutative. By Theorem 16 the absolute values $\bar{u}_i$ and $\overline{w} \circ \phi$ of $L_{i_0}$ coincide, hence so do their restrictions $u_i$ and $w$ to $L$.

We shall now show that $b$ is injective. Because $L/K$ is finite and separable, we may write $L = K(\alpha)$. We know from the discussion of (I.1.50-51) (see in particular (I.1.54) to (I.1.57)), that there are polynomials $e_i(X) \in K_u[X]$ so that, for each $i$, $e_i(1 \otimes \alpha) = e_i(\alpha) = \eta_i$ is the idempotent of $K_u \otimes_K L$ of Theorem 17. We then choose a sequence $\{e_n^{(i)}(X)\}$ of polynomials in $K[X]$ of constant degree such that in $K_u[X]$

$$\lim_{n \to \infty, \overline{u}} e_n^{(i)}(X) = e_i(X).$$

This limit is to be understood as the polynomial obtained by taking the limit of coefficients. It is clear that if $(K_u, \overline{u}) \hookrightarrow (N, v)$ is an embedding of valued fields, then

$$\lim_{n \to \infty, v} e_n^{(i)}(\beta) = e_i(\beta)$$

for $\beta \in N$. In particular for $N = (K_u \otimes_K L)\eta_j$ we have

$$\lim_{n \to \infty, \overline{u}_j} e_n^{(i)}(\alpha\eta_j) = e_i(\alpha\eta_j).$$

The right side is the image of $e_i(\alpha)\eta_j = \eta_i\eta_j = \delta_{ij}\eta_j$ under $K_u \otimes_K L \to (K_u \otimes_K L)\eta_j$. By the definition of $u_j$ on $L$, we now have

$$\begin{aligned} u_j(e_n^{(i)}(\alpha) - \delta_{ij}) &= \overline{u}_j(e_n^{(i)}(\alpha)\eta_j - \delta_{ij}\eta_j) \\ &= \overline{u}_j(e_n^{(i)}(\alpha\eta_j) - \delta_{ij}\eta_j) \\ &\to 0. \end{aligned}$$

In other words for $j \neq i$

$$u_j(e_n^{(i)}(\alpha)) \to 0$$
$$u_i(e_n^{(i)}(\alpha) - 1) \to 0$$

and so $u_i \neq u_j$. We thus have established the injectivity of $b$.

□

It follows from (II.2.21) that for an algebraic number field $L$, all the equivalence classes of absolute values are obtained by the procedure of Theorem 17, with $K = \mathbb{Q}$. Thus by Theorem 9 the completions of $L$ are finite algebraic extensions of $\mathbb{Q}_p$, or of $\mathbb{R}$ – the only ones in the latter case being $\mathbb{R}$ itself or $\mathbb{C}$.

Thanks to Theorem 17, we shall now also write the decomposition (1.6) as

$$(1.6.a) \qquad K_u \otimes_K L = \prod_{w|u} L_w$$

where the product extends over the distinct extensions $w$ of $u$ to $L$. Reading off $K_u$-dimensions we deduce that

$$(1.7) \qquad (L : K) = \sum_{w|u} (L_w : K_u).$$

Analogously, note that if $\mathfrak{p}$ is a prime ideal of a Dedekind domain $\mathfrak{o}$ with field of fractions $K$, and if $u$ is a discrete absolute value of $K$ which corresponds to $\mathfrak{p}$, then, writing $\mathfrak{o}_{K_u}$ for the valuation ring in $K_u$, we show that there is a decomposition

$$(1.8) \qquad \begin{aligned} \mathfrak{o}_{K_u} \otimes_{\mathfrak{o}_K} \mathfrak{o}_L &= \prod_i (\mathfrak{o}_{K_u} \otimes_{\mathfrak{o}_K} \mathfrak{o}_L)\eta_i \\ &= \prod_{w|u} \mathfrak{o}_{L_w}. \end{aligned}$$

To see this we note that the left-hand side is certainly contained in the right-hand side. To show that we have equality, we remark that on the one hand, by the Chinese Remainder Theorem, we know that $\mathfrak{o}_L$ is dense in the right-hand side; on the other hand, since $\mathfrak{o}_{K_u} \otimes_{\mathfrak{o}_K} \mathfrak{o}_L$ is $\mathfrak{o}_{K_u}$-free of finite rank it is closed; for, given a free basis $\{a_k\}$ we may, by (1.3), define the topology of $K_u \otimes_K L$ via the norm $\|.\|$, where $\|\sum c_k a_k\| = \sup u(c_k)$ $(c_k \in K_u)$.

We can obtain this decomposition in another way: as explained in the alternative proof of Theorem 16, we can lift all idempotents of $\mathfrak{o}_L/\mathfrak{p}\mathfrak{o}_L$ to $\mathfrak{o}_{K_u} \otimes_{\mathfrak{o}_K} \mathfrak{o}_L$, and so obtain an isomorphism

$$\varprojlim \mathfrak{o}_L/\mathfrak{p}^n\mathfrak{o}_L \cong \prod_{\mathfrak{P}|\mathfrak{p}} \varprojlim \mathfrak{o}_L/\mathfrak{P}^n\mathfrak{o}_L.$$

If we take $K = \mathbb{Q}$ and $u$ the ordinary absolute value, then $K_u = \mathbb{R}$, and so $L_w$ will be either $\mathbb{R}$ or $\mathbb{C}$; we shall call $w$ real or imaginary accordingly. In this case (1.7) shows that

$$(L : \mathbb{Q}) = s(L) + 2t(L) \tag{1.9}$$

where $s(L)$ resp. $t(L)$ is the number of real resp. imaginary absolute values. If $w$ is real, then the embedding $\theta_w: L \to L_w = \mathbb{R}$ is real, and conversely if $\theta_w: L \to L_w$ factorises through $\mathbb{R}$, then by Theorem 10 we must have $L_w = \mathbb{R}$. Therefore if $L_w = \mathbb{C}$, then $\theta_w: L \to L_w = \mathbb{C}$ is what we call imaginary, i.e. takes some non-real values. Hence, composing $\theta_w$ with complex conjugation, we get a second distinct embedding $\overline{\theta}_w: L \to L_w = \mathbb{C}$ with $w(x) = |\overline{\theta}(x)|_{\mathbb{C}}$. All these embeddings are clearly distinct. We thus have at least $s(L)$ real embeddings and $2t(L)$ imaginary embeddings. But the total number of embeddings $L \to \mathbb{C}$ is precisely $(L : \mathbb{Q})$. It follows therefore from (1.9) that

$$\begin{aligned} s(L) &= \text{number of real embeddings } L \to \mathbb{R} \\ t(L) &= \text{number of conjugate pairs of imaginary} \\ &\qquad \text{embeddings } L \to \mathbb{C}. \end{aligned} \tag{1.9.a}$$

*Remark.* By introducing the algebraic closure $\overline{\mathbb{Q}}_p$ of $\mathbb{Q}_p$ and using Galois theory of infinite extensions one can similarly describe the different extensions $w$ of $|.|_p$ to $L$ in terms of Galois orbits of embeddings $L \to \overline{\mathbb{Q}}_p$; this, however, lies beyond the scope of this book.

Next we apply a number of results for products of algebras to (1.6). Given $a \in L$, we write $a = \sum a_w$ for its decomposition under (1.6.$a$); from (I.1.37) we can immediately deduce that

$$\begin{aligned} c_{L/K,a}(X) &= \prod_{w|u} c_{L_w/K_u,a_w}(X) \\ N_{L/K}(a) &= \prod_{w|u} N_{L_w/K_u}(a_w) \\ t_{L/K}(a) &= \sum_{w|u} t_{L_w/K_u}(a_w). \end{aligned} \tag{1.10}$$

Next suppose that $\mathfrak{o}$ is a Dedekind domain with field of fractions $K$; let $\mathfrak{p}$ denote a prime ideal of $\mathfrak{o}$; let $L/K$ denote a finite separable extension of $K$, let $\mathfrak{o}_L$ denote the integral closure of $\mathfrak{o}$ in $L$, and let $\mathfrak{P}$ denote a prime ideal of $\mathfrak{o}_L$ above $\mathfrak{p}$, i.e. so that $\mathfrak{o} \cap \mathfrak{P} = \mathfrak{p}$. Write $K_\mathfrak{p}$ for the completion of $K$ with respect to a discrete absolute value $|.|_\mathfrak{p}$ associated with $v_\mathfrak{p}$. Let $\mathfrak{o}_\mathfrak{p}$ denote the closure of $\mathfrak{o}$ in $K_\mathfrak{p}$ and let $\overline{\mathfrak{p}}$ denote the closure of $\mathfrak{p}$ in $K_\mathfrak{p}$. In the remainder of this section we refer to this as the Dedekind domain situation. We define the *ramification index* of $\mathfrak{P}$ in $L/K$ to be

$$e_\mathfrak{P}(L/K) = v_\mathfrak{P}(\mathfrak{p}\mathfrak{o}_L) \tag{1.11}$$

where $v_{\mathfrak{P}}$ denotes the valuation associated with $\mathfrak{P}$. This is equivalent to saying

(1.11.$a$) $$\mathfrak{p}\mathfrak{o}_L = \mathfrak{P}^e\mathfrak{a}$$

for some $\mathfrak{o}_L$-ideal $\mathfrak{a}$ which is coprime to $\mathfrak{P}$, where we abbreviate $e_{\mathfrak{P}}(L/K)$ to $e$. Clearly $e \geq 1$. The prime ideal $\mathfrak{P}$ is said to ramify in $L/K$ when $e > 1$.

The embedding $\mathfrak{o} \hookrightarrow \mathfrak{o}_L$ gives rise to an embedding $\mathfrak{o}/\mathfrak{p} \hookrightarrow \mathfrak{o}_L/\mathfrak{P}$, and $\mathfrak{o}_L/\mathfrak{P}$, viewed as an extension of $\mathfrak{o}/\mathfrak{p}$ is of finite degree, as $\mathfrak{o}_L$ is finitely generated over $\mathfrak{o}$ by Theorem 5. The *residue class degree* of $\mathfrak{P}$ in $L/K$ is then defined as

(1.12) $$f_{\mathfrak{P}}(L/K) = \dim_{\mathfrak{o}/\mathfrak{p}}(\mathfrak{o}_L/\mathfrak{P}) = (\mathfrak{o}_L/\mathfrak{P} : \mathfrak{o}/\mathfrak{p}).$$

We remark that this invariant has already been introduced in an *ad hoc* manner in §4 of Chapter II (see Theorem 15). We claim that both $e$ and $f$ are invariant under completion with respect to $|.|_{\mathfrak{P}}$. Indeed, writing $\overline{\phantom{x}}$ for closure in this completion and using the fact that $\overline{\mathfrak{P}}$ is the unique prime ideal of $\overline{\mathfrak{o}}_L$, by the Corollary to Theorem 11, (1.11.$a$) gives $\overline{\mathfrak{p}} \cdot \overline{\mathfrak{o}}_L = \overline{\mathfrak{P}}^e$ which proves the result for ramification indices. The result for residue class extension degrees follows from the isomorphism of $\mathfrak{o}_L$-modules

$$\mathfrak{o}_L/\mathfrak{P} \cong \overline{\mathfrak{o}}_L/\overline{\mathfrak{P}}$$

shown in Theorem 11.

Let $N \supset L \supset K$ now denote a tower of finite separable extensions, and let $\mathfrak{q}$ denote a prime ideal of $\mathfrak{o}_N$ above $\mathfrak{P}$. Then we have the so-called tower formulae

(1.13.$a$) $$e_{\mathfrak{q}}(N/K) = e_{\mathfrak{q}}(N/L)e_{\mathfrak{P}}(L/K)$$

(1.13.$b$) $$f_{\mathfrak{q}}(N/K) = f_{\mathfrak{q}}(N/L)f_{\mathfrak{P}}(L/K)$$

*Proof.* The result for ramification indices is an immediate consequence of the characterisation (1.11.$a$). The result for residue class extension degrees follows from the corresponding tower formula for field extension degrees when applied to the residue class fields in question.

□

When both the prime ideal and the field extension are clear from the context, we shall write $e$, $f$ in place of $e_{\mathfrak{P}}(L/K)$, $f_{\mathfrak{P}}(L/K)$.

**(1.14)** *Suppose that $K$ is complete with respect to $|.|_{\mathfrak{p}}$, so that $\mathfrak{p}$ (resp. $\mathfrak{P}$) is the unique prime ideal of $\mathfrak{o}$ (resp $\mathfrak{o}_L$). Then*

($a$) $ef = (L : K)$;

(*b*) *The following diagram commutes*

$$\begin{array}{ccc} K^* & \xrightarrow{v_{\mathfrak{p}}} & \mathbb{Z} \\ \downarrow & & \downarrow e \\ L^* & \xrightarrow{v_{\mathfrak{P}}} & \mathbb{Z}; \end{array}$$

(*c*) *The following diagram commutes*

$$\begin{array}{ccc} L^* & \xrightarrow{v_{\mathfrak{P}}} & \mathbb{Z} \\ \downarrow N_{L/K} & & \downarrow f \\ K^* & \xrightarrow{v_{\mathfrak{p}}} & \mathbb{Z}. \end{array}$$

*Proof.* Since $\mathfrak{o}_L$ is $\mathfrak{o}$-free of rank $(L:K)$, it follows that $\mathfrak{o}_L/\mathfrak{p}\mathfrak{o}_L$ is an $\mathfrak{o}/\mathfrak{p}$ vector space of dimension $(L:K)$. By (II.1.16) we know that the $\mathfrak{o}/\mathfrak{p}$ vector spaces $\mathfrak{P}^j/\mathfrak{P}^{j+1}$ all have the same dimension, namely $f$. Since $\mathfrak{P}^e = \mathfrak{p}\mathfrak{o}_L$, it follows immediately that $ef = (L:K)$. This proves (*a*). From (1.11.*a*) it follows that for any $x \in K^*$, $v_{\mathfrak{P}}(x) = ev_{\mathfrak{p}}(x)$, and this then establishes part (*b*).

To prove part (*c*), note that, by (1.1), $v_{\mathfrak{P}}(a) = 0$ iff $v_{\mathfrak{p}}(N_{L/K}(a)) = 0$ for $a \in L^*$. So from (II.2.4) we know that for some integer $f'$, the following diagram commutes

$$\begin{array}{ccc} L^* & \xrightarrow{v_{\mathfrak{P}}} & \mathbb{Z} \\ \downarrow N_{L/K} & & \downarrow f' \\ K^* & \xrightarrow{v_{\mathfrak{p}}} & \mathbb{Z} \end{array}$$

In order to determine $f'$, we choose a $\mathfrak{o}$-generator $\pi$ of $\mathfrak{p}$; then $v_{\mathfrak{p}}(\pi) = 1$, while $N_{L/K}(\pi) = \pi^{(L:K)}$ and so $v_{\mathfrak{p}}(N_{L/K}(\pi)) = (L:K)$. On the other hand, by definition $v_{\mathfrak{P}}(\pi) = e$; thus $ef' = (L:K)$, and so, by part (*a*), $f' = f$.

□

We next consider more special properties for absolute values of a field $K$ which is a finite separable extension of either $\mathbb{Q}$ or $\mathbb{F}_p(X)$. First suppose that $K$ is a number field, and let $u$ be an absolute value of $K$; then by Theorem 9 $u|_{\mathbb{Q}}$ is either a discrete absolute value coming from a prime number $p$ or else $u|_{\mathbb{Q}}$ is equivalent to the ordinary absolute value. In the latter case $u$ is either real or complex, as described after (1.7). If, however, $u|_{\mathbb{Q}}$ is discrete, then by Theorem 16 $u$ must be discrete, and so $u$ comes from a prime ideal of $\mathfrak{o}_K$. If $K$ is a finite separable extension of $\mathbb{F}_p(X)$, then we have seen in (II.2.8) that all absolute values of $\mathbb{F}_p(X)$ are discrete, and so again by Theorem 16 all absolute values of $K$ are discrete.

If $K$ is a number field and $u$ is not discrete, then, by the above, $u$ corresponds to either a real or an imaginary embedding $j$ of $K$. We define the *normalised absolute value* corresponding to $u$, $\|.\|_u$, by

$$\|a\|_u = \begin{cases} |j(a)|_{\mathbf{R}} & j \text{ real} \\ |j(a)|^2_{\mathbf{C}} & j \text{ imaginary,} \end{cases}$$

cf. (II.2.10). Recall that if $u$ is imaginary, then in fact the 'normalised absolute value' $\|.\|_u$ is not actually an absolute value, since the triangle inequality fails to hold in general.

We now suppose that $u$ is a discrete absolute value which corresponds to the valuation $v$ on $K$. Let $\mathfrak{o}$ (resp. $\mathfrak{p}$) denote the valuation ring (resp. ideal) of $v$ in $K$. Then in both the arithmetic and geometric cases, we know (see (II.1.37)) that the index $\mathbf{N}\mathfrak{p} = [\mathfrak{o} : \mathfrak{p}]$ is finite. The normalised absolute value associated to $u$ is defined to be

$$\|a\|_u = \mathbf{N}\mathfrak{p}^{-v(a)}.$$

In this case $\|.\|_u$ is itself an absolute value and it is equivalent to $u$.

**(1.15)** *Let $K$ be as above and let $L/K$ denote a finite separable extension; then, for $a \in L^*$,*

$$\prod_{w|u} \|a\|_w = \|N_{L/K}(a)\|_u.$$

*Proof* It is readily checked that for both Archimedean and discrete $u$:

(1.16) $$\|b\|_w = \|b\|_u^{(L_w:K_u)} \quad (b \in K_u).$$

The result then follows immediately since by (1.10)

$$\|N_{L/K}(a)\|_u = \prod_w \|N_{L_w/K_u}(a_w)\|_u$$

by (1.16)

$$= \prod_w \|N_{L_w/K_u}(a_w)\|_w^{(L_w:K_u)^{-1}}$$

by (1.1), (1.14),

$$= \prod_w \|a_w\|_w.$$

□

The following beautiful and fundamental result explains why we introduced the normalised absolute values:

**Theorem 18 (The Product Rule).** *Let $K$ be a finite separable ex-*

*tension of either* $\mathbb{Q}$ *or* $\mathbb{F}_p(X)$*; then for* $a \in K^*$

$$\prod_u \|a\|_u = 1$$

*where the product extends over all equivalence classes of absolute values of* $K$.

*Proof.* For $K = \mathbb{Q}$ we proved this in (II.2.15). For any finite field $\mathbb{F}_q$, the only absolute value $|.|$ is the trivial one, as $a \in \mathbb{F}_q^*$ implies $|a|^{q-1} = 1$. Therefore (II.2.8) and (II.2.8.$a$) give the result for $\mathbb{F}_p(X)$. The argument leading to (II.2.21) applies to finite separable extensions of $\mathbb{F}_p(X)$, as well as to those of $\mathbb{Q}$. The general result then follows from (1.15).

□

*Remark.* The deeper reason for defining the normalised absolute value lies in the theory of Haar integrals on complete fields, which lies beyond the scope of this book. Note for instance that $\|a\|_{\mathbb{C}}$ is the Jacobian of the co-ordinate transformation of real variables $(x, y) \mapsto (x', y')$ where $x' + iy' = a(x + iy)$. Similarly, if $K$ is a finite extension of $\mathbb{Q}_p$, then

$$\|a\|_K = |[\mathfrak{o}_K : a\mathfrak{o}_K]|_p .$$

We conclude this section by considering Galois action on the various extensions of a given absolute value. With this in mind, unless stated to the contrary, for the remainder of this section we suppose that $L/K$ is a finite Galois extension with $\Gamma = \mathrm{Gal}(L/K)$. As previously, $u$ denotes an absolute value of $K$, and $K_u$ is the completion of $K$ with respect to $u$. We let $\Gamma$ act on the algebra $A = K_u \otimes_K L$ via the second factor, so that $(k \otimes l)^\gamma = k \otimes l^\gamma$. Thus $\Gamma$ acts as $K_u$-automorphisms of $A$, and therefore $\Gamma$ permutes the primitive idempotents $\{\eta_i\}$ of $A$. An element $\gamma$ of $\Gamma$ will thus give rise to an isomorphism $(K_u \otimes_K L)\eta_i \cong (K_u \otimes_K L)\eta_i^\gamma$, $a\eta_i \mapsto a^\gamma \eta_i^\gamma$. By Theorem 16 this is an isomorphism of valued fields.

If $w$ denotes an extension of $u$ to $L$, then $w^\gamma(a) = w(a^{\gamma^{-1}})$ defines a further extension of $u$ to $L$. If in addition $w = |.|_v$ for some valuation $v$, then $|.|_v^\gamma = |.|_{v^\gamma}$, where $v^\gamma$ is the valuation $v^\gamma(a) = v(a^{\gamma^{-1}})$and $|a|_v^\gamma = \left|a^{\gamma^{-1}}\right|_v$. If $\mathfrak{a}$ is a subset of $L$ and $\gamma \in \Gamma$, we put

$$\mathfrak{a}^\gamma = \{a^\gamma \in L \mid a \in \mathfrak{a}\}.$$

Now let $\mathfrak{o}_L$ be a Dedekind domain with field of fractions $L$, such that $\mathfrak{o}_L^\gamma = \mathfrak{o}_L$ for all $\gamma \in \Gamma$; then the map $\mathfrak{a} \mapsto \mathfrak{a}^\gamma$ permutes fractional $\mathfrak{o}_L$-ideals, and $(\mathfrak{a}\mathfrak{b})^\gamma = \mathfrak{a}^\gamma \mathfrak{b}^\gamma$, $(\mathfrak{a}^\gamma)^\delta = \mathfrak{a}^{\gamma\delta}$. Thus the group $I_{\mathfrak{o}_L}$ of fractional ideals becomes a $\Gamma$-module. Moreover, $\mathfrak{o}_L$-ideals get mapped into $\mathfrak{o}_L$-ideals, and prime ideals map to prime ideals. For the valuation associated with $\mathfrak{P}$ we then clearly have

(1.17) $$v_{\mathfrak{P}}^{\gamma} = v_{\mathfrak{P}^{\gamma}}.$$

We now return to the general situation where $u$ is an arbitrary absolute value on $K$; let $\{u_i\}$ denote the absolute values of $L$ which extend $u$. The action of $\Gamma$ on the $\{\eta_i\}$ induces an action on the subscripts via the rule $\eta_i^{\gamma} = \eta_{i\gamma}$. We claim that the bijection of Theorem 17 respects this $\Gamma$-action, that is

**(1.17.*a*)** $u_i^{\gamma} = u_{i\gamma}$.

*Proof.* Since $a\eta_i \mapsto \overline{u}_{i\gamma}\left((a\eta_i)^{\gamma}\right)$ defines an absolute value on $L_i$, by Theorem 16, $\overline{u}_i(a\eta_i) = \overline{u}_{i\gamma}\left((a\eta_i)^{\gamma}\right))$ for all $a \in A$. Now let $b \in L$, then $u_i(b) = \overline{u}_i(b\eta_i)$, and so

$$\begin{aligned} u_{i\gamma}(b) &= \overline{u}_{i\gamma}\left(b(\eta_i^{\gamma})\right) \\ &= \overline{u}_i\left((b\eta_i^{\gamma})^{\gamma^{-1}}\right) \\ &= \overline{u}_i(b^{\gamma^{-1}} \cdot \eta_i) \\ &= u_i(b^{\gamma^{-1}}) \\ &= u_i^{\gamma}(b) \end{aligned}$$

□

We define $\Delta_i = \mathrm{Stab}_{\Gamma}(u_i)$ to be the stabiliser of $u_i$ under action by $\Gamma$, i.e. the subgroup of $\Gamma$ of elements $\delta$ with $u_i^{\delta} = u_i$; we call $\Delta_i$ the *decomposition group* of the absolute value $u_i$ in $L/K$. By the standard theory of group actions on sets, and by (1.17.*a*), we note that

(1.18) $$\Delta_{i\gamma} = \gamma^{-1}\Delta_i\gamma.$$

If $\mathfrak{o}_L$ is a Dedekind domain and if $u_i = |.|_{\mathfrak{P}}$ for a prime ideal $\mathfrak{P}$ of $\mathfrak{o}_L$, then by (1.17) it follows that $u_i^{\gamma} = u_i$ iff $v_{\mathfrak{P}^{\gamma}} = v_{\mathfrak{P}}$ iff $\mathfrak{P}^{\gamma} = \mathfrak{P}$. Thus, in this case,

$$\Delta_i = \mathrm{Stab}_{\Gamma}(\mathfrak{P}).$$

In the sequel we shall usually write $\Delta(\mathfrak{P})$ for $\mathrm{Stab}_{\Gamma}(\mathfrak{P})$ and call this the *decomposition group* of $\mathfrak{P}$. We now prove the following basic result

**Theorem 19.** *Let $L/K$ be Galois with Galois group $\Gamma$. Then $\Gamma$ permutes transitively the primitive idempotents $\eta_i$ of $K_u \otimes_K L$, the field components $L_i$ of $K_u \otimes_K L$, and the extensions $u_i$ of $u$ to $L$. Moreover, restricting the action of $\gamma$ on $K_u \otimes_K L$ for each $i$, we obtain a* *commutative diagram*

$$\begin{array}{ccc} (L_i, \bar{u}_i) & \xrightarrow[\sim]{\gamma} & (L_i^\gamma, \bar{u}_i^\gamma) \\ \uparrow & & \uparrow \\ (L, u_i) & \xrightarrow[\sim]{\gamma} & (L, u_i^\gamma) \end{array}$$

*of valued fields.*

*If $K$ is the field of fractions of a Dedekind domain $\mathfrak{o}$, then $\Gamma$ permutes transitively the prime ideals of $\mathfrak{o}_L$ lying above a given prime ideal of $\mathfrak{o}$.*

By the preceding discussion it will suffice to show that the action of $\Gamma$ on the $\eta_i$ is transitive. First we show

**(1.19)** *Let $V$ be a finite-dimensional $K$-vector space, let $\Gamma$ act on $V$ by $K$-linear automorphisms, and let $E$ be a field extension of $K$. Then $\Gamma$ acts on $V \otimes_K E$ via $V$ and $(V \otimes_K E)^\Gamma = V^\Gamma \otimes_K E$.*

*Proof.* Given a $K$-linear map $l: V \to U$, then

$$\ker(l \otimes \mathrm{id}_E) = \ker(V \otimes_K E \to U \otimes_K E) = \ker(l) \otimes_K E.$$

Now set $U = \prod_{\gamma \in \Gamma} V$, and define $l$ to be the map $l(x) = \prod_\gamma x^\gamma - x$. Clearly $\ker l = V^\Gamma$, and the result follows.

□

*Proof of Theorem 19.* Consider the sum $\sum_\sigma \eta_i^\sigma$ where the sum extends over a set of coset representatives $\sigma \in \Delta_i \setminus \Gamma$. This is a sum of distinct primitive idempotents, and is fixed by $\Gamma$; so by (1.19) it lies in $A^\Gamma = K_u$. This shows that $\sum \eta_i^\sigma = 1$; so by (I.1.35) each primitive idempotent of $A$ must occur in this sum and therefore be of the form $\eta_i^\sigma$. Finally, the result for prime ideals follows immediately on taking $u = |.|_\mathfrak{p}$.

□

**Corollary.** *If $L/\mathbb{Q}$ is Galois, then the embedding $L \to \mathbb{C}$ will either all be real ($L$ is totally real) or they are all imaginary ($L$ is totally imaginary).*

Returning to the Dedekind domain situation, when $K$ is the field of fractions of a Dedekind domain $\mathfrak{o}$, Theorem 19, together with the invariance of $e$ and $f$ under completion, implies,

**Corollary.** *If $\mathfrak{P}$, $\mathfrak{P}'$ are prime ideals of $\mathfrak{o}_L$ above the same prime ideal*

*of* $\mathfrak{o}$, *then*

$$(L_{\mathfrak{P}'} : K_{\mathfrak{p}}) = (L_{\mathfrak{P}} : K_{\mathfrak{p}})$$
$$e_{\mathfrak{P}'}(L/K) = e_{\mathfrak{P}}(L/K)$$
$$f_{\mathfrak{P}'}(L/K) = f_{\mathfrak{P}}(L/K).$$

In the number field situation we can now show:

**Theorem 20.** *Let* $L/K$ *denote an extension of number fields, let* $\mathfrak{p}$ *denote a prime ideal of* $\mathfrak{o}$ *and suppose* $\mathfrak{p}\mathfrak{o}_L = \prod_{i=1}^{g} \mathfrak{P}_i^{e_i}$ *for distinct prime ideals* $\mathfrak{P}_i$. *Writing* $f_i = f_{\mathfrak{P}_i}(L/K)$,

$$(L : K) = \sum_{i=1}^{g} e_i f_i.$$

*Moreover, if* $L/K$ *is Galois, then the* $e_i$ *and* $f_i$ *are independent of* $i$. *Denoting them by* $e$ *and* $f$ *gives* $(L : K) = efg$.

*Proof.* The first part follows from (1.7) and (1.14). The above corollary then gives the second part.

□

We shall say that $\mathfrak{p}$ *splits* in $L$ if $f_i = e_i = 1$ for all $i$, i.e. if $\mathfrak{p}\mathfrak{o}_L$ is the product of $(L : K)$ distinct prime ideals; and we shall say that $\mathfrak{p}$ is *inert* in $L$ if $\mathfrak{p}\mathfrak{o}_L$ is a prime ideal, i.e. if $g = 1$ and $f = (L : K)$.

**Corollary.** *If* $\mathfrak{a}$ *is a fractional* $\mathfrak{o}$*-ideal, then*

$$\mathcal{N}_{L/K}(\mathfrak{a}\mathfrak{o}_L) = \mathfrak{a}^{(L:K)}.$$

*Proof.* It suffices to prove this for $\mathfrak{a} = \mathfrak{p}$, a prime ideal. Using Theorem 15 we get

$$\begin{aligned}\mathcal{N}_{L/K}(\mathfrak{p}\mathfrak{o}_L) &= \mathcal{N}_{L/K}\Big(\prod_i \mathfrak{P}_i^{e_i}\Big)\\ &= \prod_i (\mathcal{N}_{L/K}\mathfrak{P}_i)^{e_i}\\ &= \mathfrak{p}^{(\sum_i e_i f_i)} = \mathfrak{p}^{(L:K)}.\end{aligned}$$

□

*Remark.* For an alternative approach one can prove the equation in the corollary directly from the definition of the module index, and then use this to derive the last theorem.

**Theorem 21.** *Let $L/K$ be a finite Galois extension; let $u$ denote an absolute value of $K$, and let $w$ denote an extension of $u$ to $L$. Then $L_w$ is a Galois extension of $K_u$, and $\Delta_w$ identifies naturally with* $\mathrm{Gal}(L_w/K_u)$.

*Proof.* $\Delta_w$ acts as $K_u$-automorphisms of $L_w$ via the rule that for $a \in A$

$$(a\eta)^\delta = a^\delta \eta^\delta = a^\delta \eta$$

where $\eta$ is the idempotent associated with $w$. Note that given $\delta \in \Delta_w \backslash 1$, we can find $l \in L$ such that $l^\delta \neq l$, and so $(1 \otimes l)\eta$ is not fixed by $\delta$. This shows that $\Delta_w$ injects into the group of $K_u$ field automorphisms of $L_w$. It will therefore suffice to show $|\Delta_w| = (L_w : K_u)$. Now by Theorem 19

$$(\Gamma : \Delta_w) = \mathrm{Card}\{\eta_i\} = \frac{\dim_{K_u}(A)}{(L_w : K_u)}.$$

The result then follows since $\dim_{K_u}(A) = (L : K) = |\Gamma|$.

□

**Corollary.** *If $L/K$ is Galois with Galois group $\Gamma$, then*

$$(\mathcal{N}_{L/K}(\mathfrak{a}))\mathfrak{o}_L = \prod_{\gamma \in \Gamma} \mathfrak{a}^\gamma.$$

*Proof.* It suffices to prove this for a prime ideal $\mathfrak{P}$ of $\mathfrak{o}_L$. In the notation of the second part of Theorem 20, $ef$ is the order of $\Delta(\mathfrak{P})$ and $g$ is its index in $\Gamma$, i.e. the number of distinct prime ideals $\mathfrak{P}_i$ in $L$ above $\mathfrak{p} = \mathfrak{P} \cap \mathfrak{o}$. Therefore

$$\prod_\gamma \mathfrak{P}^\gamma = \left(\prod_i \mathfrak{P}_i^e\right)^f = (\mathfrak{p}\mathfrak{o}_L)^f = (\mathcal{N}_{L/K}(\mathfrak{P}))\mathfrak{o}_L.$$

□

Next we wish to consider the behaviour of absolute values with respect to an intermediate extension $L \supset M \supset K$, with $\mathrm{Gal}(L/M) = \Omega \subset \Gamma$. With this in mind it is important to make our notation a little more precise and write

$$\Delta_{L/K}(w) = \mathrm{Stab}_\Gamma(w)$$
$$\Delta_{L/M}(w) = \mathrm{Stab}_\Omega(w).$$

As previously $u = w|_K$; we now consider the absolute values $t$ of $M$ which extend $u$. Of course each $t$ can be extended to some $w'$ on $L$.

**(1.20)** *For $\gamma, \theta \in \Gamma$,*

$$w^\gamma|_M = w^\theta|_M$$

*iff* $\gamma$ *and* $\theta$ *lie in the same double coset of* $\Delta_w \backslash \Gamma/\Omega$.

*Proof.* Suppose first that $\theta = \delta\gamma\omega$ for some $\delta \in \Delta_w$, $\omega \in \Omega$. By definition $w^\delta = w$ and so $w^{\delta\gamma} = w^\gamma$; moreover for $x \in M$, $w^{\delta\gamma\omega}(x) = w^{\delta\gamma}(x^{\omega^{-1}}) = w^{\delta\gamma}(x) = w^\gamma(x)$; hence $w^\theta|_M = w^\gamma|_M$.

Conversely, suppose that $w^\gamma|_M = w^\theta|_M$. Applying Theorem 19 to the extension $L/M$, we can find $\omega \in \Omega$ such that $w^{\gamma\omega} = w^\theta$; it therefore follows that $\gamma\omega\theta^{-1} \in \Delta_w$ and so $\theta = \delta\gamma\omega$ for some $\delta \in \Delta_w$.

□

Suppose now that $\Omega$ is a normal subgroup of $\Gamma$, so that $M/K$ is now a Galois extension; then the double cosets $\Delta_w \backslash \Gamma/\Omega$ identify with the left cosets $\Delta_w\Omega \backslash \Gamma$, and so we deduce:

**Corollary.** *Identifying* $\mathrm{Gal}(M/K)$ *with* $\Gamma/\Omega$, *we have*

$$\Delta_{M/K}(w|_M) = \frac{\Delta_{L/K}(w)\cdot\Omega}{\Omega}.$$

Lastly we consider some aspects of the action of $\Gamma = \mathrm{Gal}(L/K)$ on the group $I_{\mathfrak{o}_L}$, of fractional $\mathfrak{o}_L$-ideals, in the case when $K$ is the field of fractions of a Dedekind domain $\mathfrak{o}$. We can use the above work to describe $I_{\mathfrak{o}_L}^\Gamma$, the subgroup of fractional ideals which are fixed by $\Gamma$. We know that

$$I_{\mathfrak{o}_L}^\Gamma = \bigoplus_{\mathfrak{p}} I_{\mathfrak{o}_L,\mathfrak{p}}^\Gamma$$

where $I_{\mathfrak{o}_L,\mathfrak{p}}$ denotes the subgroup of $I_{\mathfrak{o}_L}$ generated by the $\mathfrak{o}_L$ prime ideals above a prime ideal $\mathfrak{p}$ of $\mathfrak{o}$. Since $\Gamma$ acts transitively on the prime ideals above $\mathfrak{p}$, it follows that $I_{\mathfrak{o}_L,\mathfrak{p}}^\Gamma$ is the infinite cyclic group generated by

$$\bar{\mathfrak{p}} = \prod_{\mathfrak{P}|\mathfrak{p}} \mathfrak{P}. \tag{1.21}$$

Clearly $\bar{\mathfrak{p}}$ is characterised by the property that $\bar{\mathfrak{p}}^e = \mathfrak{p}\mathfrak{o}_L$, where $e = e_{\mathfrak{P}}(L/K)$ (which we recall is independent of the choice of $\mathfrak{P}|\mathfrak{p}$). Keeping this notation, we can also determine the kernel of the ideal norm $\mathcal{N}_{L/K}: I_{\mathfrak{o}_L} \to I_{\mathfrak{o}}$.

**(1.22)** $\mathcal{N}_{L/K}(\mathfrak{a}) = \mathfrak{o}$ *iff the fractional* $\mathfrak{o}_L$*-ideal* $\mathfrak{a}$ *can be written in the form*

$$\prod_{\gamma\in\Gamma} \mathfrak{b}_\gamma^{(1-\gamma)},$$

*for fractional* $\mathfrak{o}_L$*-ideals* $\mathfrak{b}_\gamma$.

*Proof.* If $\mathfrak{P}$ denotes a prime ideal of $\mathfrak{o}_L$, then $\mathfrak{P}$ and $\mathfrak{P}^\gamma$ both lie over the same prime ideal $\mathfrak{p}$ of $\mathfrak{o}$, and both have the same residue class degree, $f$ say, in $L/K$. Hence $\mathcal{N}_{L/K}(\mathfrak{P}) = \mathfrak{p}^f = \mathcal{N}_{L/K}(\mathfrak{P}^\gamma)$, and so $\mathcal{N}_{L/K}(\mathfrak{b}^{1-\gamma}) = \mathfrak{o}$, for any fractional $\mathfrak{o}_L$-ideal $\mathfrak{b}$.

Conversely, suppose $\mathcal{N}_{L/K}(\mathfrak{a}) = \mathfrak{o}$. Since $\Gamma$ acts transitively on the prime ideals of $\mathfrak{o}_L$ above a given prime ideal of $\mathfrak{o}$, we can write

$$\mathfrak{a} = \prod_i {\prod}' \mathfrak{P}_i^{n(i,\gamma)\gamma}$$

where $n(i,\gamma) \in \mathbb{Z}$, the $\mathfrak{P}_i \cap \mathfrak{o} = \mathfrak{p}_i$ are distinct, and where for given $i$, the second product runs over a transversal $\gamma \in \Delta(\mathfrak{P}_i) \setminus \Gamma$. Since

$$\mathcal{N}_{L/K}(\mathfrak{a}) = \prod_i {\prod}' \mathfrak{p}_i^{n(i,\gamma)} = \mathfrak{o}_L$$

it follows that for each $i$, $\sum' n(i,\gamma) = 0$. Thus we can write $\mathfrak{a}$ as

$$\mathfrak{a} = \prod_i {\prod}' \mathfrak{P}_i^{n(i,\gamma)\gamma - n(i,\gamma)}$$

which is of the required form.

□

## §2 Discriminants and differents

We have already seen how important the notion of discriminant can be in (I,§1) and (II,§2). Our initial objective here is to extend its definition to any extension of rings of algebraic integers $\mathfrak{o}_N/\mathfrak{o}_K$; this is not entirely trivial since, in general, $\mathfrak{o}_N$ is not free over $\mathfrak{o}_K$. Our main tools here will be completion, together with some of the more elementary module theory of (II,§4). The remainder of this section will then be devoted to considering various arithmetic applications. By far the most important such application is Dedekind's theorem, which shows that a prime ideal ramifies in an extension iff it divides the discriminant.

Throughout this section $\mathfrak{o}$ denotes a Dedekind domain with field of fractions $K$. For brevity we shall frequently write $[:]$ for the module index $[:]_\mathfrak{o}$, and $[:]_\mathfrak{p}$ for $[:]_{\mathfrak{o}_\mathfrak{p}}$, for a given prime ideal $\mathfrak{p}$ of $\mathfrak{o}$, c.f. (II.4.14).

We begin by considering a symmetric, non-degenerate $K$-bilinear form $b$ on a finite-dimensional $K$-vector space $V$

$$b: V \times V \to K$$

and we let $M$ always be an $\mathfrak{o}$-lattice in $V$ as defined in (II,§4). In the special case when $M$ is $\mathfrak{o}$-free on a basis $m_1, \ldots, m_n$, we define

$$\text{(2.1)} \qquad \mathfrak{d}(M) = \det(b(m_i, m_j)) \cdot \mathfrak{o}.$$

Note that this ideal is non-zero, since $b$ is non-degenerate; furthermore the right-hand term is independent of the particular choice of basis; a change of basis is given by a matrix over $\mathfrak{o}$, whose inverse also has elements in $\mathfrak{o}$. The determinant is thus in $\mathfrak{o}^*$.

Let $\mathfrak{p}$ denote a prime ideal of $\mathfrak{o}$, and let $K_\mathfrak{p}$, $\mathfrak{o}_\mathfrak{p}$ denote the completion of $K$, $\mathfrak{o}$ with respect to a discrete absolute value corresponding to $\mathfrak{p}$. We write $b_\mathfrak{p}$ for the $K_\mathfrak{p}$-bilinear form on $V_\mathfrak{p} = V \otimes_K K_\mathfrak{p}$ given by the rule

$$b_\mathfrak{p}(\sum x_i \otimes \lambda_i, \sum y_j \otimes \mu_j) = \sum_{i,j} b(x_i, y_j)\lambda_i\mu_j.$$

Then $b_\mathfrak{p}$ is again symmetric and non-degenerate. Since $M_\mathfrak{p} = M \otimes_\mathfrak{o} \mathfrak{o}_\mathfrak{p}$ is free over $\mathfrak{o}_\mathfrak{p}$, by (2.1) $\mathfrak{d}(M_\mathfrak{p})$ is well-defined.

If $M$ is not $\mathfrak{o}$-free, we know from (II.4.1) that we can find a free $\mathfrak{o}$-lattice $F$ in $V$ which contains $M$. By (II.4.12) $F_\mathfrak{p} = M_\mathfrak{p}$ for almost all $\mathfrak{p}$; hence $\mathfrak{d}(M_\mathfrak{p}) = \mathfrak{d}(F_\mathfrak{p})$ for almost all $\mathfrak{p}$. Moreover clearly $\mathfrak{d}(F)_\mathfrak{p} = \mathfrak{d}(F_\mathfrak{p})$. Thus if $\mathfrak{p}$ does not occur in the prime ideal decomposition of the fractional ideal $\mathfrak{d}(F)$ then $\mathfrak{d}(F_\mathfrak{p}) = \mathfrak{o}_\mathfrak{p}$. This then shows that $\mathfrak{d}(M_\mathfrak{p}) = \mathfrak{o}_\mathfrak{p}$ for almost all $\mathfrak{p}$. It therefore follows that there is a unique $\mathfrak{o}$-ideal which we call the *discriminant* of $M$ and which we denote by $\mathfrak{d}(M)$, such that for all prime ideals $\mathfrak{p}$ of $\mathfrak{o}$

(2.2) $$\mathfrak{d}(M)_\mathfrak{p} = \mathfrak{d}(M_\mathfrak{p}).$$

If $M$ is actually free, then we may take $F = M$ in the new definition (2.2), and so verify that (2.1) and (2.2) coincide.

We write $M = N_1 \perp N_2$ if $M$ is the internal direct sum of $\mathfrak{o}$-modules $N_1$ and $N_2$ with the property that $b(N_1, N_2) = 0$. Such $N_1$ and $N_2$ are said to be "orthogonal" to each other. If $N_1$ and $N_2$ are free $\mathfrak{o}$-modules, then we can find a basis of $M$ which consists of an $\mathfrak{o}$-basis of $N_1$ and an $\mathfrak{o}$-basis of $N_2$. Since these two subsets are mutually orthogonal

(2.3) $$\mathfrak{d}(M) = \mathfrak{d}(N_1)\mathfrak{d}(N_2)$$

where $\mathfrak{d}(N_i)$ denotes the discriminant of $N_i$ with respect to the form obtained by restricting $b$ to $N_iK$. Indeed, by applying the above reasoning to $M_\mathfrak{p}$ and then using the definition (2.2), we see that (2.3) holds for any orthogonal decomposition $M = N_1 \perp N_2$.

**(2.4)** *Given two $\mathfrak{o}$-lattices $M$, $N$ in $V$*

$$\mathfrak{d}(N) = \mathfrak{d}(M)[M : N]^2.$$

*Proof.* Since $\mathfrak{d}(N)_\mathfrak{p} = \mathfrak{d}(N_\mathfrak{p})$ and since $([M : N])_\mathfrak{p} = [M_\mathfrak{p} : N_\mathfrak{p}]_\mathfrak{p}$, we may without loss of generality suppose that $\mathfrak{o} = \mathfrak{o}_\mathfrak{p}$. Let $\{m_i\}$, $\{n_i\}$ denote $\mathfrak{o}$-bases of $M$, $N$ respectively. If $m_i = \sum a_{ij}n_j$ with $a_{ij} \in K$; then by bilinearity

$$\det(b(m_i, m_j)) = \det(a_{ij})^2 \det(b(n_i, n_j)).$$

The result then follows, since by definition (see II.4.13)

$$[M : N] = \det(a_{ij})\mathfrak{o}.$$

□

Since we can always find some free $\mathfrak{o}$-module $F \supset M$ spanning the same vector space, and since $\mathfrak{d}(F)$ is a principal $\mathfrak{o}$-ideal by (2.1), we have shown

**Corollary.** *The class of* $\mathfrak{d}(M)$ *in the ideal class group* $Cl(\mathfrak{o})$ *is a square.*

If $S$ is a subset of $V$, and $x \in V$ then $b(x, S)$ stands for the set of elements $b(x, s)$, all $s \in S$. For a given form $b$, the dual of the $\mathfrak{o}$-lattice $M$ in $V$, denoted by $M^D$, is defined by

$$M^D = \{x \in V \mid b(x, M) \subset \mathfrak{o}\}.$$

Clearly $M^D$ is an $\mathfrak{o}$-module, and also

$$M \subset N \Rightarrow N^D \subset M^D.$$

**(2.5)** *$M^D$ is an $\mathfrak{o}$-lattice in $V$.*

*Proof.* If $M$ is a free $\mathfrak{o}$-module with basis $m_1, \ldots, m_n$, let $m_1^*, \ldots, m_n^*$ denote the dual basis of $M$ with respect to $b$. Reasoning as in the proof of Theorem 5, we see that $M^D = \sum m_i^* \mathfrak{o}$, and the result is therefore shown when $M$ is free.

More generally from (II.4.1) and (II.4.1.$a$) we can find free $\mathfrak{o}$-lattices $F$, $G$ in $V$ such that $F \subset M \subset G$; hence $G^D \subset M^D \subset F^D$. Since $\mathfrak{o}$ is Noetherian, $M^D$ is finitely generated and $M^D \cdot K \supset G^D \cdot K = V$.

□

We observe that given an orthogonal decomposition $M = N_1 \perp N_2$, then

$$(N_1 \perp N_2)^D = N_1^D \perp N_2^D. \tag{2.6}$$

Next we show that the formation of duals commutes with completion:

$$(M^D)_{\mathfrak{p}} = (M_{\mathfrak{p}})^D. \tag{2.7}$$

(Note that here $(M_{\mathfrak{p}})^D$ means the dual of $M_{\mathfrak{p}}$ with respect to $b_{\mathfrak{p}}$).

*Proof.* If $M$ is $\mathfrak{o}$-free on a basis $\{m_i\}$, then $M^D$ is $\mathfrak{o}$-free on the dual basis $\{m_i^*\}$; therefore $(M^D)_{\mathfrak{p}}$ and $(M_{\mathfrak{p}})^D$ both coincide with the free $\mathfrak{o}_{\mathfrak{p}}$-module on $\{m_i^*\}$. This establishes the result when $M$ is $\mathfrak{o}$-free. Obviously

$(M^D)_\mathfrak{p}$ and $(M_\mathfrak{p})^D$ are both $\mathfrak{o}_\mathfrak{p}$-submodules of $V_\mathfrak{p}$. If $x \in (M^D)_\mathfrak{p}$, then $x = \sum y_i \otimes \lambda_i$ for $y_i \in M^D$, $\lambda_i \in \mathfrak{o}_\mathfrak{p}$; therefore

$$\begin{aligned} b_\mathfrak{p}(x, M_\mathfrak{p}) &\subset \sum b_\mathfrak{p}(y_i, M_\mathfrak{p}) \\ &= \sum b(y_i, M)\mathfrak{o}_\mathfrak{p} \\ &\subset \mathfrak{o}_\mathfrak{p}. \end{aligned}$$

This shows $(M^D)_\mathfrak{p} \subset (M_\mathfrak{p})^D$. To prove the reverse inclusion, note that because $M$ maps onto $M_\mathfrak{p}/M_\mathfrak{p}\mathfrak{p}$, we can find $m_1, \ldots, m_n$ in $M$ which are a basis of $M_\mathfrak{p}/M_\mathfrak{p}\mathfrak{p}$ over $\mathfrak{o}/\mathfrak{p}$. Hence, by Nakayama's Lemma (see (II.4.4.*a*)), $\{m_i\}$ is an $\mathfrak{o}_\mathfrak{p}$ basis of $M_\mathfrak{p}$. Let $F$ denote the free $\mathfrak{o}$-lattice on $\{m_i\}$. Then $M \supset F$ with $M_\mathfrak{p} = F_\mathfrak{p}$ and so $[M : F]$ is coprime to $\mathfrak{p}$. We can therefore find $a \in \mathfrak{o}$, $a \notin \mathfrak{p}$ such that $a^{-1}F \supset M$. Set $G = a^{-1}F$ and note $G_\mathfrak{p} = M_\mathfrak{p}$. We then have a chain of inclusions

$$(G^D)_\mathfrak{p} \subset (M^D)_\mathfrak{p} \subset (M_\mathfrak{p})^D \subset (G_\mathfrak{p})^D.$$

Since $G$ is free the two end terms coincide and hence the required equality is shown.

□

We can now prove the following fundamental result:

**(2.8)** $\mathfrak{d}(M) = [M^D : M]$.

*Proof.* By (2.7) it suffices to prove the result when $\mathfrak{o} = \mathfrak{o}_\mathfrak{p}$. We then let $\{m_i\}$ denote an $\mathfrak{o}$-basis of $M$, and we let $\{m_i^*\}$ denote the dual basis, which is therefore an $\mathfrak{o}$-basis of $M^D$. By definition, $b(m_i, m_j^*) = \delta_{ij}$; hence if $m_i = \sum m_j^* a_{ij}$, then

$$\mathfrak{d}(M) = \mathfrak{o} \cdot \det(b(m_i, m_j)) = \mathfrak{o}\det(a_{ij}) = [M^D : M].$$

□

As an immediate corollary we note

**(2.9)**

(*a*) $[M : N] = [N^D : M^D]$;

(*b*) $M^{DD} = M$.

*Proof.* To prove (*a*), note that by (2.4) and (2.8)

$$[N^D : N][M^D : M]^{-1} = \mathfrak{d}(N)\mathfrak{d}(M)^{-1} = [M : N]^2,$$

while on the other hand

$$[N^D : N][M^D : M]^{-1} = [M : N][N^D : M^D].$$

Since by definition $M \subset M^{DD}$, in order to prove $(b)$ we need only show $[M : M^D] = [M^{DD} : M^D]$; this, however, follows from part $(a)$ on setting $N = M^D$.

□

If $L$ is a finite separable extension of $K$, let $\mathfrak{o}_L$ denote the integral closure of $\mathfrak{o}$ in $L$. Then $b$ induces a form $b_L: V \otimes_K L \times V \otimes_K L \to L$ by the rule $b_L(x \otimes l, x' \otimes l') = b(x, x')ll'$. We claim that

$$\mathfrak{d}(M \otimes_{\mathfrak{o}} \mathfrak{o}_L) = \mathfrak{d}(M)\mathfrak{o}_L, \tag{2.10}$$

where the left-hand side is the discriminant with respect to $b_L$.

*Proof.* Equation (2.10) certainly holds when $M$ is a free $\mathfrak{o}$-module, say on a basis $\{m_i\}$: for then $\{m_i \otimes 1\}$ is a free basis of the $\mathfrak{o}_L$-module $M \otimes_{\mathfrak{o}} \mathfrak{o}_L$. The general result then follows by the corresponding equation for module indices (see (II.4.17)) together with (2.4).

□

We now apply the above general theory to the particular case where $A$ is a finite dimensional commutative algebra over a perfect field $K$, with the property that $\mathrm{Rad}(A) = 0$. By Theorem 1 we know that $d(x_1, \ldots, x_n) \neq 0$ for any $K$-basis $\{x_1, \ldots, x_n\}$ of $A$; hence, by (I.1.28), it follows that the map $(a, b) \mapsto t_{A/K}(ab)$ is a non-degenerate symmetric bilinear form $t_{A/K}: A \times A \to K$. Needless to say we shall principally be interested in the case when $A$ is a field. Indeed, if $L/K$ denotes a finite separable extension of fields, then we define $\mathfrak{d}(L/K)$ to be the discriminant of the $\mathfrak{o}$-lattice $\mathfrak{o}_L$ with respect to $t_{L/K}$.

If $K = \mathbb{Q}$, then $\mathfrak{o}_L$ has a $\mathbb{Z}$ basis, $\{w_i\}$ say, and so by (2.1)

$$\mathfrak{d}(L/\mathbb{Q}) = \det(t_{L/\mathbb{Q}}(w_i w_j))\mathbb{Z} = d_L\mathbb{Z} \tag{2.11}$$

where $d_L$ denotes the absolute discriminant defined in (II.1.25). Thus we see that our new, more general notion of discriminant does indeed extend the original one – provided we only consider the ideal generated by $d_L$. (In particular, note that we thereby lose the sign of $d_L$.)

From the definition of $\mathfrak{d}(L/K)$, together with the corollary to (2.4) and (II.1.23), we note that

**(2.12)** $\mathfrak{d}(L/K)$ *is an* $\mathfrak{o}$*-ideal whose class lies in* $\mathrm{Cl}(\mathfrak{o})^2$.

Even at this early stage, we can use the discriminant as a considerable aid in calculating rings of integers:

**(2.13)** *If $L$, $N$ are linearly disjoint finite separable extensions of $K$,* *in an algebraic closure of $K$, and if $(\mathfrak{d}(L/K), \mathfrak{d}(N/K)) = \mathfrak{o}$, then $\mathfrak{o}_{LN} = \mathfrak{o}_L \cdot \mathfrak{o}_N$.*

Here $\mathfrak{o}_L \cdot \mathfrak{o}_N$ denotes the $\mathfrak{o}$-module spanned by elements of the form $xy$ with $x \in \mathfrak{o}_L$, $y \in \mathfrak{o}_N$. In the case when $L$, $N$ are algebraic number fields which satisfy the conditions of (2.13), we say that $L$ and $N$ are *arithmetically disjoint.*

*Proof.* Set $F = LN$. Via the identification $L \otimes_K N \cong F$ of (I.1.49), we identify $\mathfrak{o}_L \otimes_\mathfrak{o} \mathfrak{o}_N$ with $\mathfrak{o}_L \cdot \mathfrak{o}_N$. In the same way, we view $\mathfrak{o}_F$ as an $\mathfrak{o}$-lattice in $L \otimes_K N$ which contains $\mathfrak{o}_L \otimes_\mathfrak{o} \mathfrak{o}_N$. Clearly it suffices to show $[\mathfrak{o}_F : \mathfrak{o}_L \otimes_\mathfrak{o} \mathfrak{o}_N]_\mathfrak{p} = \mathfrak{o}_\mathfrak{p}$ for each prime ideal $\mathfrak{p}$ of $\mathfrak{o}$. Given such a $\mathfrak{p}$, by hypothesis, $\mathfrak{p}$ is coprime to at least one of $\mathfrak{d}(L/K)$, $\mathfrak{d}(N/K)$; so suppose $\mathfrak{p} \nmid \mathfrak{d}(L/K)$. By the index formula (2.4)

(2.13.$a$) $$[\mathfrak{o}_F : \mathfrak{o}_L \otimes_\mathfrak{o} \mathfrak{o}_N]^2_{\mathfrak{o}_N} = \mathfrak{d}_{F/N}(\mathfrak{o}_L \otimes_\mathfrak{o} \mathfrak{o}_N)\mathfrak{d}_{F/N}(\mathfrak{o}_F)^{-1}$$

where $\mathfrak{d}_{F/N}$ denotes the discriminant with respect to $t_{F/N}$. Since $t_{F/N}$ identifies with $t_{L/K} \otimes \mathrm{id}$ on $L \otimes_K N$, we can apply (2.10) to deduce that $\mathfrak{d}_{F/N}(\mathfrak{o}_L \otimes_\mathfrak{o} \mathfrak{o}_N) = \mathfrak{d}_{L/K}(\mathfrak{o}_L) \cdot \mathfrak{o}_N$, which is coprime to $\mathfrak{p}$. However, by (2.12), $\mathfrak{d}_{F/N}(\mathfrak{o}_F) \subset \mathfrak{o}_N$. Therefore, because the left-hand term in (2.13.$a$) is integral, it must be coprime to $\mathfrak{p}$. Since this holds for all $\mathfrak{p}$, the result is shown.

□

Next we consider in detail $\mathfrak{o}_L^D$ – the dual of $\mathfrak{o}_L$ with respect to $t_{L/K}$.

**(2.14)** *$\mathfrak{o}_L^D = \mathcal{D}^{-1}(L/K)$ for some $\mathfrak{o}_L$-ideal $\mathcal{D}(L/K)$ and*

$$\mathfrak{a}^D = \mathfrak{a}^{-1}\mathcal{D}^{-1}(L/K)$$

*for any fractional $\mathfrak{o}_L$-ideal $\mathfrak{a}$. Furthermore $\mathcal{N}_{L/K}(\mathcal{D}(L/K)) = \mathfrak{d}(L/K)$.*

*Proof.* By definition

(2.14.$a$) $$a \in \mathfrak{o}_L^D \iff t_{L/K}(a \cdot \mathfrak{o}_L) \subset \mathfrak{o}.$$

From (2.5) it is therefore clear that $\mathfrak{o}_L^D$ is an $\mathfrak{o}_L$ fractional ideal which contains $\mathfrak{o}_L$; hence it is the inverse of an $\mathfrak{o}_L$-ideal. Obviously

$$t_{L/K}(\mathfrak{a} \cdot \mathfrak{a}^{-1}\mathcal{D}^{-1}(L/K)) = t_{L/K}(\mathcal{D}^{-1}(L/K)) \subset \mathfrak{o},$$

and so $\mathfrak{a}^{-1}\mathcal{D}^{-1}(L/K) \subset \mathfrak{a}^D$. Conversely, if $b \in \mathfrak{a}^D$, then $t(b \cdot \mathfrak{a}) \subset \mathfrak{o}$; hence $b\mathfrak{a} \subset \mathcal{D}^{-1}(L/K)$; and so $b \in \mathfrak{a}^{-1}\mathcal{D}^{-1}(L/K)$. By (2.8) $\mathfrak{d}(L/K) = [\mathfrak{o}_L^D : \mathfrak{o}_L]_\mathfrak{o} = [\mathfrak{o}_L : \mathcal{D}(L/K)]_\mathfrak{o} = \mathcal{N}_{L/K}(\mathcal{D}(L/K))$.

□

In the sequel we shall refer to $\mathcal{D}(L/K)$ as the *different of $L/K$*; the inverse $\mathcal{D}^{-1}(L/K)$ is called the *co-different.*

**(2.15)** *If $N \supset L \supset K$ denotes a finite tower of separable extensions, then*

(*a*) $\mathcal{D}(N/K) = \mathcal{D}(N/L)\mathcal{D}(L/K)$
(*b*) $\mathfrak{d}(N/K) = \mathcal{N}_{L/K}(\mathfrak{d}(N/L))\mathfrak{d}(L/K)^{(N:L)}$.

Note that the right-hand term in part (*a*) is the $\mathfrak{o}_N$-ideal spanned by elements of the form $xy$ for $x \in \mathcal{D}(N/L)$, $y \in \mathcal{D}(L/K)$.

*Proof.* Clearly (*b*) will follow from (*a*) by Theorem 15 and the corollary to Theorem 20 on applying $\mathcal{N}_{L/K}$ and using (2.14). By (2.14.*a*) and by the transitivity of the trace

$$\begin{aligned} x \in \mathcal{D}^{-1}(N/K) &\iff t_{L/K}(t_{N/L}(x\mathfrak{o}_N)) \subset \mathfrak{o} \\ &\iff t_{N/L}(x\mathfrak{o}_N) \subset \mathcal{D}^{-1}(L/K) \\ &\iff t_{N/L}(x\mathfrak{o}_N\mathcal{D}(L/K)) \subset \mathfrak{o}_L \\ &\iff x\mathfrak{o}_N\mathcal{D}(L/K) \subset \mathcal{D}^{-1}(N/L) \\ &\iff x \in \mathcal{D}^{-1}(L/K)\mathcal{D}^{-1}(N/L). \end{aligned}$$

□

From now on $L$ is a fixed finite separable extension of $K$. From (1.8) we know that the $K_\mathfrak{p}$-algebra decomposition $K_\mathfrak{p} \otimes_K L = \prod_{\mathfrak{P}|\mathfrak{p}} L_\mathfrak{P}$ induces a decomposition

(2.16) $$\mathfrak{o}_{K_\mathfrak{p}} \otimes_{\mathfrak{o}_K} \mathfrak{o}_L = \prod_{\mathfrak{P}|\mathfrak{p}} \mathfrak{o}_{L_\mathfrak{P}}.$$

In the sequel we write $\mathfrak{o}_{L,\mathfrak{p}}$ for this ring. The corresponding decomposition for the trace given in (1.10), together with (2.7), shows

(2.17) $$\mathcal{D}(L/K)_\mathfrak{p} = \bigoplus_{\mathfrak{P}|\mathfrak{p}} \mathcal{D}(L_\mathfrak{P}/K_\mathfrak{p}).$$

Thus, on reading off indices of (2.17) against (2.16)

(2.18) $$\mathfrak{d}(L/K)_\mathfrak{p} = \prod_{\mathfrak{P}} \mathfrak{d}(L_\mathfrak{P}/K_\mathfrak{p}).$$

Recall (see (1.11)) that a prime ideal $\mathfrak{p}$ of $\mathfrak{o}$ is said to ramify in $L$ iff $e_\mathfrak{P}(L/K) > 1$ for some prime ideal $\mathfrak{P}$ of $\mathfrak{o}_L$ above $\mathfrak{p}$. The following theorem of Dedekind is an absolutely fundamental property of discriminants.

**Theorem 22.** *Suppose that all residue class fields of $\mathfrak{o}$ are perfect; then $\mathfrak{p}$ ramifies in $L$ iff $\mathfrak{p}$ divides $\mathfrak{d}(L/K)$.*

Since $\mathfrak{d}(L/K)$ has only finitely many prime divisors, this implies

**Corollary.** *Only finitely many prime ideals of $\mathfrak{o}$ ramify in $L$.*

*Proof.* By (2.18) together with the invariance of ramification indices with respect to completions, we may suppose, without loss of generality, that $K$ is complete with valuation ring $\mathfrak{o}$ and valuation idoal $\mathfrak{p}$.

Let $k = \mathfrak{o}/\mathfrak{p}$ and let $B$ denote the $k$-algebra $\mathfrak{o}_L/\mathfrak{p}\mathfrak{o}_L$. For $a \in \mathfrak{o}_L$, we write $\overline{a}$ for its image in $B$. Let $\{b_i\}$ denote an $\mathfrak{o}$-basis of $\mathfrak{o}_L$; then $\{\overline{b}_i\}$ is clearly a $k$-basis of $B$. For any $a \in \mathfrak{o}_L$, recall that $t_{L/K}(a)$ is the trace of the endomorphism given by multiplication by $a$ on the basis $\{b_i\}$. It therefore follows that $\overline{t_{L/K}(a)} = t_{B/k}(\overline{a})$, and so

$$(2.19) \qquad \overline{\det(t_{L/K}(b_i b_j))} = \det(t_{B/k}(\overline{b}_i\overline{b}_j)).$$

Since $\mathfrak{p}\mathfrak{o}_L = \mathfrak{P}^e$, it follows that $e > 1$ iff $\mathrm{Rad}(B) \neq (0)$; however, by Theorem 1 we know $\mathrm{Rad}(B) = 0$ iff $\det(t_{B/k}(\overline{b}_i\overline{b}_j)) \neq 0$. The result then follows from (2.19).

□

We now provide some examples of the above theory in practice. First we give a useful technique, going back to Euler, for calculating discriminants and differents.

**(2.20)** *Let $L = K(\alpha)$ and suppose $g(X)$, the minimal polynomial of $\alpha$ over $K$, lies in $\mathfrak{o}[X]$. Then*

(*a*) $\mathfrak{o}[\alpha]^D = g'(\alpha)\mathfrak{o}[\alpha]$

(*b*) $\mathfrak{d}(\mathfrak{o}[\alpha]) = N_{L/K}(g'(\alpha))\mathfrak{o} = \mathrm{Disc}(g)\mathfrak{o}$.

*Proof.* Given (*a*), the first equality in (*b*) follows since

$$\mathfrak{d}(\mathfrak{o}[\alpha]) = [\mathfrak{o}[\alpha]^D : \mathfrak{o}[\alpha]] = [g'(\alpha)^{-1}\mathfrak{o}[\alpha] : \mathfrak{o}[\alpha]] = N_{L/K}(g'(\alpha))\mathfrak{o},$$

on using the definition of the norm as a determinant (see (I.1.27.*a*)). The second equality in (*b*) comes from the definition of $\mathrm{Disc}(g)$ in (I.1.5).

We now prove (*a*). Let $g(X) = \prod_{i=1}^n (X - \alpha_i)$ in a splitting field. Expanding $1/g(X)$ as a partial fraction gives

$$\frac{1}{g(X)} = \sum_i \frac{1}{g'(\alpha_i)(X - \alpha_i)}.$$

We then expand both sides in terms of the parameter $Y = 1/X$; this gives an equality in $\mathfrak{o}[[Y]]$:

$$Y^n + \sum_{k>n} b_k Y^k = \sum_{k=0}^{\infty} \left( \sum_i \frac{\alpha_i^k}{g'(\alpha_i)} \right) Y^{k+1}.$$

Equating coefficients then tells us that

$$(2.21) \qquad t_{L/K}\left(\frac{\alpha^k}{g'(\alpha)}\right) = \begin{cases} 0 & k < n-1 \\ 1 & k = n-1 \\ b_k & k > n-1. \end{cases}$$

Note that $\{\alpha^k\}_0^{n-1}$ is a $\mathfrak{o}$-basis of $\mathfrak{o}[\alpha]$. Let $M$ denote the $\mathfrak{o}$-module spanned by the elements $c_k = \alpha^k/g'(\alpha)$, $k = 0, \ldots, n-1$. Since all $b_k \in \mathfrak{o}$, from (2.21) $M \subset \mathfrak{o}[\alpha]^D$. However, the matrix $t_{L/K}(\alpha^i c_k)$ is 1 along the skew-diagonal and is zero above, and therefore has determinant $\pm 1$. Thus

$$[M : \mathfrak{o}[\alpha]] = [\mathfrak{o}[\alpha]^D : \mathfrak{o}[\alpha]]$$

and so

$$M = \mathfrak{o}[\alpha]^D.$$

□

By way of a complement to the above result, we briefly show how to determine the discriminant of an irreducible, separable trinomial $g(X) = X^n + aX + b$. Here $g'(X) = nX^{n-1} + a$ and so, if $\alpha$ is a root of the polynomial $g$ and if $L = K(\alpha)$, then

$$N_{L/K}(g'(\alpha)) = N_{L/K}(n\alpha^{n-1} + a).$$

Since $n\alpha^{n-1} + na + nb\alpha^{-1} = 0$, it follows that

$$n\alpha^{n-1} + a = -(n-1)a - nb\alpha^{-1}.$$

However, since $\alpha$ has minimal polynomial over $K$ equal to $g$, it follows that

$$-(n-1)a - nb\alpha^{-1}$$

will have minimal polynomial given by the monic numerator of

$$g\left(\frac{-nb}{X + (n-1)a}\right)$$

namely

$$(X + (n-1)a)^n - na(X + (n-1)a)^{n-1} + (-1)^n b^{n-1} n^n,$$

and so, reading off the constant term of this polynomial and using (I.1.5), we have shown

**(2.22)** *If $g = X^n + aX + b$ is an irreducible, separable polynomial over $K$, then*

$$\mathrm{Disc}(g) = (-1)^{n(n-1)/2}(n^n b^{n-1} + (-1)^{n+1}(n-1)^{n-1}a^n).$$

The following key result ("Kummer's criterion") essentially says that a prime ideal factors in the same way as the minimal polynomial of a generator factors in the residue field of that prime. This result therefore gives us a powerful new perspective on the central problem of describing decomposition laws of prime ideals.

**Theorem 23.** *Let $L = K(\alpha)$ denote a separable extension of $K$; suppose that $\alpha \in \mathfrak{o}_L$ and let $g(X)$ denote the minimal polynomial of $\alpha$ over $K$. Let $\mathfrak{p}$ denote a prime ideal of $\mathfrak{o}$ and assume that $k = \mathfrak{o}/\mathfrak{p}$ is perfect. Suppose that on taking residue classes mod $\mathfrak{p}$, $\overline{g}(X) = \prod_1^t \overline{g}_i(X)$ with the $\overline{g}_i$ distinct monic irreducible polynomials in $k[X]$. Then*

(*a*) *$\mathfrak{p}$ does not ramify in $L$;*
(*b*) *$\mathfrak{o}[\alpha]_\mathfrak{p} = \mathfrak{o}_{L,\mathfrak{p}}$;*
(*c*) *Let $\mathfrak{P}_i = (\mathfrak{p}, g_i(\alpha))$ with $g_i \in \mathfrak{o}[X]$ any polynomial whose image is $\overline{g}_i$ in $k[X]$; then the $\mathfrak{P}_i$ are the distinct prime ideals above $\mathfrak{p}$, with $f_{\mathfrak{P}_i}(L/K) = \deg(\overline{g}_i)$, and $\mathfrak{p}\mathfrak{o}_L = \prod_1^t \mathfrak{P}_i$.*

*Generalisation.* Our proof of part (*c*) can easily be extended to the case where $\overline{g}$ has repeated factors, $\overline{g} = \prod \overline{g}_i^{e_i}$, provided one supposes that $\mathfrak{p} \nmid [\mathfrak{o}_L : \mathfrak{o}[\alpha]]$. The definition of $\mathfrak{P}_i$ remains unaltered and one then shows $\mathfrak{p}\mathfrak{o}_L = \prod \mathfrak{P}_i^{e_i}$.

*Proof.* By the localisation properties of discriminants and indices

(2.23) $$\mathfrak{d}(L/K)_\mathfrak{p}[\mathfrak{o}_{L,\mathfrak{p}} : \mathfrak{o}[\alpha]_\mathfrak{p}]_\mathfrak{p}^2 = \mathfrak{d}(\mathfrak{o}_L)_\mathfrak{p}[\mathfrak{o}_L : \mathfrak{o}[\alpha]]_\mathfrak{p}^2$$

by (2.4)

$$= \mathfrak{d}(\mathfrak{o}[\alpha])_\mathfrak{p}$$

by (2.20.*b*)

$$= \mathrm{Disc}(g)\mathfrak{o}_\mathfrak{p}.$$

By hypothesis $\overline{g}$ is separable, hence $\mathrm{Disc}(\overline{g}) \in k^*$, and so $\mathrm{Disc}(g) \in \mathfrak{o}_\mathfrak{p}^*$. Since all terms on the left of (2.23) are $\mathfrak{o}_\mathfrak{p}$-ideals, we deduce that

$$\mathfrak{o}_\mathfrak{p} = \mathfrak{d}(L/K)_\mathfrak{p} = \prod_{\mathfrak{P}|\mathfrak{p}} \mathfrak{d}(L_\mathfrak{P}/K_\mathfrak{p})$$

$$\mathfrak{o}_{L,\mathfrak{p}} = \mathfrak{o}[\alpha]_\mathfrak{p}$$

Thus (*b*) is shown, and (*a*) follows from Theorem 22.

To show part (*c*), from (II.4.6) and using (*b*), we deduce that

(2.24) $$\frac{\mathfrak{o}_L}{\mathfrak{p}\mathfrak{o}_L} = \frac{\mathfrak{o}_{L,\mathfrak{p}}}{\mathfrak{p}\mathfrak{o}_{L,\mathfrak{p}}} = \frac{\mathfrak{o}[\alpha]_\mathfrak{p}}{\mathfrak{p}\mathfrak{o}[\alpha]_\mathfrak{p}}.$$

However reduction mod $\mathfrak{p}$ gives an isomorphism

(2.25) $$\frac{\mathfrak{o}[\alpha]_\mathfrak{p}}{\mathfrak{p}\mathfrak{o}[\alpha]_\mathfrak{p}} = \frac{\mathfrak{o}_\mathfrak{p}[X]}{(\mathfrak{p}, g(X))} \cong \frac{k[X]}{(\overline{g}(X))}.$$

Since $\overline{g}$ is separable, $k[X]/(\overline{g})$ splits as a product of fields $k[X]/(\overline{g}_i)$ each of which has degree $\deg(\overline{g}_i)$ over $k$. The result then follows on noting that the ideal $\mathfrak{P}_i = (\mathfrak{p}, g_i(\alpha))$ is the kernel of the map $\mathfrak{o}_L \to k[X]/(\overline{g}_i)$ obtained by composing $\mathfrak{o}_L \to \mathfrak{o}_L/\mathfrak{p}\mathfrak{o}_L$ with the maps (2.24), (2.25).

□

The remainder of this section is devoted to considering various applications of the above theory. In addition to the existing notation and hypotheses, we now suppose that all residue class fields of $\mathfrak{o}$ are finite, so that the absolute norm of ideals is defined.

For a given positive integer $m$, we let $\mu_m$ denote the group of $m$th roots of unity contained in a fixed separable closure of $K$. Throughout $\mathfrak{p}$ is a prime ideal of $\mathfrak{o}$. Recall from (II.3.26):

**(2.26)** *Let $m$ be coprime to $\mathbf{N}\mathfrak{p}$. If $\mu_m \subset K$, then $\mathbf{N}\mathfrak{p} \equiv 1 \bmod (m)$. Conversely, if $K$ is complete with respect to $|.|_\mathfrak{p}$, then $\mu_m \subset K$ if $\mathbf{N}\mathfrak{p} \equiv 1 \bmod (m)$.*

*Remark.* In fact if $K$ is an algebraic number field and if $\mathbf{N}\mathfrak{p} \equiv 1 \bmod (m)$ for all prime ideals $\mathfrak{p}$ of $\mathfrak{o}$ which are coprime to $m$, then it can be shown that $\mu_m \subset K$; however, we cannot prove this result here.

**(2.27)** *Suppose $(m, \mathbf{N}\mathfrak{p}) = 1$ and let $L = K(\mu_m)$. Then $\mathfrak{p}$ is non-ramified in $L$; furthermore for each prime $\mathfrak{P}$ of $\mathfrak{o}_L$ above $\mathfrak{p}$, $f_\mathfrak{P}(L/K)$ is the least positive integer $f$ such that $\mathbf{N}\mathfrak{p}^f \equiv 1 \bmod (m)$.*

*Proof.* Since $X^m - 1$ is separable in $k[X]$, we can apply Theorem 23, and deduce that $\mathfrak{p}$ does not ramify in $L$; furthermore, if $\zeta$ is a primitive $m$th root of unity in $L_\mathfrak{P}$, then by part (b) of the theorem $\mathfrak{o}_{L,\mathfrak{P}} = \mathfrak{o}_\mathfrak{p}[\zeta]$; hence $k_L = k[\overline{\zeta}]$, where $\overline{\zeta}$ is still a primitive $m$th root of unity, and therefore $(k_L : k) = f$, since $\{\overline{\zeta}, \overline{\zeta}^{\mathbf{N}\mathfrak{p}}, \ldots, \overline{\zeta}^{\mathbf{N}\mathfrak{p}^{f-1}}\}$ is the full set of conjugates of $\overline{\zeta}$ under powers of the Frobenius automorphism.

□

**(2.28)** *Let $l$ denote a prime number with $l \nmid \mathbf{N}\mathfrak{p}$. Let $a \in K^*$, $a \notin K^{*l}$, and let $L = K(a^{1/l})$. Then*

(*a*) *$\mathfrak{p}$ is non-ramified in $L$ iff $v_\mathfrak{p}(a) \equiv 0 \bmod (l)$; moreover, in this case we can assume without loss of generality that $v_\mathfrak{p}(a) = 0$.*

(*b*) *If $v_\mathfrak{p}(a) = 0$ and if $\mathbf{N}\mathfrak{p} \not\equiv 1 \bmod (l)$, then $\mathfrak{p}\mathfrak{o}_L = \mathfrak{P}_0\mathfrak{P}_1 \cdots \mathfrak{P}_r$ where the $\mathfrak{P}_i$ are distinct prime ideals of $\mathfrak{o}_L$ with residue class extension* *degree*

$$f_{\mathfrak{P}_0}(L/K) = 1$$
$$f_{\mathfrak{P}_i}(L/K) = f \quad \textit{for } i > 0$$

*where again $f$ denotes the order of* $\mathbf{N}\mathfrak{p}$ mod $(l)$.

(c) *If $v_{\mathfrak{p}}(a) = 0$ and if* $\mathbf{N}\mathfrak{p} \equiv 1$ mod $(l)$, *then either $\mathfrak{p}\mathfrak{o}_L$ splits into a product of $l$ distinct primes $\mathfrak{P}_i$ with $f_{\mathfrak{P}_i}(L/K) = 1$ or else $\mathfrak{p}\mathfrak{o}_L$ is itself prime, depending on whether or not $X^l - \bar{a}$ is soluble in $k[X]$.*

*Proof.* Writing $a = a_1/a_2$ with $a_1, a_2 \in \mathfrak{o}$, it is clear that $K(a^{1/l}) = K(b^{1/l})$ where $b = a_1 a_2^{l-1}$; thus, without loss of generality, we can assume $a \in \mathfrak{o}$. If $v_{\mathfrak{p}}(a) = lr$ for some integer $r$, then we can choose an $\mathfrak{o}$-ideal $\mathfrak{a}$ which lies in the class of $\mathfrak{p}^r$, but which is coprime to $\mathfrak{p}$ (see (II.1.19)); thus $\mathfrak{p}^{-r}\mathfrak{a} = b\mathfrak{o}$ for some $b \in K^*$. Therefore

$$v_{\mathfrak{p}}(ab^l) = lr + lv_{\mathfrak{p}}(b) = 0$$

and $ab^l \in \mathfrak{o}$. Hence we can now assume $v_{\mathfrak{p}}(a) = 0$. Since $X^l - \bar{a}$ is separable in $k[X]$, we can use Theorem 23 to deduce that $\mathfrak{p}$ is non-ramified in $L$.

On the other hand, if $v_{\mathfrak{p}}(a) = s$ with $l \nmid s$, then by multiplying by a suitable $l$th power of an element in $\mathfrak{p}^{-1}\backslash\mathfrak{o}$, we can assume $0 < s < l$. If $\mathfrak{P}$ denotes any prime of $L$ over $\mathfrak{p}$, we can write $a^{1/l}\mathfrak{o}_L = \mathfrak{P}^r\mathfrak{a}$ where $\mathfrak{P}$ and $\mathfrak{a}$ are coprime and $r > 0$. Hence $v_{\mathfrak{P}}(a) - rl = se$ where $e = e_{\mathfrak{P}}(L/K)$. Since $l \nmid s$, we must have $l|e$ and so $e > 1$ which completes the proof of $(a)$.

Assume now that $a \in \mathfrak{o}$ with $v_{\mathfrak{p}}(a) = 0$. If $\mathbf{N}\mathfrak{p} \equiv 1$ mod $(l)$, then $k$ contains the $l$th roots of unity; thus by standard Kummer theory (see below) $X^l - \bar{a}$ is completely split/irreducible according as $\bar{a}$ is/is not an $l$th power in $k^*$. The result is then a direct consequence of Theorem 23.

Lastly suppose $\mathbf{N}\mathfrak{p} \not\equiv 1$ mod $(l)$. Then, raising to the $l$th power induces a group automorphism of $k^*$, and so there exists a unique $\bar{b} \in k^*$ such that $\bar{b}^l = \bar{a}$. Mapping $X \mapsto X/\bar{b}$, we see that $X^l - \bar{a}$ factorises in $k[X]$ in exactly the same way as $X^l - 1$; however, $X^l - 1$ factorises as $X - 1$ times a product of distinct irreducible polynomials, each of which has degree $f$ (by (2.27)). The result $(b)$ now follows from Theorem 23. □

For the benefit of the reader who is unfamiliar with Kummer theory we prove the result quoted above. Let $F$ be a field containing a group $\mu_l$ of order $l$ ($l$-th roots of unity). If $b \in F^*$ and $\beta$ is a root of $X^l - b$ in some extension field $E$ of $F$, then the $\beta\eta$ ($\eta \in \mu_l$) are precisely the roots of $X^l - b$. Thus if $\beta \in F$, then $X^l - b$ splits completely in $F[X]$.

If $F(\beta) \neq F$ then the conjugates $\beta\eta$ will still lie in $F(\beta)$ and $\gamma \mapsto \eta_\gamma$, where $\beta^\gamma = \beta\eta_\gamma$, is a non-trivial homomorphism $\mathrm{Gal}(F(\beta)/F) \to \mu_l$, hence an isomorphism. Thus $(F(\beta) : F) = l$ and $X^l - b$ is irreducible.

We shall now specialise to $K = \mathbb{Q}$ and $l = 2$. Let $\left(\frac{\cdot}{p}\right)$ denote the Legendre symbol, so that for an integer $a$, which is coprime to the odd prime $p$,

$$\left(\frac{a}{p}\right) = \begin{cases} 1 & \text{if } X^2 \equiv a \bmod (p) \text{ is soluble for integral } X, \\ -1 & \text{if } X^2 \equiv a \bmod (p) \text{ is insoluble for integral } X. \end{cases}$$

Recall that if the integer $b$ is also coprime to $p$, then

$$\left(\frac{ab}{p}\right) = \left(\frac{a}{p}\right)\left(\frac{b}{p}\right).$$

We can now apply Theorem 23 to obtain an extremely practical criterion for determining the behaviour of a prime number $p$ in a quadratic number field $K = \mathbb{Q}(\sqrt{d})$ (with $d$ a square-free integer, $d \neq 1$).

**(2.29)** *If $p > 2$ and if $p \nmid d_K$, then $p$ splits (as a product of two distinct ideals) in $K$ iff*

$$\left(\frac{d_K}{p}\right) = 1.$$

*Also if $d_K$ is odd, then 2 splits in $K$ iff $d_K \equiv 1 \bmod (8)$.*

*Remark.* We shall obtain a generalisation of this result to other Kummer extensions in the next section.

*Proof.* Suppose first that $p$ is odd. Then from (II.1.32,33) $\mathfrak{o}_K \supset \mathbb{Z}[\sqrt{d}] \supset 2\mathfrak{o}_K$, and so $\mathfrak{o}_{K,p} = \mathbb{Z}[\sqrt{d}] \otimes \mathbb{Z}_p$. Setting $g(X) = X^2 - d$ and applying Theorem 23, we deduce that $p$ splits in $K$ iff $g(X)$ is reducible in $\mathbb{F}_p[X]$, i.e. iff

$$1 = \left(\frac{d}{p}\right) = \left(\frac{d_K}{p}\right).$$

Now take $p = 2$. From (II.1.32,33) $\mathfrak{o}_K = \mathbb{Z}[\frac{1+\sqrt{d}}{2}]$, since $d_K$ is odd. The result now follows from Theorem 23, since $\frac{1+\sqrt{d}}{2}$ has minimal polynomial $X^2 - X + \left(\frac{1-d}{4}\right)$, and because this is reducible in $\mathbb{F}_2[X]$ iff $\frac{d-1}{4} \equiv 0 \bmod (2)$.

□

## §3 Non-ramified and tamely ramified extensions

Throughout this section we let $K$ denote a field which is complete with respect to a discrete absolute value $u$ associated with a valuation $v$. We write $\mathfrak{o}$ for the valuation ring of $K$, $\mathfrak{p}$ for the valuation ideal, and we suppose $k = \mathfrak{o}/\mathfrak{p}$ always to be finite. We shall call any element $\pi \in \mathfrak{p}$ with the property that $\pi\mathfrak{o} = \mathfrak{p}$ a *uniformising parameter* of $K$.

Recall the equation $[L : K] = e(L/K)f(L/K)$ (cf (1.14)). A finite separable extension $L/K$ is said to be totally ramified if $e(L/K) = (L : K)$; thus, in this case, $f(L/K) = 1$, and so $k_L$, the residue class field of $\mathfrak{o}_L$, is equal to $k$.

A monic polynomial

$$g(X) = X^m + a_{m-1}X^{m-1} + \cdots + a_1X + a_0 \in \mathfrak{o}[X]$$

is called an *Eisenstein polynomial* over $K$ if $a_j \in \mathfrak{p}$, all $j$, and $a_0 \notin \mathfrak{p}^2$, i.e. $a_0$ is a uniformising parameter of $K$. The name derives from Eisenstein's Irreducibility Criterion for polynomials: indeed, applying this criterion to the principal ideal domain $\mathfrak{o}$, shows that $g$ is irreducible in $\mathfrak{o}[X]$; whence, by the Gauss Lemma, it is irreducible in $K[X]$. We shall give an independent proof of this fact below:

**Theorem 24.** *Let $L/K$ denote a finite separable extension.*

(*a*) *The following conditions are equivalent*
  (*i*) $L = K(\lambda)$ *for $\lambda$ a root of some Eisenstein polynomial $g(X)$.*
  (*ii*) $L/K$ *is totally ramified.*
  (*iii*) $\mathfrak{o}_L = \mathfrak{o}[\lambda]$ *for a uniformising parameter $\lambda$ of $L$.*

(*b*) *If condition (i) is satisfied, then $\lambda$ is a uniformising parameter and $\deg(g) = (L : K)$, and so $g$ is irreducible in $K$.*

(*c*) *The minimal polynomial over $K$ of a uniformising parameter of a totally ramified separable extension $L$ of $K$ is an Eisenstein polynomial over $K$.*

*Proof.* We begin by assuming (*i*) and deduce both (*b*) and (*ii*). Let $g(X) = X^n + b_{n-1}X^{n-1} + \cdots + b_0$. Since $g(X) \in \mathfrak{o}[X]$ we know that $\lambda \in \mathfrak{o}_L$. Thus

$$\lambda^n = -\sum_{j=0}^{n-1} b_j\lambda^j \in \mathfrak{p}\mathfrak{o}_L,$$

therefore if $v_L$ is the valuation on $L$ (associated with the prime ideal $\mathfrak{P}$ of $\mathfrak{o}_L$ above $\mathfrak{p}$), we have $v_L(\lambda) \geq 1$. As

$$v_L(\lambda^n + b_0) = v_L\left(-\sum_{j=1}^{n-1} b_j\lambda^j\right) \geq 1 + e(L/K)$$

and $v_L(b_0) = e(L/K)$ we must have $v_L(\lambda^n) = e(L/K)$, (use (II.2.2d)) and we get inequalities

$$(L:K) \geq e(L/K) = nv_L(\lambda) \geq n \geq (L:K).$$

We must thus have equality throughout. $e(L/K) = (L:K)$ implies that $L/K$ is totally ramified. $n = (L:K)$ implies that $g$ is irreducible. $nv_L(\lambda) = n$ implies that $\lambda$ is a uniformising parameter.

Next we show that (*ii*) implies (*iii*). More precisely we prove that $\mathfrak{o}_L = \mathfrak{o}[\lambda]$ for any uniformising parameter $\lambda$ of $L$. We shall use (II.3.24*c*). Since $k_L = k$ we may choose the set $\mathcal{R}$ of representatives of $k_L$ to be in $\mathfrak{o}$. Given a uniformising parameter $\lambda$ of $L$, for $h = i + ej$, $(e = e(L/K))$ $0 \leq i < e$, put $\lambda_h = \lambda^i \pi^j$, where $\pi$ is some uniformising parameter of $K$. Collecting terms together we have shown all elements of $\mathfrak{o}_L$ to be of the form $\sum_{i=0}^{e-1} a_i \lambda^i \in \mathfrak{o}[\lambda]$. On the other hand, since $\lambda \in \mathfrak{o}_L$ it is obvious that $\mathfrak{o}_L \supset \mathfrak{o}[\lambda]$ and so (*iii*) is established.

Finally we assume (*iii*) to hold. Reducing mod the prime ideal $\mathcal{P}$ of $\mathfrak{o}_L$ it is clear that $\mathfrak{o}_L/\mathcal{P} = \mathfrak{o}/\mathfrak{p}$, hence $f(L/K) = 1$ and so $e = e(L/K) = (L:K)$. Thus (*iii*) implies (*ii*) and also implies that $L = K(\lambda)$. Assuming (*ii*) and (*iii*) we shall show that this $\lambda$ is the root of an Eisenstein polynomial over $K$. This will then give us the implication (*iii*)$\Rightarrow$(*i*). As the original choice of uniformising parameter $\lambda$ in the proof of (*ii*)$\Rightarrow$(*iii*) was arbitrary we shall also have established (*c*).

By hypothesis we can write $\lambda^e = \sum_{i=0}^{e-1} c_i \lambda^i$ with $c_i \in \mathfrak{o}$. It then follows that $v_L(\lambda^e - c_0) \geq 1$, whence $v_L(c_0) \geq 1$; this implies $v_K(c_0) \geq 1$; hence $v_L(c_0) \geq e$, and so $v_L(\lambda^e - c_0) \geq e$. Similarly $v_L(\lambda^e - c_0 - c_1\lambda) \geq 2$ and so $v_K(c_1) \geq 1$. Continuing in this way we see that $v_K(c_j) \geq 1$ for $j = 0, 1, \ldots, e-1$. Thus we conclude that $v_L(\lambda^e - c_0) \geq e + 1$ and so $e = ev_L(\lambda) = v_L(\lambda^e) = v_L(c_0)$; therefore $v_K(c_0) = 1$ and we have shown $g = X^e - \sum_{j=0}^{l-1} c_i X^i$ to be an Eisenstein polynomial.

□

Note the following arithmetic application:

**Corollary.** *Let $F$ denote an algebraic number field, and let $g \in \mathfrak{o}_F[X]$ have the property that $g$ is Eisenstein in $F_\mathfrak{p}[X]$ for some prime ideal $\mathfrak{p}$ of $\mathfrak{o}_F$. Then $g$ is irreducible in $F[X]$.*

*Proof.* By the above $g$ is irreducible in $F_\mathfrak{p}[X]$; it is therefore irreducible in $F[X]$.

□

(3.1). As an illustration of the above ideas, for a prime number $p$ and for a positive integer $n$ consider the cyclotomic polynomial in $\mathbb{Q}_p[X]$

$$\Phi_{p^n}(X) = \frac{X^{p^n} - 1}{X^{p^{n-1}} - 1} = X^{p^{n-1}(p-1)} + X^{p^{n-1}(p-2)} + \cdots + X^{p^{n-1}} + 1.$$

Clearly the roots of $\Phi_{p^n}$ are the primitive $p^n$th roots of unity in the algebraic closure of $\mathbb{Q}_p$.

We let $g(X) = \Phi_{p^n}(X+1)$; since in $\mathbb{F}_p[X]$, $(X+1)^{p^n} - 1 \equiv X^{p^n}$, and $(X+1)^{p^{n-1}} - 1 \equiv X^{p^{n-1}}$, $\overline{g}(X) \equiv X^{p^{n-1}(p-1)}$. Thus in order to show that $g$ is an Eisenstein polynomial, it suffices to observe that $g(0) = \Phi_{p^n}(1) = p$. Writing $K = \mathbb{Q}_p(\zeta)$ for a primitive $p^n$th root of unity $\zeta$ and applying Theorem 24, we conclude that $K/\mathbb{Q}_p$ is totally ramified of degree $p^{n-1}(p-1)$, $\mathfrak{o}_K = \mathbb{Z}_p[\zeta]$, and $1 - \zeta$ is a uniformising parameter of $K$.

Next we consider the opposite case to the above: namely, we consider extensions $L/K$ which are non-ramified; thus in this case $e(L/K) = 1$, and so $f(L/K) = (L : K)$.

For any field $F$ and for any positive integer $m$ which is coprime to the characteristic of $F$, we denote by $F[m]$ the field obtained by adjoining the $m$th roots of unity (in a given separable closure of $F$) to $F$.

**Theorem 25.**

(a) *Given a positive integer $f$, there exists a non-ramified extension $K\{f\}$ of $K$ of degree $f$, and moreover this extension is unique within a given algebraic closure of $K$ and is unique up to isomorphism. In fact $K\{f\} = K[m]$ for any $m$ with the property that $\mathbf{N}\mathfrak{p}$ has order $f$ in $(\mathbb{Z}/m\mathbb{Z})^*$; thus, in particular, $K\{f\}$ is generated over $K$ by a primitive $\mathbf{N}\mathfrak{p}^f - 1$st root of unity.*

(b) *If $L/K$ is a finite separable extension, then $L \supset K\{f\} \supset K$ for $f = f(L/K)$; furthermore $L/K\{f\}$ is totally ramified, and $K\{f\}$ is the unique maximal non-ramified extension of $K$ in $L$.*

*Proof.* First we show

**(3.2.a)** *Let $M$ be a finite separable extension of $K$ and let $k_M$ denote the residue class field of $\mathfrak{o}_M$. Let $m$ be a positive integer, $(m, \mathbf{N}\mathfrak{p}) = 1$. Then the irreducible factors $\overline{g(X)}$ of $X^m - 1$ in $k_M[X]$ are precisely the reductions of the irreducible factors of $X^m - 1$ in $M[X]$.*

This follows from Hensel's Lemma, $X^m - 1$ being separable. Indeed, let $g(X)$ be an irreducible factor of $X^m - 1$ in $M[X]$ and $h(X)$ an irreducible (monic) factor of the residue class polynomial $\overline{g(X)}$ in $k_M[X]$. Choose $f(X)$ monic in $M[X]$ with $\overline{f(X)} = h(X)$; $f(X)$ must be irreducible.

Adjoin a root $\alpha$ of $f(X)$ to $M$; write $N = M(\alpha)$. In $k_N$ the polynomial $h(X)$ thus has the root $\overline{\alpha}$, hence so has $\overline{g(X)}$. By Hensel's Lemma $g(X)$ has a root in $N$. Thus $\deg g(X) \geq \deg h(X) = [N : M] \geq \deg g(X)$. Therefore $\overline{g(X)} = h(X)$, i.e. $\overline{g(X)}$ is irreducible.

(3.2.*a*) implies that

$$(K[m] : K) = (k[m] : k), \tag{3.2.b}$$

and that $k_{K[m]} \supset k[m]$. As always $(K[m] : K) \geq (k_{K[m]} : k)$, so we conclude that

$$k[m] = k_{K[m]}. \tag{3.2.c}$$

Thus $K[m]/K$ is non-ramified. (The above generalises (II.3.26) slightly.)

If now $L$ is a finite separable extension of $K$ it also follows from (3.2.*a*) that $L \supset K[m]$ iff $k_L \supset k[m]$. Choose $m_1$ so that $k_L = k[m_1]$. If first $L$ is non-ramified, then $(L : K) = (k[m_1] : k)$, hence by (3.2.*b*) $L = K[m_1]$. Thus

(3.2.*d*). *The fields* $K[m]$ *with* $(m, \mathbf{N}\mathfrak{p}) = 1$, *are precisely the non-ramified extensions of* $K$, $K[m]$ *being contained in* $L$ *iff* $k[m] \subset k_L$.

We know that a finite field $k$ has an extension of given degree $f$, unique to within isomorphism; we denote it by $k\{f\}$. If $|k| = \mathbf{N}\mathfrak{p}$ this is the field $k[m]$ for any $m$, such that $f$ is the order of $\mathbf{N}\mathfrak{p}$ modulo $m$, for instance $m = \mathbf{N}\mathfrak{p}^f - 1$. Choosing such an $m$, $K[m]$ will be of degree $f$ over $K$, by (3.2.*b*). As $m | \mathbf{N}\mathfrak{p}^f - 1$, $K[m]$ will be contained in $K[\mathbf{N}\mathfrak{p}^f - 1]$, hence will coincide with it, since they have the same degree over $K$. Thus $K$ has a non-ramified extension $K\{f\}$, unique within a given algebraic closure of $K$, by (3.2.*d*), and unique to within isomorphism over $K$. Moreover if $f_1 | f$ then $\mathbf{N}\mathfrak{p}^{f_1} - 1 | \mathbf{N}\mathfrak{p}^f - 1$ so $K\{f_1\} \subset K\{f\}$.

If now $L$ is a finite separable extension of $K$ then $k_L = k\{f(L/K)\}$. Hence by (3.2.*d*), $K\{f(L/K)\} = L_0$, say, is a non-ramified extension of $K$ in $L$ containing all non-ramified extensions of $K$ in $L$. By (3.2.*b*), $f(L/K) = f(L_0/K)$, hence $f(L/L_0) = 1$, i.e. $L/L_0$ is totally ramified.

□

If $M/K$ is a finite, separable, non-ramified extension, then, by the above result, $M = K(\eta)$ where $\eta$ is a primitive root of unity of order $m = |k_M^*|$. Since $X^m - 1$ is separable in $k[X]$, we can apply Theorem 23 to deduce that

$$\mathfrak{o}_M = \mathfrak{o}[\eta]. \tag{3.3}$$

More generally, if $L/K$ is any finite separable extension, we let $\lambda$ denote a uniformising parameter of $L$, and we let $M$ denote the maximal non-ramified extension of $K$ in $L$. We then claim that, with $\eta$ as in (3.3),

$$\mathfrak{o}_L = \mathfrak{o}[\lambda + \eta]. \tag{3.4}$$

First note that since $\mathfrak{o}[\lambda + \eta]$ is $\mathfrak{o}$-free, $\mathfrak{o}[\lambda + \eta]$ is closed and hence complete in $\mathfrak{o}_L$. Thus

$$\mathfrak{o}[\lambda + \eta] \ni \lim_{n\to\infty} (\lambda + \eta)^{|k_L|^n} = \eta$$

and so

$$\begin{aligned} \mathfrak{o}[\lambda + \eta] &\supset \mathfrak{o}[\eta][\lambda + \eta] \\ &= \mathfrak{o}_M[\lambda + \eta] \\ &= \mathfrak{o}_M[\lambda] \\ &= \mathfrak{o}_L, \quad \text{by Theorem 24.} \end{aligned}$$

The property (3.4) does not, in general, go over to algebraic number fields. To see how this can go wrong, let $E/F$ denote a Galois extension of number fields with $\Gamma = \mathrm{Gal}(E/F)$, and suppose that there is a non-ramified prime ideal $\mathfrak{P}$ of $\mathfrak{o}_E$ with the property that $\mathbf{N}\mathfrak{P} < (E : F)$. We then show that there is no element $a \in E$ such that

$$\mathfrak{o}_E = \mathfrak{o}_F[a]. \tag{3.5}$$

Indeed, suppose for contradiction that (3.5) holds; then by the pigeon-hole principle we can find distinct $\gamma$, $\delta \in \Gamma$ with $a^\gamma \equiv a^\delta \bmod \mathfrak{P}$ and $\gamma \neq \delta$. Thus, by (2.20.*b*), $\mathfrak{p} = \mathfrak{P} \cap \mathfrak{o}_F$ must divide $\mathfrak{d}(\mathfrak{o}_F[a])$. However, by Theorem 22, $\mathfrak{p} \nmid \mathfrak{d}(E/F)$, since $\mathfrak{p}$ is non-ramified in $E/F$; therefore

$$\mathfrak{p} \supset [\mathfrak{o}_E : \mathfrak{o}_F[a]]_{\mathfrak{o}}^2 = \mathfrak{d}(\mathfrak{o}_F[a])\mathfrak{d}(E/F)^{-1}$$

which contradicts (3.5).

*Example.* Let $K$ denote the cubic subfield of $\mathbb{Q}[31]$ (see VI §1 for its existence). Moreover 2 is non-ramified in $\mathbb{Q}[31]$; furthermore, since $2^5 \equiv 1 \bmod (31)$ we deduce that 2 is completely split in $K/\mathbb{Q}$ (see V §1 for details). Therefore, by the above, $\mathfrak{o}_K$ cannot be written in the form $\mathbb{Z}[a]$. The same reasoning applies to the cubic subfields of $\mathbb{Q}[43]$, $\mathbb{Q}[109]$, $\mathbb{Q}[127]$...

Next we consider the construction of certain non-ramified extensions for number fields. In the course of the next chapter we shall see that there are no non-ramified extensions of $\mathbb{Q}$. This then leads us to consider the case of a quadratic field $F = \mathbb{Q}(\sqrt{m})$, where as previously $m \neq 1$ is square-free. Note that since $d_F$ is either $m$ or $4m$, we can write $F = \mathbb{Q}(\sqrt{d_F})$.

With the construction of non-ramified extensions in mind, we define the notion of a *prime discriminant*: an integer $q$ is called a prime discriminant if $q$ is divisible by only one prime number and if $q = d_G$ for some quadratic extension $G/\mathbb{Q}$; $q$ is said to belong to its unique prime divisor. Thus the prime discriminants are $-4$, $\pm 8$ and $p^*$ for odd primes $p$, where $p^* = \pm p \equiv 1 \bmod (4)$.

**(3.6)** *Let $q_j$ for $j = 1, \ldots, n$ be prime discriminants which belong to distinct primes, and let $E = \mathbb{Q}(\sqrt{q_1}, \ldots, \sqrt{q_n})$. Then $E/\mathbb{Q}$ is a Galois extension whose Galois group is abelian, has exponent 2 and order $2^n$; moreover*

$$e_p(E/\mathbb{Q}) = \begin{cases} 2 & \text{if some } q_i \text{ belongs to } p \\ 1 & \text{otherwise.} \end{cases}$$

*Proof.* Let $q_i$ belong to $p_i$. We argue by induction on $n$; the case $n = 1$ is clear. Let $E' = \mathbb{Q}(\sqrt{q_1}, \ldots, \sqrt{q_{n-1}})$, $G = \mathbb{Q}(\sqrt{q_n})$. Since $\prod q_i^{r_i}$ is a square in $\mathbb{Q}^*$ iff the $r_i$ are all even, we deduce that $\mathrm{Gal}(E/\mathbb{Q})$ is an elementary abelian group of order $2^n$ (see below). By induction the only primes which ramify in $E'$ are the $p_i$, $i = 1, \ldots, n-1$; while by Theorem 22, $p_n$ is the only prime to ramify in $G$. Therefore $E'$ and $G$ are arithmetically disjoint over $\mathbb{Q}$.

We can therefore apply (2.13): under the identification of $E$ with $E' \otimes_{\mathbb{Q}} G$, $\mathfrak{o}_E$ identifies with $\mathfrak{o}_{E'} \otimes_{\mathbb{Z}} \mathfrak{o}_G$, and so by (2.10) $\mathfrak{d}(E/E') = \mathfrak{d}(G/\mathbb{Q})\mathfrak{o}_{E'}$. The result for the ramification indices then follows from Theorem 22 together with the tower formula (2.15).

□

The assertion on the Galois group of $E/\mathbb{Q}$ is a particular case of the following result

**(3.6.*a*)** *Let $E_0$ be a field of characteristic $\neq 2$ and let $a_j$ $(j = 1, \ldots, n)$ be elements of $E_0^*$ so that $\prod_j a_j^{n_j} \in E_0^{*2}$ only if the $n_j$ are all even. Then $E = E_0(\sqrt{a_1}, \ldots, \sqrt{a_n})$ is a Galois extension of $E_0$ whose Galois group is abelian, has exponent 2 and order $2^n$.*

*Proof.* As the conjugates of $\sqrt{a_j}$ are $\pm\sqrt{a_j}$, $E/E_0$ is clearly Galois. If $\sigma \in \mathrm{Gal}(E/E_0)$ then $(\sqrt{a_j})^{\sigma^2} = \sqrt{a_j}$. Thus $\mathrm{Gal}(E/E_0)$ has exponent 1 or 2, hence is abelian. We establish the equation $(E : E_0) = 2^n$ by induction on $n$, the case $n = 1$ being taken for granted. Put $E_1 = E_0(\sqrt{a_1})$. Then for $x, y \in E_0$, we have $(x + y\sqrt{a_1})^2 \in E_0$ iff $xy = 0$. Thus $\prod_{j=2}^n a_j^{n_j} \in E_1^{*2}$ only if the $n_j$ are all even. By induction hypothesis $(E : E_1) = 2^{n-1}$ and so $(E : E_0) = 2^n$.

□

**(3.7)** *For any quadratic field $F$, $d_F$ can be written uniquely as a product of prime discriminants $d_F = \prod q_i$. Furthermore*

$$e_p(F/\mathbb{Q}) = \begin{cases} 2 & \text{if some } q_i \text{ belongs to } p; \\ 1 & \text{otherwise.} \end{cases}$$

*Proof.* The description of $e_p(F/\mathbb{Q})$ will follow immediately from the factorisation together with Theorem 22.

Since, for odd $p$, there is exactly one prime discriminant, namely $p^*$, it is clear that any factorisation will necessarily be unique.

To show the existence of such a factorisation we consider separate cases. We write $m = \delta \prod p_i$ with $\delta = \pm 1$, and we let $N$ denote the number of primes $p_i$ which are congruent to 3 mod (4).

If $m \equiv 1 \bmod (4)$: Then $d_F = m$, and $m = \delta(-1)^N \prod p_i^*$. Reducing mod(4) gives $1 \equiv \delta(-1)^N \bmod (4)$; hence $\delta(-1)^N = 1$, and so $m = \prod p_i^*$.

If $m \equiv 3 \bmod (4)$: Then $d_F = 4m$, and $m = \delta(-1)^N \prod p_i^*$ again. Reducing mod(4) gives $\delta(-1)^N = -1$; whence we obtain the factorisation $d_F = -4 \prod p_i^*$.

If $m \equiv 2 \bmod (4)$: We write $p_1 = 2$, then

$$d_F = 4m = 8\delta(-1)^N \prod_{i>1} p_i^*,$$

and the result follows since both $\pm 8$ are prime discriminants.

□

Let $F$ be as in (3.7). From the formulae in (3.6) and (3.7) and using the tower formula (1.13$a$), we have shown the following result, which we shall encounter again in the genus theory of quadratic fields:

**(3.8)** *The extension $\mathbb{Q}(\sqrt{q_1}, \ldots, \sqrt{q_n})/F$ is non-ramified for all prime ideals of $\mathfrak{o}_F$ the $q_i$ being as in (3.7).*

(3.9) We now turn to tame ramification. Let $F$ denote the field of fractions of a Dedekind domain $\mathfrak{o}_F$, let $E/F$ denote a finite separable extension, and let $\mathfrak{o}_E$ denote the integral closure of $\mathfrak{o}_F$ in $\mathfrak{o}_E$. Observe that for any fractional $\mathfrak{o}_E$-ideal $\mathfrak{a}$, $t_{E/F}(\mathfrak{a})$ is a fractional $\mathfrak{o}_F$-ideal; furthermore, since $t_{E/F}(\mathfrak{o}_E) \subset \mathfrak{o}_F$, we note that $t_{E/F}(\mathfrak{o}_E)$ is an $\mathfrak{o}_F$-ideal. From (1.10) we recall that for any prime ideal $\mathfrak{p}$ of $\mathfrak{o}_F$

$$t_{E/F}(\mathfrak{o}_E)_{\mathfrak{p}} = t_{E/F}(\mathfrak{o}_{E,\mathfrak{p}}) = \sum_{\mathfrak{P}|\mathfrak{p}} t_{E_{\mathfrak{P}}|F_{\mathfrak{p}}}(\mathfrak{o}_{E,\mathfrak{P}}).$$

Thus, in seeking to determine $t_{E/F}(\mathfrak{o}_E)$, we may, without loss of generality, place ourselves in the situation of a finite separable extension $L/K$ of fields which are complete with respect to a discrete absolute value.

We shall call $L/K$ a *tame extension* or say that $L/K$ is *tamely ramified* iff $(e(L/K), p) = 1$, where as previously $p$ denotes the characteristic of $\mathfrak{o}/\mathfrak{p}$, assumed to be non-zero.

**Theorem 26.** *If $\mathfrak{P}$ denotes the unique prime ideal of $\mathfrak{o}_L$ above $\mathfrak{p}$, then $\mathfrak{P}^{e-1}$ divides the different $\mathcal{D}(L/K)$. Furthermore the following conditions are equivalent*

(*a*) *$L$ is a tame extension of $K$;*
(*b*) *$t_{L/K}(\mathfrak{o}_L) = \mathfrak{o}_K$;*
(*c*) *$\mathcal{D}(L/K) = \mathfrak{P}^{e-1}$.*

*Proof.* Choose $N \supset L$ with $N/K$ finite and Galois, and let $\mathfrak{q}$ denote the unique prime ideal above $\mathfrak{P}$ in $\mathfrak{o}_N$. Then $t_{L/K}(a) = \sum a^\sigma$ where the sum runs over the full set of embeddings $\sigma$ of $L$ into $N$ over $K$. If $a \in \mathfrak{P}$, then each $a^\sigma$ lies in $\mathfrak{q}$, and so $t_{L/K}(a) \in \mathfrak{q} \cap K = \mathfrak{p}$; this then shows

$$t_{L/K}(\mathfrak{P}) \subset \mathfrak{p}. \tag{3.10}$$

It therefore follows that $t_{L/K}(\mathfrak{P}\mathfrak{p}^{-1}) \subset \mathfrak{o}$, and this in turn implies that $\mathfrak{P}^{1-e} = \mathfrak{P}\mathfrak{p}^{-1} \subset \mathcal{D}^{-1}(L/K)$. We therefore deduce that $\mathcal{D}(L/K) \subset \mathfrak{P}^{e-1}$, which proves the first part of the theorem.

Next we show that (*b*) and (*c*) are equivalent. If $t_{L/K}(\mathfrak{o}_L) = \mathfrak{o}$; then by $K$-linearity $t_{L/K}(\mathfrak{p}^{-1}\mathfrak{o}_L) = \mathfrak{p}^{-1}$, and so $\mathfrak{p}^{-1}\mathfrak{o}_L = \mathfrak{P}^{-e} \supsetneqq \mathcal{D}^{-1}(L/K) \supset \mathfrak{P}^{1-e}$. Since $\mathcal{D}(L/K)$ is an $\mathfrak{o}_L$-ideal, this implies that $\mathcal{D}(L/K) = \mathfrak{P}^{e-1}$. Conversely, if $t_{L/K}(\mathfrak{o}_L) \subset \mathfrak{p}$, then $t_{L/K}(\mathfrak{p}^{-1}\mathfrak{o}_L) \subset \mathfrak{o}$, which implies $\mathcal{D}^{-1}(L/K) \supset \mathfrak{P}^{-e}$, i.e. $\mathfrak{P}^e|\mathcal{D}(L/K)$.

Now let $M$ denote the maximal non-ramified extension of $K$ in $L$. Trivially we know that $t_{M/K}(\mathfrak{o}_M) \subset \mathfrak{o}$; however, using the interpretation of the trace as the trace of the associated linear map, we know that for $u \in \mathfrak{o}_M$, $\overline{t_{M/K}(a)} = t_{k_M/k}(\bar{a})$; since $k_M/k$ is separable $t_{k_M/k}(k_M) = k$, and it follows that $t_{M/K}(\mathfrak{o}_M) = \mathfrak{o}$. Therefore, by (3.10) together with the tower formula for ramification indices, it suffices to establish the equivalence of (*a*) and (*b*) in the case when $L/K$ is totally ramified. Assuming this to be the case, then, since $\mathfrak{o}_L = \mathfrak{o} + \mathfrak{P}$,

$$t_{L/K}(\mathfrak{o}_L) = t_{L/K}(\mathfrak{o}) + t_{L/K}(\mathfrak{P}) = e\mathfrak{o} + t_{L/K}(\mathfrak{P})$$

and the right-hand term is $\mathfrak{o}$, resp. is contained in $\mathfrak{p}$, if $p \nmid e$, resp. if $p|e$.

For future reference we record

**Corollary.** *If $L/K$ is tame, then $t_{L/K}(\mathfrak{P}) = \mathfrak{p}$.*

*Proof.* If $\mathfrak{p} = \pi\mathfrak{o}$, then

$$t_{L/K}(\mathfrak{P}) \supset t_{L/K}(\pi\mathfrak{o}_L) = \pi t_{L/K}(\mathfrak{o}_L) = \pi\mathfrak{o} = \mathfrak{p}.$$

The result then follows from (3.10).

□

We conclude this section by considering a suggestive example of a "potentially wild" Kummer extension.

**(3.11)** *Let $K = \mathbb{Q}_p[p]$ and let $\mathfrak{p}$ denote the unique prime ideal of $\mathfrak{o}$. Suppose $a \in \mathfrak{o}^*$ and let $L = K(a^{1/p})$. Writing $\mathfrak{P}$ for the unique prime ideal of $\mathfrak{o}_L$:*

(*i*) $L = K$ *if* $v_\mathfrak{p}(a^{p-1} - 1) > p$.
(*ii*) $e_\mathfrak{P}(L/K) = 1$ *and* $f_\mathfrak{P}(L/K) = p$ *if* $v_\mathfrak{p}(a^{p-1} - 1) = p$.
(*iii*) $e_\mathfrak{P}(L/K) = p$ *and* $f_\mathfrak{P}(L/K) = 1$ *if* $v_\mathfrak{p}(a^{p-1} - 1) < p$.

*Proof.* Let $\zeta$ denote a primitive $p$th root of unity in $K$. From (3.1) we know that $K/\mathbb{Q}_p$ is totally ramified of degree $p-1$ and that $\mathfrak{p} = (1-\zeta)\mathfrak{o}$. As shown in the remark on Kummer theory after the proof of (2.28), $(L : K) = p$ or 1, and so (*i*)–(*iii*) are the only possibilities. Replacing $a$ by $a^{p-1}$ we may assume, without loss of generality, that $a = 1 \bmod \mathfrak{p}$. Consider a solution $x$ in $L$ to

$$\text{(3.11.}a\text{)} \qquad a = (1+x)^p = 1 + x^p + \sum_{i=1}^{p-1} \binom{p}{i} x^i.$$

For notational simplicity we extend $v_\mathfrak{p}$ from $K^*$ to $L^*$ by the rule $v_\mathfrak{p}(b) = \frac{1}{e}v_\mathfrak{P}(b)$ for $b \in L^*$; so that $\mathfrak{p}$ will be non-ramified in $L/K$ iff $v_\mathfrak{p}(L^*) = \mathbb{Z}$. We now return to (3.11.$a$). Since $a \equiv 1 \bmod \mathfrak{P}$ we conclude that $x \in \mathfrak{P}$, and so we may re-write (3.11.$a$) as

$$\text{(3.11.}b\text{)} \qquad a = 1 + x^p + pxu$$

with $u \in 1 + \mathfrak{P}$.

If $v_\mathfrak{p}(a-1) < p$ then, since $v_\mathfrak{p}(p) = p - 1$, we deduce immediately from (3.11.$b$) that $pv_\mathfrak{p}(x) = v_\mathfrak{p}(a-1)$ and so $0 < v_\mathfrak{p}(x) < 1$. This then shows that $\mathfrak{p}$ ramifies in $L/K$ and more precisely $e_\mathfrak{P}(L/K) = (L : K) = p$.

Suppose now that $v_\mathfrak{p}(a - 1) \geq p$. Using the transformation $X = (1-\zeta)Y$ for indeterminates $X$ and $Y$, we let $g$ be the polynomial over $K$ such that

$$\begin{aligned}(1+X)^p - a &= (1+(1-\zeta)Y)^p - a \\ &= (1-\zeta)^p g(Y).\end{aligned}$$

Then, by the binomial theorem, we have in $k = \mathfrak{o}/\mathfrak{p}$

$$\overline{g}(Y) = Y^p + sY - t \tag{3.12}$$

where $t$ denotes the residue class of $(a-1)/(1-\zeta)^p$ in $k$, and $s$ is the residue class of the unit $p(1-\zeta)/(1-\zeta)^p$. We shall show below that

$$s = -1. \tag{3.12.a}$$

If now $v_{\mathfrak{p}}(a-1) > p$ then $t = 0$ and so $Y^p - Y$ splits completely over $k$ and is separable. By Hensel's Lemma $g(Y)$ splits completely over $K$, hence so does $(1+X)^p - a$. This establishes (*ii*). Next, if $v_{\mathfrak{p}}(a-1) = p$ then $t \in k^*$. Now $Y^p - Y - t$ is irreducible, separable over $k$, and is called an Artin-Schreier polynomial. Indeed, if $y$ is a root, then $y^p = y + t$, so that $y^{p^i} = y + it$. In other words the roots of $\overline{g}(Y)$ form a simple orbit of size $p$ under Frobenius. The result follows now from Theorem 23.

To prove (3.12.$a$) let

$$f(X) = X^{p-1} + X^{p-2} + \cdots + X + 1.$$

This has the powers $\zeta^j$ $(j = 1, \ldots, p-1)$ as its roots. Therefore

$$p = f(1) = \prod_j (1 - \zeta^j). \tag{3.13}$$

Thus

$$\frac{p(1-\zeta)}{(1-\zeta)^p} = \prod_{j=1}^{p-1} \left( \frac{(1-\zeta^j)}{(1-\zeta)} \right) = \prod_{j=1}^{p-1} \left( \sum_{k=0}^{j-1} \zeta^k \right).$$

But $\zeta^k \equiv 1 \bmod \mathfrak{p}$. Hence

$$\frac{p(1-\zeta)}{(1-\zeta)^p} \equiv (p-1)! \bmod \mathfrak{p}$$

and, of course,

$$(p-1)! \equiv -1 \bmod (p)$$

which gives (3.12.$a$).

□

## §4 Ramification in Galois extensions

We keep the notation of the previous section; thus, in particular, $K$ is complete with respect to a discrete absolute value $u$, which is associated with the valuation $v$; the residue class field $k$ of the valuation ring $\mathfrak{o}$ is always assumed to be finite with $p$ denoting its characteristic. Throughout this section $L$ denotes a finite Galois extension of $K$, $\Delta = \mathrm{Gal}(L/K)$, and we write $v_L$ for the valuation associated with the unique prime ideal $\mathfrak{P}$ of $\mathfrak{o}_L$ above the prime ideal $\mathfrak{p}$ of $\mathfrak{o}$.

The principal theme of this section is the study of certain naturally defined subgroups of $\Delta$; these subgroups will be seen to contain a considerable amount of information about the ramification of $\mathfrak{P}$ in $L/K$. We shall then turn the situation around and use our results to obtain detailed information on the group structure of $\Delta$.

We define a group homomorphism $\rho: \Delta \to \mathrm{Gal}(k_L/k)$ by the rule $\bar{a}^{\rho(\delta)} = \bar{a}^\delta$ for $a \in \mathfrak{o}_L$, where as previously $\bar{a}$ denotes $a \bmod \mathfrak{P}$. Note that this action is well-defined since $\mathfrak{P}^\delta = \mathfrak{P}$ for all $\delta \in \Delta$. We let $\Delta_0$ denote $\ker(\rho)$, so that

(4.1) $$\Delta_0 = \{\delta \in \Delta \mid a^\delta \equiv a \bmod \mathfrak{P} \quad \text{for all } a \in \mathfrak{o}_L\}.$$

$\Delta_0$ will be called the *inertia group* of $L/K$.

**Theorem 27.** *$\rho$ is surjective, and so the following sequence is exact*

$$1 \to \Delta_0 \to \Delta \to \mathrm{Gal}(k_L/k) \to 0.$$

*Also the maximal non-ramified extension $L_0$ of $K$ in $L$ is the fixed field of $\Delta_0$.*

*Proof.* The fixed field $k_L^\Delta$ is of the form $k[m]$, $(m, \mathbf{N}\mathfrak{p}) = 1$. As $k[m] \subset k_L$ also $K[m] \subset L$ (cf. (3.2.*d*)), and we know that $k[m]$ is the residue class field of $K[m]$ (cf. (3.2.*c*)). Let $\eta$ be a primitive $m$th root of unity. Then $\bar{\eta} \in k_L^\Delta$, i.e. $\bar{\eta}^\delta = \bar{\eta}$, all $\delta \in \Delta$. As $X^m - 1$ is separable over $k$, this implies that $\eta^\delta = \eta$ for all $\delta \in \Delta$, i.e. $K[m] \subset L^\Delta$, so that $K[m] = K$. But (cf. (3.2.*b*)), $(K[m] : K) = (k[m] : k)$. Thus $k_L^\Delta = k$, whence $\rho$ is surjective.

Apply this to $L_0$ in place of $L$. $L_0$ is always Galois over $K$, being of the form $K[m]$ (by Theorem 25). As $(k_{L_0} : k) = (L_0 : K)$ (cf. (3.2.*b*)) and $k_{L_0} = k_L$, $\rho_0$ will be an isomorphism

$$\mathrm{Gal}(L_0/K) \simeq \mathrm{Gal}(k_L/k).$$

This then implies that $L_0 = L^{\Delta_0}$.

□

**(4.2) Corollary.**

(*a*) $|\Delta_0| = e(L/K)$;

(*b*) *$L/K$ is non-ramified iff $|\Delta_0| = 1$.*

*Proof.* (*a*) follows at once from the three equalities $e(L/K)f(L/K) = (L : K)$ (cf. (1.14)); $f(L/K) = |\mathrm{im}(\rho)|$; $|\Delta_0||\mathrm{im}(\rho)| = (L : K)$.

(*b*) is immediate from the Theorem.

□

From Galois theory of finite fields we know that the field extension $k_L/k$ is always Galois and that the field automorphism $x \mapsto x^{|k|}$ is a distinguished generator of $\mathrm{Gal}(k_L/k)$; we call this the *Frobenius automorphism* of $k_L/k$. By Theorem 27 there exists $\delta \in \Delta$ with the property that

$$a^\delta \equiv a^{|k|} \bmod \mathfrak{P}$$

for all $a \in \mathfrak{o}_L$. When $L/K$ is non-ramified, this is unique in $\Delta$, since $\rho$ is then injective; we call it the Frobenius element of $L/K$, and we denote it by $\phi(L/K)$.

Let $\lambda$ denote a uniformising parameter of $L$; from Theorem 24, recall that

$$\mathfrak{o}_L = \mathfrak{o}_{L_0}[\lambda]. \tag{4.3}$$

If $\delta \in \Delta$, then $\lambda^\delta$ is also a uniformising parameter for $L$, and so $\lambda^{1-\delta} \in \mathfrak{o}_L^*$ where $\lambda^{1-\delta}$ is short for $\lambda/\lambda^\delta$. We write $\theta(\delta)$ for the residue class $\lambda^{1-\delta} \bmod \mathfrak{P}$.

We assert that the restriction of $\theta$ to $\Delta_0$ is a group homomorphism

$$\theta\colon \Delta_0 \to k_L^*. \tag{4.4}$$

First note that from the definition of $\Delta_0$: $\theta(\sigma)^\delta = \theta(\sigma)$ for $\sigma \in \Delta$ and $\delta \in \Delta_0$; the multiplicativity of $\theta$ on $\Delta_0$ then follows from the congruences mod $\mathfrak{P}$

$$\theta(\sigma\delta) \equiv \frac{\lambda}{\lambda^{\sigma\delta}} \equiv \frac{\lambda}{\lambda^\delta}\left(\frac{\lambda}{\lambda^\sigma}\right)^\delta \equiv \theta(\delta)\theta(\sigma)^\delta \equiv \theta(\sigma)\theta(\delta).$$

In fact $\theta$ is independent of the choice of uniformising parameter, since for $b \in \mathfrak{o}_L^*$, $\delta \in \Delta_0$, $b^{1-\delta} \equiv 1 \bmod \mathfrak{P}$, and so

$$(\lambda b)^{1-\delta} \equiv \lambda^{1-\delta} b^{1-\delta} \equiv \lambda^{1-\delta} \bmod \mathfrak{P}.$$

We write $\Delta_1$ for $\ker\theta$, and we refer to $\Delta_1$ as the wild inertia group of $L/K$. It is also called the ramification group, or the first ramification group. There is in fact a whole hierarchy of such groups (see [Se]).

**Theorem 28.**

(*i*) *For $\sigma \in \Delta_0$ the following statements are equivalent.*

- (*a*) $\sigma \in \Delta_1$;
- (*b*) $\lambda^\sigma \equiv \lambda \bmod \mathfrak{P}^2$ *for any uniformising parameter of $L$;*
- (*c*) *For all $i \geq 0$ and for all $\lambda_i \in L$ with $\lambda_i \mathfrak{o}_L = \mathfrak{P}^i$,*

$$\lambda_i^\sigma \equiv \lambda_i \bmod \mathfrak{P}^{i+1},$$

- (*d*) *For all $a \in \mathfrak{o}_L$, $a^\sigma \equiv a \bmod \mathfrak{P}^2$,*

(e) *For all* $b \in L^*$, $b^\sigma/b \equiv 1 \bmod \mathfrak{P}$.

(ii) *Furthermore* $\Delta_1$ *is the unique, hence normal, p-Sylow subgroup of* $\Delta_0$, *and* $\Delta_0/\Delta_1$ *is a cyclic group whose order divides* $\mathbf{N}\mathfrak{P} - 1$; $\Delta_1$ *is normal in* $\Delta$.

In the sequel we shall write $L_1$ for $L^{\Delta_1}$. From part (*ii*) and (1.14) $(ef = n)$ we have

**Corollary.**

(a) *Writing* $e(L/K) = e'p^r$, *with* $(e', p) = 1$, *we have*

$$(4.5)\quad (L_1 : L_0) = e(L_1/L_0) = e(L_1/L) = e' \quad (L : L_1) = e(L/L_1) = p^r.$$

(b) *In particular* $L_1/K$ *is tamely ramified,* $L/L_1$ *is totally and wildly ramified.*

(c) $L/K$ *is tamely ramified iff* $\Delta_1 = \{1\}$.

(When $p|e(L/K)$ we say that $L/K$ is *wildly* ramified.)

*Proof of Theorem 28.* Suppose $\sigma \in \Delta_1$. We shall first establish (*e*) from (*a*). We can write $b$ in the form $\eta\lambda^i$, $\lambda$ a uniformising parameter of $L$, $i \in \mathbb{Z}$, $\eta \in \mathfrak{o}_L^*$. Then $\eta^{\sigma-1} \equiv 1 \bmod \mathfrak{P}$ as $\sigma \in \Delta_0$. Also $(\lambda_i)^{\sigma-1}$ $(\lambda^{\sigma-1})^i \equiv 1 \bmod \mathfrak{P}$, as $\sigma \in \ker\theta$.

Next note that (*e*) trivially implies (*c*) by putting $b = \lambda_i$ and by multiplying the congruence $\lambda_i^\sigma/\lambda_i \equiv 1 \bmod \mathfrak{P}$ through by $\lambda_i$. (*b*) is a special case of (*c*). But if $\lambda^\sigma = \lambda + \beta$, $\beta \in \mathfrak{P}^2$, then $\lambda^{\sigma-1} = 1 + \beta\lambda^{-1} \in 1 + \mathfrak{P}$. Thus $\sigma \in \Delta_1$. We have established the equivalence of all conditions except (*d*).

To derive (*d*) from the rest recall (4.3) which implies in particular that any element $a$ of $\mathfrak{o}_L$ can be written as $a = c + y$, $y \in \mathfrak{P}$, $c \in \mathfrak{o}_{L_0}$. As $\sigma \in \Delta_0$, $c^\sigma - c = 0$. By condition (*c*) $y^\sigma - y \in \mathfrak{P}^2$. So (*d*) is verified. Conversely, (*d*) trivially implies (*b*).

From the very definition of $\theta$ it follows that $\Delta_0/\Delta_1 \cong \mathrm{Im}\ \theta \subset k_L^*$, and hence $[\Delta_0 : \Delta_1]\ |\mathbf{N}\mathfrak{P} - 1$. Since $\Delta_1$ is normal in $\Delta_0$ by definition, it will suffice to show that $\Delta_1$ is a $p$-group. To this end we suppose for contradiction that there is an element $\sigma \in \Delta_1$ with order $r > 1$, $p \nmid r$. By (*b*) $\lambda^\sigma = \lambda(1 + \alpha)$ for some $\alpha \in \mathfrak{P}$. Since $\alpha \neq 0$, we may suppose $\alpha \in \mathfrak{P}^t$, $\alpha \notin \mathfrak{P}^{t+1}$ with $t \geq 1$. Then $\lambda^{\sigma^2} = \lambda(1 + \alpha)(1 + \alpha^\sigma)$, and similarly

$$\lambda^{\sigma^p} = \lambda \prod_{i=0}^{p-1} (1 + \alpha^{\sigma^i}).$$

By $(c)$, $\alpha^{\sigma^i} \equiv \alpha \bmod \mathfrak{P}^{t+1}$ for each such $i$, and so

$$\lambda^{\sigma^p} \equiv \lambda(1+\alpha)^p \equiv \lambda \bmod \mathfrak{P}^{t+1}.$$

Hence for any integer $s$, $\lambda^{\sigma^{sp}} = \lambda(1+\beta_s)$ for some $\beta_s \in \mathfrak{P}^{t+1}$. On the other hand $\lambda^{\sigma^r} = \lambda$, and so $\lambda^{\sigma^{rR}} = \lambda$ for any integer $R$. Choosing $R$ and $s$ such that $rR + sp = 1$, we conclude that

$$\lambda(1+\alpha) = \lambda^{\sigma} = \lambda^{\sigma^{rR+sp}} = \lambda^{\sigma^{sp}} = \lambda(1+\beta_s).$$

This, however, is absurd since $\alpha \notin \mathfrak{P}^{t+1}$, while $\beta_s \in \mathfrak{P}^{t+1}$.

The fact that $\Delta_1$ is normal in $\Delta$ then follows trivially from the above. □

(4.7) As an illustrative example of the two intermediate extensions $L_0$, $L_1$, consider the case $K = \mathbb{Q}_p$, $L = \mathbb{Q}_p[rp^n]$ with $p \nmid r$. From Theorem 25 we know that $\mathbb{Q}_p[r]$ is non-ramified over $\mathbb{Q}_p$; on the other hand we have seen in (3.1) that $\mathbb{Q}_p[p^n]/\mathbb{Q}_p$ is totally ramified of degree $p^{n-1}(p-1)$; thus $\mathbb{Q}_p[p]/\mathbb{Q}_p$ is totally ramified of degree $p-1$. In summary, in this case we have shown $L_0 = \mathbb{Q}_p[r]$, $L_1 = \mathbb{Q}_p[rp]$.

Next we prove a result which gives us a very strong grip on the structure of totally, tamely ramified Galois extensions of $K$.

**(4.8)** *Let $L/K$ be a Galois extension with $L_0 = K$ and $L = L_1$. Put $e = e(L/K)$; so $(p, e) = 1$. Then $K = K[e]$, i.e. $K$ contains the primitive* $e$-th *roots of unity, and $L = K(\alpha^{1/e})$ for some uniformising parameter $\alpha$ of $K$.*

*Proof.* As $\Delta_1 = 1$ and $\Delta = \Delta_0$, the homomorphism $\theta$ of (4.4) is now an isomorphism of $\Delta$ with the subgroup $\mu_{k,e}$ of $k^*$ of order $e$. By (II.3.26) $K^*$ contains a group $\mu_{K,e}$ mapping isomorphically onto $\mu_{k,e}$ under the residue class map $x \mapsto \overline{x}$. In particular there exists an isomorphism

(4.8.$a$) $$\psi \colon \Delta \cong \mu_{K,e} \quad \text{with } \overline{\psi} = \theta.$$

By Kummer theory $L^*$ contains an element $\beta$ with

(4.8.$b$) $$\beta^{1-\delta} = \psi(\delta), \quad \forall \delta \in \Delta.$$

This is easy to see by considering the "Lagrange resolvents"

$$\beta_i = \sum_{\delta \in \Delta} a_i^{\delta} \psi(\delta)^{-1}$$

where $\{a_i\}$ is a basis of $L/K$. As $\det(a_i^{\delta}) \neq 0$ some $\beta_{i_0}$ is non-zero, and (4.8.$b$) will then be satisfied by $\beta = \beta_{i_0}$. Moreover, as $\beta^{\delta} = \beta$ implies $\delta = 1$ it follows that

(4.8.$c$) $$L = K(\beta)$$

Next observe that $(\beta^e)^{1-\delta} = \psi(\delta)^e = 1$. Thus $\beta = \alpha^{1/e}$, $\alpha \in K^*$. We may replace $\beta$ by $\beta c$, $c \in K^*$. In other words we may replace $\alpha$ by $\alpha c^e$. By a suitable choice of $c$, we may assume that

$$\text{(4.8.}d\text{)} \qquad 0 \le v < e, \quad v = v_K(\alpha).$$

Thus $\beta = \alpha^{1/e} = \lambda^v \eta$, $\lambda$ a uniformising parameter of $L$, $\eta \in \mathfrak{o}_L^*$. Hence from the definition of $\theta$,

$$\beta^{1-\delta} \bmod \mathfrak{P} = \theta(\delta)^v,$$

i.e. by (4.8.$a$),

$$\beta^{1-\delta} = \psi(\delta)^v,$$

while by (4.8.$b$),

$$\beta^{1-\delta} = \psi(\delta).$$

Hence $v \equiv 1 \bmod (e)$ which in conjunction with (4.8.$d$) gives $v = 1$. Thus indeed $\beta$ is a uniformising parameter of $L$, and so the result follows by (4.8.$c$).

□

We shall next study the behaviour of the groups $\Delta_i = \Delta_i(L/K)$ under change of extension.

**Theorem 29.**

($a$) *Let $\Sigma$ be a subgroup of $\Delta = \mathrm{Gal}(L/K)$, with $N = L^\Sigma$ as fixed field. Identify $\Sigma = \mathrm{Gal}(L/N)$. The inertia group $\Sigma_0$ resp. the wild inertia group $\Sigma_1$ of $L/N$ is given by*

$$\Sigma_i = \Sigma \cap \Delta_i \quad (i = 0, 1)$$

*If $L/K$ is non-ramified then $\phi(L/N) = \phi(L/K)^{f(N/K)}$.*

($b$) *Suppose now that $\Sigma$ is normal in $\Delta$, and write $\Omega = \Delta/\Sigma$. Identify $\Omega = \mathrm{Gal}(N/K)$. Then*

$$\Omega_i = \Delta_i\Sigma/\Sigma \quad \textit{for i=0,1}$$

*and if $L/K$ is non-ramified then $\phi(L/K)$ has image $\phi(N/K)$ in $\Omega$.*

*Proof.* For $\sigma \in \Sigma$ we have $\sigma \in \Sigma_i$ iff $a^\sigma \equiv a \bmod \mathfrak{P}^{i+1}$ for all $a \in \mathfrak{o}_L$ iff $\sigma \in \Delta_i \cap \Sigma$. The result on Frobenius elements follows immediately from the fact that $|k_N| = |k|^{f(N/K)}$.

Now we prove ($b$). Let $\mathfrak{q} = \mathfrak{P} \cap N$, and let $\delta \in \Delta_0$. Then $a^\delta - a \in \mathfrak{P}$ for all $a \in \mathfrak{o}_L$; hence $a^\delta - a \in \mathfrak{P} \cap N = \mathfrak{q}$ when $a \in \mathfrak{o}_N$; this implies that $\delta|_N \in \Omega_0$, and so we have shown $\Delta_0\Sigma/\Sigma \subset \Omega_0$. However, by ($a$) $\Delta_0 \cap \Sigma = \Sigma_0$; thus, on comparing orders, we have

$$\text{(4.9)} \qquad \begin{aligned}(\Delta_0\Sigma : \Sigma) = (\Delta_0 : \Sigma_0) &= e(L/K)e(L/N)^{-1} \\ &= e(N/K) = |\Omega_0|\end{aligned}$$

which implies that $\Delta_0\Sigma/\Sigma = \Omega_0$.

Next we use the characterisation of the wild inertia group as the $p$-Sylow group of the inertia group. Clearly $\Delta_1\Sigma/\Sigma$ is a $p$-subgroup of $\Omega_0 = \Delta_0\Sigma/\Sigma$, hence is contained in $\Omega_1$. But $\Delta_0\Sigma/\Delta_1\Sigma$ is of order dividing $[\Delta_0 : \Delta_1]$, i.e. prime to $p$. Hence $\Delta_1\Sigma/\Sigma = \Omega_1$.

The result for Frobenius elements follows at once from the fact that the Frobenius automorphism of $k_L/k$ restricts to the Frobenius automorphism of $k_N/k$.

□

By putting together Theorems 28 and 29, we can now deduce a powerful result on the group structure of $\Delta$.

**(4.10)** *The subgroups $\Delta_1 \subset \Delta_0$ are both normal in $\Delta$; $\Delta/\Delta_0$ is cyclic, generated by the coset which identifies with the Frobenius element of $L_0/K$; $\Delta_0/\Delta_1$ is cyclic of order prime to $p$, and $\Delta_1$ is a $p$-group. In particular $\Delta$ is a soluble group.*

We now seek to apply the preceding local results to a Galois extension $E/F$ of number fields, whose Galois group we denote by $\Gamma$. Let $\mathfrak{p}$ denote a prime ideal of $\mathfrak{o}_F$, and let $\mathfrak{P}$ denote a prime ideal of $\mathfrak{o}_E$ above $\mathfrak{p}$. We set $K = F_\mathfrak{p}$, $L = E_\mathfrak{P}$ and we take $E$ embedded in $L$. From §1 recall that $\Gamma_\mathfrak{P}$, the decomposition group of $\mathfrak{P}$ in $L/K$, i.e.

$$\Gamma_\mathfrak{P} = \{\gamma \in \Gamma \mid \mathfrak{P}^\gamma = \mathfrak{P}\}$$

identifies with $\Delta = \mathrm{Gal}(L/K)$ (cf. Theorem 21). For $i = 0, 1$ we write

$$\Gamma_{\mathfrak{P},i}(E/F) = \Delta_i(L/K).$$

We shall usually abbreviate the left-hand term to $\Gamma_{\mathfrak{P},i}$. The subgroups $\Gamma_{\mathfrak{P},i}$ have the following useful characterisation:

**(4.11)** $$\Gamma_{\mathfrak{P},i} = \{\gamma \in \Gamma_\mathfrak{P} \mid a^\gamma \equiv a \bmod \mathfrak{P}^{i+1} \quad \text{for all } a \in \mathfrak{o}_E\}.$$

*Proof.* First observe that if $\delta \in \Delta_i$, then for any $a \in \mathfrak{o}_E$,

$$a^\delta - a \in \mathfrak{P}^{i+1}\mathfrak{o}_L \cap \mathfrak{o}_E = \mathfrak{P}^{i+1}.$$

Conversely (see II.4.4.$a$) we can find an $\mathfrak{o}_K$ basis $\{x_j\}$ of $\mathfrak{o}_L$, which consists of elements of $\mathfrak{o}_E$. If, for given $i = 0, 1$, $\gamma \in \Gamma$ belongs to the right-hand side of (4.11), then for $a = \sum b_j x_j$, $b_j \in \mathfrak{o}_K$,

$$a^\gamma - a = \sum b_j(x_j^\gamma - x_j) \in \mathfrak{P}^{i+1}\mathfrak{o}_K \subset \mathfrak{P}^{i+1}\mathfrak{o}_L$$

and so indeed $\gamma$ lies in $\Delta_i$.

□

The ramification index $e_{\mathfrak{p}}(E/F)$, the residue class degree $f_{\mathfrak{p}}(E/F)$ – both for prime ideals of $\mathfrak{o}_E$ lying above a prime ideal $\mathfrak{p}$ of $\mathfrak{o}_F$ – as well as the number $g_{\mathfrak{p}}(E/F)$ of prime factors of $\mathfrak{p}$ in $\mathfrak{o}_L$ are now given in terms of $\Gamma$ and its subgroups.

**(4.12) Corollary.**

$$|\Gamma_{\mathfrak{P},0}| = e_{\mathfrak{p}}(E/F); \;\; [\Gamma_{\mathfrak{P}} : \Gamma_{\mathfrak{P},0}] = f_{\mathfrak{p}}(E/F); \;\; (\Gamma : \Gamma_{\mathfrak{P}}) = g_{\mathfrak{p}}(E/F).$$

**(4.12.*a*) Corollary.** *If $p$ is wildly ramified in $E/F$ (i.e. if $p | e_{\mathfrak{p}}(E/F)$ for some prime ideal $\mathfrak{p}$ of $\mathfrak{o}_E$ above $p$) then $p \mid |\Gamma|$.*

If $\mathfrak{p}$ does not ramify in $E/F$, then we write $(\mathfrak{P}, E/F)$ for the element $\phi(E/K)$, when it is viewed as an element of $\Gamma$. We call $(\mathfrak{P}, E/F)$ the *Frobenius element* of $\mathfrak{P}$ in $E/F$; the definition of this element may easily be reformulated as:

**(4.13)** *If $\mathfrak{p}$ is non-ramified in $E/F$, then there exists a unique element $(\mathfrak{P}, E/F) \in \Gamma_{\mathfrak{P}}$ such that for all $a \in \mathfrak{o}_E$*

$$a^{(\mathfrak{P},E/F)} \equiv a^{\mathbf{N}\mathfrak{p}} \bmod \mathfrak{P}.$$

*Moreover $(\mathfrak{P}, E/F)$ generates $\Gamma_{\mathfrak{P}}$.*

In due course we shall see that the notion of the Frobenius element of a prime ideal is an absolutely fundamental and key tool in algebraic number theory. Amongst other things, it plays an important role in some of the nicest proofs of the Law of Quadratic Reciprocity and in the definition of $L$-functions.

The following result describes the main properties of Frobenius elements with respect to change of extension.

**Theorem 30.**

(*a*) $\gamma^{-1}\Gamma_{\mathfrak{P}}\gamma = \Gamma_{\mathfrak{P}^\gamma}$; $\gamma^{-1}\Gamma_{\mathfrak{P},i}\gamma = \Gamma_{\mathfrak{P}^\gamma,i}$ *for* $i = 0, 1$; *and*

$$\gamma^{-1}(\mathfrak{P}, E/F)\gamma = (\mathfrak{P}^\gamma, E/F).$$

*Thus the Frobenius elements of the prime ideals above $\mathfrak{p}$ determine a conjugacy class of $\Gamma$. If $\Gamma$ is abelian, then all these entities depend only on $\mathfrak{p}$ and not on the particular prime ideal $\mathfrak{P}$ above $\mathfrak{p}$. [In this situation we write $\Gamma_{\mathfrak{p}}$, $(\mathfrak{p}, E/F)$ etc.].*

(*b*) *Let $\Sigma$ denote a subgroup of $\Gamma$, let $G$ denote the fixed field $E^\Sigma$ and identify $\Sigma$ with* $\mathrm{Gal}(E/G)$. *Then $\Sigma_{\mathfrak{P}} = \Sigma \cap \Gamma_{\mathfrak{P}}$, $\Sigma_{\mathfrak{P},i} = \Sigma \cap \Gamma_{\mathfrak{P},i}$ for $i = 0, 1$; moreover, if $\mathfrak{q} = \mathfrak{P} \cap G$, then*

$$(\mathfrak{P}, E/G) = (\mathfrak{P}, E/F)^{f_{\mathfrak{q}}(G/F)}.$$

(c) *Suppose further that $\Sigma$ is normal in $\Gamma$, and identify $\Omega = \Gamma/\Sigma$ with* $\mathrm{Gal}(G/F)$. *Then $\Gamma_{\mathfrak{P}}\Sigma/\Sigma = \Omega_{\mathfrak{q}}$; $\Gamma_{\mathfrak{P},i}\Sigma/\Sigma = \Omega_{\mathfrak{q},i}$ for $i = 0, 1$; and* $(\mathfrak{P}, E/F)|_G = (\mathfrak{q}, G/F)$.

*Proof.* $(a)$ is just a straightforward application of the theory of groups acting on sets; $(b)$ follows from Theorem 29$(a)$; $(c)$ follows from Theorem 29$(b)$.

□

We shall conclude this section by using some of the above material to describe the arithmetic of the splitting fields of certain trinomials; the result which we give is due to Uchida.

Let $r$ denote a prime number, and let $f(X) = X^r - aX - b$ be an irreducible polynomial in $\mathbb{Z}[X]$ and suppose also that

(4.14) $$((r-1)a, rb) = 1.$$

Let $N$ denote the splitting field of $f$ over $\mathbb{Q}$. We shall show:

**(4.15)**

(a) $\mathrm{Gal}(N/\mathbb{Q}) \cong S_r$, *the symmetric group on $r$ letters.*

(b) *If $A_r$ denotes the alternating subgroup of $S_r$ and $K = N^{A_r}$, then $N/K$ is a non-ramified extension. (Elementary group theory of $S_r$ tells us of course that $K$ is the unique quadratic extension of $\mathbb{Q}$ in $N$).*

(c) *If $p$ ramifies in $N/\mathbb{Q}$; then $e_p(N/\mathbb{Q}) = 2$.*

Before proving this result we briefly consider possible conditions which guarantee the irreducibility of $f$. On the one hand, by (4.14) note that $f$ cannot be an Eisenstein polynomial for any prime number; on the other hand, if $a \equiv 1 \bmod (r)$ and if $b \not\equiv 0 \bmod (r)$ then the image of $f$ in $\mathbb{F}_r[X]$ is irreducible (see the proof of (3.11)). This condition will therefore guarantee that $f$ is irreducible in $\mathbb{Q}[X]$.

In order to prove (4.15) we first note that

$$Xf'(X) - rf(X) = (r-1)aX + rb.$$

Hence for any prime number $p$, the monic HCF of $\overline{f}$ and $\overline{f'}$ in $\mathbb{F}_p[X]$, $h(X)$ say, must divide the polynomial $\overline{(r-1)a}X + \overline{rb}$ (which is non-zero by (4.14)). Thus there are two possibilities: either $h = 1$, in which case $\overline{f}$ is separable; or $h$ is linear, and then

(4.16) $$\overline{f} = gh^2$$

with $gh$ separable in $\mathbb{F}_p[X]$.

We put $\Gamma = \mathrm{Gal}(N/\mathbb{Q})$. We view $\Gamma$ as a subgroup of $S_n$ via its action on the roots of $f$. In particular, since $r$ is prime and since $f$ is irreducible $\Gamma$ must contain an $r$-cycle.

As an intermediate step towards the proof of (4.15) we first show:

**(4.17)** *If $\mathfrak{P}$ is a ramified prime ideal of $N/\mathbb{Q}$ above the prime number $p$, then $\Gamma_{\mathfrak{P},0}$ has order 2, and is generated by a transposition (under the embedding $\Gamma \hookrightarrow S_r$).*

*Proof.* By Theorem 23 $\overline{f}$ cannot be separable in $\mathbb{F}_p[X]$, and so (4.16) must obtain. If $\alpha_1, \ldots, \alpha_r$ denote the distinct roots of $f$ in $N$; then

$$\prod_{i=1}^{r}(X - \alpha_i) = gh^2 \bmod \mathfrak{P}.$$

Reordering the roots if necessary, we may therefore write

$$h^2 \equiv (X - \alpha_1)(X - \alpha_2) \equiv (X - \alpha_1)^2 \bmod \mathfrak{P}$$

$$g \equiv \prod_{i=3}^{r}(X - \alpha_i) \bmod \mathfrak{P}.$$

Now choose $\sigma \in \Gamma_{\mathfrak{P},0}$, $\sigma \neq 1$; from (4.16) we know that $gh$ is separable and so the $\alpha_2, \alpha_3, \ldots, \alpha_r$ are all pairwise non-congruent mod $\mathfrak{P}$. Thus we must have $\alpha_i^\sigma = \alpha_i$ if $i \geq 3$; however, since $\sigma \neq 1$, it follows that $\alpha_1^\sigma = \alpha_2$, $\alpha_2^\sigma = \alpha_1$, thus we have shown that, as a permutation, $\sigma = (1\,2)$. □

The proof of (4.15) now follows at once. As has already been remarked, in the next chapter we shall show that $N/\mathbb{Q}$ must ramify for some prime $p$. Thus $\Gamma$, viewed as a subgroup of $S_r$, contains both a transposition and an $r$-cycle; since an arbitrary $r$-cycle and transposition always generate $S_r$, it follows that $\Gamma = S_r$, which proves $(a)$. To prove $(b)$ we note that by (4.17) $A_\Gamma \cap \Gamma_{\mathfrak{P},0} = \{1\}$; and so by Theorem 4$(b)$, $N/K$ is non-ramified. This also shows $(c)$.

□

# IV

# Classgroups and Units

In this chapter we study in detail both the classgroup and the unit group of a ring of algebraic integers: in §3 we shall determine a practical procedure for calculating classgroups; while in §4 we shall completely describe the $\mathbb{Z}$-module structure of the unit group. These two results are absolutely basic in algebraic number theory, and they will be used repeatedly in subsequent chapters.

## §1 Elementary results

This section mainly serves as a preparation and introduction for the remainder of the chapter. We start by briefly recalling a number of standard results concerning Euclidean domains.

Recall that an integral domain $\mathfrak{o}$ is said to be *Euclidean* if there exists a function $N$ from $\mathfrak{o} \setminus 0$ to the non-negative integers such that for all $a, b \in \mathfrak{o} \setminus 0$

(1.1.$a$) $N(b) \leq N(ab)$

(1.1.$b$) $a$ can be written as $qb+r$, with $q, r, \in \mathfrak{o}$ where either $N(r) < N(b)$ or $r = 0$.

The function $N$ is usually referred to as a Euclidean norm.

**(1.2)** *A Euclidean domain is a principal ideal domain.*

*Proof.* Let $\mathfrak{a}$ denote a non-zero $\mathfrak{o}$-ideal and choose $a \in \mathfrak{a} \setminus 0$ with $N(a)$

minimal. We assert that $\mathfrak{a} = a\mathfrak{o}$: indeed, the inclusion $\mathfrak{a} \supset a\mathfrak{o}$ is clear; conversely, for $b \in \mathfrak{a}$ we write $b = qa + r$ where either $r = 0$ or $N(r) < N(a)$ (by (1.1.$b$)). Since $r \in \mathfrak{a}$, by the minimality condition on $a$, it follows that $r$ must be zero, and so $b \in a\mathfrak{o}$.

□

The great advantage of Euclidean domains is that their definition is of a highly constructive nature: for instance $\mathbb{Z}$ is a Euclidean domain because $a \mapsto |a|_{\mathbb{R}}$ defines a Euclidean norm; similarly, $f \mapsto \deg(f)$ is seen to define a Euclidean norm on $K[X]$, the polynomial ring over $X$ with coefficients in a field $K$. Thus we may immediately deduce that both $\mathbb{Z}$ and $K[X]$ are principal ideal domains.

The case of the Gaussian integers is a little bit more complicated:

**(1.3)** $\mathbb{Z}[i]$ *is a Euclidean domain.*

*Proof.* For $a \in \mathbb{Z}[i]$, define $N(a) = |a|_{\mathbb{C}}$ (the modulus of $a$); we assert that $N$ is a Euclidean norm. Since $N(ab) = N(a)N(b)$, condition (1.1.$a$) is clear. Next suppose $b \in \mathbb{Z}[i]$, $b \neq 0$, and let $q$ denote a nearest point in $\mathbb{Z}[i]$ to the complex number $a/b$ in the Argand diagram.

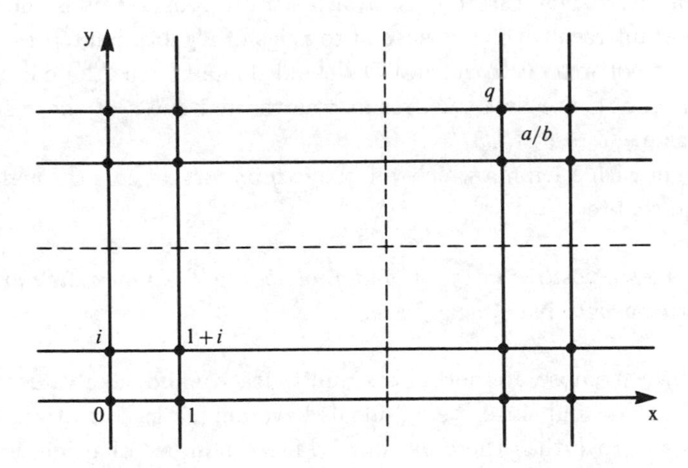

Then, by geometric considerations, we have

$$\left|q - \frac{a}{b}\right|_{\mathbf{C}} \leq \frac{\sqrt{2}}{2} \Rightarrow N(a-qb) < N(b)$$

and so $a = qb + (a - qb)$ gives a representation of $a$ which satisfies (1.1.$b$). □

The above argument can be applied to a number of rings of algebraic integers of quadratic fields. In fact it is known that for square-free negative $d$, the ring of integers of $\mathbf{Q}(\sqrt{d})$ is a principal ideal domain precisely for the nine values $d = -1, -2, -3, -7, -11, -19, -47, -67, -163$. Of these values, the ring of integers is a Euclidean domain only when $d = -1, -3, -7, -11$.

Henceforth we restrict our attention to the arithmetic situation. So now suppose that $K$ is an algebraic number field, with ring of integers $\mathfrak{o}_K$. Recall that the classgroup of $\mathfrak{o}_K$, $\mathrm{Cl}(\mathfrak{o}_K)$, is defined to be the group of fractional $\mathfrak{o}_K$-ideals modulo the subgroup of principal fractional $\mathfrak{o}_K$-ideals; thus $\mathrm{Cl}(\mathfrak{o}_K) = \{1\}$ precisely when $\mathfrak{o}_K$ is a principal ideal domain.

In the arithmetic situation it is natural to view $\mathrm{Cl}(\mathfrak{o}_K)$ as a function of $K$, and so for brevity we shall write $C_K$ for $\mathrm{Cl}(\mathfrak{o}_K)$. As we have seen, in certain special circumstances we may be able to define a Euclidean norm on $\mathfrak{o}_K$, and so show that $C_K = \{1\}$; however, this is an extremely limited method of attack.

We shall now show that $C_K$ is always a finite group; this is the second important result which is special to rings of algebraic integers, and which does not extend to general Dedekind domains (see the corollary to Theorem 55); the first such result was the finiteness of the residue class rings $\mathfrak{o}_K/\mathfrak{a}$ (see II.1.37).

We begin with a lemma which will play an important role throughout the whole chapter:

**(1.4)** *Given a positive integer $n$ and a number field $K$; then the number of $\mathfrak{o}_K$-ideals $\mathfrak{a}$ with $\mathbf{N}\mathfrak{a} \leq n$ is finite.*

*Proof.* Since the absolute norm $\mathbf{N}$ is multiplicative and take values $> 1$ on prime ideals, and since the $\mathfrak{o}_K$-ideals have unique factorisation, it is sufficient to prove that there are only a finite number of prime ideals $\mathfrak{p}$ with $\mathbf{N}\mathfrak{p} \leq n$. This follows from the fact that there are only a finite number of (rational) prime numbers $p \leq n$, that by Theorem 5 there are only a finite number of prime ideals of $\mathfrak{o}_K$ above each $p$, and that the norm of each prime ideal above $p$ is a power of $p$. □

**Theorem 31.** *The ideal classgroup $C_K$ of a number field $K$ is finite.*

*Proof.* We choose a $\mathbb{Z}$-basis $v_1, \ldots, v_n$ of $\mathfrak{o}_K$; without loss of generality we may assume $n > 1$, since $\mathbb{Z}$ has been shown to be a principal ideal domain.

Let $c$ be a given class in $C_K$; by (II.1.18) we can choose an $\mathfrak{o}_K$-ideal $\mathfrak{a}$ with class $c^{-1}$. Let $\mathcal{S}$ denote the set

$$\mathcal{S} = \{s \in \mathfrak{o}_K \mid s = \sum_{i=1}^{n} m_i v_i,\ m_i \in \mathbb{Z},\ 0 \le m_i < \mathbf{N}\mathfrak{a}^{1/n} + 1\}.$$

Then $|\mathcal{S}| \ge (\mathbf{N}\mathfrak{a} + 1)$. Since $\mathbf{N}\mathfrak{a} = [\mathfrak{o}_K : \mathfrak{a}]$, by the pigeonhole principle we can find distinct $a$, $b$ in $\mathcal{S}$ such that $a \equiv b \bmod \mathfrak{a}$; hence $(a - b) = \mathfrak{a}\mathfrak{b}$, for some $\mathfrak{o}_K$-ideal $\mathfrak{b}$; furthermore we observe that $\mathfrak{b}$ has class $c$. Denote by $\sigma_1, \ldots, \sigma_n$ the distinct embeddings of $K$ into $\mathbb{C}$. We define $\kappa$ by

$$\kappa = \prod_{j=1}^{n}\left(\sum_{i=1}^{n}|v_i^{\sigma_j}|\right);$$

thus $\kappa$ depends not only on $K$, but also on the given choice of basis.

Writing $a - b = \sum p_i v_i$, with $|p_i| \le \mathbf{N}\mathfrak{a}^{1/n} + 1$, we have

$$\begin{aligned} |N_{K/\mathbb{Q}}(a-b)| &= \left|\prod_{j=1}^{n}\left(\sum_{i=1}^{n} p_i v_i^{\sigma_j}\right)\right| \\ &\le \prod_j \left(\sum |p_i||v_i^{\sigma_j}|\right) \\ &\le (\mathbf{N}\mathfrak{a}^{1/n} + 1)^n \kappa. \end{aligned} \tag{1.5}$$

From (II.1.35) and (II.1.38) we know that $|N_{K/\mathbb{Q}}(a-b)| = \mathbf{N}\mathfrak{a} \cdot \mathbf{N}\mathfrak{b}$; and so by (1.5) we conclude that given an $\mathfrak{o}_K$-ideal $\mathfrak{a}$ with class $c^{-1}$, there exists an $\mathfrak{o}_K$-ideal $\mathfrak{b}$ with class $c$ such that

$$\mathbf{N}\mathfrak{b} \le (1 + \mathbf{N}\mathfrak{a}^{-1/n})^n \kappa.$$

However, observe that we can replace $\mathfrak{a}$ by the ideal $a\mathfrak{a}$ for any $a \in \mathfrak{o}_K \backslash 0$, with $|N_{K/\mathbb{Q}}(a)|$ arbitrarily large; we can therefore make $(1 + \mathbf{N}\mathfrak{a}^{-1/n})^n$ arbitrarily close to 1, and so, because the $\mathbf{N}\mathfrak{b}$ lie in the discrete set of positive integers, we can always find $\mathfrak{b}$ such that $\mathbf{N}\mathfrak{b} \le \kappa$. The theorem now follows immediately from (1.4).

□

The general principle of showing that any class is represented by some $\mathfrak{o}_K$-ideal whose norm is less than some bound is fundamental in the calculation of classgroups. Unfortunately, as we shall see presently, the particular bound $\kappa$ given in the proof of Theorem 31 rapidly becomes quite unworkable.

As a first example of a use of this bound consider the case $K = \mathbb{Q}(\omega)$ with $\omega$ a non-trivial cube root of unity. From (II.1.33) we know that $d_K = -3$ and that we can take $\{1, \omega\}$ as a $\mathbb{Z}$-basis of $\mathfrak{o}_K$; so in this case $\kappa = 4$. We shall show that $C_K = \{1\}$, by showing that each $\mathfrak{o}_K$-ideal with norm less than or equal to 4 is principal: indeed, since every such ideal is a product of prime ideals with the same property, it suffices to restrict attention to prime ideals with this property.

Since $-3\omega = (1-\omega)^2$, it follows that 3 is ramified, and that the unique prime ideal of $\mathfrak{o}_K$ above 3 is principal. Since $2 \nmid d_K$, 2 is unramified in $K$ and so 2 is inert by (III.2.29); hence the unique prime ideal of $\mathfrak{o}_K$ above 2 is also principal. We have therefore shown all the prime ideals in question to be principal.

To illustrate the point that the bound $\kappa$ of Theorem 31 is in general too weak we might consider the case $K = \mathbb{Q}(\sqrt{102})$, with $v_1 = 1$, $v_2 = \sqrt{102}$, so that $\kappa = |1 + \sqrt{102}|^2 = 123.2\cdots$ and so we should have to determine the behaviour of all primes with norm less than 123. It therefore becomes apparent that we now need to develop tools which will dramatically improve upon the constant $\kappa$. To this end, in the next section we introduce a number of geometric techniques concerning the intersection of lattices with symmetric convex bodies. We shall then use these methods in the third section to derive a vastly improved constant: for the case $K = \mathbb{Q}(\sqrt{102})$ we shall see that it in fact suffices to consider prime ideals of $\mathfrak{o}_K$ whose norm is less than or equal to 10.

## §2 Lattices in Euclidean space.

This section is of an entirely geometric nature. It will provide nearly all the techniques and basic results which we require for the two final arithmetic sections of this chapter.

Throughout this section $V$ denotes an $n$-dimensional $\mathbb{R}$-vector space; once and for all we fix an $\mathbb{R}$-basis $e_1, \ldots, e_n$ of $V$. Given an element $x \in V$, we write $x = \sum_{h=1}^n e_h x_h$ with $x_h \in \mathbb{R}$. $V$ is considered as endowed with the Euclidean topology with respect to this basis; however, recall that the topology so obtained is of course independent of the particular choice of basis.

We let

$$\Omega = \sum_{h=1}^{n} e_h \mathbb{Z}$$

denote the additive subgroup of $V$ generated by the $e_h$. An additive subgroup $\Lambda$ of $V$ is said to be *discrete* if the set $\Lambda \cap C$ is finite for all bounded subsets $C$ of $V$. Thus for instance given any $r \leq n$, the subgroup $\Omega_r = \sum_{h=1}^{r} e_h \mathbb{Z}$ is a discrete subgroup of $V$, since it can only have finite intersection with any bounded subset of $V$.

The discrete subgroups of $V$ are fully described by the following result:

**(2.1)** *$\Lambda$ is a discrete subgroup of $V$ iff $\Lambda$ is freely generated over $\mathbb{Z}$ by $r$ vectors which are linearly independent over $\mathbb{R}$ (hence $r \leq n$).*

*Proof.* First suppose that $\Lambda$ is freely generated by $\mathbb{R}$-linearly independent vectors $\{v_1, \ldots, v_r\}$. We then complete this set to an $\mathbb{R}$-basis $\{v_1, \ldots, v_n\}$ of $V$. If $C$ denotes a bounded subset of $V$, then the coordinates $x_i$ of all $x \in C$ are bounded, and so $\Lambda \cap C$ is indeed bounded. This shows that $\Lambda$ is discrete.

Now suppose that $\Lambda$ is discrete. We shall prove in several stages that $V$ is freely generated over $\mathbb{Z}$ by $\mathbb{R}$-linearly independent vectors: firstly we use the discreteness of $\Lambda$ to deduce that $\Lambda$ is a finitely generated $\mathbb{Z}$-module; we then use the theory of finitely generated $\mathbb{Z}$-modules to deduce that $\Lambda$ is $\mathbb{Z}$-free; lastly we show that some $\mathbb{Z}$-basis is (and hence all $\mathbb{Z}$-bases are) linearly independent over $\mathbb{R}$.

Let $\mathfrak{B} = \{v_1, \ldots, v_r\}$ denote a maximal subset of $\Lambda$ with the property that the $\{v_i\}$ are $\mathbb{R}$-linearly independent. We then define

$$P_{\mathfrak{B}} = \{\sum_{h=1}^{r} v_h \alpha_h \mid 0 \leq \alpha_h \leq 1\}.$$

$P_{\mathfrak{B}}$ is obviously bounded. By the maximality condition, given any $x \in \Lambda$ we can write $x = \sum_{h=1}^{r} v_h \lambda_h$ with $\lambda_h \in \mathbb{R}$; thus $x$ can be rewritten as

$$x = \left(\sum v_h(\lambda_h - [\lambda_h])\right) + \left(\sum v_h[\lambda_h]\right), \tag{2.2}$$

where for a real number $\lambda$ we denote by $[\lambda]$ its integer part; that is to say $[\lambda] \in \mathbb{Z}$ and $[\lambda] \leq \lambda < [\lambda] + 1$. Clearly the first term on the right-hand side of (2.2) lies in both $P_{\mathfrak{B}}$ and $\Lambda$; thus by (2.2) $\Lambda$ is generated over $\mathbb{Z}$ by $P_{\mathfrak{B}} \cap \Lambda$ and the $\{v_h\}$. Therefore $\Lambda$ is finitely generated over $\mathbb{Z}$, since $P_{\mathfrak{B}} \cap \Lambda$ is finite; moreover, $\Lambda$ is $\mathbb{Z}$-torsion free, since $V$ contains no $\mathbb{Z}$-torsion. Hence, by the theory of finitely generated modules over $\mathbb{Z}$, we know that $\Lambda$ must be $\mathbb{Z}$-free.

Let $x \in P_{\mathfrak{B}} \cap \Lambda$, $x = \sum v_h \lambda_h$. For each positive integer $j$ we write

$$x_j = \sum_h v_h(\lambda_h j - [\lambda_h j])\,; \tag{2.3}$$

thus $x_j$ lies in the finite set $P_{\mathfrak{B}} \cap \Lambda$; we can therefore find distinct integers $j$, $k$ such that $x_j = x_k$, that is to say

$$\sum_h v_h \lambda_h (j - k) = \sum_h v_h([j\lambda_h] - [k\lambda_h]).$$

Since the $v_h$ are $\mathbb{R}$-linearly independent, we may equate coefficients and deduce that for all $h$, $\lambda_h(j-k) = [j\lambda_h]-[k\lambda_h]$, which shows that $\lambda_h \in \mathbb{Q}$. Since $P_{\mathfrak{B}} \cap \Lambda$ is finite, we conclude that $\Lambda' = \sum_i^r v_h\mathbb{Z}$ has finite index in $\Lambda$. By construction the $v_1, \ldots, v_r$ are linearly independent over $\mathbb{R}$. However, any $\mathbb{Z}$-basis $u_1, \ldots, u_r$ of $\Lambda$ is the transform of $v_1, \ldots, v_r$ by a non-singular matrix over $\mathbb{Q}$; the matrix is therefore non-singular over $\mathbb{R}$ and so $u_1, \ldots, u_r$ are linearly independent over $\mathbb{R}$.

□

At a number of points in both this chapter and also in Chapter VIII we shall need the notion of the volume of a bounded set $S$ in $\mathbb{R}^n$. For this we appeal to the Riemann integral and we say that $S$ is *Jordan measurable* if the characteristic function of $S$ is Riemann integrable; we call the value of this integral the volume of $S$. From time to time we shall use the result that any bounded convex set in $V$ is always Jordan measurable. (See Exercise 10 for this chapter.)

In the sequel we shall be particularly interested in discrete subgroups of maximal rank, that is to say in the case $r = n$ in the above result: we shall call such subgroups *lattices* of $V$. If $\Lambda$ is a lattice of $V$, we choose a $\mathbb{Z}$-basis $\mathfrak{B} = \{v_1, \ldots, v_n\}$ of $\Lambda$ and we define the fundamental parallelepiped of $\Lambda$, with respect to $\mathfrak{B}$, to be

$$P_{\mathfrak{B}} = \{\sum v_h\alpha_h \mid 0 \leq \alpha_h \leq 1\};$$

the volume of $P_{\mathfrak{B}}$ (with respect to the coordinate system $\{e_h\}$) is then defined to be

$$\mathrm{vol}(P_{\mathfrak{B}}) = \left| \int_{x \in P_{\mathfrak{B}}} dx_1 \ldots dx_n \right|.$$

**(2.4)** *If* $v_h = \sum \mu_{hj}e_j$ *with* $\mu_{hj} \in \mathbb{R}$, *then*

$$\mathrm{vol}(P_{\mathfrak{B}}) = |\det(\mu_{hj})|.$$

*Proof.* Put $\mathfrak{E} = \{e_1, \ldots, e_n\}$ and let $g\colon V \to V$ denote the linear transformation induced by setting $g(e_h) = v_h$. Then by the Jacobian transformation formula

$$\left| \int_{P_{\mathfrak{B}}} dx_1 \ldots dx_n \right| = \left| \int_{P_{\mathfrak{E}}} |\det(\mu_{hj})| dx_1 \ldots dx_n \right|$$
$$= |\det(\mu_{hj})|.$$

□

If $\mathfrak{B}'$ is a further basis of $\Lambda$, then there is a matrix $m \in GL_n(\mathbb{Z})$ which transforms $\mathfrak{B}'$ to $\mathfrak{B}$. Thus by (2.4)

$$\mathrm{vol}(P_{\mathfrak{B}'}) = |\det(m)||\det(\mu_{hj})|.$$

Since $\det(m) = \pm 1$, we conclude that $\mathrm{vol}(P_{\mathfrak{B}})$ is in fact independent of the particular choice of $\mathbb{Z}$-basis; therefore, from now on, we call this invariant the volume of $\Lambda$, and we denote it by $\mathrm{vol}(\Lambda)$.

Now that we have dealt with the above preliminaries, we can begin to consider the main theme of this section, and address the question of how a given lattice meets a given subset of $\mathbb{R}^n$. Our first result in this direction is due to Minkowski:

**(2.5)** *Suppose that $S$ is a bounded Jordan measurable subset of $V$, let $\Lambda$ denote a lattice in $V$ and suppose $\mathrm{vol}(S) > \mathrm{vol}(\Lambda)$; then there exist distinct elements $x, y \in S$ such that $x - y$ lies in $\Lambda$.*

*Proof.* Let $\mathfrak{B}$ denote a $\mathbb{Z}$-basis of $\Lambda$; thus $V = \bigcup_{\lambda\in\Lambda}(\lambda + P_{\mathfrak{B}})$, and furthermore for distinct $\lambda, \mu \in \Lambda$, $(\lambda + P_{\mathfrak{B}}) \cap (\mu + P_{\mathfrak{B}})$ has zero volume. We deduce that

$$S = \bigcup_{\lambda}((\lambda + P_{\mathfrak{B}}) \cap S)$$

and

$$\begin{aligned}\mathrm{vol}(S) &= \sum_{\lambda} \mathrm{vol}((\lambda + P_{\mathfrak{B}}) \cap S) \\ &= \sum_{\lambda} \mathrm{vol}(P_{\mathfrak{B}} \cap (S - \lambda)).\end{aligned}$$

(Note that both sums on the right are finite since $S$ is bounded). Thus, if all the $\{S - \lambda\}_{\lambda\in\Lambda}$ were disjoint, then we could conclude that

$$\mathrm{vol}(S) = \sum \mathrm{vol}(P_{\mathfrak{B}} \cap (S - \lambda)) \leq \mathrm{vol}(P_{\mathfrak{B}}) = \mathrm{vol}(\Lambda).$$

□

We shall call $S \subset V$ *symmetric* if given $v \in S$ then $-v \in S$, i.e. if $-S = S$. We call $S$ *convex* if given $x, y \in S$, then $xt + y(1-t) \in S$ for all $t \in [0, 1]$, that is to say the line segment connecting any two points in $S$, lies entirely within $S$.

The key lemma which lies at the heart of all subsequent results in this section is:

**(2.6) (Blichfeldt's Lemma).** *Suppose that $S \subset V$ is bounded, convex and symmetric. Let $\Lambda$ denote a lattice in $V$, and suppose that $\mathrm{vol}(S) > 2^n\mathrm{vol}(\Lambda)$; then $S \cap \Lambda$ contains a non-zero point. If, in addition, $S$ is* *closed and we are now only given* $\mathrm{vol}(S) = 2^n \mathrm{vol}(\Lambda)$, *then again* $S \cap \Lambda$ *contains a non-zero point.*

*Proof.* Initially suppose $\mathrm{vol}(S) > 2^n \mathrm{vol}(\Lambda)$, and define $S'$ to be the contracted set $\frac{1}{2}S$; then $\mathrm{vol}(S') = 2^{-n}\mathrm{vol}(S) > \mathrm{vol}(\Lambda)$. We can thus apply (2.5) to $S'$ to find distinct points $y, z \in S'$ such that $x = y - z \in \Lambda$. We rewrite

$$x = \frac{1}{2}(2y - 2z).$$

Then $2y$, $2z$ lie in $S$; so $-2z \in S$ by symmetry; and so by convexity we conclude that $x$ lies in $S$, as required.

Lastly suppose that $S$ is closed and thus complete, and that $\mathrm{vol}(S) = 2^n \mathrm{vol}(\Lambda)$. We consider the dilation $S_k = (1 + \frac{1}{k})S$ for each positive integer $k$. Each $S_k$ satisfies the conditions of the first part, and so we can find non-zero $x_k \in S_k \cap \Lambda$. However, since $2S$ is compact with $\Lambda$ discrete, there must be an infinite number of repeats in the sequence $\{x_k\}$, i.e. we can find some $x_m$ such that $x_m \in S_k$ for all $k$. Since $S$ is closed, $S = \bigcap_k S_k$, and therefore $x_m$ lies in $S$.

□

Although the above result will be used in some of our arithmetic applications, for a number of other applications we really need to reinterpret Blichfeldt's Lemma in terms of linear forms, and then study in some detail the particular symmetric, convex regions which we shall later require.

A *linear form* $f$ on $V$ is just an $\mathbb{R}$-linear map $f: V \to \mathbb{R}$. For $x = \sum e_j x_j$, we then have $f(x) = \sum a_j x_j$, where $a_j = f(e_j)$ for each $j$. Thus a map $f: V \to \mathbb{R}$ is a linear form iff for some given constants $a_i$ we have $f(\sum e_j x_j) = \sum a_j x_j$ for all $\{x_j\}$.

**Theorem 32.** *Let* $f_h$ $(h = 1, \dots, n)$ *denote* $n$ *linear forms on* $V$, $f_h(x) = \sum a_{hj} x_j$, *and suppose that* $\Delta = |\det(a_{hj})|$ *is non-zero; let* $S \subset V$ *denote a compact, symmetric, convex region; if* $\mathrm{vol}(S) \geq 2^n \Delta$ *then there exists a non-zero* $x \in \Omega = \sum_1^n e_h \mathbb{Z}$ *with* $\sum e_h f_h(x) \in S$.

Before proving the theorem we note that the region

$$R = \{x \in V \mid |x_h| \leq r_h\},$$

for positive real numbers $r_h$, is a Riemann integrable, compact, symmetric, convex region with volume $2^n \prod_1^n r_h$. We can therefore deduce the following result on integral solutions:

**Corollary.** *If $\prod r_h \geq \Delta$, then we can find integers $x_h$, not all zero, such that*

$$|\sum_j a_{hj}x_j| \leq r_h \quad \textit{for } h = 1, \ldots, n.$$

We now prove Theorem 32. Define $\theta_f \colon V \to V$ by

$$\theta_f(x) = \sum e_h f_h(x). \tag{2.7.a}$$

Thus, in full, for $x = \sum e_j x_j$

$$\theta_f(x) = \sum_{h,j} e_h a_{hj} x_j. \tag{2.7.b}$$

Let $\Lambda$ denote the set $\theta_f(\Omega)$. Since $\Delta$ is non-zero, $\theta_f$ is invertible, and so $\Lambda$ is a lattice in $V$; furthermore, by (2.4), we know that $\mathrm{vol}(\Lambda) = \Delta$. We can therefore immediately apply (2.6), and thereby deduce the existence of a non-zero point $y \in S \cap \Lambda$. From the very definition of $\Lambda$, $y = \theta_f(x)$ for some non-zero $x \in \Omega$; writing $x = \sum e_h x_h$ then yields the required integral solution.

□

In arithmetic applications we shall also need a version of Theorem 32 where we admit the possibility of complex valued $\mathbb{R}$-linear forms $f \colon V \to \mathbb{C}$, i.e. $\mathbb{R}$-linear maps $f \colon V \to \mathbb{C}$. We impose the special condition that in any given system $\{f_h\}$, $(h = 1, \ldots, n)$ of such forms, the non-real forms occur in complex conjugate pairs

$$f(x) = \sum a_j x_j \qquad \bar{f}(x) = \sum \bar{a}_j x_j.$$

The associated system of real linear forms $\{g_h\}$ is then defined as follows

$$g_h = f_h \qquad \text{if } f_h = \overline{f}_h \tag{2.8.a}$$

$$\left.\begin{aligned} g_h &= \frac{1}{2}(f_h + f_k) \\ g_k &= \frac{1}{2i}(f_h - f_k) \end{aligned}\right\} \text{if } \overline{f}_h = f_k \neq \overline{f}_k \text{ with } h < k. \tag{2.8.b}$$

As we shall see presently, it then suffices to apply Theorem 32 to this new system of real linear forms. With this in mind we let $\Lambda = \theta_g(\Omega)$, i.e. $\Lambda$ is the lattice in $V$ associated with the $\{g_h\}$. By (2.8), it follows that if $f_h(x) = \sum a_{hj} x_j$ and $\Delta = |\det(a_{hj})|$, then

$$\mathrm{vol}(\Lambda) = 2^{-t}\Delta \tag{2.9}$$

where for all applications we may thus assume $\Delta \neq 0$ and where $t$ is the number of complex conjugate pairs of non-real forms $f_h$. For simplicity of notation we assume that $f_h$, $(h = 1, \ldots, s)$ are the real forms (whence of course $n = s + 2t$), and that $f_{s+h}$, $(h = 1, \ldots, t)$ are representatives of the pairs of complex conjugate forms with $\overline{f}_{s+h} = f_{s+t+h}$. We consider $\{f_h\}$, together with such a numbering, as being fixed for the remainder of this section.

In the next section we shall see that it can be extremely advantageous to work with a more subtly chosen region than the straightforward box used in the corollary to Theorem 32: indeed, the choice of region can yet further substantially reduce the amount of work required in calculating a class number. With just this advantage in mind, define the region $R_d$, for a positive real number $d$, by

$$R_d = \{x \in V \mid \sum_{h=1}^{s} |x_h| + 2 \sum_{h=s+1}^{s+t} |x_h^2 + x_{h+t}^2|^{1/2} \leq d\}. \tag{2.10}$$

**(2.11)** *$R_d$ is a compact, symmetric, convex subset of $V$.*

*Proof.* It is clear from (2.10) that $R_d$ is both symmetric and bounded. To show that $R_d$ is convex, choose $\lambda \in [0,1]$ and $x, y \in R_d$; then, by the triangle inequality, for $1 \leq h \leq s$

$$|\lambda x_h + (1-\lambda)y_h| \leq \lambda|x_h| + (1-\lambda)|y_h|$$

and for each $h$, $s < h \leq s+t$, by the triangle inequality in $\mathbb{R}^2$,

$$\begin{aligned}|(\lambda x_h + (1-\lambda)y_h)^2 + (\lambda x_{h+t} + (1-\lambda)y_{h+t})^2|^{1/2}& \\ \leq \lambda|x_h^2 + x_{h+t}^2|^{1/2} + (1-\lambda)|y_h^2 + y_{h+t}^2|^{1/2}&.\end{aligned}$$

Summing over $h$ we obtain

$$\begin{aligned}&\sum_{h=1}^{s} |\lambda x_h + (1-\lambda)y_h| \\ &\quad + 2\sum_{h=s+1}^{s+t} |(\lambda x_h + (1-\lambda)y_h)^2 + (\lambda x_{h+t} + (1-\lambda)y_{h+t})^2|^{1/2} \\ &\leq \lambda\left(\sum_{1}^{s} |x_h| + 2\sum_{s+1}^{s+t} |x_h^2 + x_{h+t}^2|^{1/2}\right) \\ &\qquad + (1-\lambda)\left(\sum_{1}^{s} |y_h| + 2\sum_{s+1}^{s+t} \left|y_h^2 + y_{h+t}^2\right|^{1/2}\right) \\ &\leq \lambda d + (1-\lambda)d = d.\end{aligned}$$

Hence $\lambda x + (1-\lambda)r \in R_d$. Since $R_d$ is defined as the inverse image of a continuous function on a closed set, it follows that $R_d$ itself be closed. Because $R_d$ is closed and bounded, we deduce that it is compact.

□

Lastly, in order to be able to apply Theorem 32 to $R_d$, we now need to calculate its volume. With this in mind define

$$S_d^{(s,t)} = \{x \in R_d \mid x_j \geq 0 \quad \text{for } j = 1, \ldots, s\}.$$

We shall show that

$$\text{vol}(S_d^{(s,t)}) = \frac{d^n}{n!}\left(\frac{\pi}{2}\right)^t. \tag{2.12}$$

It then follows trivially that

$$\text{vol}(R_d) = \frac{2^s d^n (\pi/2)^t}{n!}. \tag{2.13}$$

We shall prove (2.12) by induction on $n$. Firstly we suppose that $t = 0$, and that (2.12) holds for $(s-1, 0)$. Then

$$\begin{aligned}
\text{vol}(S_d^{(s,t)}) &= \left| \int_0^d \text{vol}(S_{d-x_s}^{(s-1,0)}) dx_s \right| \\
&= \left| \int_0^d \frac{(d-x_s)^{n-1}}{(n-1)!} \cdot dx_s \right| = \frac{d^n}{n!}.
\end{aligned}$$

So now suppose $t > 0$, and suppose (2.12) to hold for $(s, t-1)$. Given the nature of the domain of integration, it is natural to write $x_{s+t} = r\cos\phi$, $x_{s+2t} = r\sin\phi$; then

$$\begin{aligned}
\text{vol}(S_d^{(s,t)}) &= \left| \int_{r=0}^{d/2} \int_{\phi=0}^{2\pi} \text{vol}(S_{d-2r}^{(s,t-1)}) \cdot r dr d\phi \right| \\
&= \left| \int_0^{d/2} \int_0^{2\pi} \frac{(d-2r)^{n-2}}{(n-2)!} \cdot (\pi/2)^{t-1} \cdot r dr d\phi \right| \\
&= \frac{(\pi/2)^t}{n!} \cdot d^n.
\end{aligned}$$

□

Recall that $\{f_h\}$ denotes our chosen system of linear forms with $f_h(e_j) = a_{hj}$, i.e. $f_h(x) = \sum a_{hj} x_j$. Then $\{g_h\}$ is the associated system of real linear forms (see (2.8)) and $\theta_g \colon V \to V$ is the map of (2.7).

For future reference we now draw together all the above work and show

**Theorem 33.** *Let $\{f_h\}$, $(h = 1, \ldots, n)$ denote a system of complex valued linear forms on $V$ as above, and suppose $\Delta = |\det(a_{hj})| \neq 0$; let $R_d$ denote the region given in (2.10), and suppose that $d$ is chosen so that*

$$\frac{\pi^t d^n}{n!} \geq 4^t \Delta. \tag{2.14}$$

*Then there is a non-zero* $x \in \Omega = \sum_1^n e_j \mathbb{Z}$ *such that*

$$0 < \sum_{h=1}^{s} |f_h(x)| + 2 \sum_{h=s+1}^{s+t} |f_h(x)| \leq d.$$

*Proof.* Let $\Lambda = \theta_g(\Omega)$. Because $\Delta \neq 0$, we know from (2.9) that $\Lambda$ is a lattice. We now show that there is a non-zero point $y \in R_d \cap \Lambda$. By Theorem 32 and (2.11) the existence of such a point is guaranteed provided $\mathrm{vol}(R_d) \geq 2^n \mathrm{vol}(\Lambda)$. Since $\mathrm{vol}(\Lambda) = 2^{-t}\Delta$ by (2.9), this condition is satisfied thanks to (2.13) and (2.14).

Let $x$ denote the unique element in $\Omega$ such that $\theta_g(x) = y$. By the definition of $\theta_g$ in (2.7), together with the fact that $\theta_g(x) \in R_d$, it follows that

$$\sum_1^s |g_h(x)| + 2 \sum_{s+1}^{s+t} |g_h^2(x) + g_{h+t}^2(x)|^{1/2} \leq d,$$

so from (2.8) we conclude that

$$\sum_1^s |f_h(x)| + 2 \sum_{s+1}^{s+t} |f_h(x)| \leq d.$$

□

## §3 Classgroups

We shall now apply the geometric methods developed in the preceding section in order to obtain a powerful bound for the calculation of the classgroup of a given number field $K$; we then conclude this section with a number of worked examples.

Throughout this section and the next we let $n$ denote the degree of $K/\mathbb{Q}$. From (I.1.48) we know that $x \mapsto x \otimes 1$ induces an injective $K$-homomorphism $\phi: K \to K \otimes_{\mathbb{Q}} \mathbb{R}$. By Theorem 17 we know that there is a product decomposition of $\mathbb{R}$-algebras

(3.1) $$K \otimes_{\mathbb{Q}} \mathbb{R} = \mathbb{R}^{(s)} \times \mathbb{C}^{(t)}$$

where $s$ denotes the number of real embeddings of $K$, and $2t$ the number of imaginary embeddings of $K$ into $\mathbb{C}$. We number the $n$ embeddings $\sigma_i: K \to \mathbb{C}$ in such a way that $\sigma_i$ is real iff $i \leq s$, and $\sigma_{s+i} \circ \rho = \sigma_{s+i+t}$ for $1 \leq i \leq t$, where $\rho: \mathbb{C} \to \mathbb{C}$ denotes complex conjugation. Then for each $i$, the map $\sigma_i \otimes \mathrm{id}: K \otimes \mathbb{R} \to \mathbb{C}$ induced by $x \otimes r \mapsto x^{\sigma_i} r$ is an $\mathbb{R}$-linear form on the $\mathbb{R}$-vector space $K \otimes_{\mathbb{Q}} \mathbb{R}$. We now employ the notation of the previous section, and set $V = K \otimes_{\mathbb{Q}} \mathbb{R}$, $f_i = \sigma_i \otimes \mathrm{id}$. Clearly, given

$f_i \in \{f_h\}$, the complex conjugate $\overline{f}_i$ also belongs to $\{f_h\}$, and so we write $\{g_i\}$ for the associated system of real forms (see (2.8)).

If $a_1, \ldots, a_n$ denotes a $\mathbb{Q}$-basis of $K$, then $\{e_i = a_i \otimes 1\}$ is an $\mathbb{R}$-basis of $V$. We now wish to determine the coefficients of the $f_i$ with respect to this chosen basis: for an element $x = \sum a_j \otimes x_j$ of $V$, we have

$$f_i(x) = \sum a_j^{\sigma_i} x_j \tag{3.2.a}$$

thus, in particular, for $a \in K$

$$f_i(\phi(a)) = a^{\sigma_i}. \tag{3.2.b}$$

However, from (I.1.18.$b$) we know that

$$\Delta = |\det(a_{ij})| = |\det(a_j^{\sigma_i})| \tag{3.2.c}$$

is non-zero, since $K/\mathbb{Q}$ is a separable extension. This then gives us a new and interesting way of viewing discriminants as volumes: for, if we now choose $\{a_i\}$ to be a $\mathbb{Z}$-basis of an $\mathfrak{o}_K$-fractional ideal $\mathfrak{a}$, and if we let $\theta_g \colon V \to V$ denote the automorphism of $V$ associated with the $\{g_i\}$, via $\theta_g(x) = \sum_i e_i g_i(x)$; then by (3.2.$c$), (2.9) and (II.1.39)

$$\mathbf{N}\mathfrak{a}|d_K|^{1/2} = \Delta \quad \text{and} \quad \operatorname{vol}(\theta_g(\Omega)) = 2^{-t}\mathbf{N}\mathfrak{a}|d_K|^{1/2}. \tag{3.3}$$

**Theorem 34.** *Let $\mathfrak{a}$ denote a fractional $\mathfrak{o}_K$-ideal. Then there is a non-zero element $y$ of $\mathfrak{a}$ with the property that*

$$|N_{K/\mathbb{Q}}(y)| \leq \left(\frac{4}{\pi}\right)^t \frac{n!}{n^n} |d_K|^{1/2}\mathbf{N}\mathfrak{a}.$$

*Proof.* Again choose $\{a_i\}$ to be a $\mathbb{Z}$-basis of $\mathfrak{a}$, and choose $d$ to be the positive real number such that

$$\frac{\pi^t d^n}{n!} = 4^t\Delta = 4^t|d_K|^{1/2}\mathbf{N}\mathfrak{a}. \tag{3.4}$$

Then, by Theorem 33, we can find a non-zero $x \in \Omega$ such that $\theta_g(x) \in \theta_g(\Omega) \cap R_d$, where $R_d$ denotes the region in $V$ cut out by the inequalities given in (2.10). Thus, in summary, if $x = \sum a_i \otimes x_i$ ($x_i$ all integral), then $y = \sum a_i x_i$ lies in $\mathfrak{a}$, it is non-zero, $\phi(y) = x$, and furthermore by (3.2)

$$\sum_{i=1}^n |f_i(\phi(y))| = \sum_{i=1}^n |y^{\sigma_i}| \leq d.$$

In conclusion, because the geometric mean of a finite set of positive numbers is bounded above by their arithmetic mean, we note that

$$|N_{K/\mathbb{Q}}(y)| = \prod_{i=1}^n |y^{\sigma_i}| \leq \left(\frac{d}{n}\right)^n.$$

The result then follows immediately from (3.4).

□

Now let $c$ denote an arbitrary class in the classgroup $C_K$; from (II.1.18) we know that we can find an $\mathfrak{o}_K$-ideal $\mathfrak{a}$ which represents $c^{-1}$. By Theorem 34, there is a non-zero element $x$ in $\mathfrak{a}$ with the property that

$$|N_{K/\mathbb{Q}}(x)| \leq \left(\frac{4}{\pi}\right)^t \frac{n!}{n^n}|d_K|^{1/2}\mathbf{N}\mathfrak{a}.$$

Since $x$ lies in $\mathfrak{a}$, we see that $x\mathfrak{a}^{-1}$ is an $\mathfrak{o}_K$-ideal with class $c$, and by (II.1.35, 38) $\mathbf{N}(x\mathfrak{a}^{-1}) = |N_{K/\mathbb{Q}}(x)|(\mathbf{N}\mathfrak{a})^{-1}$, so that

$$\mathbf{N}(x\mathfrak{a}^{-1}) \leq \left(\frac{4}{\pi}\right)^t \frac{n!}{n^n}|d_K|^{1/2}.$$

We have therefore shown

**Theorem 35.** *Given a class $c \in C_K$, there exists an $\mathfrak{o}_K$-ideal $\mathfrak{b}$ with class $c$ such that*

$$\mathbf{N}\mathfrak{b} \leq \left(\frac{4}{\pi}\right)^t \frac{n!}{n^n}|d_K|^{1/2}.$$

It is interesting to note that in the case when $n = s$, had we worked only with a simple box region, in place of the more cunning region $R_d$ and not used the arithmetic mean of conjugates, then we should have only achieved the bound $|d_K|^{1/2}$ in the above.

As an illustration of the strength of our bound, consider briefly the case $K = \mathbb{Q}(b)$, where $b$ is the unique real root of $X^3 - X - 1$. By (III.2.22) we know that $\mathbb{Z}[b]$ has discriminant $-23$. Thus by (II.1.28) we know $s = 1$, $t = 1$; moreover by (III.2.4)

$$\mathbb{Z}d_K \cdot [\mathfrak{o}_K : \mathbb{Z}[b]]^2 = \mathfrak{d}(\mathbb{Z}[b])\mathbb{Z}.$$

Therefore, because $d_K$ is an integer, while 23 is square-free, we conclude $\mathfrak{o}_K = \mathbb{Z}[b]$ and $d_K = -23$. By Theorem 35, in determining $C_K$, we need only consider $\mathfrak{o}_K$-ideals $\mathfrak{b}$ with

$$\mathbf{N}\mathfrak{b} \leq \frac{4}{\pi} \cdot \frac{3!}{27} \cdot \sqrt{23} = 1.357\cdots$$

The only possibility for $\mathfrak{b}$ is $\mathfrak{o}_K$, which is clearly principal: we have therefore shown $\mathfrak{o}_K$ to be a principal ideal domain with virtually no effort at all.

Before proceeding to apply the above so-called Minkowski bound to some further examples, we first show how Theorem 35 can be used to obtain the following classical result of Hermite and Minkowski:

**Theorem 36.** *If $K$ is a finite, non-trivial extension of $\mathbb{Q}$ then $|d_K| > 1$, and hence, by Theorem 22, some rational prime number must ramify in $K$.*

*Proof.* Since $\mathbf{N}\mathfrak{b} \geq 1$ in Theorem 35, we deduce that

$$1 \leq \left(\frac{4}{\pi}\right)^t \frac{n!}{n^n}|d_K|^{1/2} \leq \left(\frac{4}{\pi}\right)^n \frac{n!}{n^n}|d_K|^{1/2}.$$

We set $b_n = \left(\frac{\pi}{4}\right)^n \frac{n^n}{n!}$, and we show that $b_n > 1$ for all $n \geq 2$. The case $n = 2$ is clear, since $b_2 = \pi^2/8 > 1$. More generally, we observe that

$$\frac{b_{n+1}}{b_n} = \frac{\pi}{4}(1 + \frac{1}{n})^n \geq \frac{\pi}{2}.$$

□

We now conclude this section by using Theorem 35 to determine two classgroups.

*Example 1.* $K = \mathbb{Q}(\sqrt{-6})$. By (II.1.32) we know that $\mathfrak{o}_K = \mathbb{Z}[\sqrt{-6}]$ and $d_K = -24$; thus by Theorem 35 we need only examine the classes of $\mathfrak{o}_K$-ideals $\mathfrak{b}$ with absolute norm

$$\mathbf{N}\mathfrak{b} \leq \frac{4}{\pi} \cdot \frac{1}{2} \cdot \sqrt{24} = 3.119\cdots.$$

Since 2 and 3 both ramify in $K$, the only ideals which we need to consider are $\mathfrak{p}_2$, $\mathfrak{p}_3$, the unique $\mathbb{Z}[\sqrt{-6}]$-prime ideals above 2 and 3 respectively. As $\mathfrak{p}_2^2 = 2\mathfrak{o}_K$, $\mathfrak{p}_3^2 = 3\mathfrak{o}_K$, we immediately conclude that the only possible non-trivial classes can have order 2 and because there are at most two such classes, $C_K$ is either trivial or of order 2. However, in the Introduction we showed that $C_K \neq \{1\}$, and therefore $C_K$ must have order 2.

*Example 2.* $K = \mathbb{Q}(\sqrt{-26})$. From (II.1.32), $\mathfrak{o}_K = \mathbb{Z}[\sqrt{-26}]$ and $d_K = -104$. By Theorem 35 we know that each class is represented by an $\mathfrak{o}_K$-ideal $\mathfrak{b}$ with

$$\mathbf{N}\mathfrak{b} \leq \frac{2}{\pi} \cdot \sqrt{104} = 6.492\cdots.$$

Since the ideals with norm $\leq 6$ are products of prime ideals with the same property, we begin by considering only prime ideals of $\mathfrak{o}_K$.

(a) 2 ramifies, $\mathfrak{p}_2^2 = 2\mathfrak{o}_K$. If $\mathfrak{p}_2 = \alpha\mathfrak{o}_K$ with $\alpha = a + b\sqrt{-26}$, then on taking norms we have $2 = a^2 + 26b^2$, which is absurd, since $26 > 2$ implies $b = 0$ and $a^2 = 2$. Thus $\mathfrak{p}_2$ represents a class of order 2.

(b) Since $\left(\frac{-26}{3}\right) = 1$, from (III.2.29) we know 3 splits: $3\mathfrak{o}_K = \mathfrak{p}_3\overline{\mathfrak{p}}_3$ where $\mathfrak{p}_3 = (1 - \sqrt{-26}, 3)$, $\overline{\mathfrak{p}}_3 = (1 + \sqrt{-26}, 3)$ (as per Theorem 23). If

$\mathfrak{p}_3 = \alpha\mathfrak{o}_K$, then arguing as before we should have $3 = a^2 + 26b^2$ which is absurd since, $26 > 3$ implies $b = 0$ and so $3 = a^2$. Next we suppose $\mathfrak{p}_3^2 = \beta\mathfrak{o}_K$, $\beta = c + d\sqrt{-26}$; again this would imply that $9 = c^2 + 26b^2$; hence $b = 0$ and so $\beta = \pm 3$; this implies $\mathfrak{p}_3^2 = \mathfrak{p}_3\overline{\mathfrak{p}}_3$ which is absurd. However $N_{K/\mathbb{Q}}(1 - \sqrt{-26}) = 27$, and clearly $3 \nmid 1 - \sqrt{-26}$, so that $\mathfrak{p}_3^3 = (1 - \sqrt{-26})\mathfrak{o}_K$. Thus $\mathfrak{p}_3$ represents a class of order 3.

(c) Again 5 splits, since $\left(\frac{-26}{5}\right) = 1$; we write $5\mathfrak{o}_K = \mathfrak{p}_5\overline{\mathfrak{p}}_5$. Since $30 = N_{K/\mathbb{Q}}(2 - \sqrt{-26})$, we see that, interchanging $\mathfrak{p}_5$ and $\overline{\mathfrak{p}}_5$ if necessary, $\mathfrak{p}_2\mathfrak{p}_3\mathfrak{p}_5$ has trivial class and so, from the above $\mathfrak{p}_5$ must have order 6.

In conclusion, we have shown $C_K$ to be cyclic of order 6, and that it is generated by the class of $\mathfrak{p}_5$.

## §4 Units

Recall that $U_K$, the group of units of $K$, is the multiplicative group of invertible elements of $\mathfrak{o}_K$. We conclude this chapter by applying the geometric methods of §2 to the study of $U_K$. Here we shall be content with a description of $U_K$ as an abstract group; in the next chapter we shall consider the problem of determining explicit generators for $U_K$ for certain fields $K$ of small degree.

Let $\mu_K$ denote the group of roots of unity in $K$. Any $\zeta \in \mu_K$ must be an algebraic integer since it satisfies a polynomial of the form $X^m - 1$; indeed, as $\zeta \cdot \zeta^{m-1} = \zeta^m = 1$, we deduce that $\zeta \in U_K$, and more precisely we have $\mu_K \subset U_{K,\mathrm{tor}}$, the $\mathbb{Z}$-torsion submodule of $U_K$. Conversely, given $u \in U_{K,\mathrm{tor}}$ there must exist a positive integer $n$ such that $u^n = 1$; thus we have shown that

(4.1) $$\mu_K = U_{K,\mathrm{tor}}.$$

For the most part we shall be concerned with the torsion-free quotient $\mathfrak{U}_K = U_K/\mu_K$.

We begin by giving a general criterion for determining whether a given algebraic integer is a unit or not.

**(4.2)** *An element $a \in \mathfrak{o}_K$ is a unit iff $N_{K/\mathbb{Q}}(a) = \pm 1$.*

*Proof.* Recall that $N_{K/\mathbb{Q}}(a) = \prod_\sigma a^\sigma$ where the product extends over the distinct embeddings $\{\sigma\colon K \hookrightarrow \mathbb{C}\}$. If $a$ is a unit; then so are all the $a^\sigma$; hence $N_{K/\mathbb{Q}}(a)$ must be a unit in $\mathbb{Z}$, and so is $\pm 1$. Conversely, if $a \in \mathfrak{o}_K$ and $N_{K/\mathbb{Q}}(a) = \pm 1$; then $a\left(\prod_{\sigma \neq \mathrm{id}} a^\sigma\right)(\pm 1) = 1$, and so $a$ is a unit as all terms here are algebraic integers.

□

By way of motivation for what follows, we briefly review some already observed phenomena for quadratic number fields. Suppose first that $K$ is imaginary quadratic, $K = \mathbb{Q}(\sqrt{d})$, with $d$ a square-free negative integer. If $a + b\sqrt{d}$ is a unit; then $2a, 2b \in \mathbb{Z}$, and by (4.2), $(2a)^2 - (2b)^2 d = \pm 4$; in fact, since the term on the left is clearly positive, only the positive sign can occur. Obviously such an equation can have only finitely many integer solutions $(a, b)$ for given $d$; hence we have shown

**(4.3)** *If $K$ is a quadratic imaginary number field, then $U_K$ is finite and so $U_K = \mu_K$.*

The case of real quadratic number fields is considerably richer. We readily observe that $\mathbb{Q}(\sqrt{2})$ contains the unit $(1+\sqrt{2})$ (by (4.2)), and this is clearly not a root of unity; in the same way $\mathbb{Q}(\sqrt{3})$ contains the unit $2 + \sqrt{3}$. Thus we begin to suspect that the existence of real embeddings of a number field $K$ *vis a vis* the existence of imaginary embeddings strongly influences the structure of $U_K$.

In the previous section, in order to use the geometric techniques of §2, we considered the vector space $V = K \otimes_{\mathbb{Q}} \mathbb{R}$ together with linear forms induced by the embeddings $\sigma_h: K \hookrightarrow \mathbb{C}$. For the study of units, we set $W = \mathbb{R}^{(s+t)}$ where $s$, and $2t$, denote the number of real, and imaginary, embeddings respectively of $K$ into $\mathbb{C}$. We define a group homomorphism

$$\psi: U_K \to W \tag{4.4}$$

by

$$\psi(u) = \sum_{h \leq s} d_h \log |u^{\sigma_h}| + \sum_{s<h\leq s+t} d_h 2 \log |u^{\sigma_h}|$$

where $\{d_h\}$ always denotes the canonical basis of $W$. Here, as previously, the $\sigma_h$ are ordered so that the first $s$ are real, and the $\sigma_h$, $s < h \leq s+t$ are a set of representatives of the imaginary embeddings under the action of complex conjugation. The behaviour of the homomorphism $\psi$ lies right at the heart of our understanding of $U_K$. In due course we shall show that $\ker \psi = \mu_K$ is finite, and that $\operatorname{im} \psi$ is a lattice in a certain hyperplane of $W$. This done, we shall then have established

**Theorem 37 (Dirichlet).** *$U_K$ is a finitely generated $\mathbb{Z}$-module and*

$$U_K \cong \mu_K \times \mathbb{Z}^{(s+t-1)}.$$

In the sequel we put $r = s + t - 1$; we call $r$ the Dirichlet rank.

As a first step in our understanding of $\psi$ we show

**(4.5)** *Let $\beta$ denote a positive real number; then the set*

$$\mathcal{S} = \{a \in \mathfrak{o}_K \mid |a^{\sigma_h}| \leq \beta \ \forall h\}$$

*is finite.*

*Proof.* Since the coefficients of the characteristic polynomial $c_{K/\mathbb{Q},a}$ are all symmetric functions in the $a^{\sigma_h}$, we deduce that if $a \in \mathcal{S}$, then $c_{K/\mathbb{Q},a}$ has bounded coefficients in $\mathbb{Z}$; therefore there are only a finite number of possibilities for the characteristic polynomials, e.g. $2^{n^2}(\beta+1)^{n^2}$ is such a bound.

□

We can immediately deduce two corollaries:

**(4.5.*a*) Corollary** $\ker \psi = \mu_K$ *and* $\mu_K$ *is a finite cyclic group.*

*Proof.* From the very definition of $\psi$ in (4.4) it follows that $\mu_K \subset \ker \psi$, since roots of unity all have absolute value 1. Conversely, if $u \in \ker \psi$, then the $u^{\sigma_i}$ all have absolute value 1, and so by (4.5) $\ker \psi$ is a finite subgroup of $K^*$; as such it must be cyclic and must consist entirely of roots of unity.

□

**(4.5.*b*) Corollary** $\mathrm{im}\,(\psi)$ *is a discrete subgroup of* $W$.

*Proof.* Let $R$ denote a bounded region in $W$, and let $u \in U_K$ be such that $\psi(u) \in R$; thus the $\log |u^{\sigma_h}|$ and hence the $|u^{\sigma_h}|$ are all bounded, and so by (4.5) we conclude that $\psi(U_K) \cap R$ is finite.

□

Following (4.5.*b*), it is natural to ask what the $\mathbb{Z}$-rank of $U_K/\mu_K$ is. Clearly it is at most $s+t$ by (2.1). In fact it is easy to see that it is bounded above by $s+t-1$: for, if $\mathcal{H}$ denotes the hyperplane in $W$

$$\mathcal{H} = \{x = \sum d_h x_h \in W \mid \sum_{h=1}^{s} x_h + 2 \sum_{h=s+1}^{s+t} x_h = 0\},$$

then we claim

$$\mathrm{im}\,\psi \subset \mathcal{H}. \tag{4.6}$$

Indeed, if $u \in U_K$, then by (4.2)

$$0 = \log(|N_{K/\mathbb{Q}}(u)|) = \log\left(\prod_{1}^{n} |u^{\sigma_h}|\right)$$
$$= \sum_{h=1}^{s} \log|u^{\sigma_h}| + 2 \sum_{s<h\leq s+t} \log|u^{\sigma_h}|.$$

□

In the light of the above discussion we see that, in order to prove Theorem 37, it will suffice to show that $\mathrm{im}\,(\psi)$ contains $s+t-1$ $\mathbb{R}$-linearly independent vectors, i.e.

**(4.7)** $\mathrm{im}(\psi)$ *is an* $\mathcal{H}$ *lattice.*

Before embarking on the proof of this result, we first need to introduce some further notation. For the time being we fix an integer $h$: $1 \leq h \leq s+t$, and a positive real number $\epsilon$. We then define the $h$th $\epsilon$-tube in $V$, $T_\epsilon$, to be

$$\{x \in V \mid \text{for } j \neq h\text{: } |x_j| < \epsilon \text{ for } j \leq s; |x_j^2 + x_{j+t}^2| < \epsilon^2 \text{ for } s < j \leq s+t\}.$$

Given a positive real number $\alpha$, we define the $h$th $\epsilon$-box of volume $\alpha$ to be

if $h \leq s$

$$B_{\epsilon,\alpha} = \{x \in T_\epsilon \mid |x_h| \leq 2^{-s}\pi^{-t}\alpha\epsilon^{1-n}\}$$

if $h > s$

$$B_{\epsilon,\alpha} = \{x \in T_\epsilon \mid |x_h^2 + x_{h+t}^2| \leq 2^{-s}\pi^{-t}\alpha\epsilon^{2-n}\}.$$

It follows at once that indeed

if $h \leq s$ then

$$\mathrm{vol}(B_{\epsilon,\alpha}) = [2 \cdot 2^{-s}\pi^{-t}\alpha\epsilon^{1-n}] \prod_{\substack{j=1 \\ j\neq h}}^{s} (2\epsilon) \prod_{j>s}^{s+t} \pi\epsilon^2 = \alpha$$

if $h > s$ then

$$\mathrm{vol}(B_{\epsilon,\alpha}) = [\pi 2^{-s}\pi^{-t}\epsilon^{2-n}\alpha] \prod_{j=1}^{s} (2\epsilon) \prod_{\substack{j>s \\ j\neq h}}^{s+t} \pi\epsilon^2 = \alpha.$$

Moreover, it is trivial to check that $B_{\epsilon,\alpha}$ is a compact, symmetric, convex subset of $V$. For future reference we note that if $b = (b_1, \ldots, b_n) \in B_{\epsilon,\alpha}$, then of course

$$\prod_{k=1}^{s} 2|b_k| \prod_{k>s}^{s+t} \pi|b_k^2 + b_{k+t}^2| \leq \mathrm{vol}(B_{\epsilon,\alpha}) = \alpha. \tag{4.8}$$

*Proof of* (4.7). Our strategy will be to show that each $\epsilon$-tube contains an infinitude of units.It will then be relatively straightforward to show that if we choose $\epsilon < 1$, and a unit in each $\epsilon$-tube for each $h \leq r$, then the resulting units have $\mathbb{R}$-linearly independent images in $W$ under $\psi$.

Set $\alpha = 2^{n-t}|d_K|^{1/2}$, and let $\{a_i\}$ denote a $\mathbb{Z}$-basis of $\mathfrak{o}_K$. As in §3 $\{e_k = a_k \otimes 1\}$ is an $\mathbb{R}$-basis of $V$. We put $\theta_g(y) = \sum e_k g_k(y)$ where $\{g_k\}$ is the associated system of real linear forms associated with the linear forms derived from the $\{\sigma_k\}$. Then from (3.3), by Blichfeldt's Lemma (2.6), there is a non-zero element $x \in \mathfrak{o}_K$ such that

$$\theta_g(\phi(x)) \in B_{\epsilon,\alpha}.$$

Since $\theta_g(y)$ has $i$th coordinate $g_i(y)$, it follows from the definition of $B_{\epsilon,\alpha}$ that

$$|N_{K/\mathbb{Q}}(x)| = \prod_{k=1}^{n} |f_k(\phi(x))| =$$

by (2.8) $$\prod_{k=1}^{s} |g_k(x)| \prod_{k=s+1}^{s+t} |g_k^2(x) + g_{k+t}^2(x)|$$

by (4.8) $$\leq 2^{-s}\pi^{-t}\alpha.$$

We now repeat the process with $\epsilon$ replaced by $2^{-m}\epsilon$, for all positive integers $m$, still keeping $k$ fixed. We thereby construct an infinite sequence of elements of $\mathfrak{o}_K$, $\{x(i)\}$ with $|N_{K/\mathbb{Q}}(x(i))| \leq 2^{-s}\pi^{-t}\alpha$. A given element of the sequence can only be repeated a finite number of times. Therefore, by discarding repeats, we may assume the elements of the sequence to be distinct. Thus by (1.4) there are infinitely many associates in the sequence; and so, for some $j$, $x(i)x(j)^{-1}{}_{i\in\mathbb{N}}$ contains an infinite sequence of units $\{y_l\}$. Furthermore, as $|x(i)^{\sigma_k}| \to 0$ as $i \to \infty$ for each $k \neq h$, $k \leq s+t$ given $\eta > 0$ we can find $N$ such that $\phi(y_l) \in T_\eta$ for all $l \geq N$. This then completes the first part of our proof.

By the above work, for each integer $h$, $1 \leq h \leq s+t$, we can find a unit $u_h$ with $\phi(u_h)$ in the $h$th $\epsilon$-tube. Choosing $\epsilon < 1$, we have

$$|u_h^{\sigma_j}| \begin{cases} > 1 & \text{if } h = j \\ < 1 & \text{if } h \neq j \end{cases}$$

and so

(4.9) $$\psi_j(u_h) \begin{cases} > 0 & \text{if } h = j \\ < 0 & \text{if } h \neq j \end{cases}$$

where $\psi_j \colon U_K \to \mathbb{R}$ denotes composition of $\psi$ with projection onto the $j$th coordinate. It therefore now suffices to show that

**(4.10)** *The matrix $(\psi_j(u_h))$ has rank greater than or equal to $r$.*

*Proof.* Let $\Psi$ denote the matrix $(\psi_j(u_h))$; by (4.6) the sum over each row is zero. Suppose for contradiction that we can find a vector $\lambda$ such that $\lambda \cdot \Psi = \mathbf{0}$, but such that $\lambda$ is not a multiple of the vector all of whose entries are 1. Thus in particular

$$\sum_{j=1}^{s+t} \lambda_j \psi_j(u_h) = 0 \quad \text{for all } h. \tag{4.11}$$

Re-ordering the embeddings $\{\sigma_j\}$ and multiplying by $-1$ if necessary, we may suppose that $\lambda_1$ is the largest coefficient and is positive; we then apply the same permutation to the $\{u_h\}$ so that (4.9) still holds, and in particular

$$\begin{aligned} &\psi_1(u_1) > 0 \\ &\psi_j(u_1) < 0 \quad \text{for } j > 1. \end{aligned} \tag{4.12}$$

From (4.6) we know that

$$\sum_{j=1}^{s+t} \psi_j(u_1) = 0 \tag{4.13}$$

However, by (4.11,12,13),

$$0 = \sum_{j=1}^{s+t} \lambda_j \psi_j(u_1) > \sum_{j=1}^{s+t} \lambda_1 \psi_j(u_1) = 0$$

which is absurd.

□

We shall postpone using the Dirichlet theorem for the explicit determination of unit groups of number fields until the next chapter: there we shall consider various techniques for calculating generators of quadratic, biquadratic, cubic and sextic fields. Many of these techniques will be at their strongest when combined with the $L$-function results of Chapter VIII. There we shall encounter, as a basic invariant of $K$, the volume of $\psi(U_K)$ in the hyperplane $\mathcal{H}$ (viewed as endowed with the chosen basis $(d_h - d_{s+t} : 1 \leq h \leq r)$); this value is called the regulator and is denoted by $R_K$. Thus

$$R_K = |\det(\psi_k(u_i))| \tag{4.14}$$

where $1 \leq k, i \leq r$ and where $u_i$, $i = 1, \ldots, r$, have the property that their images in $\mathfrak{U}_K$ are a $\mathbb{Z}$-basis. We shall refer to such a set $\{u_i\}$ as a system of *fundamental units* for $U_K$. It is clear that $R_K$ is independent of the choice of the system of fundamental units.

We conclude this section by establishing an important generalisation of Dirichlet's theorem. Let $S$ denote a finite set of inequivalent absolute values of $K$, which includes all the Archimedean absolute values; we write $S_f$ for the subset of discrete absolute values in $S$, and we put $m = |S_f|$. The group of $S$-units of $K$, denoted $U_K(S)$, is defined to be

$$U_K(S) = \{a \in K^* \mid |a|_v = 1 \ \forall v \notin S\}.$$

Clearly $U_K \subset U_K(S)$; moreover, in the special case when $S_f = \emptyset$, we have equality $U_K = U_K(S)$.

**Theorem 38.**

$$U_K(S) \cong \mu_K \times \mathbb{Z}^{|S|-1}.$$

*where, as usual, $|S|$ denotes the number of elements in the set $S$.*

*Proof.* Let $v_1, \ldots, v_m$ denote the valuations corresponding to the absolute values in $S_f$, and let $h = |C_K|$. If $\mathfrak{p}_i$ is the prime ideal of $\mathfrak{o}_K$ corresponding to $v_i$ under Theorem 7, then we know that $\mathfrak{p}_i^h$ is principal, and so we may choose a generator $\pi$ such that $\pi_i \mathfrak{o}_K = \mathfrak{p}_i^h$.

Next define the group homomorphism

$$\nu \colon U_K(S) \to \prod_1^m \mathbb{Z}.$$

where $\nu(a) = (v_1(a), \ldots, v_m(a))$. We claim $\ker \nu = U_K$: indeed, $a \in \ker \nu$ iff $a\mathfrak{o}_K = \mathfrak{o}_K$, and this occurs iff $a \in U_K$.

In order to show that the rank of $U_K(S)$ actually achieves the stated value, it will suffice to show that $\nu$ has finite cokernel; this, however, is now immediate since each $\pi_i \in U_K(S)$ and $\nu(\pi_i)$ is $h$ in the $i$th position, and 0 elsewhere.

□

# V

# Fields of low degree

In the preceding chapter, we obtained an abstract description of the unit group of a number field, and we also described a powerful technique for determining classgroups. The present chapter is intended as a complement to the previous one, with the main aim being to gain explicit information on both fundamental units and classgroups for number fields of small degree; more precisely, we wish to derive a detailed and intimate knowledge of the arithmetic of quadratic fields, biquadratic fields, as well as certain cubic and sextic fields. The knowledge which we accumulate concerning these fields – both in this chapter and in subsequent chapters – will later be used to obtain results on ordinary integers.

We mention that we shall return to the theme of explicit methods when we consider cyclotomic extensions in Chapter VI. In the same vein, it is important to emphasise that many of the techniques introduced here will only be partially successful in this chapter; they will only realise their full potential when combined with the $L$-function results of Chapter VIII.

## §1 Quadratic fields

An algebraic number field $K$ with $(K : \mathbb{Q}) = 2$ is called a *quadratic field.* These fields correspond bijectively to the square-free integers $d \neq 1$, under $K \leftrightarrow d$, with $K = \mathbb{Q}(\sqrt{d})$.

From (II.1.32,33) recall that

$$\begin{cases} \mathfrak{o}_K = \mathbb{Z}[\sqrt{d}],\ d_K = 4d, & \text{if } d \equiv 2,3 \bmod (4) \\ \mathfrak{o}_K = \mathbb{Z}[\frac{1+\sqrt{d}}{2}],\ d_K = d, & \text{if } d \equiv 1 \bmod (4). \end{cases}$$

From Theorem 20 we get the equation $efg = 2$ for the ramification index $e$, the residue class degree $f$ and the number $g$ of prime ideal factors in $\mathfrak{o}_K$ of a given prime $p$. (More precisely, of a given prime ideal $p\mathbb{Z}$ of $\mathbb{Z}$.) As every factor is $\geq 1$, we have three possible cases:
(i) The ramified case $f = g = 1$, $e = 2$. By Theorem 22 and equation (III.2.1), this is the case iff $p|d_K$, i.e. by (II.1.32), (II.1.33) iff $p|d$ or $d \equiv -1 \bmod (4)$, with $p = 2$;
(ii) The split case $e = f = 1$, $g = 2$. Here there is a prime ideal $\mathfrak{p}$ of $\mathfrak{o}_K$ distinct from $\mathfrak{p}^\tau$ (where $\langle\tau\rangle = \mathrm{Gal}(K/\mathbb{Q})$), with $p\mathfrak{o}_K = \mathfrak{p}\mathfrak{p}^\tau$, hence $p\mathbb{Z} = \mathcal{N}_{k/\mathbb{Q}}(\mathfrak{p})$ (see Corollary to Theorem 21).
(iii) The inert case $g = e = 1$, $f = 2$, when $p\mathfrak{o}_K$ is still a prime ideal of $\mathfrak{o}_K$.

There is an explicit criterion to distinguish between cases (ii) and (iii). By (III.2.29)

**(1.1)** *Suppose $p \nmid d_K$. Then $p\mathfrak{o}_K$ is the product of two distinct prime ideals iff*

$$\begin{cases} \left(\frac{d_K}{p}\right) = 1 & \textit{for } p > 2, \\ d_K \equiv 1 \bmod (8) & \textit{for } p = 2. \end{cases}$$

Note also that by Theorem 15 we have $p^f = \mathbf{N}\mathfrak{p}$, where $\mathfrak{p}$ is any prime ideal divisor of $p$ in $\mathfrak{o}_K$. So in cases (i) and (ii), $p = \mathbf{N}\mathfrak{p}$.

We complement this criterion for prime ideals with an obvious one for completions "at infinity", i.e. at the Archimedean absolute values: If $d > 0$, then for any embedding $\sigma: K \hookrightarrow \mathbb{C}$, $\sqrt{d}^\sigma$ is real, i.e. $\sigma$ embeds $K$ into $\mathbb{R}$; in this case we call $K$ a real quadratic field. (At this stage it is advantageous not to think of $\sqrt{d}$ as meaning a positive real number, but instead to view it as one of the roots to $X^2 - d$.) If $d < 0$, then $\sqrt{d}^\sigma$ must be imaginary, so that $K$ cannot be embedded in $\mathbb{R}$; in this case we call $K$ an imaginary quadratic field. From the point of view of the general theory of (III.§1), in the first case $K \otimes_\mathbb{Q} \mathbb{R} = \mathbb{R} \times \mathbb{R}$, i.e. the ordinary absolute value $|.|$ of $\mathbb{Q}$ has two extensions to $K$. In the second case $K \otimes_\mathbb{Q} \mathbb{R}$ is a field and so must be $\mathbb{C}$, i.e. $|.|$ possesses a unique extension to $K$. We also know that, in the former case, the two embeddings of $K$ into $\mathbb{R}$ are interchanged by the generator $\tau$ of $\mathrm{Gal}(K/\mathbb{Q})$: $\sqrt{d}^\tau = -\sqrt{d}$. (See the discussion following (III.1.9)).

Our first step in penetrating the arithmetic of the quadratic field $K$ is to understand its unit group $U_K$. We know from Dirichlet's Unit Theorem that

$$(1.2)\qquad \begin{cases} U_K \cong \mathbb{Z} \times \mu_K & \text{if } K \text{ is real} \\ U_K = \mu_K & \text{if } K \text{ is imaginary} \end{cases}$$

where, as previously, $\mu_K$ denotes the group of roots of unity in $K$. Of course the group $\mu_{\mathbb{R}}$ of real roots of unity is $\{\pm 1\} = \mu_{\mathbb{Q}}$. As any embedding $K \hookrightarrow \mathbb{R}$ maps $\mu_K$ into $\mu_{\mathbb{R}}$, we have

**(1.3)** $\mu_K = \{\pm 1\}$ *if* $K$ *is real.*

To find $\mu_K$ for imaginary $K$, we introduce a useful explicit description of $U_K$ which is valid in both the real and imaginary cases. For a square-free integer $d \neq 1$, a number $\alpha = \frac{1}{2}(x + y\sqrt{d})$, $x, y \in \mathbb{Q}$ is a unit of $K$ iff $x, y \in \mathbb{Z}$ and $N_{K/\mathbb{Q}}(\alpha) = \frac{1}{4}(x^2 - y^2 d) = \pm 1$. Indeed, this follows almost immediately from (II.1.31) and (IV.4.2). We restate this criterion, fixing once and for all a square root $\sqrt{d}$ of $d$.

**(1.4)** *The units* $\alpha$ *of* $K$ *correspond bijectively to pairs* $(x, y)$ *of integers such that* $x^2 - y^2 d = \pm 4$, *under the map* $(x, y) \mapsto \frac{1}{2}(x + y\sqrt{d})$.

Let $d = -m$, $m > 0$; then clearly $-4$ cannot occur in the above, so that now, in the imaginary quadratic case with $K = \mathbb{Q}(\sqrt{-m})$, the units in $K$ correspond to the integral solutions of

$$(1.4.a)\qquad x^2 + my^2 = 4.$$

We quickly verify that for $m > 4$, the only solutions are $x = \pm 2$, $y = 0$, i.e. $\mu_K = \{\pm 1\}$. For $m = 3$ we get six solutions: $(\pm 1, \pm 1)$, $(\pm 2, 0)$; so that $\mu_K$ has order 6. For $m = 2$, there are only the two solutions $(\pm 2, 0)$. For $m = 4$ we have the four solutions $\pm 1$, $\pm i$, given by $(\pm 2, 0)$, $(0, \pm 2)$.

There is an alternative, slightly more sophisticated approach to this problem, which also works for fields of degree greater than 2. We anticipate the fact to be proved in (VI.§1), that the degree of the field $\mathbb{Q}(\mu_n)$, generated by the $n$th roots of unity $\mu_n$, is $\phi(n)$ (the Euler function). As $\mu_{\mathbb{Q}} = \{\pm 1\}$ and $\phi(n) > 1$ for $n > 2$, we have to find all $n > 2$ with $\phi(n) \mid 2$, i.e. $\phi(n) \leq 2$. There are three such values: $n = 3, 6$ which both give $|\mu_K| = 6$, $K = \mathbb{Q}(\mu_3) = \mathbb{Q}(\mu_6)$; and $n = 4$ which gives $|\mu_K| = 4$, $K = \mathbb{Q}(\mu_4)$.

We now turn to a detailed analysis of $U_K$ for real quadratic $K = \mathbb{Q}(\sqrt{d})$. Let

$$\eta = x + y\sqrt{d} \qquad (x, y \in \mathbb{Q})$$

denote a unit of $K$ which is different from $\pm 1$. Then the four elements

$$\{\eta, \eta^{-1}, -\eta, -\eta^{-1}\}$$

are all distinct. They coincide, in some order, with

$$\{\eta, \eta^{\tau}, -\eta^{\tau}, -\eta\}$$

where $\tau$ generates $\mathrm{Gal}(K/\mathbb{Q})$, and they also coincide in some order with $\{\pm x \pm y\sqrt{d}\}$. Moreover, the set of all units of $K$ of infinite order is the disjoint union of such quadruplets. Recall that $\eta$ is said to be a fundamental unit if $U_K = \eta^{\mathbb{Z}} \times \{\pm 1\}$. Elementary group theory tells us that the fundamental units form such a quadruplet. Now fix an embedding $\sigma: K \hookrightarrow \mathbb{R}$. We then view the elements of $K$ as real numbers and let $\sqrt{d}$ be the positive square root of $d$ under this embedding. One easily sees that each quadruplet has precisely one unit in each of the intervals $(-\infty, -1)$, $(-1, 0)$, $(0, 1)$, $(1, \infty)$. In particular, there is precisely one fundamental unit $u$ with $u > 1$ (with respect to $\sigma$). We call $u$ the *normalised* fundamental unit of $K$ (with respect to $\sigma$). Each unit $\eta \neq \pm 1$ then has a unique representation $\eta = u^{\delta m}\delta'$, $\delta = \pm 1$, $\delta' = \pm 1$, $m \in \mathbb{N}$, and of course any such element is a unit. It follows that, in the quadruplet $\{u^{\delta m}\delta'\}$ $(\delta, \delta' \in \{\pm 1\})$, $u^m$ is the only unit which is greater than 1. In other words the units $v > 1$ are precisely the $u^m$, $m \in \mathbb{N}$. Clearly $u^{m+1} > u^m$, and therefore

**(1.5)** *The normalised fundamental unit of $K$ is the minimal unit $u$ with $u > 1$.*

We now wish to derive a criterion for $v = a + b\sqrt{d}$ to be fundamental, in terms of the rationals $a, b$. Clearly the units $c + e\sqrt{d} > 1$ are precisely those in each quadruplet with $c > 0$, $e > 0$.

**(1.6)** *The unit $v = a + b\sqrt{d}$ $(a, b > 0)$ is the normalised fundamental unit iff $a < c$ for all other units of $K$, $\eta = c + e\sqrt{d}$, $\eta > 1$.*

Note: this criterion is independent of the choice of embedding $K \hookrightarrow \mathbb{R}$.

Before proving the result, we briefly note that we cannot replace $a$ by $b$ in the above, since $\frac{1}{2}(1 + \sqrt{5})$ is fundamental in $\mathbb{Z}[\sqrt{5}]$ but $\left(\frac{1}{2}(1 + \sqrt{5})\right)^2 = \frac{1}{2}(3 + \sqrt{5})$.

*Proof.* Let $u = p + q\sqrt{d}$ denote the normalised fundamental unit of $K$, and let $v$ denote a unit of $K$, which is greater than 1; then, as has been shown, $v = u^n$ for some $n \in \mathbb{N}$. If we write $u^m = p_m + q_m\sqrt{d}$, for $m \in \mathbb{N}$, then it suffices to show that $\{p_m\}$ is a strictly monotonic sequence. Since $u^{m+1} = u \cdot u^m$, we obtain the recursion formula

$$\begin{aligned} p_{m+1} + q_{m+1}\sqrt{d} &= (p + q\sqrt{d})(p_m + q_m\sqrt{d}) \\ &= (pp_m + dqq_m) + (pq_m + qp_m)\sqrt{d}. \end{aligned}$$

If $d \not\equiv 1 \bmod (4)$, then $p$ and $q$ are positive integers, so by induction $p_m$ and $q_m$ are positive integers and it is clear from this formula that $p_{m+1} = pp_m + dqq_m$; hence $\{p_m\}$ is indeed strictly monotonic. If $d \equiv 1 \bmod (4)$, then we can argue as before, provided $p \geq 1$. However, if $p = \frac{1}{2}$ then we would need $\frac{1}{4} - q^2 d = \pm 1$; hence $q = \frac{1}{2}$, $d = 5$. In this case we suppose that $v = \frac{1}{2} + b\sqrt{5}$ is a unit of $K$, greater than 1; then $N_{K/\mathbb{Q}}(v) = \pm 1$, implies that $\frac{1}{4} - 5b^2 = \pm 1$, and this shows that $v = u$.

□

We now briefly consider two illustrative examples:

$K = \mathbb{Q}(\sqrt{2})$: $1+\sqrt{2}$ is a unit and clearly the first coefficient is minimal; thus by (1.6), it must be fundamental.

$K = \mathbb{Q}(\sqrt{3})$: $2+\sqrt{3}$ is a unit. To show that 2 is minimal, we note that for $b \neq 0$, $|N_{K/\mathbb{Q}}(1+b\sqrt{3})| = |3b^2 - 1| \geq 2$. Hence $2+\sqrt{3}$ is fundamental.

This technique shows how to check whether a given unit is fundamental or not; it does not, however, provide a practical method for constructing fundamental units. Indeed the construction of units by *ad hoc* means rapidly becomes impracticable: for instance $\mathbb{Q}(\sqrt{46})$ has normalised fundamental unit $24335 + 3588\sqrt{46}$. There is an efficient algorithm for constructing a fundamental unit of a real quadratic field by use of continued fractions. The reader is referred to Davenport's book [D] for an account of this topic.

We saw that the norm $N_{K/\mathbb{Q}}(\eta)$ of a unit in the quadratic field $K$ is $+1$ or $-1$. One important question, both for the problem of solving the equation $x^2 - y^2 d = -1$ in integers (see VII, §2) and in the determination of certain classgroups (see below), is: does there exist a unit $\eta$ of a real quadratic field $K$, with $N_{K/\mathbb{Q}}(\eta) = -1$? Clearly if a fundamental unit $u$ has norm 1, then so does every unit in $K$. Therefore the problem comes down to the determination of the sign of the norm of a fundamental unit. A first, easy result is:

**(1.7)** *If $d_K$ is divisible by a prime $p \equiv -1 \bmod (4)$, then $N_{K/\mathbb{Q}}(u) = 1$.*

*Proof.* By (II.1.34), the norm of any algebraic integer of $K$ is congruent to an integer square mod$(d)$, hence mod$(p)$. But for $p \equiv -1 \bmod (4)$, $\mathbb{F}_p^*$ has no element of order 4, therefore -1 is not a square mod $(p)$.

The question of the sign of the norm of the fundamental unit leads us to some new concepts which will also provide us with a finer analysis of both the unit group and the classgroup. For the time being let $N$ denote an algebraic number field, which is not necessarily quadratic. An element $a$ of $N$ is said to be *totally positive* if $a^\sigma$ is positive for all embeddings $\sigma: N \to \mathbb{R}$. The totally positive elements of $N$ form a subgroup $N_+$ of $N^*$, and we write $U_N^+ = U_N \cap N_+$ for the group of totally positive units. Note that certainly $N^{*2} \subset N_+$, $U_N^2 \subset U_N^+$. A principal ideal of $\mathfrak{o}_N$ is said to be totally positive if it can be written in the form $a\mathfrak{o}_N$ for $a \in N_+$. The totally positive principal ideals form a subgroup $P_N^+$ of $P_N$,the principal fractional $\mathfrak{o}_N$-ideals. We define the *narrow* ideal class group of $N$ to be $C_N^+ = I_N/P_N^+$, with $I_N$ the group of all fractional $\mathfrak{o}_N$-ideals. Note that, if $N$ is totally imaginary, i.e. has no real embeddings, then total positivity is no restriction, and so $N_+ = N^*$ etc. In general we have a commutative diagram with exact rows:

$$\begin{array}{ccccccccc} 0 & \longrightarrow & P_N^+ & \longrightarrow & I_N & \longrightarrow & C_N^+ & \longrightarrow & 0 \\ & & \downarrow{\scriptstyle \theta_N} & & \| & & \downarrow{\scriptstyle \pi_N} & & \\ 0 & \longrightarrow & P_N & \longrightarrow & I_N & \longrightarrow & C_N & \longrightarrow & 0 \end{array} \tag{1.8}$$

which can be completed by the map $\pi_N$; it follows that $\pi_N$ is surjective and

$$\ker \pi_N \cong \operatorname{coker} \theta_N. \tag{1.9}$$

We also have a commutative diagram with exact rows and injective columns:

$$\begin{array}{ccccccccc} 0 & \longrightarrow & U_N^+ & \longrightarrow & N_+ & \longrightarrow & P_N^+ & \longrightarrow & 0 \\ & & \downarrow & & \downarrow & & \downarrow{\scriptstyle \theta_N} & & \\ 0 & \longrightarrow & U_N & \longrightarrow & N^* & \longrightarrow & P_N & \longrightarrow & 0 \end{array} \tag{1.10}$$

Recall that for any non-zero real number $x$

$$\operatorname{sign}(x) = \frac{x}{|x|} = \pm 1.$$

Let $s$ denote the number of embeddings $N \hookrightarrow \mathbb{R}$. To each $a \in N^*$ we associate the signature vector

$$(\operatorname{sign}(a^{\sigma_1}), \operatorname{sign}(a^{\sigma_2}), \ldots, \operatorname{sign}(a^{\sigma_s}))$$

where the $\sigma_i$ run through the real embeddings of $N$, and where each $\operatorname{sign}(a^{\sigma_i}) = \pm 1$. This gives a homomorphism $N^* \to < \pm 1 >^{(s)}$. The kernel is clearly $N_+$ and by (II.2.14), the homomorphism is surjective. Thus

$$N^*/N_+ \cong < \pm 1 >^{(s)} \tag{1.11}$$

and from (1.9) and (1.10) we get an exact sequence

$$0 \to U_N/U_N^+ \to < \pm 1 >^{(s)} \to \ker \pi_N \to 0. \tag{1.12}$$

Note that $U_N/U_N^+$ contains the "canonical" non-trivial element represented by $-1$. In particular

**(1.13)** $\ker \pi_N$ *is a 2-group of order dividing* $2^{s-1}$.

We now turn to the case of a real quadratic field $K$ – so that now $s = 2$. We have

**(1.14)** $\ker \pi_K$ *has order* 1 *or* 2 *according as whether the norm of a fundamental unit* $u$, $N_{K/\mathbf{Q}}(u)$, *is* $-1$ *or* $+1$.

*Proof.* We already know that $\ker \pi_K$ is of order dividing 2. By the exactness of (1.12) $\ker \pi_K = 1$ precisely when $U_N/U_N^+$ has order 4, i.e. if for each signature vector $(\delta_1, \delta_2) \in <\pm 1>^{(s)}$, there is a unit $\eta$ with $\mathrm{sign}(\eta^{\sigma_i}) = \delta_i$ $(i = 1, 2)$. If this is the case, let $\mathrm{sign}(\eta^{\sigma_1}) = 1$, $\mathrm{sign}(\eta^{\sigma_2}) = -1$; then $N_{K/\mathbf{Q}}(\eta) = \eta^{\sigma_1}\eta^{\sigma_2} = -1$. Conversely, if $N_{K/\mathbf{Q}}(\eta) = -1$; then $\mathrm{sign}(\eta^{\sigma_1}) \neq \mathrm{sign}(\eta^{\sigma_2})$, $\mathrm{sign}(\eta^{\sigma_1}) = 1$, $\mathrm{sign}(\eta^{\sigma_2}) = -1$ say; hence $1, -1, \eta, -\eta$ are four units with distinct signature vectors.

□

Now let $K$ be an arbitrary quadratic field. If $K$ is imaginary, then of course $C_K^+ = C_K$. Let $\mathcal{S}$ denote the group of $\mathfrak{o}_K$-ideals of the form $\prod_j \mathfrak{p}_j^{v_j}$, where the $\mathfrak{p}_j$ denote the ramified prime ideals of $K$. Mapping $\prod \mathfrak{p}_j^{v_j} \mapsto c^+(\prod \mathfrak{p}_j^{v_j})$, where $c^+(\mathfrak{a})$ denotes the (narrow) class of $\mathfrak{a}$ in $C_K^+$, gives a homomorphism

$$\nu_K : \mathcal{S} \to C_K.$$

**Theorem 39.**

(a) $\mathrm{Im}\,\nu_K = C_{K,2}^+$ *(the subgroup of* $C_K^+$ *of elements of order dividing* 2*)*.

(*b*) $\ker \nu_K \supset \mathcal{S}^2$ *and* $[\ker \nu_K : \mathcal{S}^2] = 2$, *i.e.* $\nu_K$ *induces a surjection*

$$\nu' : \mathcal{S}/\mathcal{S}^2 \to C_{K,2}^+$$

*whose kernel has order* 2.

Applying (1.14) we have

**Corollary 1.** *If* $K$ *is imaginary, or if* $K$ *is real and the norm of the fundamental unit is* $-1$, *then* $C_{K,2}$ *is an elementary* 2*-group of order* $2^{g-1}$, *where* $g$ *denotes the number of distinct primes dividing* $d_K$. *If* $K$ *is real and the norm of the fundamental unit is* $+1$, *then* $C_{K,2}$ *is an elementary* 2*-group of order* $2^{g-2}$.

For brevity we write $h_K = |C_K|$ and $h_K^+ = |C_K^+|$.

**Corollary 2.** *If $d_K$ has only one prime divisor then $h_K$ is odd; furthermore, if $K$ is real, then the norm of the fundamental unit is $-1$.*

Note that, in particular, this corollary applies to $\mathbb{Q}(\sqrt{p})$, if $p \equiv 1 \bmod (4)$; $\mathbb{Q}(\sqrt{-p})$, if $p \equiv -1 \bmod (4)$.

First we need two preliminary results.

**(1.15)** *Let $E = F(\sqrt{d})/F$ denote a quadratic extension of fields of characteristic zero, and suppose $a \in E$ has the property that $N_{E/F}(a) = 1$. Then there exists $b \in E^*$ such that $a = b^\sigma/b$, where $\sigma$ denotes the non-trivial automorphism of $E/F$.*

*Proof.* If $a = -1$, we may take $b = \sqrt{d}$. Otherwise we set $b = (1+a)^{-1}$, for then

$$\frac{1+a}{1+a^\sigma} = \frac{(1+a)a}{(1+a^\sigma)a} = \frac{(1+a)a}{a+aa^\sigma} = \frac{(1+a)a}{1+a} = a.$$

□

This lemma is in fact a particular case of a more general result, known as Hilbert's Theorem 90, which describes the elements of norm 1 for any finite cyclic field extension.

The next result which we need follows at once from (III.1.21).

**(1.16)** *Let $\tau$ denote the non-trivial automorphism of $K/\mathbb{Q}$, and let $\mathfrak{a}$ denote an $\mathfrak{o}_K$-ideal with the property that $\mathfrak{a}^\tau = \mathfrak{a}$, then $\mathfrak{a} = r\mathfrak{q}$, where $r$ is a positive rational and $\mathfrak{q}$ is a square-free $\mathfrak{o}_K$-ideal which is divisible only by ramified primes of $K/\mathbb{Q}$.*

We can now prove Theorem 39. If $\mathfrak{p}$ denotes the $\mathfrak{o}_K$ prime ideal dividing a ramified prime $p$, then $\mathfrak{p}^2 = p\mathfrak{o}_K$ and so we immediately conclude that $\mathcal{S}^2 \subset \ker \nu_K$, and hence

$$\operatorname{im} \nu_K \subset C_{K,2}^+.$$

Again we let $\tau$ denote the non-trivial automorphism of $K/\mathbb{Q}$, and we let $c$ denote a class in $C_K^+$. By Theorem 15 and the Corollary to Theorem 21 we know that for any non-zero $\mathfrak{o}_K$-ideal $\mathfrak{a}$

$$\mathfrak{a}\mathfrak{a}^\tau = \mathcal{N}_{K/\mathbb{Q}}(\mathfrak{a})\mathfrak{o}_K = \mathbf{N}\mathfrak{a}\mathfrak{o}_K$$

and so $c^{1+\tau} = 1$; thus

$$c^{1-\tau} = c^{1+\tau}c^{-2\tau} = c^{-2\tau} \tag{1.17}$$

which implies the equalities

$$\begin{cases} C_{K,2}^+ = \ker(1-\tau\colon C_K^+ \to C_K^+) \\ (C_K^+)^2 = (C_K^+)^{1-\tau} \end{cases} \tag{1.18}$$

Now suppose that $c^2 = 1$, and choose an ideal $\mathfrak{a}$ with narrow class $c$. Then by (1.18) $\mathfrak{a}^{1-\tau} = a\mathfrak{o}_K$ for some $a \in K_+$, and so

$$N_{K/\mathbb{Q}}(a)\mathfrak{o}_K = a^{1+\tau}\mathfrak{o}_K = \mathfrak{a}^{(1-\tau)(1+\tau)} = \mathfrak{o}_K.$$

It follows that $N_{K/\mathbb{Q}}(a) = \pm 1$; moreover, because $a \in K_+$, $N_{K/\mathbb{Q}}(a) = 1$. Hence by (1.15) we can find $b \in K^*$ such that $a = b^{\tau-1}$; indeed, because $a \in K_+$, $b$ and $b^\tau$ must have the same sign under any real embedding of $K$; thus, on replacing $b$ by $-b$ if necessary, we may suppose $b$ to be totally positive. Since $\mathfrak{a}^{1-\tau} = a\mathfrak{o}_K = b^{\tau-1}\mathfrak{o}_K$, we conclude that $(\mathfrak{a}b)^\tau = \mathfrak{a}b$, and so, by (1.16), $\mathfrak{a}b = \mathfrak{a}_1 r$ where $\mathfrak{a}_1 \in \mathcal{S}$ and $r \in \mathbb{Q}_+$. We have therefore shown that $\nu_K(\mathfrak{a}_1) = c$. Thus indeed $\operatorname{im}\nu_K = C^+_{K,2}$.

Prior to our proving $[\ker\nu_K : \mathcal{S}^2] = 2$, we first note that

$$(U_K^+)^{1-\tau} = (U_K^+)^2. \tag{1.19}$$

Indeed, if $v \in {U_K}^+$, then $v^{1+\tau} = N_{K/\mathbb{Q}}(v) = 1$ and so (1.19) follows from the identity

$$v^{1-\tau} = v^{1+\tau+1-\tau} = v^2.$$

Suppose now that $\mathfrak{a} \in \ker\nu_K$, so that $\mathfrak{a} = a\mathfrak{o}_K$ for some $a \in K_+$. Since $\mathfrak{a} \in \mathcal{S}$, $\mathfrak{a}^\tau = \mathfrak{a}$, and so $a^{1-\tau} \in U_K$; moreover, because $a \in K_+$, we know $a^{1-\tau} \in {U_K}^+$. Given $\mathfrak{a}$, $a$ is only unique up to a totally positive multiplicative unit; however, from (1.19) we note that for $v \in {U_K}^+$

$$a^{1-\tau} = (av)^{1-\tau} \bmod ({U_K}^+)^2.$$

Thus the map $\mathfrak{a} \mapsto a^{1-\tau} \bmod ({U_K}^+)^2$ induces a group homomorphism

$$\rho\colon \ker\nu_K \to {U_K}^+/({U_K}^+)^2. \tag{1.20}$$

As we have already noted, for $\mathfrak{a} \in \mathcal{S}$, $\mathfrak{a}^2 = r\mathfrak{o}_K$ with $r \in \mathbb{Q}_+$, and so

$$\mathcal{S}^2 \subset \ker\rho. \tag{1.21}$$

Conversely, if $a \in K_+$ with $a\mathfrak{o}_K \in \ker\rho$, then by (1.19) we must have $a^{1-\tau} = v^{1-\tau}$ for some $v \in {U_K}^+$; this then implies that $(av^{-1})^\tau = av^{-1} \in \mathbb{Q}_+$, and so $a\mathfrak{o}_K = av^{-1}\mathfrak{o}_K \in \mathcal{S}^2$. This shows that $\ker\rho \subset \mathcal{S}^2$ and hence by (1.21)

$$\ker\rho = \mathcal{S}^2. \tag{1.22}$$

We now show that $\rho$ is surjective. Let $v \in {U_K}^+$ so that $N_{K/\mathbb{Q}}(v) = 1$; thus by (1.15) we can find $a \in K^\times$ such that $a^{1-\tau} = v$. Since $a^{1-\tau}$ is totally positive, $a$ and $a^\tau$ must have the same sign under any real embedding of $K$; hence, on multiplying $a$ by $-1$ if necessary, we may suppose $a$ to be totally positive. We then apply (1.16) and write $a = a_1 r$, where $r \in \mathbb{Q}_+$ and $a_1\mathfrak{o}_K \in \mathcal{S}$. Clearly $a_1^{1-\tau} = a^{1-\tau} = v$ and so $\rho(a_1\mathfrak{o}_K) = v \bmod (U_K^+)^2$. This now proves that

$$\ker\nu_K/\mathcal{S}^2 \cong U_K^+/(U_K^+)^2.$$

We complete our proof of the theorem by showing that in all cases

(1.23) $$[U_K^+ : (U_K^+)^2] = 2.$$

First, if $K$ is quadratic imaginary, then $U_K^+ = U_K = \mu_K$. Since $-1 \in \mu_K$ clearly $[\mu_K : \mu_K^2] = 2$. On the other hand if $K$ is real, then $U_K^+$ is an infinite cyclic group, and so again (1.23) holds.

□

Methods such as those used here occur repeatedly in algebraic number theory in many places, and they can be formalised in terms of what is known as the cohomology theory of groups.

We shall frequently want to study norm properties of quadratic fields. These are closely related to quadratic forms (over $\mathbb{Q}$ and over $\mathbb{Z}$) – as we have already indicated in the introduction and as we shall see here and again in Chapter VII. With $K$ still a quadratic field, let $\omega_1, \omega_2$ denote a $\mathbb{Z}$-basis of $\mathfrak{o}_K$. Thus, as $(x, y)$ runs over $\mathbb{Z} \times \mathbb{Z}$, $\omega_1 x + \omega_2 y$ will run over $\mathfrak{o}_K$ and

(1.24) $$N_{K/\mathbb{Q}}(\omega_1 x + \omega_2 y) = q(x, y)$$

where $q(x, y) = a_1 x^2 + bxy + a_2 y^2$ is the quadratic form with $a_i = N_{K/\mathbb{Q}}(\omega_i)$, $b = t_{K/\mathbb{Q}}(\omega_1 \omega_2^{\tau})$. We call $q = q_K$ the norm form (with respect to the basis $\omega_1, \omega_2$). We shall usually choose $\omega_1 = 1$ and

$$\omega_2 = \begin{cases} \sqrt{d} & \text{if } d \not\equiv 1 \bmod (4) \\ \frac{1}{2}(1 + \sqrt{d}) & \text{if } d \equiv 1 \bmod (4). \end{cases}$$

There are three ways in which a given prime number $p$ can occur as a norm; this hierarchy of norm properties is given by:

(1.24.$a$) $\quad p = \mathbf{N}\mathfrak{p} \quad$ for some prime ideal $\mathfrak{p}$ of $\mathfrak{o}_K$

(1.24.$b$) $\quad p = q_K(x, y) \quad x, y \in \mathbb{Q}$

(1.24.$c$) $\quad p = q_K(x, y) \quad x, y \in \mathbb{Z}$.

Presently we shall see that these norm properties are intimately related to the Kummer decomposition law for $p$ in $K$ in (1.1). The norm form $q_K$ has particularly nice properties when $C_K^+ = \{1\}$. By Theorem 39 this can only happen if $|d_K|$ is a prime power. There are precisely nine imaginary quadratic fields with class number $h_K^+ = 1$, i.e. with $C_K^+ = \{1\}$. In the real quadratic case one does not even know whether there are infinitely many fields $K$ with $h_K = 1$.

*Examples.*

(1) $K = \mathbb{Q}(i)$ $(i^2 = -1)$. Then $\mathfrak{o}_K = \mathbb{Z}[i]$ and from (IV.1.2) we know that $\mathfrak{o}_K$ is a Euclidean domain, and so $h_K = 1$. Clearly in this case

$$q_K(x, y) = x^2 + y^2.$$

(2) $K = \mathbb{Q}(\omega)$ ($\omega^3 = 1$, $\omega \neq 1$). As in (1) one shows that $C_K^+ = C_K = \{1\}$. In this case

$$q_K(x,y) = x^2 - xy + y^2.$$

(3) $K = \mathbb{Q}(\sqrt{2})$. Using the Minkowski bound of Theorem 35, we know that every class in $C_K$ is represented by an $\mathfrak{o}_K$-ideal with absolute norm less than or equal to $\sqrt{2}$. Therefore $h_K = 1$; in fact $h_K^+ = 1$ since the norm of the fundamental unit $1 + \sqrt{2}$ is $-1$ (which agrees with the Corollary 2 to Theorem 39). The norm form here is

$$q_K(x,y) = x^2 - 2y^2.$$

Our main result here for quadratic forms will, in particular, give famous classical theorems on the forms mentioned above.

**(1.25)** *Suppose that for $K = \mathbb{Q}(\sqrt{d})$, we have $h_K^+ = 1$.*

(*a*) *A prime number $p$ is of the form $p = q_K(x,y)$ for some $(x,y) \in \mathbb{Z} \times \mathbb{Z}$ iff*

$$(*) \quad \text{either} \quad \left(\frac{d_K}{p}\right) = 1 \textit{ if } p > 2, \textit{ resp. } d \equiv 1 \bmod (8) \textit{ if } p = 2$$
$$\text{or} \quad p \mid d_K.$$

(*b*) *Let $a$ be a positive integer and let $a = \prod_i p_i^{r_i}$ be its prime factorisation. Then $a = q_K(x,y)$ for some $(x,y) \in \mathbb{Z} \times \mathbb{Z}$ iff for each $i$, with $r_i$ odd we have*

$$p_i = q_K(x_i, y_i)$$

*for some $(x_i, y_i) \in \mathbb{Z} \times \mathbb{Z}$.*

For the proof of this result we need a simple lemma.

**(1.26)** *Let $E/F$ denote a Galois extension of number fields. Let $\mathfrak{p}$ denote a prime ideal of $\mathfrak{o}_F$ with $\mathfrak{p}^r = \mathcal{N}_{E/F}(\mathfrak{a})$ for some $\mathfrak{o}_E$-fractional ideal $\mathfrak{a}$; then $f_\mathfrak{p}(E/F) \mid r$.*

*Proof.* Write $\mathfrak{a} = \mathfrak{a}' \prod_j \mathfrak{P}_j^{t_j}$, where the $\mathfrak{P}_j$ are the prime ideals above $\mathfrak{p}$, and $\mathfrak{a}'$ is coprime to all such prime ideals. Then $\mathcal{N}_{E/F}(\mathfrak{a}') = \mathfrak{o}_F$, $\mathcal{N}_{E/F}(\prod_j \mathfrak{P}_j^{t_j}) = \mathfrak{p}^{f \sum t_j}$, where $f = f_\mathfrak{p}(E/F)$. □

*Proof of* (1.25). Part (*a*). By the Kummer decomposition law (see (1.1)), the condition (*) holds precisely when $p\mathbb{Z} = \mathcal{N}_{K/\mathbb{Q}}(\mathfrak{P})$ for some prime ideal $\mathfrak{P}$ of $\mathfrak{o}_K$. If this is the case, then as $\mathfrak{P} = \alpha\mathfrak{o}_K$ with $\alpha$ totally positive, we have $p\mathbb{Z} = \mathcal{N}_{K/\mathbb{Q}}(\alpha\mathfrak{o}_K)$, i.e. $p = \pm N_{K/\mathbb{Q}}(\alpha)$. As $p > 0$ and

$N_{K/\mathbb{Q}}(\alpha) > 0$, we must obtain the positive sign, and so $p = q_K(x, y)$ where $\alpha = \omega_1 x + \omega_2 y$.

Conversely, if $p = q_K(x, y)$, then $p = N_{K/\mathbb{Q}}(\alpha)$ for some totally positive $\alpha \in \mathfrak{o}_K$; hence $p\mathbb{Z} = \mathcal{N}_{K/\mathbb{Q}}(\alpha\mathfrak{o}_K)$. By (1.26) $f(K/\mathbb{Q}) = 1$, i.e. $p\mathbb{Z} = \mathcal{N}_{K/\mathbb{Q}}\mathfrak{P}$, as above.

(*b*) Trivially, any integral square is represented as $q_K(x', y')$, with $(x', y') \in \mathbb{Z} \times \mathbb{Z}$. Next suppose $a = q_K(x, y)$ for $(x, y) \in \mathbb{Z} \times \mathbb{Z}$; then $a\mathbb{Z} = \mathcal{N}_{K/\mathbb{Q}}(\mathfrak{a})$ for some fractional $\mathfrak{o}_K$-ideal $\mathfrak{a}$, and so for each $i$, $p^{r_i}\mathbb{Z} = \mathcal{N}_{K/\mathbb{Q}}(\mathfrak{a}_i)$ for some $\mathfrak{o}_K$-fractional ideal $\mathfrak{a}_i$. If $r_i$ is odd, then by (1.26) $f_{p_i}(K/\mathbb{Q})$ divides $(2, r_i) = 1$; hence by part (*a*): $p_i = q_K(x_i, y_i)$, where $(x_i, y_i) \in \mathbb{Z} \times \mathbb{Z}$. Conversely, if each $p_i = q_K(x_i, y_i)$, then we have $p_i = N_{K/\mathbb{Q}}(\beta_i)$, for some $\beta_i \in \mathfrak{o}_K$ for each $i$, where $r_i$ is odd. Trivially, if $r_i = 2s_i$, then $p_i^{r_i} = N_{K/\mathbb{Q}}(p_i^{s_i})$, and so

$$a = N_{K/\mathbb{Q}}\left(\prod_{r_i \text{ odd}} \beta_i^{r_i} \prod_{r_i \text{ even}} p_i^{s_i}\right) = q_K(x, y)$$

as required.

□

It is interesting to see just how neatly the above result solves the classical problems of determining which prime numbers can be written in the form $x^2 + y^2$, $x^2 - xy + y^2$, $x^2 - 2y^2$ (with $x, y \in \mathbb{Z}$ of course).

Applying (1.25) to $\mathbb{Q}(\sqrt{-1})$, where $q_K(x, y) = x^2 + y^2$, we see that a prime number $p = x^2 + y^2$ iff $p = 2$ or if $p > 2$ and $\left(\frac{-4}{p}\right) = 1$. However, $-1$ is a square mod $(p)$ iff $p \equiv 1 \bmod (4)$ and so, as shown in the Introduction, we see that $p = x^2 + y^2$ iff $p = 2$ or $p \equiv 1 \bmod (4)$.

When we apply (1.25) to $\mathbb{Q}(\omega)$ and $\mathbb{Q}(\sqrt{-2})$, we find that:

$$p = x^2 - xy + y^2 \quad \text{iff } p = 3 \text{ or } p \equiv 1 \bmod (3);$$

$$p = x^2 - 2y^2 \quad \text{iff } p = 2 \text{ or } p \equiv 1, 7 \bmod (8).$$

These results of (1.25) do not extend when $C_K^+ \neq \{1\}$. For the connection with quadratic forms over $\mathbb{Z}$ see (VII,§2). Here we shall be concerned with a weakening of the above situation, where we consider forms over $\mathbb{Q}$ for arbitrary quadratic $K$. Observe that any norm $b = N_{K/\mathbb{Q}}(\beta)$, $\beta \in K$ can be written as $b = x^2 - dy^2$ with $(x, y) \in \mathbb{Q} \times \mathbb{Q}$. Note also that if $b$ is positive , then $\beta$ may be chosen to be totally positive. For, if $K$ is real with embeddings $\sigma_i \colon K \hookrightarrow \mathbb{R}$ $(i = 1, 2)$, then $\beta^{\sigma_1}\beta^{\sigma_2}$ is positive; hence either $\beta^{\sigma_1}$ and $\beta^{\sigma_2}$ are both positive, as required, or $(-\beta)^{\sigma_1}$, $(-\beta)^{\sigma_2}$ are both positive, and we can replace $\beta$ by $-\beta$.

**(1.27)** *Let $p$ be a prime number. The following three conditions are equivalent:*

(a) $p = \mathbf{N}\mathfrak{p}$ *for some prime ideal* $\mathfrak{p}$ *of* $\mathfrak{o}_K$ *(i.e. by* (1.1) *either* $p \mid d_K$, *or if* $p \neq 2$ *(resp.* $p = 2$*) then* $\left(\frac{d}{p}\right) = 1$ *(resp.* $d \equiv 1 \bmod (8)$*))* *and moreover the class* $c^+(\mathfrak{p})$ *of* $\mathfrak{p}$ *in* $C_K^+$ *is a square, i.e.* $c^+(\mathfrak{p}) \in (C_K^+)^2$.

(*b*) $p = N_{K/\mathbb{Q}}(\alpha)$ *for some totally positive* $\alpha \in K$, *i.e.* $p = x^2 - dy^2$ *for some* $(x, y) \in \mathbb{Q} \times \mathbb{Q}$.

(*c*) *There exist integers* $u, v, w$ *with* $(u, v, w) = 1$ *such that* $u^2 - dv^2 - pw^2 = 0$.

*Remark.* The choice of $\mathfrak{p}$ in (*a*) clearly does not affect the relation on $c^+(\mathfrak{p})$.

*Proof.* Recall from (1.18) that

$$(C_K^+)^2 = (C_K^+)^{1-\tau}. \tag{1.28}$$

(*a*)$\Rightarrow$(*b*) From (1.28), we can write $\mathfrak{p} = \alpha\mathfrak{a}^{1-\tau}$ with $\alpha$ totally positive. By (III.1.22) $\mathcal{N}_{K/\mathbb{Q}}(\mathfrak{a}^{1-\tau}) = \mathbb{Z}$, and so $p\mathbb{Z} = \mathcal{N}_{K/\mathbb{Q}}(\mathfrak{p}) = \mathcal{N}_{K/\mathbb{Q}}(\alpha\mathfrak{o}_K)$; hence $p = \pm N_{K/\mathbb{Q}}(\alpha)$. However, $p$ and $N_{K/\mathbb{Q}}(\alpha)$ are both positive, whence $p = N_{K/\mathbb{Q}}(\alpha)$.

(*b*)$\Rightarrow$(*a*) In this case $p\mathbb{Z} = \mathcal{N}_{K/\mathbb{Q}}(\alpha\mathfrak{o}_K)$, and so by (1.26) $p = \mathbf{N}\mathfrak{p}$ for some prime ideal $\mathfrak{p}$. Therefore $\mathcal{N}_{K/\mathbb{Q}}(\alpha^{-1}\mathfrak{p}) = \mathbb{Z}$, and so by (III.1.22) $\alpha^{-1}\mathfrak{p} = \mathfrak{a}^{1-\tau}$, i.e. $c^+(\mathfrak{p}) = c^+(\mathfrak{a})^{1-\tau}$; and the result then follows from (1.28).

The equivalence of (*b*) and (*c*) is completely obvious.

□

By Theorem 39 we know certain classes in $C_K^+$ – namely those in $C_{K,2}^+$ – and by the last result it is of substantial interest to know whether they are squares in $C_K^+$. In other words, by Theorem 39, $2^{g-1} \mid h_K^+$, where $g$ denotes the number of prime divisors of $d_K$; the question now is whether a higher power of 2 divides $h_K^+$? There are indeed criteria for this. We shall deal here with the special case when $g = 2$, so that $C_{K,2}^+$ is a cyclic group of order 2.

The theorem that we shall establish (Theorem 41) is very closely connected with the quadratic reciprocity law, and in fact we shall both state and prove this here (Theorem 40). This will be the first of four proofs that we shall give.

We briefly recall some properties of the Legendre Symbol, and also define certain residue class characters mod (4) and mod (8).

If $p$ is an odd prime, and if $a$ is an integer which is not divisible by $p$, then we define

$$\left(\frac{a}{p}\right) = \pm 1 \quad \text{and} \quad \left(\frac{a}{p}\right) \equiv a^{\frac{p-1}{2}} \bmod (p). \tag{1.29}$$

It follows that

$$\left(\frac{ab}{p}\right) = \left(\frac{a}{p}\right)\left(\frac{b}{p}\right), \quad \left(\frac{a}{p}\right) = 1 \iff a \equiv x^2 \bmod (p),$$

$$\text{and} \sum_{b=1}^{p-1} \left(\frac{b}{p}\right) = 0. \tag{1.30}$$

For an odd integer $a$, we define

$$\lambda_4(a) = \begin{cases} 1 & \text{if } a \equiv 1 \bmod (4) \\ -1 & \text{if } a \equiv -1 \bmod (4) \end{cases} \tag{1.31}$$

i.e. $\lambda_4(a) \equiv a \bmod (4)$.

$$\lambda_8(a) = \begin{cases} 1 & \text{if } a \equiv \pm 1 \bmod (8) \\ -1 & \text{if } a \equiv \pm 3 \bmod (8) \end{cases} \tag{1.32}$$

i.e. $\lambda_8(a) = (-1)^{(a^2-1)/8}$.

**Theorem 40 (Quadratic Reciprocity Law).**
*For distinct odd primes $p, q$*

$$\left(\frac{p}{q}\right)\left(\frac{q}{p}\right) = (-1)^{\frac{p-1}{2} \cdot \frac{q-1}{2}}.$$

*Also, for a prime $p > 2$*

$$\left(\frac{-1}{p}\right) = \lambda_4(p) \qquad (\textit{1st subsidiary law}),$$

$$\left(\frac{2}{p}\right) = \lambda_8(p) \qquad (\textit{2nd subsidiary law}).$$

*Remark 1.* The Quadratic Reciprocity Law can be viewed as a "two prime interchange" rule for the behaviour of one prime in a quadratic field whose only ramified prime is the other one. Thus, to take the simplest case, if $p \equiv 1 \equiv q \bmod (4)$, so that $\left(\frac{p}{q}\right) = \left(\frac{q}{p}\right)$; then the law says that $p$ will split in $\mathbb{Q}(\sqrt{q})$ iff $q$ splits in $\mathbb{Q}(\sqrt{p})$.

*Remark 2.* The first subsidiary law is of course entirely elementary: by (1.29)

$$\left(\frac{-1}{p}\right) \equiv (-1)^{(p-1)/2} \bmod (p).$$

But $\left(\frac{-1}{p}\right)$ and $(-1)^{(p-1)/2}$ are both $\pm 1$; since $+1 \not\equiv -1 \bmod (p)$ it follows that $\left(\frac{-1}{p}\right) = (-1)^{(p-1)/2}$. We can, however, give a more "complicated" proof, which shows how this law reflects the comparison of norms in two quadratic fields. Indeed, if $\left(\frac{-1}{p}\right) = 1$, then by (1.25) $p = x^2 + y^2$ $(x, y \in \mathbb{Z})$. But $p \equiv x^2 + y^2 \bmod (4)$ implies $p \equiv 1 \bmod (4)$. This in turn tells us, by Corollary 2 to Theorem 39, that the fundamental unit $a + b\sqrt{p}$ has norm $a^2 - b^2 p = -1$, i.e. $4a^2 \equiv -4 \bmod (p)$, and so $\left(\frac{-1}{p}\right) = 1$.

*Proof of Theorem 40.* First let $p, q$ denote distinct odd primes, with $p \equiv 1 \bmod (4)$, and put $q^* = (-1)^{(q-1)/2} \cdot q$, so that $q^* \equiv 1 \bmod (4)$. If $\left(\frac{q}{p}\right) = 1$, then $\left(\frac{q^*}{p}\right) = 1$, and so, by (1.1), in the field $K = \mathbb{Q}(\sqrt{q^*})$, we have $p = \mathbf{N}\mathfrak{p}$ for a prime ideal $\mathfrak{p}$ of $\mathfrak{o}_K$. Moreover, by Corollary 2 to Theorem 39, the class $c^+(\mathfrak{p})$ in $C_K^+$ is a square; hence by (1.27) we have integers $u, v, w$, with $(u, v, w) = 1$, such that $u^2 - q^* v^2 - p w^2 = 0$. Here $q \nmid w$, as otherwise $q \mid (u, v, w)$. Therefore $\left(\frac{p}{q}\right) = 1$. Next if $\left(\frac{p}{q}\right) = 1$, then, since $p^* = p$, in the same way we get $u_1^2 - p v_1^2 - q w_1^2 = 0$ with $(u_1, v_1, w_1) = 1$; this then implies $\left(\frac{q}{p}\right) = 1$ In summary, if $p \equiv 1 \bmod (4)$, then we have shown that if one of $\left(\frac{p}{q}\right)$, $\left(\frac{q}{p}\right)$ is 1, then so is the other, i.e. $\left(\frac{p}{q}\right) = \left(\frac{q}{p}\right)$.

Next, with $q$ and $q^*$ as above, assume that $\lambda_8(q) = 1$, i.e. $\lambda_8(q^*) = 1$. Since 2 splits in $\mathbb{Q}(\sqrt{q^*})$, the same argument as above tells us that $u^2 - q^* v^2 - 2w^2 = 0$ for some coprime integers $u, v, w$; hence $\left(\frac{2}{q}\right) = 1$. Conversely, if $\left(\frac{2}{q}\right) = 1$, then $q = \mathbf{N}\mathfrak{q}$ for some prime ideal in $K = \mathbb{Q}(\sqrt{2})$. By Theorem 39 $h_K^+$ is odd, and so by (1.27) we obtain an equation $u_1^2 - 2v_1^2 - q w_1^2 = 0$ with coprime integers $u_1, v_1, w_1$. In particular $q w_1^2 \equiv u_1^2 - 2v_1^2 \bmod (8)$, which implies that $q \equiv \pm 1 \bmod (8)$, i.e. $\lambda_8(q) = 1$.

Finally let $p, q$ be distinct primes with $p \equiv q \equiv -1 \bmod (4)$. We now work in the field $K = \mathbb{Q}(\sqrt{pq})$. Since $p$ and $q$ both ramify in $K$, $p = \mathbf{N}\mathfrak{p}$, $q = \mathbf{N}\mathfrak{q}$ for prime ideals $\mathfrak{p}, \mathfrak{q}$ of $\mathfrak{o}_K$. We prove

**(1.33)** *If* $c^+(\mathfrak{p})$ *is a square, then* $\left(\frac{p}{q}\right) = 1$ *and* $\left(\frac{q}{p}\right) = -1$.

*Proof.* Indeed, by (1.27) we get an equation

$$u^2 - pqv^2 - pw^2 = 0$$

with coprime integers $u, v, w$. This implies that $u = pu_1$ and so

$$pu_1^2 - qv^2 - w^2 = 0 \qquad (u_1, v, w) = 1.$$

Now $(u_1, q) = 1$ and $(v, p) = 1$, so that we can deduce the required result (as $\left(\frac{-q}{p}\right) = -\left(\frac{q}{p}\right)$).

□

By Theorem 39 there is a unique class $c_0 \in C_K^+$ of order 2, and the classes $c^+(\mathfrak{p})$, $c^+(\mathfrak{q})$ both have order dividing 2. As $p \equiv -1 \bmod (4)$, by (1.7) the norm of the fundamental unit of $K$ is 1; therefore, as $\sqrt{pq}$ is positive under one real embedding and negative under the other, $\sqrt{pq}\mathfrak{o}_K \notin P_K^+$, i.e. $c_0 = c^+(\sqrt{pq}\mathfrak{o}_K)$. As $c^+(\mathfrak{p})$ and $c^+(\mathfrak{q})$ both have order at most two and since $c^+(\mathfrak{p})c^+(\mathfrak{q}) = c_0$, it follows that, after reordering $p$ and $q$ if necessary, $c^+(\mathfrak{p}) = 1$ and $c^+(\mathfrak{q}) = c_0$. By (1.33) we have $\left(\frac{p}{q}\right)\left(\frac{q}{p}\right) = -1$, and this completes the proof of Theorem 40. For future reference we note that by (1.33) $c^+(\mathfrak{q})$, and hence $c_0$, is not a square.

□

For the next theorem we consider quadratic fields $K$ with exactly two distinct prime divisors of $d_K$. By Theorem 39 $h_K^+$ is even, and there is a unique class $c_0$ of order 2. Hence $4 \mid h_K^+$ iff $c_0$ is a square in $C_K^+$.

**Theorem 41.**

(*a*) *Let* $K = \mathbb{Q}(\sqrt{\delta pq})$ *for distinct odd primes* $p, q$ *with* $\delta = \pm 1$ *such that* $\delta pq \equiv 1 \bmod (4)$. *Then* $4 \mid h_K^+$ *iff* $\left(\frac{p}{q}\right) = \left(\frac{q}{p}\right) = 1$.

(*b*) *Let* $K = \mathbb{Q}(\sqrt{\delta p})$ *for* $p$ *an odd prime,* $\delta = \pm 1$ *such that* $\delta p \equiv -1 \bmod (4)$. *Then* $4 \mid h_K^+$ *iff* $\lambda_4(p) = \lambda_8(p) = 1$.

(*c*) *Let* $K = \mathbb{Q}(\sqrt{2\delta p})$ *where* $\delta = \pm 1$, *and* $p$ *is an odd prime. Then* $4 \mid h_K^+$ *iff either*

(*i*) $\delta = 1$ *and* $\lambda_4(p) = \lambda_8(p) = 1$

*or*

(*ii*) $\delta = -1$ *and* $\lambda_8(p) = 1$.

Observe that by Theorem 39, if $\mathfrak{p}, \mathfrak{q}$ are the two ramified prime ideals of $\mathfrak{o}_K$, then the three groups

$$C_{K,2}^+ = \langle c_0 \rangle = \langle c^+(\mathfrak{p}), c^+(\mathfrak{q}) \rangle \tag{1.34}$$

coincide. This observation will be used throughout. Before proving Theorem 41, we first show

**(1.35)** *If* $p$ *is a prime* $\equiv 1 \bmod (4)$ *or if* $p = 2$, *and if* $a$ *is a positive rational; then there is a solution in coprime integers* $u, v, w$ *for*

$$u^2 - pv^2 - aw^2 = 0$$

*precisely when there is a solution to*

$$u^2 - pv^2 + aw^2 = 0.$$

*Proof.* Let $L = \mathbb{Q}(\sqrt{p})$. By (1.27) we have to show that if $b = N_{L/\mathbb{Q}}(\beta)$, for some $\beta \subset L^*$, then also $-b = N_{L/\mathbb{Q}}(\beta')$ for some $\beta' \in L^*$. This is immediate: for we can put $\beta' = \beta u$ with $u$ a unit with norm $-1$ (which exists by Corollary 2 of Theorem 39).

□

At the end of the proof of Theorem 40, we have already seen that when $p, q$ are two odd primes $\equiv -1 \bmod (4)$, then for $K = \mathbb{Q}(\sqrt{pq})$ we know that $c_0$ is not a square, and of course $\left(\frac{p}{q}\right) \neq \left(\frac{q}{p}\right)$.

We next consider a field $K = \mathbb{Q}(\sqrt{\delta rq})$, with $r, q$ distinct primes, not both $\equiv -1 \bmod (4)$ and $\delta = \pm 1$ as in the statement of the theorem. We show

**(1.36)** *If* $\mathfrak{r}$ *is the prime ideal of* $\mathfrak{o}_K$ *above* $r$ *and* $c^+(\mathfrak{r})$ *is a square, then*

(*a*) $\left(\frac{r}{q}\right) = 1$ *for* $q > 2$

(*b*) *for* $q = 2$ *and* $\delta = 1$, $\lambda_4(r)\lambda_8(r) = 1$

(*c*) *for* $q = 2$ *and* $\delta = -1$, $\lambda_8(r) = 1$.

*Proof.* By (1.27) $c^+(\mathfrak{r})$ being a square implies

$$u^2 - \delta qrv^2 - rw^2 = 0.$$

It is always to be understood, both here and in the sequel, that symbols $u, v, w$, with various subscripts, denote coprime integers. We can therefore write $u = ru_1$, and so

$$ru_1^2 - \delta qv^2 - w^2 = 0$$

with $q \nmid u_1$, which gives the result.

□

We have a converse (still for the same field $K$):

**(1.37)** *If* $r \not\equiv -1 \bmod (4)$ *and if*

$$\lambda_8(r) = 1 \quad \textit{when } q = 2$$

*or*

$$\left(\frac{r}{q}\right) = 1 \quad \textit{when } q > 2,$$

*then* $c^+(\mathfrak{r})$ *is a square.*

*Proof.* Since only $r$ ramifies in $\mathbb{Q}(\sqrt{r})$, by Corollary 2 to Theorem 39 $C^+_{\mathbb{Q}(\sqrt{r})}$ has odd order. So by (1.27), applied to $\mathbb{Q}(\sqrt{r})$, we deduce that

$$u^2 - rv^2 - qw^2 = 0$$

hence by (1.35)

$$u_1^2 - rv_1^2 - \delta q w_1^2 = 0.$$

Thus

$$ru_1^2 - (rv_1)^2 + \delta q r w_1^2 = 0$$

whence, by (1.27) applied to $K$, $c^+(\mathfrak{r})$ must be a square.

□

We now piece the above results together to prove the three parts of the theorem.

*Part* (*a*). As explained after the proof of (1.35) we may suppose here that $p \equiv 1 \bmod (4)$. If $c_0$ is a square, then by (1.34) $c^+(\mathfrak{p})$ must be a square; so, by (1.36), $\left(\frac{p}{q}\right) = 1$, and by Theorem 40 also $\left(\frac{q}{p}\right) = 1$. Next suppose that conversely $\left(\frac{p}{q}\right) = 1 = \left(\frac{q}{p}\right)$. If $q \equiv 1 \bmod (4)$, then by (1.37), $c^+(\mathfrak{p})$ and $c^+(\mathfrak{q})$ are squares; hence so is $c_0$. On the other hand, if $q \equiv -1 \bmod (4)$, then $K$ is imaginary and $c^+(\sqrt{-pq}\mathfrak{o}_K) = 1$, i.e. $c^+(\mathfrak{pq}) = 1$. By (1.37) $\left(\frac{p}{q}\right) = 1$ implies that $c^+(\mathfrak{p})$ is a square and so $c^+(\mathfrak{q})$ is a square. This therefore shows that $c_0$ is a square.

*Part* (*c*). Let $\mathfrak{s}$ denote the prime ideal above 2 in $\mathfrak{o}_K$, and, as previously, let $\mathfrak{p}$ denote the prime ideal of $\mathfrak{o}_K$ above $p$. First let $\delta = 1$. If $c_0$ is a square, i.e. $c^+(\mathfrak{s})$ and $c^+(\mathfrak{p})$ are squares, then, by (1.36), $\lambda_4(p)\lambda_8(p) = 1$ (taking $r = p$, $q = 2$) and $\left(\frac{2}{p}\right) = 1$ (taking $q = p$, $r = 2$); so, by Theorem 40, $\lambda_8(p) = 1$.

Conversely, if $\lambda_4(p) = \lambda_8(p) = \left(\frac{2}{p}\right) = 1$, we apply (1.37) with $q = 2$, $r = p$ to deduce that $c^+(\mathfrak{p})$ is a square; then apply it again with $r = 2$, $q = p$ to show that $c^+(\mathfrak{s})$ is a square. Therefore $c_0$ is a square.

Now take $\delta = -1$. Then $c^+(\mathfrak{sp}) = c^+(\sqrt{-2p}\mathfrak{o}_K) = 1$, i.e. $c^+(\mathfrak{p}) = c^+(\mathfrak{s}) = c_0$. If $c_0$ is a square, then, by (1.36), $\lambda_8(p) = 1$. Conversely, if $\lambda_8(p) = 1$, then apply (1.37) with $r = p$ and $q = 2$, to give us $c^+(\mathfrak{p})$ is a square in the case $p \equiv 1 \bmod(4)$. If $p \equiv -1 \bmod(4)$ but $\lambda_8(p) = 1$, then, working in the field $\mathbb{Q}(\sqrt{-p})$ and using Corollary 2 to Theorem 39 and

(1.27), we get an equation $u^2 + pv^2 - 2w^2 = 0$. By (1.35) we also have an equation

$$u_1^2 - pv_1^2 - 2w_1^2 = 0$$

whence

$$(pv_1)^2 - pu_1^2 + 2pw_1^2 = 0.$$

This implies that $c^+(\mathfrak{p})$ is a square.

$\square$

*Part* (*b*). First let $p \equiv 1 \bmod (4)$. Then $K$ is imaginary and

$$c^+(\mathfrak{p}) = c^+(\sqrt{-p}\mathfrak{o}_K) = 1.$$

Hence $c_0 = c^+(\mathfrak{s})$, with $\mathfrak{s}$ again the prime ideal of $\mathfrak{o}_K$ above 2. This will be a square iff

$$u^2 + pv^2 - 2w^2 = 0.$$

Clearly this implies $\lambda_8(p) = 1$.

Conversely, if $\lambda_4(p) = \lambda_8(p) = 1$, then $\left(\frac{2}{p}\right) = 1$. As $h^+_{\mathbb{Q}(\sqrt{2})}$ is odd, by Corollary 2 to Theorem 39, we get from (1.27)

$$u_1^2 - pv_1^2 - 2w_1^2 = 0$$

and hence by (1.35)

$$u^2 + pv^2 - 2w^2 = 0.$$

Next if $p \equiv -1 \bmod (4)$; then $K$ is real, and the norm of a fundamental unit must be 1. Hence $c^+(\sqrt{p}\mathfrak{o}_K) = c_0$. By (1.27) this is a square precisely when $u^2 - pv^2 - pw^2 = 0$, which obviously implies $p \equiv 1 \bmod (4)$ – a contradiction.

$\square$

In Chapter VI, §3 we shall give three further proofs of the quadratic reciprocity law, by entirely different means. In the meantime we shall not have to make use of this law – with the exception of the elementary first subsidiary law.

## §2 Biquadratic fields

A number field $K$ is called *biquadratic* here if it is the compositum of two distinct quadratic extensions of $\mathbb{Q}$; so that $K/\mathbb{Q}$ is a Galois extension whose Galois group is an elementary 2-group of order 4; we do, though, warn the reader that some sources use the term biquadratic for cyclic extensions of degree 4. In this section we shall mainly be concerned with describing methods for determining the units of $K$. We shall make some remarks concerning class numbers; however, we defer a full treatment of this matter to Chapter VIII, when we shall have the required $L$-function results to hand.

We begin with a few initial observations concerning the unit group $U_K$. By Dirichlet's Unit Theorem we know that $\mathfrak{U}_K = U_K/\mu_K$ has $\mathbb{Z}$-rank three or one according as $K$ is real or imaginary. In the previous section we have seen that it is relatively easy to calculate the unit group of a quadratic number field. It is therefore of great interest to determine how close the group generated by the units of the three quadratic subfields is to the full unit group $U_K$.

Let us now fix some notation. Given a biquadratic field $K$, we shall write $k_1, k_2, k_3$ for the three quadratic subfields of $K$; we let $\sigma_i$ denote the non-trivial Galois automorphism of $K/k_i$, and we let $s_i$ denote the non-trivial Galois automorphism of $k_i/\mathbb{Q}$. We let $v_i$ denote a fundamental unit of $k_i$ if $k_i$ is real, and it is to be understood that $v_i$ is taken to be 1 if $k_i$ is imaginary. If $K$ itself is imaginary, then we always take $k_1$ to be the (unique) real quadratic subfield: more precisely, number the $\sigma_i$ such that $\sigma_1$ generates $\mathrm{Gal}(K_v/\mathbb{Q}_v)$ for some Archimedean absolute value of $K$; because $K/\mathbb{Q}$ is abelian, $\sigma_1$ is the unique generator of $\mathrm{Gal}(K_w/\mathbb{Q}_w)$ for all Archimedean absolute values $w$ of $K$ (see (III.1.18)).

We begin our study of $U_K$ by describing the possibilities of $\mu_K$; we write $W_K = |\mu_K|$.

**(2.1)** *For a biquadratic field $K$, $W_K$ is either* 2, 4, 6, 8 *or* 12.

*Proof.* We reason as in the more sophisticated approach to the same problem for quadratic fields. Suppose that $n = W_K$; then $K \supset \mathbb{Q}(\mu_K) \supset \mathbb{Q}$, while in (VII,§1) we shall show $\phi(n) = [\mathbb{Q}(\mu_K) : \mathbb{Q}]$. We are therefore reduced to first determining the even $n$ for which $\phi(n) \mid 4$. Since $\phi$ is multiplicative with $\phi(p^r) = (p-1)p^{r-1}$ for prime $p$, we see that $n$ can only take on the values 2, 4, 6, 8, 10 or 12. We discard the case $n = 10$, since $\mathbb{Q}[10]/\mathbb{Q}$ is cyclic of degree 4. We conclude by noting that each of the remaining five values yields a subfield of a biquadratic field.

□

If $K$ is imaginary we again have the notion of a fundamental unit, i.e. we have a unit $u$ so that every unit $w$ in $U_K$ can be written in the form $\zeta u^r$ for some $r \in \mathbb{Z}$, $\zeta \in \mu_K$. In particular we note that $u$ mod $\mu_K$ is a generator for $\mathfrak{U}_k$. The following result, though not very hard, plays a fundamental role in all of our calculations.

**Theorem 42.** *Let $K$ denote an imaginary biquadratic field, then*

$$[U_K : U_{k_1}\mu_K] = 1 \text{ or } 2\,.$$

*Furthermore, denoting this index by $\delta_K$, we have:*

(a) $\delta_K = 1$ *if and only if $v_1$ is a fundamental unit of $K$;*

(b) *if $\delta_K = 2$ and if $u$ is a fundamental unit of $K$, then $u^2 = \zeta v_1$ for some $\zeta \in \mu_K$ with $v_1$ a fundamental unit of $k_1$.*

*Generalisation.* The proof that we give for Theorem 42 can readily be extended to show

*If $K \supset L \supset \mathbb{Q}$ with $K$ a totally imaginary number field, $K/L$ quadratic, and $L$ totally real, then $[U_K : U_L\mu_K] \leq 2$.*

Before proving Theorem 42, we observe that if in case (b) $\zeta = \xi^2$ for some root of unity $\xi$ in $\mu_K$, then $\sqrt{v_1} \in K$. However, $v_1$ is fundamental in $k_1$, so that $\sqrt{v_1}$ cannot lie in $k_1$, and therefore $K = k(\sqrt{v_1})$. We have therefore shown:

**Corollary.** *If $\delta_K = 2$ and if $v_1$ is positive under some real embedding of $k_1$, then $\zeta$ does not lie in $\mu_K^2$.*

Next we show a result which will be of great use in a number of subsequent proofs:

**(2.2)** *$\sigma_1$ acts trivially on $\mathfrak{U}_K$.*

*Proof.* By Dirichlet's Theorem we have a group isomorphism $\phi\colon \mathfrak{U}_K \cong \mathbb{Z}$. So, via $\phi$, $\sigma_1$ acts as an automorphism of $\mathbb{Z}$, that is to say for all $x \in \mathfrak{U}_K$

$$x^{\sigma_1} = sx$$

with $s = \pm 1$ independent of $x$. However, since $v_1$ is not a root of unity, $v_1\mu_K$ is a non-trivial element of $\mathfrak{U}_K$ and obviously $(v_1\mu_K)^{\sigma_1} = v_1\mu_K$. We therefore conclude that $s = 1$. □

We are now in a position to prove Theorem 42. We begin by showing that $\delta_K = 1$ or 2. By (2.2) it is clear that for $y \in U_K$, $y^{1-\sigma_1}$ must lie in $\mu_K$. We therefore let $\lambda\colon U_K \to \mu_K \bmod \mu_K^2$ be defined by $\lambda(y) = y^{1-\sigma_1}\mu_K^2$. We assert that $\ker\lambda = U_{k_1}\mu_K$. Clearly $\lambda$ is trivial on $U_{k_1}$ and also $\mu_K \subset \ker\lambda$ since for $\zeta \in \mu_K$, $\lambda(\zeta) = \zeta/\zeta^{-1}\mu_K^2 = \mu_K^2$. On the other hand, if $y \in \ker\lambda$, then $y^{1-\sigma_1} = \zeta^2$ for some $\zeta \in \mu_K$. Since $\zeta^2 = \zeta^{1-\sigma_1}$, we see that $y\zeta^{-1}$ is fixed by $\sigma_1$; hence $y\zeta^{-1} \in U_{k_1}$, and so $y = y\zeta^{-1}\zeta \in U_{k_1}\mu_K$. The result then follows from the inequalities

$$[U_K : \ker \lambda] \leq [\mu_K : \mu_K^2] \leq 2.$$

Part (*a*) of the theorem follows from the definition of a fundamental unit. Next we suppose $\delta_K = 2$; then, using the isomorphism $\phi$ in the proof of (2.2), we must have $\phi(U_{k_1}\mu_K) = 2\mathbb{Z}$; hence $\phi(u^2\mu_K) = \phi(v_1\mu_K)$ for some fundamental unit $v_1$ of $k_1$.

□

The main part of this sub-section on biquadratic imaginary fields is devoted to finding criteria to determine whether $\delta_K$ is 1 or 2. We begin by describing a result of a reasonably general nature:

**(2.3)** *Suppose that $\mu_K = <\pm 1>$ and that there is an odd prime of $k_1$ which ramifies in $K/k_1$; then $\delta_K = 1$.*

*Proof.* Assume for contradiction that $\delta_K = 2$. By the theorem we can write $u^2 = \zeta v_1$ with $\zeta = \pm 1$; so, setting $v' = \zeta v_1$, we see that $K = k(\sqrt{v'})$, and hence no odd prime can ramify in $K/k$ (by (III.2.28)). □

Next we consider two families of biquadratic fields. In the first family we show that $\delta_K$ is 1, while in the second family $\delta_K$ will always be 2.

**(2.4)** *Let $K$ denote an imaginary biquadratic field. If $N_{k_1/\mathbb{Q}}v_1 = -1$, then $\delta_K = 1$. In particular, this holds if $k_1 = \mathbb{Q}(\sqrt{p})$, $p \equiv 1 \bmod (4)$.*

*Proof.* We suppose for contradiction that $v_1$ is not fundamental in $K$. Then, by Theorem 42, we know that $\sqrt{\zeta v_1} \in K$ for some $\zeta \in \mu_K$. Since $\zeta^{1/2}$ is again a root of unity, $\mathbb{Q}(\zeta^{1/2})/\mathbb{Q}$ is an abelian extension, i.e. has abelian Galois group. The composition of two abelian extensions is again abelian, and hence the compositum $K\mathbb{Q}(\zeta^{1/2}) = \mathbb{Q}(\zeta^{1/2}, v_1^{1/2})$ is also abelian over $\mathbb{Q}$; thus we see that $\mathbb{Q}(v_1^{1/2})$ is abelian over $\mathbb{Q}$. From Corollary 2 to Theorem 39, in the previous section, we know that the fundamental unit of $\mathbb{Q}(\sqrt{p})$ has norm $-1$ when $p \equiv 1 \bmod (4)$. Since $N_{k_1/\mathbb{Q}}(v_1) = -1$, $v_1$ will be positive under one real embedding of $k_1$ and negative under the other. We therefore see that $\mathbb{Q}(v_1^{1/2})$ possesses both real and imaginary embeddings. However, this is impossible, since for a Galois extension $N/\mathbb{Q}$, the Galois group transitively permutes the embeddings $N \to \mathbb{C}$, so that they all have common image.

□

Our proof of the following result is due to Kubota:

**(2.5)** *Let $p_1$, $p_2$ denote distinct prime numbers with the property that $p_1 \equiv -1 \equiv p_2 \bmod (4)$, and let $K = \mathbb{Q}(\sqrt{-p_1}, \sqrt{-p_2})$; then $\delta_K = 2$.*

*Proof.* From (1.7) we know that $k_1 = \mathbb{Q}(\sqrt{p_1 p_2})$ has a fundamental unit $v_1$ with norm 1; and so by (1.15) we can find $\rho \in k_1$ such that $\rho^{1-s_1} = v_1$; indeed, by multiplying $\rho$ by a suitable rational integer, we may assume $\rho \in \mathfrak{o}_{k_1}$. Since $\rho = v_1 \rho^{s_1}$, we see that the ideal $\rho\mathfrak{o}_{k_1}$ is $s_1$-fixed, and so by (1.16) we may write $(\rho) = \mathfrak{a}r$ where $r \in \mathbb{Q}^*$ and $\mathfrak{a}$ is a $\mathfrak{o}_{k_1}$-ideal which divides $(\sqrt{p_1 p_2})$. Therefore, replacing $\rho$ by $\rho r^{-1}$, we may suppose that $\rho$ is an algebraic integer of $k_1$ which divides $\sqrt{p_1 p_2}$: we now consider the possible factorisations of $\rho$ and we show that $\delta_K = 2$ in all cases.

Firstly suppose that $(\rho) = \mathfrak{o}_{k_1}$; hence, as required,

$$v_1 = \rho^{1-s_1} = \frac{\rho^2}{N_{k_1/\mathbb{Q}}(\rho)} = \pm\rho^2.$$

On the other hand, if $(\rho) = (\sqrt{p_1 p_2})$, then $w = \rho/\sqrt{p_1 p_2}$ is a unit,and so

$$v_1 = \rho^{1-s_1} = \frac{w\sqrt{p_1 p_2}}{(w\sqrt{p_1 p_2})^{s_1}} = \frac{-w}{w^{s_1}} = \frac{-w^2}{w^{1+s_1}} = \pm w^2$$

which again shows that $v_1$ is not fundamental. Re-indexing the $p_i$ if necessary, we lastly suppose that $(\rho)^2 = (p_1)$; thus, in particular, the element $x = \rho/\sqrt{-p_1} \in U_K$. We then note that

$$v_1 = \frac{\rho}{\rho^{s_1}} = \frac{\rho^2}{N_{k_1/\mathbb{Q}}(\rho)} = \pm\frac{\rho^2}{p_1} = \pm x^2.$$

□

More generally we remark that the map $v_1 \to N_{k_1/\mathbb{Q}}(\rho)\mathbb{Q}^{*2}$ plays a similarly, crucial role in the case when $K$ is a real biquadratic field. However, since $\mathfrak{U}_K$ has $\mathbb{Z}$-rank 3, the details are somewhat more involved.

The question of class numbers of biquadratic fields will be considered in full in Chapter VIII; however, in the situation described in (2.5), we now show

**(2.6)** *Let $K = \mathbb{Q}(\sqrt{-p_1}, \sqrt{-p_2})$ where the $p_i$ are distinct prime numbers with each $p_i \equiv -1 \bmod (4)$, and each $p_i > 3$ for $i = 1, 2$. Then $h_K$ is odd.*

*Proof.* Let $\mathfrak{b}$ denote a non-zero $\mathfrak{o}_K$-ideal with the property that $\mathfrak{b}^2$ is principal: we show that $\mathfrak{b}$ itself is principal.

Observe that $k_1 = \mathbb{Q}(\sqrt{p_1 p_2})$. From Theorem 41 we know that

$h_{k_1}^+ \equiv 2 \bmod (4)$. By (1.7) $N_{k_1/\mathbb{Q}}(v_1) = 1$, and hence, by Corollary 1 to Theorem 39, $h_{k_1}$ must be odd. Thus $N_{K/k_1}\mathfrak{b}$ is principal and hence $\mathfrak{b}^{1+\sigma_1}$ is a principal $\mathfrak{o}_K$-ideal. It follows that $\mathfrak{b}^{1-\sigma_1} = \mathfrak{b}^{1+\sigma_1}\mathfrak{b}^{-2\sigma_1}$ is also a principal $\mathfrak{o}_K$-ideal: we let $\beta$ denote a generator of $\mathfrak{b}^{1-\sigma_1}$.

Since $N_{K/k_1}(\mathfrak{b}^{1-\sigma_1}) = \mathfrak{o}_{k_1}$, it is clear that $N_{K/k_1}\beta$ is a unit of $\mathfrak{o}_{k_1}$. From (2.1) we see immediately that $\mu_K =< \pm 1 >$, and so by (2.5) we know that $u^2 = \pm v_1$, hence $N_{K/k_1}(u) = \mp v_1$. Multiplying $\beta$ by a suitable power of $u$, we can assume that $N_{K/k_1}(\beta) = \pm 1$. However, $N_{K/k_1}(\beta)$ must be positive under each real embedding of $k_1$ since under each complex embedding of $K$ it assumes the form $z\bar{z}$. We have therefore shown that $N_{K/k_1}(\beta) = 1$, and so, by (1.15), we can find $\alpha \in K$ such that $\alpha^{1-\sigma_1} = \beta$. This gives $\mathfrak{b}^{1-\sigma_1} = (\beta) = (\alpha^{1-\sigma_1})$, and so the ideal $\mathfrak{b}\alpha^{-1}$ is $\sigma_1$-stable. However, $K/k_1$ is non-ramified (c.f. (III.3.8)); thus, by (III.1.21), $\mathfrak{b}\alpha^{-1}$ is of the form $\mathfrak{c}\mathfrak{o}_K$ for some $\mathfrak{o}_{k_1}$-ideal $\mathfrak{c}$. Since $\mathfrak{o}_{k_1}$ has odd class number and since $\mathfrak{c}^2 = N_{K/k_1}(\mathfrak{b}\alpha^{-1})$ has already been shown to be principal in $k_1$, we deduce that $\mathfrak{c}$ is a principal $\mathfrak{o}_{k_1}$-ideal.

□

## §3 Cubic and sextic fields

In this section we shall consider the arithmetic of cubic extensions $K/\mathbb{Q}$ which possess precisely one real embedding: of course such a number field is necessarily non-normal over $\mathbb{Q}$. More precisely, if $K = \mathbb{Q}(\alpha)$ with $\alpha = \alpha_1, \alpha_2, \alpha_3$ denoting the three distinct Galois conjugates of $\alpha$; then by Galois theory the normal closure $N$ of $K/\mathbb{Q}$ is a quadratic extension of $K$, with Gal$(N/K)$ being generated by the automorphism which transposes $\alpha_2$ and $\alpha_3$; hence $N$ is also generated over $K$ by $(\alpha - \alpha_2)(\alpha_2 - \alpha_3)(\alpha_3 - \alpha)$; we have therefore shown that $N = K(\sqrt{d_K})$ and that Gal$(N/\mathbb{Q})$ is isomorphic to $S_3$, the symmetric group on three letters. Since $K$ has only two complex embeddings, $d_K$ is negative by (II.1.28). We note, for future reference, that the imaginary quadratic number field $L = \mathbb{Q}(\sqrt{d_K})$ is the unique quadratic extension of $\mathbb{Q}$ which is contained in $N$.

Given $y \in K$, we shall always write $y_2$ (resp. $y_3$) for the image of $y$ under a Galois automorphism of $N$ which takes $\alpha$ to $\alpha_2$ (resp. $\alpha_3$) – there are precisely two such automorphisms, but the image of $y$ is independent of the choice involved. We let $\tau$ denote the non-trivial automorphism of $N/K$, and so, given the embedding $K \hookrightarrow \mathbb{R}$, $\tau$ acts as complex conjugation on $N$.

In working with such cubic fields, it is frequently useful to have to hand Cardan's formula for the root of a cubic $x^3 + cx^2 + ax + b$; in fact, by means of the substitution $x \to (x - c/3)$, we may, without loss of generality, take $c = 0$. We then have

**(3.1)** *Let $a, b \in \mathbb{Q}$ and suppose that $f(x) = x^3 + ax + b$ has negative discriminant. Then $f$ has a unique real root*

$$\alpha = \sqrt[3]{\frac{-b + \Delta/(3\sqrt{-3})}{2}} + \sqrt[3]{\frac{-b - \Delta/(3\sqrt{-3})}{2}}$$

*where $\Delta$ denotes a square root of the discriminant $-4a^3 - 27b^2$, and where the cube roots denote the real cube roots of the real numbers in question.*

*Proof.* We choose $\lambda, \mu \in \mathbb{R}$ such that $\lambda\omega + \mu\omega^2$ is one of the complex roots of $f$; here $\omega^3 = 1$, $\omega \neq 1$. Then $\lambda\omega^2 + \mu\omega$ is the other complex root, and so $\lambda + \mu$ is the real root, since the sum of the three roots is zero. On the one hand $-b = \lambda^3 + \mu^3$; while on the other hand we can express $\Delta^2$ as the product of the differences of the roots and obtain

$$\pm\Delta = 3\sqrt{-3}(\lambda^3 - \mu^3);$$

thus

$$2\lambda^3 = -b \pm \Delta/3\sqrt{-3}, \qquad 2\mu^3 = -b \mp \Delta/3\sqrt{-3}.$$

This determines $\lambda$ and $\mu$ since they are real. $\square$

We return again to the arithmetic of the cubic field $K = \mathbb{Q}(\alpha)$. Since $K$ has a real embedding, we note that $\mu_K = \langle \pm 1 \rangle$. Applying Dirichlet's Unit Theorem, we see that, since $s(K) = t(K) = 1$ and $s(N) = 0$, $t(N) = 3$, $\mathfrak{U}_K$ has $\mathbb{Z}$-rank one, while $\mathfrak{U}_N$ has $\mathbb{Z}$-rank two. Thus $K$ possesses a fundamental unit, i.e. a unit whose class in $\mathfrak{U}_K = U_K/\langle \pm 1 \rangle$ is a generator. As in the case of quadratic fields, one shows that there is a unique such fundamental unit which is greater than 1 under the embedding $K \to \mathbb{R}$; we denote it by $u$.

We now describe both the structure and the contents of this section. We begin by giving a lower bound for $u$ in terms of $|d_K|$, and we then show how this bound may be used to actually determine a fundamental unit of $K$. We conclude by considering the relationship between $U_N$ and the unit groups $U_K, U_{K_2}, U_{K_3}$.

**(3.2) (Artin)** *Let $u$ be as above, then $|d_K| < 4u^3 + 24$.*

Before proving this result, we note the following very useful and easily applicable corollary:

**Corollary.** *Let $v \in U_K$ with $v > 1$. If $4v^{3/2} + 24 < |d_K|$, then $v$ is a fundamental unit of $K$.*

*Proof.* Once and for all we fix an embedding $N \to \mathbb{C}$ which extends the real embedding $K \to \mathbb{R}$; so, from now on, we may view elements of $N$ as complex numbers. Because $U_K = \langle \pm 1 \rangle \times u^{\mathbb{Z}}$, we immediately see that $v = u^n$ for some positive integer $n$. If, however, $n \geq 2$, then we would have $|d_K| \leq 4u^3 + 24 = 4v^{3/n} + 24 \leq 4v^{3/2} + 24$.

□

We now prove (3.2). By (IV.4.2), we know that $N_{K/\mathbb{Q}}(u) = \pm 1$, since $u$ is a unit. However, by hypothesis $u > 0$, while $u_2 u_3 = u_2 \overline{u}_2 > 0$, so that

(3.3) $$u u_2 u_3 = +1.$$

Reordering our initial choice of $\alpha_2$ and $\alpha_3$ if necessary, we choose $r > 1$, $\theta \in (0, \pi)$ such that

$$u = r^2, \qquad u_2 = r^{-1} e^{i\theta}, \qquad u_3 = r^{-1} e^{-i\theta}.$$

Letting $D$ denote the discriminant of $\mathbb{Z}[u]$, we have

$$D^{1/2} = \begin{vmatrix} 1 & r^2 & r^4 \\ 1 & r^{-1} e^{i\theta} & r^{-2} e^{2i\theta} \\ 1 & r^{-1} e^{-i\theta} & r^{-2} e^{-2i\theta} \end{vmatrix}$$

and so

(3.4) $$D^{1/2} = D^{1/2}(r, \theta) = -2i((r^3 + r^{-3}) \sin\theta - \sin(2\theta)).$$

We now consider $D = D(r, \theta)$ for a fixed real number $r$ and for a real variable $\theta$; we set $\xi = \frac{1}{2}(r^3 + r^{-3})$. Then $|D|$ has a maximum only when $\frac{\partial D^{1/2}}{\partial \theta} = 0$; that is to say we must have

$$2\xi \cos\theta - 2\cos 2\theta = 0.$$

So, writing $x_0 = \cos\theta$, we have

(3.5) $$\xi x_0 - 2x_0^2 + 1 = 0.$$

In the sequel we set $g(X) = 2X^2 - \xi X - 1$. We note that from (3.4)

$$|D| \leq 16(\xi - \cos\theta)^2 \sin^2\theta = 16(\xi - x_0)^2(1 - x_0^2)$$

hence

(3.6) $$|D| \leq 16(\xi^2 - 2\xi x_0 + x_0^2)(1 - x_0^2).$$

From the maximality condition (3.5), we know that $\xi x_0 = 2x_0^2 - 1$, and so $\xi^2 x_0^2 = 4x_0^4 - 4x_0^2 + 1$. So, by (3.6), we conclude that

$$\begin{aligned} |D| &\leq 16(\xi^2 + 1 - x_0^4 - x_0^2) \\ &\leq 4(r^6 + 6 + (r^{-6} - 4x_0^2 - 4x_0^4)). \end{aligned}$$

Therefore it now suffices to show

$$4x_0^4 + 4x_0^2 - r^{-6} > 0. \tag{3.7}$$

To see this, we first consider the roots of $g(X)$: we note that since $2\xi = r + r^{-1} > 2$,

$$g(1) = 2 - \xi - 1 = 1 - \xi < 0.$$

On the other hand

$$g\left(-\frac{1}{2r^3}\right) = \frac{2}{4r^6} + \frac{r^3 + r^{-3}}{4r^3} - 1 = \frac{3}{4}(r^{-6} - 1) < 0.$$

Summarising, we see that $g(X)$ has two roots: one with $x_0 > 1$ (which we ignore, since $x = \cos\theta$); and a negative root $x_0 < -1/2r^3$. Thus $x_0^2 > 1/4r^6$, and so $4x_0^2 - r^{-6} > 0$, which proves (3.7).

□

Next we give two worked examples of the above result in action.

*Example 1.* $K = \mathbb{Q}(\sqrt[3]{2})$. We note that $\sqrt[3]{2} - 1$ is obviously a unit of $K$ with inverse $v = \sqrt[3]{4} + \sqrt[3]{2} + 1$, and $v = 3.85\cdots$. We have already seen in Exercise 2 in Chapter II that $\mathfrak{o}_K = \mathbb{Z}[\sqrt[3]{2}]$ and so $d_K = -108$; thus, noting that $4(3.85)^{3/2} + 24 < 108$, we conclude that $v$ is fundamental for $K$. It is worth remarking that the complexity of fundamental units of pure cubic fields can increase very quickly: for instance $\mathbb{Q}(\sqrt[3]{23})$ has a fundamental unit

$$2166673601 + 761875860\sqrt[3]{23} + 267901370\sqrt[3]{529}.$$

The fundamental unit for $\mathbb{Q}(\sqrt[3]{239})$ is yet more impressive; H. Wada has calculated this unit, and his result shows that the coefficient of 1 is greater than $10^{187}$.

*Example 2.* $K = \mathbb{Q}(\alpha)$ where $\alpha$ is the real solution to $X^3 - X + 2 = 0$. From (III.2.22) we know $\mathbb{Z}[\alpha]$ has discriminant $-104$; thus $d_K$ is either $-104$ or $-26$, and the latter is impossible by (II.1.26); therefore we conclude that $d_K = -104$ and that $\mathfrak{o}_K = \mathbb{Z}[\alpha]$. To construct a unit of infinite order we try to locate elements with the same norm, and then check to see whether they are associates.

Since $X^3 - X + 2 \equiv X(X+1)^2 \bmod (2)$, by the generalisation of Theorem 23, we see that $2\mathfrak{o}_K = \mathfrak{p}^2\mathfrak{q}$ with $\mathfrak{p} \neq \mathfrak{q}$. Directly from the minimal equation we see that $\alpha(\alpha-1)(\alpha+1) = -2$, while of course $(\alpha, \alpha+1) = 1 = (\alpha, \alpha-1)$. Since $N_{K/\mathbb{Q}}(\alpha) = -2$, we see that $(\alpha+1) = \mathfrak{p} = (\alpha - 1)$ and so $\alpha + 1$ and $\alpha - 1$ are associates.

We put $v = \frac{\alpha-1}{\alpha+1}$ and now try to apply the above corollary. From Cardan's formula in (3.1), we see that $\alpha = -1.521\cdots$ and so $v = 4.839\cdots$, thus

$$4v^{3/2} + 24 \leq 4 \cdot 5^{3/2} + 24 < 104.$$

We have therefore shown $v$ to be fundamental in $K$.

□

Next we describe a situation where the above Artin bound can be used to give a systematic determination of the fundamental unit of $K$. Let $l \geq 2$ denote an integer with the property that $4l^3+27$ is square-free, and let $f(X) = X^3 + lX - 1$. Since $\frac{df}{dX}$ is positive-definite, $f$ has a unique real root, which we shall denote by $v$. We put $K = \mathbb{Q}(v)$, and note that, because $f(X)$ has discriminant $-(4l^3 + 27)$ which is square-free, by (III.2.4), $\mathfrak{o}_K = \mathbb{Z}[v]$ and $d_K = -(4l^3 + 27)$.

Since $v^{-1} = l + v^2$, we note that $v^{-1} > 1$, and so $v^{-1} = u^n$ for some $n \geq 1$. We wish to show that $n = 1$, and hence that $v^{-1}$ is fundamental; we therefore suppose for contradiction that $n \geq 2$. Applying (3.2), we see that

$$4l^3 + 27 = |d_K| < 4u^3 + 24$$

so that $l < u$. It therefore follows that

$$l^2 < u^2 \leq u^n = v^{-1} = l + v^2 < l + 1.$$

This is impossible since $l \geq 2$, and so we have shown

**(3.8) (Ishida).** *Suppose that the integer $l \geq 2$ has the property that $4l^3+27$ is square-free, and let $v$ denote the unique real root of $X^3+lX-1$; then $v^{-1}$ is a fundamental unit of $\mathbb{Q}(v)$ which is greater than* 1.

We now consider the construction of fundamental units for the normal closure $N$ of a general cubic field $K$ with negative discriminant. Let $V_N = \langle \pm 1, u, u_2, u_3 \rangle$ and note that since $uu_2u_3 = \pm 1$, $V_N = \langle \pm 1, u, u_2 \rangle$; indeed, because $\langle u \rangle \cap \langle u_2 \rangle \subset K \cap K_2 = \mathbb{Q}$ we see that the two infinite cyclic groups $\langle u \rangle$, $\langle u_2 \rangle$ have disjoint images in $\mathfrak{B}_N = V_N\mu_N/\mu_N$; hence $\mathfrak{B}_N$ has $\mathbb{Z}$-rank two, and so $V_N$ has finite index in $U_N$. Our aim is to prove the following result of Berwick.

**Theorem 43.** *Suppose $\mu_L = \langle \pm 1 \rangle$; then $[U_N : V_N]$ is* 1 *or* 3.

Recall that $L = \mathbb{Q}(\sqrt{d_K})$. As a first step we show the following result, which does not depend on the hypothesis $\mu_L = \langle \pm 1 \rangle$.

$$\mu_N = \mu_L. \tag{3.9}$$

*Proof.* $L \supset \mathbb{Q}(\mu_N)$, since $L$ is the maximal abelian extension of $\mathbb{Q}$ in $N$.

□

In order to prove the theorem we first require some notation: let $\sigma$ denote the generator of $\Sigma = \mathrm{Gal}(N/L)$ such that $\alpha^\sigma = \alpha_2$; then $\Gamma$ is described by the generators and relations

$$\Gamma = \langle \sigma, \tau \mid \sigma^3 = 1 = \tau^2,\ \tau\sigma\tau = \sigma^2 \rangle.$$

The field isomorphisms $K \cong K_2 \cong K_3$ induce isomorphisms on their unit groups, so that, in particular, $u_2$ and $u_3$ are fundamental units of $K_2$ and $K_3$. We note that for $y \in \mathfrak{U}_N$, $y^{1+\sigma+\sigma^2} = 1$, since $N_{N/L}(U_N) \subset U_L = \mu_L$. We can therefore write $y^3$ in the form

$$y^3 = y^{1+\tau} y^{1+\sigma\tau} y^{1+\sigma^2\tau}.$$

Since $K, K_2, K_3$ is fixed by $\langle\tau\rangle, \langle\sigma\tau\rangle, \langle\sigma^2\tau\rangle$ respectively, we see that $y^{1+\tau} \in \mathfrak{U}_K$, $y^{1+\sigma\tau} \in \mathfrak{U}_{K_2}$, $y^{1+\sigma^2\tau} \in \mathfrak{U}_{K_3}$, and so $y^3 \in \mathfrak{V}_N$. As $\mathfrak{U}_N$ has $\mathbb{Z}$-rank two we deduce that $[\mathfrak{U}_N : \mathfrak{V}_N] \mid 9$; moreover if $[\mathfrak{U}_N : \mathfrak{V}_N] = 9$, then we can find $w \in U_N$ with the property that $w^3 = u\zeta$, for $\zeta \in \mu_L = \langle \pm 1 \rangle$. Therefore, changing the sign of $w$ if necessary, we have $w^3 = u$. As $u$ is fundamental, $X^3 - u$ has no linear factor in $K[X]$ and so is irreducible over $K$. Thus $3 = (K(w) : K) \mid (N : K) = 2$, which is absurd, and so we have shown $[\mathfrak{U}_N : \mathfrak{V}_N] \mid 3$. We conclude by remarking that $[U_N : V_N] = [\mathfrak{U}_N : \mathfrak{V}_N]$ because $\mu_N = \pm 1 = V_N \cap \mu_N$.

□

As an illustrative example we continue our study of the Ishida family of polynomials $f(X) = X^3 + lX - 1$ introduced in (3.8). Keeping the same notation and hypotheses we have:

**(3.10)** *If $l$ is even, then $U_N = V_N$.*

*Proof.* First note that $d_L$ is square free and $\leq -59$ and so $\mu_L = \{\pm 1\}$. Suppose, to derive a contradiction, that $U_N \neq V_N$; then by Theorem 43 $[U_N : V_N] = 3$. We can therefore find $w \in U_N \setminus V_N$ such that $w^3 \in V_N$; we then write $w^3 = \pm v^r v_2^s$ for suitable integers $r$, $s$. Changing $w$ to $-w$ if necessary and multiplying by powers of $v$ and $v_2$, we may suppose $w = v^r v_2^s$ with $0 \leq r, s < 3$. We note that the case $r = s = 0$ is excluded, by hypothesis.

Next we observe that $(w^\tau)^3 = v^r v_3^s$, and so $(ww^\tau)^3 = v^{2r} v_2^s v_3^s = v^{2r-s}$. Since $v$ is fundamental for $K$, we know that $ww^\tau = \pm v^t$; furthermore, since $v > 0$ and $ww^\tau > 0$, we see that only the positive sign is possible. We have therefore shown $2r - s \equiv 0 \bmod (3)$ and so either $(r, s) = (2, 1)$ or $(r, s) = (1, 2)$. Replacing $w$ by $w^{-1}$ if necessary, we may therefore take $(r, s) = (2, 1)$. Next we observe that, since $\mathrm{Tr}_{K/\mathbb{Q}}(v) = 0$,

$$\begin{aligned} w^3 + w^{3\tau} &= v^2 v_2 + v^2 v_3 \\ &= v^2(v_2 + v_3) \\ &= -v^3. \end{aligned} \tag{3.11}$$

Since $f \equiv X^3 + 1 \equiv (X+1)(X^2+X+1) \bmod (2)$, we know by Theorem 23 that

$$2\mathfrak{o}_K = \mathfrak{p}_1\mathfrak{p}_2 \qquad \mathbf{N}\mathfrak{p}_1 = 2,\ \mathbf{N}\mathfrak{p}_2 = 4$$
$$2\mathfrak{o}_N = \mathfrak{P}_1\mathfrak{P}_2\mathfrak{P}_3 \qquad \mathbf{N}\mathfrak{P}_i = 4.$$

We may therefore conclude that for any $z \in U_N$, $z^3 \equiv 1 \bmod \mathfrak{P}_1$; thus, by (3.11), we see that $1 + 1 \equiv 1 \bmod \mathfrak{P}_1$, which is the desired contradiction.

□

# VI

# Cyclotomic Fields

Cyclotomic fields are fields obtained by adjoining to $\mathbb{Q}$ roots of unity, i.e. roots of polynomials of the form $X^n - 1$, although the reader is warned that this terminology will be extended in §2. Geometrically these arise from the problem of dividing the unit circle into equal parts. They form a class of algebraic number fields with very beautiful and important properties. Cyclotomic fields play a fundamental rôle in a number of arithmetic problems: for instance primes in arithmetic progressions (see VIII,§4) and Fermat's Last Theorem (see VII,§1). They also throw new light on the theory of quadratic fields which we have already considered in (V,§1), and in particular provide the background to the "most natural" proof of the quadratic reciprocity law. They are Galois extensions of $\mathbb{Q}$ with abelian Galois group, and conversely, as one can show, although we shall not be able to, any finite extension of $\mathbb{Q}$ with abelian Galois group is a subfield of a cyclotomic field. Thus the theory looks forward to one of the most sophisticated aspects of algebraic number theory: the so-called class field theory.

## §1 Basic theory

We begin by considering some field theoretic aspects, where the base field is not necessarily the field of rationals. Observe that any finite subgroup $S$ of the multiplicative group of a field is necessarily cyclic: for, in a non-cyclic finite abelian group $G$ of exponent $h$, there are $|G| > h$ roots of $X^h = 1_G$, while in a field the equation $X^h - 1$ has at most $h$ solutions.

Assume from now on that $F$ is a field of characteristic zero, or that $m$ runs through natural numbers which are not multiples of the characteristic of $F$. In any extension field $E$ of $F$, the roots of $X^m - 1$ in $E$, i.e. the $m$th roots of unity in $E$, then form a cyclic group of order dividing $m$. If $F$ is big enough (for instance if $F$ is algebraically closed), then this group has order exactly $m$, since $X^m - 1$ is separable, because $(X^m - 1, mX^{m-1}) = 1$.

We say that $\zeta$ is a *primitive* $m$th root of unity, if its order is exactly $m$. We define the $m$th cyclotomic polynomial by

$$\Phi_m(X) = \prod (X - \zeta) \tag{1.1}$$

where the product runs over the primitive $m$th roots of unity $\zeta$ in a splitting field $E$ of $X^m - 1$. Note that the number of generators of a cyclic group of order $m$ is the Euler function $\phi(m)$; thus

$$\deg(\Phi_m(X)) = \phi(m). \tag{1.2}$$

Note also that for a positive divisor $d$ of $m$, any $d$th root of unity is an $m$th root of unity, and furthermore that an $m$th root of unity is a primitive $d$th root for exactly one positive divisor $d$ of $m$. Therefore, for each natural number $m$

$$X^m - 1 = \prod_{d|m} \Phi_d(X). \tag{1.3}$$

Applying the Möbius inversion formula, we get

$$\Phi_m(X) = \prod_{d|m} (X^d - 1)^{\mu(m/d)} \tag{1.4}$$

which establishes the fact that $\Phi_m(X)$ is always a polynomial over the base field $F$. In particular we note for future reference that for $r \geq 1$

$$\Phi_{p^r}(X) = \frac{X^{p^r} - 1}{X^{p^{r-1}} - 1} = \sum_{i=0}^{p-1} X^{ip^{r-1}}. \tag{1.4.$a$}$$

We still assume that $m$ is not a multiple of the characteristic of $F$. As previously we write $F[m]$ for the splitting field of $X^m - 1$ over $F$, the cyclotomic field of $m$th roots of unity over $F$.

Let $\zeta$ denote a primitive $m$th root of unity. Every $m$th root of unity is then a power of $\zeta$, i.e.

$$F(\zeta) = F[m]. \tag{1.5}$$

Moreover, if $\sigma \in \mathrm{Gal}(F[m]/F)$, then $\zeta^\sigma = \zeta^{t(\sigma)}$, for some $t(\sigma) \in \mathbb{Z}$ which is coprime to $m$. Here the residue class $t(\sigma) \bmod (m)$ is uniquely determined, and determines $\sigma$. We therefore have a map

$$\begin{aligned} r_m \colon \mathrm{Gal}(F[m]/F) &\to (\mathbb{Z}/m\mathbb{Z})^* \\ r_m(\sigma) &= t(\sigma) \bmod (m). \end{aligned} \tag{1.6}$$

It is clear that $r_m$ does not depend on the particular choice of the primitive $m$th root $\zeta$. Moreover, one verifies immediately that $r_m(\sigma\tau) = r_m(\sigma)r_m(\tau)$. Also $r_m(\sigma) = 1$ implies $\zeta^\sigma = \zeta$, and hence $\sigma = \text{id}$. We have therefore shown

**(1.6.*a*)** *$r_m$ is an injective homomorphism of groups. In particular,* $\text{Gal}(F[m]/F)$ *is abelian.*

We remark that

**(1.7)** *$r_m$ is an isomorphism iff $\Phi_m(X)$ is irreducible in $F[X]$.*

*Proof.* The $\zeta$ in (1.5) is a root of $\Phi_m(X)$. Thus $(F[m] : F) = \phi(m)$ precisely when $\Phi_m(X)$ is irreducible. On the other hand $|\text{Gal}(F[m]/F)| = \phi(m)$ precisely when $r_m$ is surjective (and hence an isomorphism). □

We now consider $F[m]$ together with $F[d]$ for $d \mid m$. Immediately from the definition, we see that the diagram

$$\text{(1.8)} \qquad \begin{array}{ccc} \text{Gal}\,(F[m]/F) & \xrightarrow{r_m} & (\mathbb{Z}/m\mathbb{Z})^* \\ \downarrow & & \downarrow \\ \text{Gal}\,(F[d]/F) & \xrightarrow{r_d} & (\mathbb{Z}/d\mathbb{Z})^* \end{array}$$

commutes. Here the vertical arrow on the left is the quotient map of Galois groups, and on the right is the quotient map of residue classes $t \bmod (m) \mapsto t \bmod (d)$. In fact we can show directly

**(1.9)** *The map $(\mathbb{Z}/m\mathbb{Z})^* \to (\mathbb{Z}/d\mathbb{Z})^*$ is surjective.*

*Proof.* Consider $t \bmod (d)$ for some $t$ coprime to $d$, and write $m = n_1 n_2$ where $(n_1, d) = 1$ and where $n_2$ is only divisible by prime numbers which divide $d$. By the Chinese Remainder Theorem we can find an integer $s$ such that $s \equiv 1 \bmod (n_1)$ and $s \equiv t \bmod (n_2)$; thus $s$ is coprime to $m$; hence $s \bmod (m)$ lies in $(\mathbb{Z}/m\mathbb{Z})^*$ and maps to $t \bmod (d)$ under the residue class quotient map. □

Next let $m = m_1 m_2$ for any coprime positive integers $m_1$ and $m_2$. From the quotient maps $\text{Gal}(F[m]/F) \to \text{Gal}(F[m_i]/F)$ $(i = 1, 2)$, we get a homomorphism

$$\text{(1.10)} \qquad \text{Gal}(F[m]/F) \to \text{Gal}(F[m_1]/F) \times \text{Gal}(F[m_2]/F).$$

This is injective: for, the primitive $m$th roots $\zeta$ are precisely the products $\zeta_1\zeta_2$, where $\zeta_i$ is a primitive $m_i$th root of unity. Thus, if for $\sigma \in \mathrm{Gal}(F[m]/F)$, we have $\zeta_1^\sigma = \zeta_1$, $\zeta_2^\sigma = \zeta_2$; then $\zeta^\sigma = \zeta$. Via the maps $r_m, r_{m_i}$ the injection (1.10) is reflected in the product isomorphism

$$(1.10.a) \qquad (\mathbb{Z}/m\mathbb{Z})^* \cong (\mathbb{Z}/m_1\mathbb{Z})^* \times (\mathbb{Z}/m_2\mathbb{Z})^*.$$

From now on, in this section our base field will always be $\mathbb{Q}$. Since $\mathbb{Q}[2m] = \mathbb{Q}[m]$ for odd $m$, in order to avoid trivialities we henceforth assume $m \not\equiv 2 \bmod (4)$.

**Theorem 44.**

(*a*) *The map $r_m$ is always an isomorphism*

$$\mathrm{Gal}(\mathbb{Q}[m]/\mathbb{Q}) \cong (\mathbb{Z}/m\mathbb{Z})^*.$$

(*b*) *A prime $p$ is non-ramified in $\mathbb{Q}[m]$ iff $p \nmid m$. In this case, the inverse of the isomorphism $r_m$ maps the class of $p \bmod (m)$ onto the Frobenius element $(p, \mathbb{Q}[m]/\mathbb{Q})$. (Note that as $\mathbb{Q}[m]/\mathbb{Q}$ is an abelian extension, there is a unique Frobenius element associated to $p$.)*

(For the definition of the Frobenius element see (III.4.13).)

*Proof.* Suppose $p \nmid m$. As the ramification index of a prime ideal is preserved by completion at that prime, (III.2.27) tells us that $p$ is non-ramified in $\mathbb{Q}[m]$. Alternatively, by (I.1.5) we know that $X^m - 1$ has discriminant $(-1)^{\frac{1}{2}(m-1)(m+2)}m^m$, and so (see exercise I.11) $p$ does not divide the discriminant of the minimal polynomial of the primitive $m$th root of unity $\zeta$ , which by (II.1.39) is $[\mathfrak{o}_{\mathbb{Q}[m]} : \mathbb{Z}[\zeta]]^2 \cdot d_{\mathbb{Q}[m]}$; thus, $p \nmid d_{\mathbb{Q}[m]}$ and the result now follows by Theorem 22. For future reference we record that we have shown

$$(1.11) \qquad p \nmid d_K \quad \text{and } p \nmid [\mathfrak{o}_{\mathbb{Q}[m]} : \mathbb{Z}[\zeta]].$$

Denote by $\mathfrak{p}$ a prime ideal above $p$ in $\mathbb{Q}[m]$ and write $\sigma = (\mathfrak{p}, \mathbb{Q}[m]/\mathbb{Q}) = (p, \mathbb{Q}[m]/\mathbb{Q})$. Then for any primitive $m$th root of unity $\zeta$

$$\zeta^\sigma \equiv \zeta^p \bmod \mathfrak{p}.$$

If $\zeta^\sigma \neq \zeta^p$, then it would follow that $\mathfrak{p}$ divides the non-zero element $\zeta^\sigma - \zeta^p$; hence $\mathfrak{p}$ divides the discriminant of $X^m - 1$, i.e. $m^m\mathbb{Z}$, which is absurd. Thus indeed $\zeta^\sigma = \zeta^p$, and so $r_m(\sigma) = p \bmod (m)$. Therefore the image of $r_m$ contains all classes of primes $p$ (not dividing $m$), and of course these generate $(\mathbb{Z}/m\mathbb{Z})^*$; hence $r_m$ is indeed surjective, and thus bijective.

Next assume that $m = np^r$ with $r > 0$ and $(n, p) = 1$. We assert that

$p$ ramifies totally in $\mathbb{Q}[p^r]$, and hence ramifies in $\mathbb{Q}[m]$. We give two proofs:

(*i*) Since the ramification index of a prime ideal is preserved by completion at that prime, the result follows at once from (III.3.1).

(*ii*) Let $K$ be any intermediate field $\mathbb{Q} \subset K \subset \mathbb{Q}[p^r]$. If $p$ is non-ramified in $K$, then by the first part of the theorem we know that $K/\mathbb{Q}$ is non-ramified at all primes; hence, by Theorem 36, $K = \mathbb{Q}$. It therefore follows that $p$ is totally ramified in $\mathbb{Q}[p^r]$.

□

We now deduce a series of corollaries:

**(1.12)** *Suppose $p \nmid m$. Then the residue class degree $f_p$ of $p$ in $\mathbb{Q}[m]/\mathbb{Q}$ is the order of $p$ mod $(m)$ (i.e. the order of the class of $p$ in $(\mathbb{Z}/m\mathbb{Z})^*$). In particular $p$ splits completely in $\mathbb{Q}[m]$ iff $p \equiv 1 \bmod (m)$.*

*Proof.* The residue class degree $f_p$ is the order of the Frobenius element of $p$ in $\mathrm{Gal}(\mathbb{Q}[m]/\mathbb{Q})$; by Theorem 44(*b*), this is the same as the order of $p$ mod $(m)$.

□

(This result could also be deduced from (III.2.27).)

**(1.13)** *The cyclotomic polynomial $\Phi_m(X)$ is irreducible in $\mathbb{Q}[X]$.*

**(1.14)** *If $m = m_1 m_2$ with $(m_1, m_2) = 1$, then*

$$\mathrm{Gal}(\mathbb{Q}[m]/\mathbb{Q}) = \mathrm{Gal}(\mathbb{Q}[m_1]/\mathbb{Q}) \times \mathrm{Gal}(\mathbb{Q}[m_2]/\mathbb{Q}).$$

*Also $\mathbb{Q}[m]$ is the compositum of arithmetically disjoint fields*

$$\mathbb{Q}[m] \cong \mathbb{Q}[m_1] \otimes_{\mathbb{Q}} \mathbb{Q}[m_2]$$

$$\mathfrak{o}_{\mathbb{Q}[m]} \cong \mathfrak{o}_{\mathbb{Q}[m_1]} \otimes_{\mathbb{Z}} \mathfrak{o}_{\mathbb{Q}[m_2]}.$$

*Proof.* The first part follows from (1.10) on comparing orders; for the second part we apply (III.2.13).

□

Using (1.12) we can give a more precise picture of the behaviour of ramified primes. Again we suppose $m = np^r$ with $r > 0$ and $p \nmid n$. Since $p$ is non-ramified in $\mathbb{Q}[n]$, but is totally ramified in $\mathbb{Q}[p^r]/\mathbb{Q}$, applying the above work and using the tower formulae for $e$ and $f$, we conclude

**(1.15)** *With the above notation* $p$ *is totally ramified in* $\mathbb{Q}[m]/\mathbb{Q}[n]$, *and* $\mathbb{Q}[n]$ *is the maximal subextension of* $\mathbb{Q}[m]$ *in which* $p$ *is non-ramified. Also* $e_p(\mathbb{Q}[m]/\mathbb{Q}) = \phi(p^r)$ *and* $f_p(\mathbb{Q}[m]/\mathbb{Q})$ *is the order of* $p \bmod (n)$.

Next we outline an alternative and independent approach to Theorem 44, which does not involve the Frobenius element, but which, on the other hand, gives more information on the ramified primes; the local version of these results for $\mathbb{Q}_p[p^r]$ has already been seen in (III.3.1).

Let $\zeta$ denote a primitive $p^r$th root of unity for a prime $p$, with $r > 0$, and for brevity we set $K = \mathbb{Q}(p^r)$. If $\zeta'$ is a further primitive $p^r$th root of unity, then $\zeta' = \zeta^t$ for some $t$; thus

$$\frac{1-\zeta'}{1-\zeta} = 1 + \zeta + \cdots + \zeta^{t-1} \in \mathfrak{o}_K.$$

Interchanging $\zeta'$ and $\zeta$, we see that $(1-\zeta)/(1-\zeta') \in \mathfrak{o}_K$ and hence is a unit. Letting $\zeta'$ run through all the primitive $p^r$th roots of unity, we get the equation in ideals

$$(1-\zeta)^{\phi(p^r)}\mathfrak{o}_K = \prod_{\zeta'}(1-\zeta')\mathfrak{o}_K = \Phi_{p^r}(1)\mathfrak{o}_K = p\mathfrak{o}_K.$$

As $(\mathbb{Q}[p^r] : \mathbb{Q}) \leq \phi(p^r)$, we again conclude that $p$ is totally ramified in $\mathbb{Q}[p^r]$, with ramification index $\phi(p^r)$, and that $(\mathbb{Q}[p^r] : \mathbb{Q}) = \phi(p^r)$. Furthermore the unique prime ideal in $\mathbb{Q}[p^r]$ above $p$ is given by

$$\mathfrak{p} = (1-\zeta)\mathfrak{o}_K.$$

Using only the first part of Theorem 44 (that if $p \nmid m$, then $p$ is non-ramified in $\mathbb{Q}[m]$), we infer that $\mathbb{Q}[mp^r] = \mathbb{Q}[m]\mathbb{Q}[p^r]$ with $\mathbb{Q}[p^r] \cap \mathbb{Q}[m] = \mathbb{Q}$. The fact that $r_m$ is an isomorphism for every $m$ now follows by induction on the number of distinct prime divisors of $m$. Restating that part of the above work which goes beyond Theorem 44, together with a further result, we have

**Theorem 45.**

(*a*) *Let* $\zeta, \zeta'$ *denote primitive* $p^r$*th roots of unity with* $p$ *a prime and* $r > 0$. *Then* $(1-\zeta')/(1-\zeta)$ *is a unit and*

$$\mathfrak{p} = (1-\zeta)\mathfrak{o}_{\mathbb{Q}[p^r]} = (1-\zeta')\mathfrak{o}_{\mathbb{Q}[p^r]}$$

*is the unique prime ideal above* $p$ *in* $\mathbb{Q}[p^r]$.

(*b*) *Also, if* $m$ *has at least two distinct prime divisors and if* $\zeta$ *is a primitive* $m$*th root of unity, then* $(1-\zeta)$ *is a unit.*

*Proof of* (*b*). Let $f_m = (X^m - 1)/(X-1)$, so that

$$f_m(1) = m \tag{1.16}$$

and

$$f_m(X) = \prod_{\substack{d|m \\ d>1}} \Phi_d(X)$$

whence

$$f_m(1) = \prod_{\substack{d|m \\ d>1}} \Phi_d(1). \tag{1.17}$$

On the other hand, by (1.4.$a$), $\Phi_{p^r}(1) = p$. Therefore, if $m = \prod_{i=1}^{l} p_i^{t_i}$, with the $p_i$ distinct primes, then

$$m = \prod_i \prod_{j=1}^{t_i} \Phi_{p^j}(1)$$

and so

$$m = \prod{}' \Phi_d(1)$$

where the product $\prod'$ runs over all positive divisors of $m$ which are prime powers (greater than 1). Comparing with (1.16) and (1.17), we deduce that

$$1 = \prod{}'' \Phi_d(1)$$

where $\prod''$ extends over all positive divisors of $m$ which have at least two distinct prime factors. But each $\Phi_d(1)$ is a non-zero integer, and so must be $\pm 1$. In particular $\Phi_m(1) = \pm 1$. On the other hand

$$\Phi_m(1) = N_{\mathbb{Q}[m]/\mathbb{Q}}(1 - \zeta)$$

and so $1 - \zeta$ is a unit.

□

We now consider the additive structure of $\mathfrak{o}_{\mathbb{Q}[m]}$.

**Theorem 46.** *Let $m$ denote a positive integer and let $\zeta$ denote a primitive mth root of unity, then*

$$\mathfrak{o}_{\mathbb{Q}[m]} = \mathbb{Z}[\zeta].$$

*Proof.* By (1.14) we are immediately reduced to proving the theorem for $m = p^r$, $p$ a prime and $r > 0$. For brevity we put $K = \mathbb{Q}[p^r]$ and we let $\mathfrak{p}$ denote the unique prime ideal of $\mathfrak{o}_K$ above $p$.

Since $\zeta$ is an algebraic integer and since $K = \mathbb{Q}[\zeta]$, $\mathfrak{o}_K$ contains $\mathbb{Z}[\zeta]$ with finite index. It will therefore suffice to show that

$$[\mathfrak{o}_K : \mathbb{Z}[\zeta]]_q = \mathbb{Z}_q \tag{1.18}$$

for each prime number $q$. If $q \neq p$, then we know that (1.18) holds by (1.11); however, when $q = p$, we know from (III.3.1) that $\mathfrak{o}_{K,p} = \mathbb{Z}_p[\zeta]$. $\square$

Given a positive integer $m > 2$ we put $K = \mathbb{Q}[m]$ and we let $\zeta$ denote a primitive $m$th root of unity in $K$. As we have seen previously in (V,§2), a field $N$ with abelian Galois group over $\mathbb{Q}$ has a unique maximal totally real subfield. For $N = K$ this is the field $L = \mathbb{Q}(\zeta + \zeta^{-1})$. This can be seen directly: clearly under each embedding of $K$ into $\mathbb{C}$, $\zeta + \zeta^{-1}$ is mapped to a real number $e^{2\pi ia/m} + e^{-2\pi ia/m}$, for some integer $a$ coprime to $m$; hence $L$ is certainly totally real. Moreover, if $L' \subset K$ has a real embedding, then so has $LL'$. Since $L \subset LL' \subsetneqq K$ and since $(K : L) = 2$, it follows that $LL' = L$, and so $L' \subset L$.

We let $\rho$ denote the generator of $\mathrm{Gal}(K/L)$, so that $\rho$ is complex conjugation in $K$. From the generalisation of Theorem 42 in (V,§2), we know that the map $u \mapsto u^{1-\rho}\mu_K^2$ induces an injection

$$\lambda' : U_K/U_L\mu_K \hookrightarrow \mu_K/\mu_K^2$$

and so in particular

$$[U_K : U_L\mu_K] \leq 2.$$

The value of this index is given by

**(1.19)**

(*a*) *If $m$ is a prime power, then $U_K = U_L\mu_K$.*

(*b*) *If $\zeta$ denotes a primitive $m$th root of unity in $\mathbb{Q}[m]$, then $-\zeta$ generates $\mu_{\mathbb{Q}[m]}$.*

(*c*) *Without loss of generality assume $m \not\equiv 2 \bmod (4)$. If $m$ is divisible by two distinct primes, then $[U_K : U_L\mu_K] = 2$.*

*Proof.* In proving (*a*) we take $p > 2$. (For the case $p = 2$ see the exercises for this chapter.) Suppose for contradiction that $[U_K : U_L\mu_K] = 2$; then we can find $u \in U_K$ such that $u^2 \in L$, but $u \notin L$; hence $K = L(\sqrt{u})$, and so by (III.2.28) $K/L$ is non-ramified at $p$; this is absurd since we know that $p$ is totally ramified in $K/\mathbb{Q}$ when $m$ is a power of $p$.

Next we show (*b*) and we suppose that $\mu_K$ has order $m'$; then $m \mid m'$ and $\mathbb{Q}[m] = \mathbb{Q}[m']$, so that

$$\phi(m) = (\mathbb{Q}[m] : \mathbb{Q}) = (\mathbb{Q}[m'] : \mathbb{Q}) = \phi(m').$$

On the one hand if $m$ is odd, all this implies that $m' = 2m$; on the other hand, if $m \equiv 0 \bmod (4)$, then this implies $m' = m$. In both cases $-\zeta$ has order $m'$.

To prove $(c)$ it suffices to show $\lambda'$ is surjective. With the notation introduced above, by Theorem 45 we know that $1-\zeta \in U_K$, since $m$ has two distinct prime factors. As $(1-\zeta)^\rho = -\zeta^{-1}(1-\zeta)$, it follows that $\lambda((1-\zeta)U_L\mu_K) = -\zeta\mu_K^2$, and therefore $\lambda'$ is surjective.

□

Keeping the above notation, we conclude this section by considering the (ordinary) classgroups $C_K$ and $C_L$. Extension of ideals $\mathfrak{a} \mapsto \mathfrak{a}\mathfrak{o}_K$ induces a group homomorphism $i: C_L \to C_K$. We show

**(1.20)** *The homomorphism $i: C_L \to C_K$ is injective, so that in particular $h_L \mid h_K$.*

*Proof.* Without loss of generality we take $m \not\equiv 2 \bmod (4)$ and $m > 2$. Suppose that $\mathfrak{a}$ is an $\mathfrak{o}_L$-ideal with the property that $\mathfrak{a}\mathfrak{o}_K$ is a principal $\mathfrak{o}_K$-ideal, $\mathfrak{a}\mathfrak{o}_K = a\mathfrak{o}_K$ say; we then show that $\mathfrak{a}$ is a principal $\mathfrak{o}_L$-ideal. Since $a^\rho\mathfrak{o}_K = \mathfrak{a}^\rho\mathfrak{o}_K = \mathfrak{a}\mathfrak{o}_K = a\mathfrak{o}_K$, it follows that $a^{1-\rho}$ is a unit of $K$; furthermore, it clearly has absolute value 1 at each Archimedean absolute value of $K$. (Note that these are all complex.) Thus, by (IV.4.5.$a$), $a^{1-\rho} \in \mu_K$.

Suppose firstly that $m$ is not a prime power. From the proof of (1.19.$c$) we know that $\lambda'$ is surjective, and so we can find a unit $u \in U_K$ such that $u^{1-\rho} = a^{\rho-1}$; hence $au = (au)^\rho$; therefore $au \in L$ and $\mathfrak{a} = au\mathfrak{o}_L$.

Lastly we suppose that $m = p^r$ for a prime $p$ and for $r > 0$. Let $\zeta$ denote a primitive $p^r$th root of unity. As $(1-\zeta)^\rho = -\zeta^{-1}(1-\zeta)$ and as $-\zeta^{-1}$ generates $\mu_K$, we can find an integer $n$ such that $a^{1-\rho} = (1-\zeta)^{n(\rho-1)}$. Let $\mathfrak{p}$ now denote the unique prime ideal of $\mathfrak{o}_K$ above $p$, and let $\mathfrak{q}$ denote the prime ideal of $\mathfrak{o}_L$ below $\mathfrak{p}$. Since $a(1-\zeta)^n \in L$, $v_\mathfrak{p}(a(1-\zeta)^n)$ is even, as is $v_\mathfrak{p}(a) = v_\mathfrak{p}(\mathfrak{a})$; thus $n$ must be even, and so $a(1-\zeta)^n\mathfrak{o}_L = \mathfrak{a}\mathfrak{q}^{n/2}$. As $\mathfrak{q} = \mathcal{N}_{K/L}(\mathfrak{p}) = \mathcal{N}_{K/L}((1-\zeta)\mathfrak{o}_L) = N_{K/L}(1-\zeta)\mathfrak{o}_K$, it follows that $\mathfrak{q}$, and hence $\mathfrak{a}$, is principal.

□

## §2 Characters

From now on a subfield $K$ of some field $\mathbb{Q}[m]$ will also be called a cyclotomic field. From (1.6.$a$) we know that such a field is always Galois over $\mathbb{Q}$ with abelian Galois group. In fact, the converse is also true: the Kronecker-Weber theorem asserts that a field $K$ which is Galois over $\mathbb{Q}$ with abelian Galois group must be cyclotomic. We shall, however, not be able to give a proof of this here.

We shall introduce the notion of (rational) residue class characters to obtain a description of the cyclotomic fields in terms of objects defined within the field $\mathbb{Q}$ of rational numbers. This description will provide us with the tools to study the intererrelations between cyclotomic fields, their ramification and their prime decomposition laws. In Chapter VIII these characters will occur in the definition of the Dirichlet $L$-functions which on the one hand will provide us with explicit class number formulae, and which on the other hand are instrumental in the proof of Dirichlet's theorem on primes in arithmetic progressions. Already, in the next section, we shall reap the rewards of our general treatment, when we obtain further deep results on quadratic fields.

For any natural number $m$, let $\Theta_m$ denote the character group of $(\mathbb{Z}/m\mathbb{Z})^*$; its elements are called *residue class characters*. Throughout this section we shall use the results in Appendix A on characters. Let $m \mid n$; then the surjection $(\mathbb{Z}/n\mathbb{Z})^* \to (\mathbb{Z}/m\mathbb{Z})^*$ will, by (A11), give rise to an injective homomorphism

(2.1) $$i_m^n \colon \Theta_m \to \Theta_n$$

and if $d \mid m$ we have

(2.1.*a*) $$i_m^n \circ i_d^m = i_d^n.$$

In the sequel, when there is no fear of confusion, we shall often omit reference to $i_m^n$ and view $\Theta_m$ as a subgroup of $\Theta_n$. Note in passing that if $m$ is odd, then $(\mathbb{Z}/2m\mathbb{Z})^* \to (\mathbb{Z}/m\mathbb{Z})^*$ is an isomorphism, i.e. $\Theta_{2m} = \Theta_m$. Similarly $\mathbb{Q}[2m] = \mathbb{Q}[m]$. As a rule we shall therefore assume

(2.1.*b*) $$m \not\equiv 2 \bmod (4).$$

**(2.2)** *Let $d$ denote the HCF of $m$ and $n$, and let $g$ denote the LCM of $m$ and $n$. Then we have equalities in $\Theta_g$*

(2.2.*a*) $$\Theta_m \Theta_n = \Theta_g$$

(2.2.*b*) $$\Theta_m \cap \Theta_n = \Theta_d.$$

First we prove a purely algebraic lemma

**(2.3)** *Let $J$ denote a finite abelian group and let $H$ be a subgroup of $J$. As per (A11), we view $X(H) = \widehat{J/H}$ as a subgroup of $\hat{J}$. If $K$ is a further subgroup of $J$, then*

(2.3.*a*) $$X(H) \cap X(K) = X(HK)$$

(2.3.*b*) $$X(H) \cdot X(K) = X(H \cap K).$$

*Proof.* Consider the exact sequence

$$1 \to \frac{J}{H\cap K} \xrightarrow{\beta} \frac{J}{H} \times \frac{J}{K} \xrightarrow{\alpha} \frac{J}{HK} \to 1$$

where $\alpha(gH, lK) = gl^{-1}HK$, $\beta(g(H\cap K)) = (gH, gK)$. Applying (A11) gives a further exact sequence

$$1 \to X(HK) \xrightarrow{\hat{\alpha}} X(H) \times X(K) \xrightarrow{\hat{\beta}} X(H\cap K) \to 1.$$

Here $\hat{\beta}(\chi, \phi) = \chi\phi$ and so $\ker\hat{\beta} \cong X(H)\cap X(K)$; clearly as subgroups of $\hat{J}$, $X(H)\cap X(K) \supset X(HK)$, and the above now shows equality, which gives (*a*). Part (*b*) on the other hand follows from the surjectivity of $\hat{\beta}$. □

We now prove (2.2). We let $G_m$ denote the kernel of the surjection $(\mathbb{Z}/g\mathbb{Z})^* \to (\mathbb{Z}/m\mathbb{Z})^*$, so $\Theta_m$ identifies as the group of characters of $(\mathbb{Z}/g\mathbb{Z})^*/G_m$. Applying (2.3) to $G_m$ and $G_n$ in $(\mathbb{Z}/g\mathbb{Z})^*$, it will suffice to prove

(2.4.*a*) $$G_m \cdot G_n = G_d$$

(2.4.*b*) $$G_m \cap G_n = G_g.$$

(*b*) is completely obvious and it is also clear that $G_m \cdot G_n \subset G_d$. It therefore suffices to show that if $x \equiv 1 \bmod (d)$, then $x \equiv yz \bmod (g)$ with $y \equiv 1 \bmod (m)$, $z \equiv 1 \bmod (n)$. This, however, is obvious; write $x = 1 + mm' + nn' = (1+mm')(1+nn') - mnm'n' \equiv yz \bmod (g)$ where $y = 1 + mm'$, $z = 1 + nn'$. □

A residue class character $\theta \in \Theta_m$ (with $m \not\equiv 2 \bmod (4)$) is called *primitive* if $\theta = i_d^m(\theta')$ (for $d \mid m$, $d \not\equiv 2 \bmod (m)$) implies that $d = m$ and $\theta = \theta'$. In this case we say that $m$ is the *conductor* of $\theta$ and we write $m = f(\theta)$.

*Example 1.* With the notation of Theorem 40, $\Theta_8 = \{\epsilon, \lambda_4, \lambda_8, \lambda_4\lambda_8\}$, where $\epsilon$ is the identity character. Here $\lambda_8$ and $\lambda_4 \cdot \lambda_8$ are primitive, but $\epsilon$ and $\lambda_4$ are not.

*Example 2.* The map $x \mapsto \left(\frac{x}{3}\right)\left(\frac{x}{5}\right)$ induces a primitive character of order two in $\Theta_{15}$; however, the characters induced by $x \mapsto \left(\frac{x}{3}\right)$ and $x \mapsto \left(\frac{x}{5}\right)$ are not primitive.

**(2.5)** *Every residue class character $\theta \in \Theta_m$ is of the form $\theta = i_f^m(\theta')$*

*wih $\theta'$ primitive in $\Theta_f$ and $f = f(\theta) \not\equiv 2 \bmod (4)$; moreover $\theta'$ is uniquely determined by $\theta$.*

*Proof.* If $\theta$ is not primitive, then $\theta = i^m_{d_1}(\theta_1)$ ($d_1 \not\equiv 2 \bmod (4)$). If $\theta_1$ is not already primitive in $\Theta_{d_1}$, then $\theta_1 = i^{d_1}_{d_2}(\theta_2)$, so that $\theta = i^m_{d_2}(\theta_2)$. After a finite number of steps, we end up with an equation $\theta = i^m_f(\theta')$ with $\theta'$ primitive. Suppose also $\theta = i^m_{f^*}(\theta^*)$ with $\theta^*$ primitive. If $f = f^*$, then, as $i^m_f$ is injective $\theta' = \theta^*$. If $f \neq f^*$, let $d = (f, f^*)$ and let $g$ denote the LCM of $f$, $f^*$. Thus $i^m_g(i^g_f(\theta')) = i^m_g(i^g_{f^*}(\theta^*))$, whence $i^g_f(\theta') = i^g_{f^*}(\theta^*)$. By (2.2.*b*), $\theta' = i^f_d(\tilde{\theta})$ for $d \mid f$, $d \neq f$ and $\tilde{\theta} \in \Theta_d$. This shows that $\theta'$ is not primitive – a contradiction.

□

Next we denote by $\Psi_K$ the character group of $\mathrm{Gal}(K/\mathbb{Q})$, for a cyclotomic field $K$. The elements of $\Psi_K$ are called *Galois characters.* If $L$ is a further cyclotomic field and if $L \supset K$, then we have a surjection $\mathrm{Gal}(L/\mathbb{Q}) \to \mathrm{Gal}(K/\mathbb{Q})$ with kernel $\mathrm{Gal}(L/K)$. By (A11) this gives rise to an injective homomorphism

(2.6) $$j^L_K \colon \Psi_K \to \Psi_L.$$

If $M \subset K \subset L$, then

(2.6.*a*) $$j^L_K \circ j^K_M = j^L_M.$$

Moreover, for $K = \mathbb{Q}[m]$, $L = \mathbb{Q}[n]$ with $m \mid n$, we now have a commutative diagram

(2.7) $$\begin{array}{ccc} \Theta_K & \xrightarrow{\hat{r}_K} & \Psi_K \\ \downarrow{\scriptstyle i^L_K} & & \downarrow{\scriptstyle j^L_K} \\ \Theta_L & \xrightarrow{\hat{r}_L} & \Psi_L \end{array}$$

Here we have written $\Theta_K$ for $\Theta_m$, $\Theta_L$ for $\Theta_n$, $i^L_K$ for $i^n_m$, and the maps $\hat{r}_K = \hat{r}_m$, $\hat{r}_L = \hat{r}_n$ are induced by the isomorphisms $r_m$, $r_n$ of Theorem 44, and are therefore both themselves isomorphisms. The commutativity follows from (1.8).

We now have

**(2.8)** *If $d$ is the HCF of $m, n$, and if $g$ is the LCM of $m$ and $n$, then*

$$\mathbb{Q}[m] \cap \mathbb{Q}[n] = \mathbb{Q}[d]$$
$$\mathbb{Q}[m] \cdot \mathbb{Q}[n] = \mathbb{Q}[g].$$

This follows at once from (2.2), (2.3) and (2.7).

□

From (2.8) it now follows that for any cyclotomic field $K$, there is a natural number $f = f(K)$, the *conductor of* $K$, with $f \not\equiv 2 \bmod (4)$, so that $K \subset \mathbb{Q}[f]$ with the property that if $K \subset \mathbb{Q}[m]$, then $f \mid m$ i.e. $\mathbb{Q}[f] \subset \mathbb{Q}[m]$. The isomorphism $\hat{r}^{-1}_{\mathbb{Q}[f]}: \Psi_{\mathbb{Q}[f]} \cong \Theta_f$, $f = f(K)$, will map $j_K^{\mathbb{Q}[f]}(\Psi_K)$ isomorphically onto a subgroup of $\Theta_f$, which we denote by $\Theta_K$. We thus get an isomorphism $\hat{r}_K: \Theta_K \cong \Psi_K$ and an injective homomorphism $i_K^L: \Theta_K \to \Theta_L$, and for these the diagram (2.7) will still commute. Also, if $M \subset K \subset L$, then $i_K^L \circ i_M^K = i_M^L$.

**(2.9)** *$p$ ramifies in $K$ iff $p \mid f(K)$.*

*Proof.* If $p$ ramifies in $K$, then $p$ ramifies in $\mathbb{Q}[f]$; hence, by Theorem 44, $p \mid f$.

By (1.15) the maximal subfield of $\mathbb{Q}[f]$ in which $p$ is non-ramified is $\mathbb{Q}[f_1]$, where $f_1$ is the greatest divisor of $f$ which is not divisible by $p$. But, if $f = f(K)$ and if $p$ is non-ramified in $K$, then $K \subset \mathbb{Q}[f_1]$ hence $f \mid f_1$ and so $f = f_1$.

□

We can describe primitive residue class characters in terms of Galois characters. For this we shall need the notion of the *kernel field* of such a character. If $\psi \in \Psi_K$, then $\ker \psi$ is a subgroup of $\mathrm{Gal}(K/\mathbb{Q})$ of the form $\mathrm{Gal}(K/\mathbb{Q}_\psi)$, with $K \supset \mathbb{Q}_\psi$; we call $\mathbb{Q}_\psi$ the kernel field of $\psi$. Note from the definition that, if $L \supset K$, with $L$ a cyclotomic field, and if $\psi' = j_K^L \psi$, then $\mathbb{Q}_{\psi'} = \mathbb{Q}_\psi$. Actually, if in the above $K = \mathbb{Q}_\psi$, then we call $\psi$ a *faithful* character of $K$. Evidently in general $\psi = j_{\mathbb{Q}_\psi}^K(\psi_0)$ for some faithful $\psi_0$ which is uniquely determined by $\psi$.

**(2.10)** *Let $f = f(K)$ and let $\psi \in \Psi_K$ be faithful. Then $\hat{r}_K^{-1}(\psi) = \chi \in \Theta_f$ is primitive; moreover, every primitive character can be written in this form for some $K$.*

*Proof.* Write $\psi^* = j_K^{\mathbb{Q}[f]}(\psi)$, and suppose that $\hat{r}_K^{-1}(\psi) = \chi$ is not primitive, i.e. $\chi = i_d^f(\chi')$ with $d \mid f$ and $d \neq f$. Then by the commutativity of (2.7), $\psi^* = j_{\mathbb{Q}[d]}^{\mathbb{Q}[f]}(\psi')$, where $\psi' = \hat{r}_{\mathbb{Q}[d]}(\chi')$. Thus $\mathbb{Q}_{\psi^*} \subset K$, $\mathbb{Q}_{\psi^*} \subset \mathbb{Q}[d]$, and therefore $\mathbb{Q}_\psi = \mathbb{Q}_{\psi^*}$ is a proper subfield of $K$, which is a contradiction.

Now let $\theta$ be a primitive residue class character mod$(f)$, and let $\hat{r}_L(\theta) = \psi$ for $L = \mathbb{Q}[f]$. Then by primitivity $\mathbb{Q}_\psi \not\subset \mathbb{Q}[d]$ for any proper divisor $d$ of $f$. Putting $K = \mathbb{Q}_\psi$ for brevity, we see that $K$ has conductor $f$ and that if $\psi = j_K^L(\psi_1)$ with $\psi_1$ faithful in $\Psi_K$, then indeed $\hat{r}_K^{-1}(\psi_1) = \theta$ in $\Theta_K \subset \Theta_f$.

□

Given a primitive residue class character $\theta \in \Theta_f$, we define a function $\tilde{\theta}$ on $\mathbb{Z}$, called an *extended residue class character*, by

$$\tilde{\theta}(a) = \begin{cases} \theta(a \bmod (f)) & \text{if } (a, f) = 1 \\ 0 & \text{if } (a, f) > 1. \end{cases} \tag{2.11}$$

We call $f(\tilde{\theta}) = f(\theta)$ the conductor of $\tilde{\theta}$. Note that the function $\tilde{\theta}$ satisfies the rules

**(2.11.a)**

(*i*) $\tilde{\theta}(ab) = \tilde{\theta}(a)\tilde{\theta}(b)$
(*ii*) $\tilde{\theta}(a) = 1$ if $a \equiv 1 \bmod (f)$
(*iii*) $\tilde{\theta}(a) = 0$ if $(a, f) > 1$
(*iv*) Suppose that $\tilde{\theta}(a) = 1$ whenever $a \equiv 1 \bmod (m)$; then $f \mid m$.

In fact one can characterise the extended primitive residue class characters by means of these properties.

Recall now that the inertia group, the decomposition group, and the Frobenius element of a prime ideal $\mathfrak{p}$ in a cyclotomic field $K$ all only depend on the prime number $p$ below $\mathfrak{p}$ (see Theorem 30). Accordingly, we now speak of the inertia group, the decomposition group and the Frobenius element of $p$ in $K$.

Similarly, if $w$ is an Archimedean absolute value of $K$ with completion $K_w$, then $\mathrm{Gal}(K_w/\mathbb{Q}_\infty)$ is independent of the choice of $w$ (where $\mathbb{Q}_\infty$ denotes the completion of $\mathbb{Q}$ with respect to the ordinary absolute value). We call $\mathrm{Gal}(K_w/\mathbb{Q}_\infty)$ the local group at infinity. A Galois character $\psi$ of $K$ is said to be *non-ramified at* $p$ (resp. *split at* $p$) if $\ker\psi$ contains the inertia group (resp. decomposition group) of $p$ in $K/\mathbb{Q}$. We say that $\psi$ is *real* if $\ker\psi$ contains the local Galois group at infinity. Note that, by Theorem 30, these properties are all preserved under the maps $j_K^L$ (for cyclotomic $L$ containing $K$).

Now let $\psi \in \Psi_K$ and suppose $\hat{r}_K^{-1}\psi = \theta \in \Theta_f$ where $f = f(K)$. Let $\chi$ denote the primitive residue class character with $i_d^f(\chi) = \theta$, so that $d = f(\chi)$. As previously, we write $\tilde{\chi}$ for the extended residue class character of $\chi$.

**(2.12)**

(*a*) *$\psi$ is non-ramified at $p$ iff $\tilde{\chi}(p) \neq 0$.*
(*b*) *$\psi$ is split at $p$ iff $\tilde{\chi}(p) = 1$.*

(c) *$\psi$ is real iff $\tilde{\chi}$ is even, i.e. if $\tilde{\chi}(-1)=1$.*
[If $\tilde{\chi}(-1) = -1$, then we say that $\tilde{\chi}$ is odd.]

*Proof.* We can replace $\psi$ by the faithful character which maps to $\psi$ (under the appropriate map $j$); this will not change the stated properties (*a*)-(*c*), nor will it change $\tilde{\chi}$. So, from now on, we assume that $\psi$ is faithful and so $f = f(\tilde{\chi}) = f(K)$.

To say that $\psi$ is non-ramified at $p$ now simply means that $p$ is non-ramified in $K$. Part (*a*) then follows immediately from (2.9).

Next assume that $p$ is non-ramified in $K$. Then, by Theorem 44, $\tilde{\chi}(p) = \psi(p, K/\mathbb{Q})$, and so this value is equal to 1 precisely when the decomposition group of $p$ in $\mathrm{Gal}(K/\mathbb{Q}) = \{1\}$.

Lastly observe that for any field $\mathbb{Q}[m]$ and for all embeddings $\mathbb{Q}[m] \hookrightarrow \mathbb{C}$, complex conjugation $\rho$ on $\mathbb{Q}[m]$ is given by $\zeta^\rho = \zeta^{-1}$ for all $m$th roots of unity, i.e. $\psi(\rho) = \chi(-1)$. This then extends to any cyclotomic field $K$ and so establishes (*c*).

□

We now come to the result which motivates the introduction of extended residue class characters. For $\theta \in \Theta_K$ we have a unique primitive residue class character $\theta'$ with $i_f^m(\theta') = \theta$, where $m = f(K)$, $f = f(\theta')$. The set of extended residue class characters $\tilde{\theta}'$ for all $\theta \in \Theta_K$ will be denoted by $\tilde{\Theta}_K$; thus $\tilde{\Theta}_K$ is in bijective correspondence with both $\Theta_K$ and $\Psi_K$.

**Theorem 47.** *Let $K$ be a cyclotomic field, let $p$ be a prime number, with ramification index $e$ and residue class degree $f$, and with $g$ distinct prime ideals in $\mathfrak{o}_K$ lying above $p$ (so that $efg = (K : \mathbb{Q})$). Then, for an algebraic indeterminate $T$, we have*

$$\prod_{\nu \in \tilde{\Theta}_K} (T - \nu(p)) = T^{fg(e-1)}(T^f - 1)^g.$$

*Proof.* Let $\Gamma = \mathrm{Gal}(K/\mathbb{Q})$ and let $\Delta$ denote the decomposition group (resp. $\Sigma$ denote the inertia group) of $p$ in $K/\mathbb{Q}$. Among the $efg$ characters in $\Psi_K$, those which are non-ramified at $p$ are precisely those whose kernel contains $\Sigma$. These form a subgroup of $\Psi_K$, the image of $j_M^K$, where $M = K^\Sigma$. There are therefore $[\Gamma : \Sigma] = (K : \mathbb{Q})/|\Sigma| = fg$ of these; the remaining $efg - fg = (e-1)fg$ characters have $\psi(\Sigma) \neq 1$; so by (2.12) there are precisely $(e-1)fg$ extended residue class characters $\nu \in \tilde{\Theta}_K$ with $\nu(p) = 0$. This accounts for the factor $T^{fg(e-1)}$ in our formula.

Next let $\psi \in \Psi_K$ be non-ramified at $p$ i.e. $\psi = j_M^K(\psi')$. If $\theta = \hat{r}_K^{-1}(\psi)$, then by (2.1.$a$) and the commutativity of (2.7), $\theta(p) = \psi'(p, M/\mathbb{Q})$. Thus, for the corresponding $\nu \in \tilde{\Theta}_K$, we have $\nu(p) = \psi'(p, M/\mathbb{Q})$. But $(p, M/\mathbb{Q})$ generates a cyclic group of order $f$ in $\Gamma/\Sigma$. Thus each $f$th root of unity is taken equally often as a value of $\psi'(p, M/\mathbb{Q})$, as $\psi$ runs over the characters in $\Psi_K$ which are non-ramified at $p$. Letting $\nu$ run over the corresponding extended residue class characters we get

$$\prod_{\nu}(T - \nu(p)) = \prod_{\zeta \mid \zeta^f = 1} (T - \zeta)^g = (T^f - 1)^g.$$

□

## §3 Quadratic fields revisited

We shall now use the general theory of §2 to gain new insights into quadratic fields, by viewing them as subfields of fields of roots of unity. In particular we shall obtain a number of further proofs of the quadratic reciprocity law.

Let $K/\mathbb{Q}$ be a quadratic extension, i.e. $(K : \mathbb{Q}) = 2$. Suppose we know that $K \subset \mathbb{Q}[f]$ for some $f$; then we may, in particular, choose the least such $f$, which by definition is $f(K)$. $\tilde{\Theta}_K$ will then consist of two extended primitive characters: one, the identity character $\epsilon$ with $\epsilon(x) = 1$ for all $x$; the other, the non-trivial element of $\tilde{\Theta}_K$, will be denoted by $\tilde{\lambda}_K$, where $\lambda_K$ is the corresponding primitive residue class character. Note that

$$\tilde{\lambda}_K(a) = 0 \qquad \text{if } (a, f(K)) \neq 1$$
$$\tilde{\lambda}_K(a)^2 = 1 \qquad \text{otherwise.}$$

**Theorem 48.**

($a$) *Every quadratic field $K$ is a subfield of some $\mathbb{Q}[f]$, and $f(K) = |d_K|$.*

($b$) *Let $p$ be an odd prime and put $p^* = \left(\frac{-1}{p}\right) p$. Then $\mathbb{Q}[p]$ contains the quadratic field $K = \mathbb{Q}(\sqrt{p^*})$, and this is also the unique quadratic subfield of $\mathbb{Q}[p^n]$ for each $n > 0$. The character $\lambda_K$ (which will often also be denoted by $\lambda_p$) is given by*

$$\lambda_K(x) = \left(\frac{x}{p}\right) \qquad \textit{for } (x, p) = 1.$$

($c$) *The quadratic subfields of $\mathbb{Q}[2^n]$ for $n \geq 3$ are precisely those of $\mathbb{Q}[8]$. Their characters, for odd $x$, are given by*

$$K = \mathbb{Q}(\sqrt{-1}) \textit{ has } \lambda_K(x) = (-1)^{\frac{1}{2}(x-1)} \ (= \lambda_4(x))$$
$$K = \mathbb{Q}(\sqrt{2}) \textit{ has } \lambda_K(x) = (-1)^{\frac{1}{8}(x^2-1)} \ (= \lambda_8(x))$$
$$K = \mathbb{Q}(\sqrt{-2}) \textit{ has } \lambda_K(x) = (-1)^{\frac{1}{2}(x-1)+\frac{1}{8}(x^2-1)} \ (= \lambda_4\lambda_8(x)).$$

(In the sequel we shall sometimes refer to $\lambda_4\lambda_8$ as $\lambda_{-8}$.)

By comparing the decomposition law of Theorem 48 with the Kummer type decomposition law of (III.2.29), we obtain an extremely elegant proof of the Quadratic Reciprocity Law.

*Second proof of the Quadratic Reciprocity Law.* To prove the main law let $p, q$ be distinct odd primes. Then $q$ splits in $\mathbb{Q}(\sqrt{p^*})$ iff $\lambda_p(q) = 1$, by (2.12). However, by (III.2.29) this occurs iff $\left(\frac{p^*}{q}\right) = 1$; hence indeed $\left(\frac{q}{p}\right) = \left(\frac{p^*}{q}\right)$.

In order to derive the two subsidiary laws, note that $p$ splits in $\mathbb{Q}(\sqrt{-1})$ iff $\lambda_K(p) = 1$ by the above; while this occurs iff $\left(\frac{-4}{p}\right) = 1$ by (III.2.29). Similarly $q$ splits in $\mathbb{Q}(\sqrt{2})$ iff $\lambda_8(p) = 1$ iff $\left(\frac{2}{p}\right) = 1$.

*Proof of Theorem 48.* (*b*) Let $p$ denote an odd prime number. For each $r \geq 1$, the Galois group $\mathrm{Gal}(\mathbb{Q}[p^r]/\mathbb{Q}) \cong (\mathbb{Z}/p^r\mathbb{Z})^*$ has a unique subgroup of index 2. Thus $\mathbb{Q}[p^r]$ has a unique quadratic subfield in which $p$ is the one and only ramified prime. But, from the list of quadratic fields and their discriminants we know that there is exactly one such quadratic field: namely $K = \mathbb{Q}(\sqrt{p^*})$. Moreover, $\lambda_K$ is a quadratic charactor mod$(p)$ i.e. has the property that $\lambda_K^2(x) = 1$ for $(x,p) = 1$; there is only one such character, namely $x \mapsto \left(\frac{x}{p}\right)$. This then proves (*b*).

To prove (*c*), observe that $\mathrm{Gal}(\mathbb{Q}[8]/\mathbb{Q}) \cong (\mathbb{Z}/8\mathbb{Z})^*$ has exactly three subgroups of index 2 (and so, more generally, does $(\mathbb{Z}/2^n\mathbb{Z})^*$ for all $n \geq 3$). There are then exactly three quadratic fields contained in $\mathbb{Q}[8]$ (and likewise for $\mathbb{Q}[2^n]$ $n \geq 3$). On the other hand, there are exactly three quadratic fields in which 2, and no other prime, is ramified, namely $\mathbb{Q}(\sqrt{-1})$, $\mathbb{Q}(\sqrt{2})$, $\mathbb{Q}(\sqrt{-2})$; thus the two sets coincide. Moreover, $\mathbb{Q}(\sqrt{-1}) = \mathbb{Q}[4]$, so that now $\lambda_K$ must be the unique primitive character mod$(4)$, namely $\lambda_4$. This leaves us to allocate the two primitive characters mod$(8)$ to the remaining two fields. As $\mathbb{Q}(\sqrt{2}) = K$ is real, i.e. is the maximal real subfield of $\mathbb{Q}[8]$, $\lambda_K(-1) = 1$, by (2.12); therefore $\lambda_K = \lambda_8$.

Note now that all quadratic fields with only one ramified prime have been taken care of. In fact parts (*b*) and (*c*) show that for each prime discriminant $q$, $\mathbb{Q}(\sqrt{q}) \subset \mathbb{Q}[|q|]$. More generally, by (III.3.7) any quadratic discriminant $d_K$ can be written uniquely as a product of prime discriminants $d_K = q_1 \cdots q_n$. Therefore

$$K = \mathbb{Q}(\sqrt{d_K}) \subset \mathbb{Q}(\sqrt{q_1}, \ldots, \sqrt{q_n}) \subset \mathbb{Q}[|d_K|]. \tag{3.1}$$

Hence $f(K) \mid |d_K|$. We now need to show $|d_K| \mid f(K)$. Consider first an odd prime $p$ dividing $|d_K|$. As $d_K = m$ or $4m$ with $m$ square-free, clearly $p$ only divides $|d_K|$ to the first power. If, however, $p \nmid f(K)$, then the inclusion $K \subset \mathbb{Q}[f]$, would imply that $p$ was non-ramified in $K$, which contradicts Theorem 22. There therefore remains the case $f(K) = m2^r$ and $|d_K| = m2^s$. Arguing as above, and using $f(K) \mid |d_K|$, we see that $2 \leq r \leq s \leq 3$. Thus if $s = 2$, $r$ must be 2 and we are done. If $s = 3$, then we must have $K = \mathbb{Q}(\sqrt{2n})$ for some odd square-free $n$. By (3.1) we know $K \subset \mathbb{Q}[8|n|]$. Suppose for contradiction that in fact $\sqrt{2n} \in \mathbb{Q}[4m]$ for some odd $m$. Increasing $m$ if necessary, we may assume $n \mid m$. By (3.1) we know $\sqrt{\delta n} \in \mathbb{Q}[m]$ for some choice of $\delta \in \{\pm 1\}$. This implies that $\mathbb{Q}[8] \subset \mathbb{Q}[4m]$ which is absurd. We therefore conclude $r = 3$, as required.

□

It is interesting to note that we can also express the sign of $d_K$ in terms of $\lambda_K$

$$d_K = f(K)\lambda_K(-1). \tag{3.2}$$

Indeed, $f(K) > 0$, whereas $d_K < 0$ iff $K$ is not real, i.e. by (2.12) iff $\lambda_K(-1) = -1$.

We shall present two more proofs that quadratic fields with exactly one ramified prime are subfields of appropriate fields of roots of unity, by providing explicit generators. This approach, via explicit computation, will also yield two more proofs of the quadratic reciprocity law.

The first attack is via *quadratic Gauss sums.* Let $\theta$ be a primitive quadratic residue character with conductor $f$. (Recall that a character is said to be quadratic if it has order 2.) Let $\zeta$ be a primitive $f$th root of unity ($f \not\equiv 2 \bmod (4)$). The corresponding Gauss sum is then defined by

$$\tau(\theta, \zeta) = \sum_{x \bmod (f)} \tilde{\theta}(x)\zeta^x \in \mathbb{Q}[f]. \tag{3.3}$$

This is clearly independent of the choice of representatives of $\mathbb{Z}$ mod $(f)$. Recall that $\tilde{\theta}(x) = 0$ if $(x, f) \neq 1$. If $(a, f) = 1$, then denote by $\sigma_a \in \mathrm{Gal}(\mathbb{Q}[f]/\mathbb{Q})$ the automorphism with the property that $\zeta^{\sigma_a} = \zeta^a$. We assert that

$$\tau(\theta, \zeta)^{\sigma_a} = \tau(\theta, \zeta^a) = \theta(a)^{-1}\tau(\theta, \zeta). \tag{3.4}$$

For the last equality observe that

$$\sum_{x \bmod (f)} \tilde{\theta}(x)\zeta^{ax} = \sum_{x \bmod (f)} \theta(a)^{-1}\tilde{\theta}(ax)\zeta^{ax}$$
$$= \theta(a)^{-1}\sum_{x} \theta(x)\zeta^{x}.$$

Here we shall only have to deal with the case where $f$ is a prime power (see the next section for the more general case).

**Theorem 49.** *Let $K$ be a quadratic field with prime discriminant $d_K$. Let $\zeta$ denote a primitive $|d_K|$th root of unity and let $\lambda = \lambda_K$ be the quadratic primitive residue class character associated with $K$; then*

$$\tau(\lambda, \zeta)^2 = d_K$$

*and so in particular $K \subset \mathbb{Q}[|d_K|]$.*

*Proof.* We begin by considering the case when $d_K$ is a power of 2. $\mathbb{Q}(\zeta)$ is isomorphic to a subfield of $\mathbb{C}$ and, of course, any such isomorphism fixes $\mathbb{Q}$; we may therefore now work inside $\mathbb{C}$. Moreover, by (3.4), we may identify $\zeta$ with a particular root of unity $e^{2\pi i/2^n}$ $(n = 2, 3)$. Later on we shall adopt this point of view more generally and find that it leads to deep additional arithmetic insights.

(3.5.$a$) If $\lambda = \lambda_4$ then

$$\tau(\lambda, \zeta) = e^{2\pi i/4}\lambda_4(1) + e^{-2\pi i/4}\lambda_4(-1) = i - i^{-1} = 2i.$$

(3.5.$b$) If $\lambda = \lambda_8$ then

$$\tau(\lambda, \zeta) = \sum_{x \bmod (8)} \lambda_8(x)e^{2\pi ix/8}$$
$$= e^{\pi i/4} + e^{-\pi i/4} - e^{5\pi i/4} - e^{-5\pi i/4} = 2(e^{\pi i/4} + e^{-\pi i/4})$$
$$= 4\cos(\pi/4) = 2\sqrt{2}.$$

(3.5.$c$) If $\lambda = \lambda_{-8}$ then

$$\tau(\lambda, \zeta) = \sum_{x \bmod (8)} \lambda_{-8}(x)e^{\pi ix/4}$$
$$= e^{\pi i/4} - e^{-\pi i/4} + e^{3\pi i/4} - e^{-3\pi i/4}$$
$$= 2(e^{\pi i/4} - e^{-\pi i/4})$$
$$= 4i\sin(\pi/4) = 2\sqrt{2}i.$$

Note that in the above three cases we get for the resulting complex numbers the equation

$$\tau(\lambda_K, e^{2\pi i/|d_K|}) = \begin{cases} \sqrt{|d_K|}, & \text{if } d_K > 0, \\ i\sqrt{|d_K|}, & \text{if } d_K < 0, \end{cases}$$

with $\sqrt{\phantom{x}}$ the positive square root. This looks forward to Theorem 50.

From now on let $d_K = p^* \; (= \left(\frac{-1}{p}\right) p)$ with $p > 2$. Then, if $p \nmid x$, $\lambda(x) = \left(\frac{x}{p}\right)$, and for brevity we write $\tau(\lambda, \zeta) = \tau_p(\zeta)$.

$$\tau_p(\zeta)^2 = \sum_{x,y} \left(\frac{x}{p}\right)\left(\frac{y}{p}\right) \zeta^{x+y}$$

where $x, y$ run through complete sets of prime residues $\bmod (p)$. We put $y \equiv xz \bmod (p)$, with $p \nmid z$, so that

$$\begin{aligned}
\tau_p(\zeta)^2 &= \sum_{z,x} \left(\frac{zx^2}{p}\right) \zeta^{x(1+z)} \\
&= \sum_{z \not\equiv -1} \left(\frac{z}{p}\right) \sum_{x} \zeta^{x(1+z)} + \left(\frac{-1}{p}\right)(p-1)
\end{aligned}$$

as $\left(\frac{x^2}{p}\right) = \left(\frac{x}{p}\right)^2 = 1$ and $\sum\limits_{\substack{x \bmod (p) \\ x \not\equiv 0}} \zeta^0 = p - 1$.

If $1 + z \not\equiv 0 \bmod (p)$, then $\zeta^{1+z}$ is a primitive $p$th root of unity, and so is a root of $X^{p-1} + \cdots + X + 1$; hence $\sum_x \zeta^{x(1+z)} = -1$. Therefore, from the above, we now obtain

$$\begin{aligned}
\tau_p(\zeta)^2 &= (-1) \sum_{z \not\equiv -1} \left(\frac{z}{p}\right) - \left(\frac{-1}{p}\right) + p\left(\frac{-1}{p}\right) \\
&= (-1) \sum_{z \bmod (p)} \left(\frac{z}{p}\right) + p\left(\frac{-1}{p}\right) \\
&= p\left(\frac{-1}{p}\right) = d_K.
\end{aligned}$$

□

In summary, we have shown directly, without any reference to the previous results of this chapter, that the fields $K$, with $d_K$ a prime discriminant, are subfields of fields of roots of unity and that $\mathbb{Q}(\sqrt{p^*}) \subset \mathbb{Q}[p]$. If we now want to introduce characters, then we can describe $\lambda_K$ as the non-trivial element of $\Theta_K$: for instance, when $d_K = p^*$, then $\lambda_K$ is the only quadratic primitive residue class character with $p$-power conductor, i.e. $\lambda_K(x) = \left(\frac{x}{p}\right)$.

We next use Gauss sums to obtain yet another proof of quadratic reciprocity. Keeping the above notation, we work in $\mathbb{Q}[f]$ and let $q$ denote an odd prime which does not divide the conductor $f = f(K)$. From (3.4)

(3.6) $$\tau(\lambda, \zeta^q) = \lambda(q)\tau(\lambda, \zeta).$$

On the other hand, taking congruences modulo the ideal $q\mathbb{Z}[\zeta]$, we obtain congruences

$$\tau(\lambda, \zeta^q) \equiv \tau(\lambda, \zeta)^q = \tau(\lambda, \zeta)\tau(\lambda, \zeta)^{2((q-1)/2)}$$
$$\equiv \tau(\lambda, \zeta) d_K^{(q-1)/2} \equiv \tau(\lambda, \zeta) \left(\frac{d_K}{q}\right).$$

As $\tau(\lambda, \zeta)^2 = d_K$ is coprime to $q$, on comparing with (3.6), we get

$$\lambda(q) \equiv \left(\frac{d_K}{q}\right) \bmod (q).$$

This is of course the Quadratic Reciprocity Law: for, on the one hand, if $d_K = p^*$ with $p > 2$, then we get

$$\left(\frac{q}{p}\right) = \left(\frac{p}{q}\right)\left(\frac{(-1)^{(p-1)/2}}{q}\right) = \left(\frac{p}{q}\right)(-1)^{((p-1)/2)((q-1)/2)}.$$

On the other hand if $d_K = -4$ or if $d_K = 8$, then we get $(-1)^{(q-1)/2} = \left(\frac{-4}{q}\right)$ and $(-1)^{(q^2-1)/2} = \left(\frac{2}{q}\right)$.

For the next approach we restrict ourselves to odd primes. We shall obtain another number which obviously lies in $\mathbb{Q}[p]$, and which actually turns out to again be a square root of $d_K$ (where $K = \mathbb{Q}(\sqrt{p^*})$).

Let $\zeta$ again denote a primitive $p$th root of unity, and define

$$\Pi_p(\zeta) = \prod_{x=1}^{(p-1)/2} (\zeta^x - \zeta^{-x}). \tag{3.7}$$

For reasons which will rapidly become clear, we call this product the "sine product". In analogy to (3.4) we have

$$\Pi_p(\zeta)^{\sigma_a} = \Pi_p(\zeta^a) = \Pi_p(\zeta)\left(\frac{a}{p}\right). \tag{3.8}$$

To prove this we use the Gauss Lemma:

**(3.9)** *Let $S = \{1, 2, \ldots, \frac{p-1}{2}\}$ and let $(a, p) = 1$. For $x \in S$, we have $ax \equiv \delta_x y \bmod (p)$ for $y = y(x) \in S$ and $\delta = \pm 1$. Then*

$$\prod_{x \in S} \delta_x = \left(\frac{a}{p}\right). \tag{3.9.a}$$

Indeed, $a^{(p-1)/2} \prod_{x \in S} x \equiv \prod_{x \in S} \delta_x \prod_{x \in S} x$. Therefore

$$\prod_{x \in S} \delta_x \equiv \left(\frac{a}{p}\right) \bmod (p) \qquad \text{with } \prod_{x \in S} \delta_x = \pm 1$$

which proves (3.9).

With the notation that we have just introduced, we get

$$\begin{aligned}\Pi_p(\zeta^a) &= \prod_{x\in S}(\zeta^{ax}-\zeta^{-ax}) = \prod_{x\in S}(\zeta^{\delta_x y(x)}-\zeta^{-\delta_x y(x)})\\ &= \prod_{x\in S}((\zeta^{y(x)}-\zeta^{-y(x)})\delta_x)\\ &= \left(\frac{a}{p}\right)\prod_{x\in S}(\zeta^x-\zeta^{-x})\\ &= \left(\frac{a}{p}\right)\Pi_p(\zeta).\end{aligned}$$

We also have a further result for $p > 2$

$$\Pi_p(\zeta)^2 = \left(\frac{-1}{p}\right)p. \tag{3.10}$$

Indeed, on the one hand

$$\begin{aligned}\Pi_p(\zeta)\Pi_p(\zeta^{-1}) &= \Pi_p(\zeta)\Pi_p(\zeta^{p-1})\\ &= \prod_{x=1}^{p-1}(\zeta^x-\zeta^{-x})\\ &= \prod_{x=1}^{p-1}\zeta^x\prod_{x=1}^{p-1}(1-\zeta^{-2x})\\ &= \zeta^{p(p-1)/2}\prod_{x=1}^{p-1}(1-\zeta^x)\\ &= p.\end{aligned}$$

On the other hand, by (3.8), $\Pi_p(\zeta^{-1}) = \left(\frac{-1}{p}\right)\Pi_p(\zeta)$, and this gives the desired result.

□

We shall now use the product $\Pi_p(\zeta)$ to obtain a further proof of the Quadratic Reciprocity Law for two distinct odd primes $p, q$. This proof is due to Eisenstein, and was shown to us by Cassels. Let $\eta$ denote a primitive $q$th root of unity. Because

$$\frac{X^q - X^{-q}}{X - X^{-1}} = \prod_{k=1}^{q-1}(X\eta^k - X^{-1}\eta^{-k}),$$

it follows from (3.8) that

$$\left(\frac{q}{p}\right) = \frac{\Pi_p(\zeta^q)}{\Pi_p(\zeta)} = \prod_{j=1}^{(p-1)/2}\prod_{k=1}^{q-1}T_{j,k} \tag{3.11.a}$$

where $T_{j,k} = (\zeta^j\eta^k - \zeta^{-j}\eta^{-k})$. By symmetry we have

$$\text{(3.11.}b\text{)}\qquad \left(\frac{p}{q}\right) = \prod_{j=1}^{p-1}\prod_{k=1}^{(q-1)/2} T_{j,k}.$$

Now observe that $T_{p-j,q-k} = -T_{j,k}$. The map $(j,k) \mapsto (p-j, q-k)$ sets up a bijection between those terms in (3.11.$a$) which are not in (3.11.$b$) (viz. $k > \frac{q-1}{2}$) and those terms in (3.11.$b$) which are not in (3.11.$a$) (viz. $j > \frac{p-1}{2}$). Thus we conclude that

$$\left(\frac{p}{q}\right) \Big/ \left(\frac{q}{p}\right) = (-1)^{\frac{p-1}{2}\frac{q-1}{2}}.$$

There is a variant of the product $\Pi_p$ which provides us with a unit in $\mathbb{Q}(\sqrt{p})$, and which is not a root of unity. For this we have to assume that $p \equiv 1 \bmod (4)$. We define

$$\text{(3.12)}\qquad \Pi_p^* = \prod_{x=1}^{(p-1)/2} (\zeta^x - \zeta^{-x})^{-\left(\frac{x}{p}\right)}$$

or in other words

$$\text{(3.12.}a\text{)}\qquad \Pi_p^* = \frac{\prod_b(\zeta^b - \zeta^{-b})}{\prod_a(\zeta^a - \zeta^{-a})}$$

where $a$, resp. $b$, runs through the quadratic residues, resp. the non-residues, in $[1, \frac{p-1}{2}]$. Since $\left(\frac{x}{p}\right) = \left(\frac{p-x}{p}\right)$, the number of residues (resp. non-residues) in $[1, \frac{p-1}{2}]$ is precisely half the number in $[1, p-1]$; hence the number of $a$-factors in $\Pi_p^*$ is the same as the number of $b$-factors, i.e.

$$(\Pi_p^*) = \mathfrak{p}^{\sum_b 1 - \sum_a 1} = (1).$$

On the other hand, the proof of (3.8) for $\Pi_p$ extends to $\Pi_p^*$ on using the definition (3.12). This then shows that

$$(\Pi_p^*)^{\sigma_a} = \left(\frac{a}{p}\right) \Pi_p^{*\left(\frac{a}{p}\right)}.$$

We may therefore conclude that $\Pi_p^*$ is a unit which lies in $\mathbb{Q}(\sqrt{p})$. Next choose $a$ coprime to $p$ such that $\left(\frac{a}{p}\right) = -1$; then $\sigma_a$ induces the non-trivial automorphism of $\mathbb{Q}(\sqrt{p})/\mathbb{Q}$ and

$$\text{(3.13)}\qquad N_{\mathbb{Q}(\sqrt{p})/\mathbb{Q}}(\Pi_p^*) = (\Pi_p^*)^{1+\sigma_a} = -1.$$

Hence, in particular, $\Pi_p^* \neq \pm 1$. It is interesting to compare this result with Corollary 2 to Theorem 39.

Up to now in this section we have only used algebraic methods. There are, however, deep problems and results which require analytic techniques, such as the use of complex variables – even if the use is only of an elementary kind. Once and for all we embed our cyclotomic and quadratic fields in the field $\mathbb{C}$ of complex numbers. In other words, we shall now define Gauss sums and sine products as complex numbers. It is crucially important that the reader understand this change in point of view, since we shall now be interested in the question as to precisely which square root of $d_K$ the Gauss sum actually is. When we use $L$-functions in Chapter VIII we shall see the Gauss sums and sine products both arise quite naturally as complex numbers. Moreover, this complex number approach will yield important explicit arithmetic results.

For an odd prime $p$, the Gauss sum is now normalised as

$$\tau_p = \sum_{x=1}^{p-1} \left(\frac{x}{p}\right) e^{2\pi i x/p}. \tag{3.14}$$

By Theorem 49 we know that $\tau_p^2 = \left(\frac{-1}{p}\right) p$. Let $\sqrt{p}$ be the positive square root of $p$ in $\mathbb{C}$, and let $i$ be the square root of $-1$ which lies in the upper half plane of $\mathbb{C}$. Then we know that

$$\tau_p = \pm\sqrt{p} \qquad \text{if } p \equiv 1 \bmod (4)$$
$$\tau_p = \pm i\sqrt{p} \qquad \text{if } p \equiv -1 \bmod (4).$$

**Theorem 50.**

$\tau_p = +\sqrt{p}$ *if* $p \equiv 1 \bmod (4)$;
$\tau_p = +i\sqrt{p}$ *if* $p \equiv -1 \bmod (4)$.

We shall see later in Chapter VIII, that this result has important consequences for the distribution of quadratic residues amongst the residue classes.

Next we define $\Pi_p$ and $\Pi_p^*$ in normalised form by

$$\Pi_p = \prod_{x=1}^{(p-1)/2} (e^{2\pi i x/p} - e^{-2\pi i x/p})$$

$$\Pi_p^* = \frac{\prod_b (e^{2\pi i b/p} - e^{-2\pi i b/p})}{\prod_a (e^{2\pi i a/p} - e^{-2\pi i a/p})}.$$

**Theorem 51.**

(*a*) $(2i)^{(1-p)/2}\Pi_p > 0$.
(*b*) $\Pi_p^* > 1$.

We shall prove (*b*) in Chapter VIII, by the use of $L$-functions. We shall not use (*b*) here, but include it for the sake of completeness.

*Proof.* In order to establish (*a*), note that

$$e^{2\pi i x/p} - e^{-2\pi i x/p} = 2i \sin\left(\frac{2\pi x}{p}\right).$$

Now for $0 < z < p/2$, $\sin\left(\frac{2\pi z}{p}\right)$ is positive which shows $(a)$.

□

Observe that, by the same reasoning, the unit $\Pi_p^*$ is clearly positive. In fact

$$\Pi_p^* = \frac{\prod_b \sin\left(\frac{2\pi b}{p}\right)}{\prod_a \sin\left(\frac{2\pi a}{p}\right)} > 0. \tag{3.15}$$

Clearly part $(b)$ of the theorem amounts to a highly non-trivial statement on the distribution of quadratic residues in $[1, \frac{p-1}{2}]$.

Note that, because $\Pi_p^2 = \left(\frac{-1}{p}\right) p$, we must have

$$\Pi_p = \begin{cases} \pm\sqrt{p} & \text{if } p \equiv 1 \bmod (4), \\ \pm i\sqrt{p} & \text{if } p \equiv -1 \bmod (4). \end{cases}$$

Part $(a)$ of Theorem 51 then tells us which sign to choose. Firstly, if $p \equiv 1 \bmod (4)$, then

$$i^{(1-p)/2} \text{ is } \begin{cases} \text{positive} & \text{if } p \equiv 1 \bmod (8) \\ \text{negative} & \text{if } p \equiv 5 \bmod (8) \end{cases}$$

and so

$$\Pi_p = \begin{cases} +\sqrt{p} & \text{if } p \equiv 1 \bmod (8), \\ -\sqrt{p} & \text{if } p \equiv 5 \bmod (8). \end{cases} \tag{3.16}$$

Similarly

$$\Pi_p = \begin{cases} i\sqrt{p} & \text{if } p \equiv 3 \bmod (8), \\ -i\sqrt{p} & \text{if } p \equiv 7 \bmod (8). \end{cases} \tag{3.16.a}$$

Using part $(a)$ of Theorem 51, we shall compare $\Pi_p$ and $\tau_p$ in order to prove Theorem 50. Write $\nu = 1 - e^{2\pi i/p}$, so that, in $\mathbb{Q}(e^{2\pi i/p})$, the element $\nu$ generates the prime ideal $\mathfrak{p}$ above $p$. We shall prove the following two congruences $\bmod \mathfrak{p}^{(p+1)/2}$ $(= \bmod \mathfrak{p}\sqrt{p^*})$.

$$\tau_p \equiv \left(\frac{p-1}{2}\right)!\,\nu^{(p-1)/2} \tag{3.17}$$

$$\left(\frac{-2}{p}\right)\Pi_p \equiv \left(\frac{p-1}{2}\right)!\,\nu^{(p-1)/2}. \tag{3.18}$$

We can restate (3.16) and (3.16.$a$) as

$$\left(\frac{-2}{p}\right)\Pi_p = \begin{cases} +\sqrt{p} & \text{if } p \equiv 1 \bmod (4) \\ +i\sqrt{p} & \text{if } p \equiv -1 \bmod (4) \end{cases} \tag{3.19}$$

where here we are assuming the subsidiary Quadratic Reciprocity Law for $p$ and $-2$. From (3.17), (3.18) and (3.19) we deduce the congruence $\bmod \mathfrak{p}^{(p+1)/2}$

$$\tau_p \equiv \begin{cases} +\sqrt{p} & \text{if } p \equiv 1 \bmod (4) \\ i\sqrt{p} & \text{if } p \equiv -1 \bmod (4). \end{cases}$$

This will finally then imply Theorem 50: indeed, writing $\tau_p = \delta\sqrt{p}$ (resp. $\delta i\sqrt{p}$) with $\delta = \pm 1$, it follows that $\bmod \mathfrak{p}^{(p+1)/2}$

$$\delta\sqrt{p} \equiv \sqrt{p} \qquad \text{resp.} \quad \delta i\sqrt{p} \equiv i\sqrt{p}.$$

As $(\sqrt{p^*}) = \mathfrak{p}^{(p-1)/2}$, the congruence $\bmod \mathfrak{p}^{(p+1)/2}$ implies that $\delta = 1$, as required.

It now remains to prove (3.17) and (3.18).

*Proof of (3.18).* Writing $\zeta = e^{2\pi i/p}$, we have

$$\begin{aligned}\Pi_p &= (-1)^{(p-1)/2} \prod_{x=1}^{(p-1)/2} (\zeta^{-x} - \zeta^{x}) \\ &= (-1)^{(p-1)/2} \prod_{x=1}^{(p-1)/2} \zeta^{-x} \prod_{x=1}^{(p-1)/2} (1 - \zeta^{2x}).\end{aligned}$$

Now $(1-\zeta^{2x}) = (1-\zeta)(1+\zeta+\cdots+\zeta^{2x-1})$, and $\zeta^t \equiv 1 \bmod \mathfrak{p}$ for all $t$; therefore

$$(1-\zeta^{2x})/(1-\zeta) \equiv 2x \bmod \mathfrak{p}$$

and

$$\zeta^{-x}(1-\zeta^{2x})/(1-\zeta) \equiv 2x \bmod \mathfrak{p}.$$

Hence

$$\Pi_p/\nu^{(p-1)/2} \equiv (-2)^{(p-1)/2} \left(\frac{p-1}{2}\right)! \bmod \mathfrak{p}$$

whence, on multiplying up, we get (3.18). □

Our proof of (3.17) is based on the following:

**(3.20)** *Let $f(T) \in \mathbb{Z}[T]$ have degree less than $p-1$; then*

$$\sum_{x=0}^{p-1} f(x) \equiv 0 \bmod (p).$$

*Proof.* It is enough to take $f(T) = T^m$ with $0 \le m < p-1$. The case when $m = 0$ is clear; we now suppose $m > 0$. If $x_0$ denotes a primitive root $\bmod(p)$ (i.e. its residue class is a generator of $(\mathbb{Z}/p\mathbb{Z})^*$), then $\bmod(p)$

$$(x_0^m - 1)\sum_{x=1}^{p-1} x^m \equiv \left(\sum_{j=0}^{p-2} x_0^{jm}\right)(x_0^m - 1) \equiv x_0^{m(p-1)} - 1 \equiv 0$$

while $(x_0^m - 1) \not\equiv 0 \bmod (p)$.

□

We now have congruences mod$\mathfrak{p}^{(p+1)/2}$

$$\tau_p = \sum_{x=1}^{p-1} \left(\frac{x}{p}\right)(1-\nu)^x$$
$$\equiv \sum_{x=0}^{p-1} \left[ x^{(p-1)/2} \sum_{j=0}^{x} (-1)^j \nu^j \binom{x}{j} \right].$$

Since $\binom{x}{j} = 0$ for $0 \leq x < j$

$$\tau_p \equiv \sum_{x=0}^{p-1} x^{(p-1)/2} \sum_{j=0}^{p-1} (-1)^j \nu^j \binom{x}{j}$$
$$\equiv \sum_{j=0}^{p-1} (-1)^j \nu^j \sum_{x=0}^{p-1} x^{(p-1)/2} \binom{x}{j} \bmod \mathfrak{p}^{(p+1)/2}.$$

By (3.20), $\sum_{x=0}^{p-1} x^{(p-1)/2} \binom{x}{j} \equiv 0 \bmod (p)$ for $j < \frac{p-1}{2}$, while again by (3.20)

$$\sum_{x=0}^{p-1} x^{(p-1)/2} \binom{x}{(p-1)/2} \equiv \sum_{x=0}^{p-1} \frac{x^{p-1}}{(\frac{p-1}{2})!} \equiv \frac{-1}{(\frac{p-1}{2})!} \bmod (p).$$

Therefore

$$\tau_p \equiv (-1)^{(p-1)/2} \nu^{(p-1)/2} \frac{-1}{(\frac{p-1}{2})!} \bmod \mathfrak{p}^{(p+1)/2}.$$

By an elementary congruence, together with Wilson's theorem,

$$(-1)^{(p-1)/2} \left[ \left(\frac{p-1}{2}\right)! \right]^2 \equiv (p-1)! \equiv -1 \bmod (p)$$

and so

$$\frac{1}{(\frac{p-1}{2})!} \equiv (-1) \cdot (-1)^{(p-1)/2} \left(\frac{p-1}{2}\right)! \bmod (p).$$

This, together with the above congruence for $\tau_p$, then establishes (3.17).

□

## §4 Gauss sums

In the previous section we have seen the importance of the quadratic Gauss sum. Here we shall define Gauss sums for all primitive residue class characters. They are important in many contexts – although some of these lie quite outside the scope of our book. We shall consider their role in connection with (i) the generators of certain number fields associated with fields of roots of unity, (ii) results that certain ideals in cyclotomic fields must be principal (this then yields a relationship on ideal classgroups in terms of operators from the Galois group), (iii) explicit class number formulae (in Chapter VIII).

By way of preparation, we first study "localisation" aspects of residue class characters. Let $f$ denote a natural number, $f \not\equiv 2 \bmod (4)$, and let

$$f = \prod_j q_j$$

be its factorisation as a product of prime powers $q_j$ of distinct primes $p_j$. By the Chinese Remainder Theorem (see Theorem 4) we have

$$(4.1) \qquad (\mathbf{Z}/f\mathbf{Z})^{*\wedge} \cong \prod_j (\mathbf{Z}/q_j\mathbf{Z})^{*\wedge} \qquad \text{(direct product)}$$

for the groups of residue class characters. Explicitly, we embed $(\mathbf{Z}/q_j\mathbf{Z})^*$ as a subgroup of $(\mathbf{Z}/f\mathbf{Z})^*$, of those residue classes which are $\equiv 1 \bmod (f_j)$, where always

$$(4.1.a) \qquad f_j = \prod_{i \neq j} q_i \qquad \text{i.e. } f = f_j q_j.$$

The component map

$$(\mathbf{Z}/f\mathbf{Z})^{*\wedge} \to (\mathbf{Z}/q_j\mathbf{Z})^{*\wedge}, \qquad \theta \mapsto \theta_j$$

is just restriction to this subgroup. Moreover, if $x \in (\mathbf{Z}/f\mathbf{Z})^*$ is of the form

$$(4.1.b) \qquad x = \prod_j x_j \qquad x_j \in (\mathbf{Z}/q_j\mathbf{Z})^* \subset (\mathbf{Z}/f\mathbf{Z})^*$$

then, for each $j$, $x \equiv x_j \bmod (q_j)$; whence $\theta_j(x_j) = \theta_j(x)$. Therefore

$$(4.1.c) \qquad \theta(x) = \prod_j \theta(x_j) = \prod_j \theta_j(x_j) = \prod_j \theta_j(x).$$

The fact that each $x$ has a unique representation of the form (4.1.$b$), then tells us that the component maps $\theta \mapsto \theta_j$ actually yield an isomorphism (4.1). We shall call $\theta_j$ the prime component of $\theta$ at $q_j$; sometimes, for greater clarity, we write $\theta_j = \theta_{p_j}$. When $p \nmid f$, we formally define the component of $\theta$ at $p$ to be $\theta_p = \epsilon$, where $\epsilon(x) = 1$ for all $x \neq 0$. From our description, it follows that $\theta$ is primitive mod$(f)$ iff each $\theta_j$ is primitive mod$(q_j)$, $q_j$ being the conductor of $\theta_j$. In the sequel we shall always assume this to be the case.

For simplicity of notation we shall write

$$e(z) = e^{2\pi i z}$$

for any complex number $z$. We define the Gauss sum associated with $\theta$

to be the complex number

(4.2)
$$\tau(\theta) = \sum_{x \bmod (f)} \theta(x)e(x/f)$$

and more generally for $a \in \mathbb{Z}$ which is coprime to $f$

(4.2.a)
$$\tau_a(\theta) = \sum_{x \bmod (f)} \theta(x)e(ax/f).$$

In both equalities the sum over $x$ runs through a full system of representatives of the units $(\mathbb{Z}/f\mathbb{Z})^*$; in the sequel, for brevity, we refer to this as a *prime system of residues* mod$(f)$. Given such a system $\{x\}$, then $\{ax\}$ will also be a prime system of residues. Therefore

$$\tau_a(\theta) = \sum_{x \bmod (f)} \theta(ax)e(ax/f)\theta(a)^{-1}$$

and so

(4.2.b)
$$\tau_a(\theta) = \theta(a)^{-1}\tau(\theta).$$

As a complement to (4.2.b) we have

(4.2.b′) If $(f, a) > 1$ then $\sum_{x \bmod (f)} \theta(x)e(ax/f) = 0.$

*Proof.* Let $d = f/(f,a)$, $b = a/(f,a)$. A prime system of residues mod$(f)$ can then be written in the form $gx$ where $x$ runs through a set of representatives of the subgroup $H$ of $(\mathbb{Z}/f\mathbb{Z})^*$ of classes $= 1$ mod $(d)$, and the classes of the elements $g$ form a set of representatives of $(\mathbb{Z}/f\mathbb{Z})^*/H$. Then

$$\tau_a(\theta) = \sum_g \theta(g)s_g$$

where

$$s_g = \sum_x \theta(x)e(axg/f) = \sum_x \theta(x)e(bxg/d).$$

But as $x \equiv 1 \bmod (d)$, we get

$$s_g = e(bg/d)\sum_x \theta(x).$$

As $f$ is the conductor of $\theta$, the restriction of $\theta$ to $H$ is non-trivial, whence by Appendix A (see (A10.a))

$$\sum_x \theta(x) = 0, \qquad \text{i.e. } s_g = 0.$$

□

Next we consider the complex conjugate of $\tau(\theta)$

$$\overline{\tau(\theta)} = \sum \theta^{-1}(x)e(-x/f)$$

whence

(4.2.c) $$\overline{\tau(\theta)} = \tau(\theta^{-1})\theta(-1).$$

The prime decomposition of the character (see (4.1)) is reflected in that of the Gauss sum as follows:

**(4.3)** *With $f_j$ as in (4.1.a)*

$$\tau(\theta) = \prod_j \tau(\theta_j) \prod_j \theta_j(f_j).$$

*Proof.* As the $\{f_k\}$ have no common factor, there exist integers $m_k$, such that

(4.3.a) $$\sum_k f_k m_k = 1$$

Write $x_k = x m_k$. Then

(4.3.b) $$x = \sum_k f_k x_k \qquad \text{and } x \equiv x_j f_j \bmod (q_j).$$

By (4.1.c)

$$\theta(x) = \prod_j \theta_j(x) = \prod_j \theta_j(x_j f_j) = \prod_j \theta_j(x_j)\theta_j(f_j)$$

and

$$x/f = \sum x_j f_j/f = \sum x_j/q_j.$$

Taking as summation index the vector $(x_k)_k$ we then have

$$\begin{aligned}\tau(\theta) &= \sum \prod_j \theta_j(x_j)e(x_j/q_j) \prod_j \theta_j(f_j)\\ &= \prod_j \theta_j(f_j) \prod_j \tau(\theta_j)\end{aligned}$$

as we had to show.

□

Of course as yet we do not know whether the $\tau(\theta)$ are non-zero.This in fact follows from the equation

(4.4) $$|\tau(\theta)| = +\sqrt{f}$$

proved below – see (4.4.a).

*Remark.* $\tau(\theta)$ is clearly an algebraic integer in $\mathbb{C}$. If $\theta$ has order $k$, then $\tau(\theta)$ obviously lies in $\mathbb{Q}(e(1/k), e(1/f))$. The various embeddings of this field into $\mathbb{C}$ are induced by maps $e(1/k) \mapsto e(r/k)$ for $(r, k) = 1$, and $e(1/f) \mapsto e(a/f)$ for $(a, f) = 1$. Such a map then takes $\tau(\theta)$ into

$\tau_a(\theta^r)$. But $f(\theta^r) = f(\theta)$ and $\tau_a(\theta^r) = \tau(\theta^r)\theta(a)^{-r}$. Hence by (4.4) $|\tau_a(\theta^r)| = |\tau(\theta)|$, i.e. all Galois conjugates of $\tau(\theta)$ possess the same absolute value in $\mathbb{C}$, or equivalently all normalised Archimedean absolute values of the field $\mathbb{Q}(e(1/k), e(1/f))$ coincide on $\tau(\theta)$.

We shall actually prove that

(4.4.*a*) $$\tau(\theta)\tau(\theta^{-1}) = \theta(-1)f.$$

(4.4) is then a consequence of (4.4.*a*) and (4.2.*c*).

For the proof of (4.4.*a*) we may assume, by (4.3), that $f = p^t$ is a power of a prime $p$, and $t > 0$. (By definition of course $\tau(\epsilon) = 1$.) The sum of all $p^r$th roots of unity is zero for $r > 0$. Subtracting two such sums for $r = s - 1$ and $r = s > 1$, we see that

**(4.4.b)** *For $s > 1$ the sum of primitive $p^s$th roots of unity is zero.*

We have

$$\tau(\theta)\tau(\theta^{-1}) = \sum_x \sum_y \theta(x)\theta(y)^{-1}e((x+y)/p^t)$$

with the sums both running over complete prime systems of residues mod$(p^t)$. On substituting $x = -yz \bmod (p^t)$ this becomes

(4.4.*c*) $$\begin{cases} \tau(\theta)\tau(\theta^{-1}) = \theta(-1)\sum_z \theta(z)A(z) \\ A(z) = \sum_y e(y(1-z)/p^t). \end{cases}$$

Let $r = v_p(1-z)$, so that $(1-z)/p^t = b/p^s$ where $r + s = t$ and $p \nmid b$. Therefore now

$$A(z) = \sum_y e(yb/p^s) = \sum_y e(y/p^s).$$

For $r < t$, i.e. $s > 0$, replace this sum by a double sum over $u$ and $w$, where $u$ runs through a complete prime system of residues mod $(p^s)$ and $w$ runs through a full set of residues mod $(p^t)$ such that $w \equiv 1 \bmod (p^s)$. There are $p^r$ values of $w$, and for each $w$ we have $e(uw/p^s) = e(u/p^s)$. Hence on writing

$$B_r = p^r\left(\sum_u e(u/p^s)\right).$$

we have

(4.4.*d*) for $v_p(1-z) = r,\ r < t,$ we have $A(z) = B_r$.

This formula is also correct for $r = t$, $s = 0$ if we put

$$B_t = p^t - p^{t-1}.$$

From (4.4.$b$), $B_r = 0$ when $s > 1$, i.e. $r < t - 1$. Also for $s = 1$ evidently $\sum_u e(u/p) = -1$, whence

$$B_{t-1} = -p^{t-1}.$$

From (4.4.$c$) we now have

$$\tau(\theta)\tau(\theta^{-1}) = \theta(-1)[B_t + B_{t-1} \sum \theta(z)]$$

(sum over $z \equiv 1 \bmod (p^{t-1})$, $z \not\equiv 1 \bmod (p^t)$). So $\sum \theta(z) = -1$ by (A10.$a$).) Hence finally we have shown that $\tau(\theta)\tau(\theta^{-1}) = \theta(-1)p^t$.

□

From now on we take $f$ to be an odd prime number $p$. Let $p - 1 = km$ and for the sake of simplicity we write $H = (\mathbb{Z}/p\mathbb{Z})^*$. We view $\mathbb{Q}[p]$ as the subfield $\mathbb{Q}(e(1/p))$ of $\mathbb{C}$. Under the isomorphism

$$H \cong \mathrm{Gal}(\mathbb{Q}[p]/\mathbb{Q})$$

we let $K$ be the subfield of $\mathbb{Q}[p]$ fixed by the image in $\mathrm{Gal}(\mathbb{Q}[p]/\mathbb{Q})$ of $H^k$, the group of $k$th powers in $H$. We also adjoin the primitive $k$th roots of unity, and thus obtain a diagram of fields

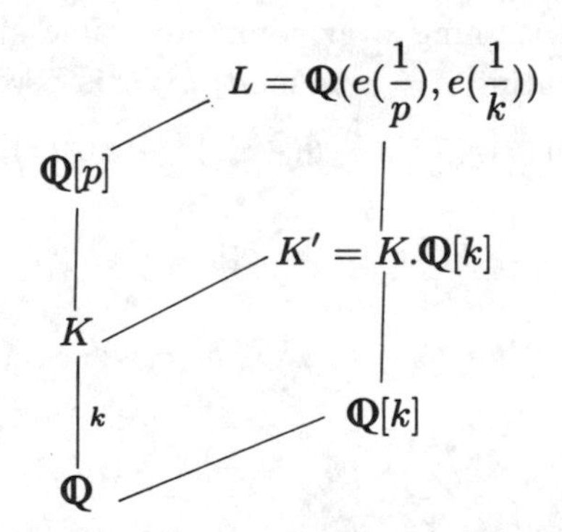


Restriction of automorphisms gives rise to an isomorphism

$$\mathrm{Gal}(L/\mathbb{Q}[k]) \cong \mathrm{Gal}(\mathbb{Q}[p]/\mathbb{Q})$$

and so, via the inverse isomorphism, we get isomorphisms

$$(4.5) \qquad \begin{cases} H \cong \mathrm{Gal}(L/\mathbb{Q}[k]) \\ H^k \cong \mathrm{Gal}(L/K') \\ H/H^k \cong \mathrm{Gal}(K'/\mathbb{Q}[k]). \end{cases}$$

We also note for the character groups that

$$(4.6) \qquad (H/H^k)^\wedge = \{\theta \in \hat{H} \mid \theta^k = \epsilon\}.$$

We shall often denote this group by $\hat{H}_k$. We have some obvious inclusions: clearly

$$(4.7) \qquad \tau(\theta) \in L \text{ if } \theta \in \hat{H}_k.$$

Next let $x$ run through a complete set of residues mod$(p)$ of the subgroup $H^k$. Then for any $a \in \mathbb{Z}$ with $(a,p)=1$, we define

(4.8)
$$S_{k,a} = \sum_x e\left(\frac{ax}{p}\right).$$

$S_{k,a}$ is called a *Gauss period.* Note that $S_{k,a} = t_{\mathbb{Q}[p]/K}(e\left(\frac{a}{p}\right))$, so that $S_{k,a} \in K$.

We then have the following strengthening of (4.7)

**Theorem 52.**

(a) *If $\theta \in \hat{H}_k$, then $\tau(\theta) \in K'$. Moreover, if in fact $\theta$ generates $\hat{H}_k$, i.e. if $\theta$ has order $k$, then*
$$K' = \mathbb{Q}[k](\tau(\theta)).$$

(b) *$K = \mathbb{Q}(S_{k,a})$ for any $a$ which is coprime to $p$.*

*Proof.* (a) Let $\theta \in \hat{H}_k$, and let $\sigma \in \mathrm{Gal}(L/K')$. Via (4.5), $e\left(\frac{1}{p}\right)^\sigma = e\left(\frac{b}{p}\right)$ for some $b$ whose class lies in $H^k$. Moreover, the values of $\theta$ are fixed under $\sigma$. Therefore $\tau(\theta)^\sigma = \tau_b(\theta) = \theta(b)^{-1}\tau(\theta)$, by (4.2.b). However, by (4.6), $\theta(b)=1$; thus $\tau(\theta)$ is fixed by $\mathrm{Gal}(L/K')$, i.e. $\tau(\theta) \in K'$. In other words

(4.8.a)
$$\mathbb{Q}[k] \subset \mathbb{Q}[k](\tau(\theta)) \subset K'.$$

Now suppose that $\theta$ has exact order $k$. Let $\omega \in \mathrm{Gal}(L/\mathbb{Q}[k](\tau(\theta)))$. Then $\omega$ fixes the values of $\theta$, and, via the first isomorphism in (4.5), $e\left(\frac{1}{p}\right)^\omega = e\left(\frac{a}{p}\right)$ for some $a$ which is coprime to $p$. Hence
$$\tau(\theta) = \tau(\theta)^\omega = \tau_a(\theta) = \theta(a)^{-1}\tau(\theta).$$

Thus $a \in \ker(\theta)$; hence $a \in H^k$, and so $\omega \in \mathrm{Gal}(L/K')$; this then shows that
$$K' \subset \mathbb{Q}[k](\tau(\theta)).$$

(b) Let $\theta$ again generate $\hat{H}_k$, and let $a$ run through a complete set of representatives for the classes of $H/H^k$. Let $x$ run through a complete set of residues mod$(p)$ of the classes in $H^k$. Then
$$\sum_a S_{k,a}\theta(a) = \sum_{a,x} e\left(\frac{ax}{p}\right)\theta(a)$$
$$= \sum_{a,x} e\left(\frac{ax}{p}\right)\theta(ax)$$

since $\theta(x)=1$. But $ax$ runs through a complete set of prime residues mod$(p)$. Therefore

$$\sum_a S_{k,a}\theta(a) = \tau(\theta). \tag{4.9}$$

Let $K_0 = \mathbb{Q}(S_{k,b})$ for some $b$. As the $S_{k,a}$ are all Galois conjugates, $K_0$ contains $S_{k,a}$ for each $a$ coprime to $p$. We know that $K_0 \subset K$. By (4.9), $\tau(\theta) \in K_0\mathbb{Q}[k]$, i.e. $K\mathbb{Q}[k] \subset K_0\mathbb{Q}[k]$. Therefore $K = K_0$.

□

Our next aim is to describe the prime factorisation of the ideal $\tau(\theta)$, where as before $\theta$ is a residue class character mod$(p)$ of order $k$. In the sequel we still keep all the field notation introduced in the above work. Directly from its definition as a sum of roots of unity, we know that $\tau(\theta)$ is an algebraic integer. Moreover, by (4.4.$a$), $\tau(\theta)$ divides $p$. Therefore the only prime divisors of $(\tau(\theta))$ are those in $K'$ lying above $p$. As $p \equiv 1 \bmod (k)$, we know by (1.12) that

$$p\mathfrak{o} = \prod_{j=1}^{\phi(k)} \mathfrak{p}_j \qquad \mathbf{N}\mathfrak{p}_j = p \tag{4.10.a}$$

where $\mathfrak{o} = \mathfrak{o}_{\mathbb{Q}[k]}$. Since $p$ is totally ramified in $\mathbb{Q}[p]$, it is totally ramified in $K/\mathbb{Q}$ with ramification index $k$. That is to say

$$\mathfrak{o}_{K'}\mathfrak{p}_i = \mathfrak{P}_i^k \qquad (i = 1, 2, \ldots, \phi(k)). \tag{4.10.b}$$

Now let $\mathfrak{p}$ denote one of the $\mathfrak{p}_j$ and write $\mathfrak{P}$ for $\mathfrak{P}_j$. By (4.10.$a$) the embedding $\mathbb{Z} \to \mathfrak{o}$ gives rise to an isomorphism $(\mathbb{Z}/p\mathbb{Z})^* \cong (\mathfrak{o}/\mathfrak{p})^*$. After taking residue classes, the map $x \mapsto \theta(x)$ ($x \in \mathbb{Z}$, $(x,p) = 1$) yields an endomorphism $x_\mathfrak{p} \mapsto \theta(x)_\mathfrak{p}$ of $(\mathfrak{o}/\mathfrak{p})^*$. Any endomorphism of this cyclic group is necessarily of the form $x_\mathfrak{p} \mapsto x_\mathfrak{p}^s$ ($s \in \mathbb{Z}$). In other words for some $s$, $0 \leq s < p-1$, we have

**(4.10.$c$)** $\theta(x) \equiv x^s \bmod \mathfrak{p}$ *for all* $x \in \mathbb{Z}$, $(x,p) = 1$.

This then extends to all $x \in \mathfrak{o}$ with $x \notin \mathfrak{p}$. Since $\theta(x)^k = 1$ and since for some $x_0$ the order of $\theta(x_0)$ is precisely $k$, we see that the order of $x_0^s$ mod $\mathfrak{p}$ is precisely $k$ and so

**(4.10.$d$)** $s = \frac{p-1}{k}t = mt$ *for some* $t$ *coprime to* $k$ *with* $0 < t < k$.

The prime factorisation of $\tau(\theta)\mathfrak{o}_L$ will be given in terms of $s$. As in §3, we again put $\nu = 1 - e\left(\frac{1}{p}\right)$. Thus $\nu\mathfrak{o}_{\mathbb{Q}[p]}$ is the unique prime ideal in $\mathbb{Q}[p]$ lying above $p$.

**Theorem 53.** *Let* $\mathfrak{P}$ *denote the prime ideal of* $\mathfrak{o}_{K'}$ *above* $\mathfrak{p}$; *then*

(*a*) $\tau(\theta) \equiv (-1)^{p-s}\frac{1}{(p-1-s)!}\nu^{p-1-s} \bmod (\mathfrak{p}, \nu^{p-s})\mathfrak{o}_L$.
(*b*) $v_{\mathfrak{P}}(\tau(\theta)) = k - t$.

*Proof.* (*a*) We remark that our proof of this result is an extension of the proof of (3.17) for quadratic Gauss sums. By (4.10.*c*)

$$\tau(\theta) \equiv \sum_{x=1}^{p-1} x^s(1-\nu)^x \bmod \mathfrak{p}\mathfrak{o}_L$$
$$\equiv \sum_{x=1}^{p-1}\left[x^s \sum_{j=0}^{x}(-1)^j\nu^j\binom{x}{j}\right] \bmod \mathfrak{p}\mathfrak{o}_L.$$

Since $\binom{x}{j} = 0$ when $j > x \geq 0$, we can rewrite this as

$$\tau(\theta) \equiv \sum_{j=0}^{p-1}\left[(-1)^j\nu^j \sum_{x=1}^{p-1} x^s\binom{x}{j}\right] \bmod \mathfrak{p}\mathfrak{o}_L.$$

We now apply (3.20) to deduce that the coefficient of $\nu^j$ is zero mod $\mathfrak{p}$ whenever $s + j < p - 1$, and that when $s + j = p - 1$

$$\sum_{x=0}^{p-1} x^s\binom{x}{j} \equiv \sum_{x=0}^{p-1}\frac{x^{p-1}}{j!} \equiv -\frac{1}{j!} \bmod (p).$$

This then gives the congruence

$$\tau(\theta) \equiv (-1)^{p-s}\frac{\nu^{p-1-s}}{(p-1-s)!} \bmod (\mathfrak{p}, \nu^{p-s})$$

which proves (*a*).

To prove (*b*), let $\tilde{\mathfrak{P}}$ denote the unique prime ideal in $\mathbb{Q}[p]$ above $p$ and let $\mathfrak{P}_L$ denote the unique prime ideal in $L$ above $\mathfrak{p}$. As $\nu\mathfrak{o}_L = \tilde{\mathfrak{P}}\mathfrak{o}_L$ and $(\mathfrak{p}, \nu^{p-s})\mathfrak{o}_L = (\mathfrak{P}_L^{p-1}, \mathfrak{P}_L^{p-s}) = \mathfrak{P}_L^{p-s}$, we deduce from part (*a*) that $v_{\mathfrak{P}_L}(\tau(\theta)) = p - 1 - s$.

Now $p-1 = km$, $s = tm$ and $\mathfrak{P}\mathfrak{o}_L = \mathfrak{P}_L^m$. Therefore indeed $\mathfrak{P}_L^{p-1-s} = \mathfrak{P}^{k-t}\mathfrak{o}_L$. The result then follows since $\tau(\theta) \in K'$.

□

In order to cast the last result into a different form, observe that as $(t, k) = 1$, we get $tt' \equiv 1 \bmod (k)$, and so $\theta^{t'}(x) \equiv x^m \bmod \mathfrak{p}$. Replacing $\theta$ by $\theta^{t'}$ if necessary, we get $\theta(x) \equiv x^m \bmod \mathfrak{p}$ and so

$$v_{\mathfrak{P}}(\tau(\theta)) = k - 1.$$

Let $(b, k) = 1$ and let $\omega_b \in \mathrm{Gal}(\mathbb{Q}[k]/\mathbb{Q})$ denote the element with the property that $e\left(\frac{1}{k}\right)^{\omega_b} = e\left(\frac{b}{k}\right)$. As $\mathfrak{P}$ is fixed under $\mathrm{Gal}(K'/\mathbb{Q}[k])$, $\mathfrak{P}^{\omega_b}$ is well-defined. We then have $\theta^{\omega_b} = \theta^b$, while for any integer $x$, $x = x^{\omega_b}$; hence

$$\theta^b(x) \equiv x^m \bmod \mathfrak{P}^{\omega_b}.$$

Let $b'$ be the unique integer in $[1, k]$ such that $bb' = 1 \bmod (k)$; then $\theta^{bb'} = \theta$. We now get

$$\theta(x) \equiv x^{mb'} \bmod \mathfrak{P}^{\omega_b}.$$

By the previous theorem

$$v_{\mathfrak{P}^{\omega_b}}(\tau(\theta)) = k - b'.$$

Replacing initially in (4.10.*c*) $\mathfrak{p}$ if necessary by one of its conjugates, we may assume that

(4.11) $$\theta(x) \equiv x^m \bmod \mathfrak{p}.$$

Then we have

**(4.12)** $$\tau(\theta)\mathfrak{o}_{K'} = p \prod_b \mathfrak{P}^{-\omega_b b'}$$

*where b runs through a prime system of residues* mod$(k)$.

Let $\sigma_a \in \mathrm{Gal}(L/\mathbb{Q}[k])$ be the Galois automorphism with $e\left(\frac{1}{p}\right)^{\sigma_a} = e\left(\frac{a}{p}\right)$. By (4.2.*b*)

$$\tau(\theta)^{k\sigma_a} = \tau(\theta)^k \theta^k(a)^{-1} = \tau(\theta)^k$$

and this will hold for all $a$ with $(a, p) = 1$. From this we deduce that $\tau(\theta)^k \in \mathbb{Q}[k]$, and so $\tau(\theta)$ is a Kummer generator of $K'/\mathbb{Q}[k]$. In particular, writing $\mathfrak{o} = \mathfrak{o}_{\mathbb{Q}[k]}$, by (4.12) we have

$$\tau(\theta)^k \mathfrak{o} = p^k \prod_b \mathfrak{p}^{-\omega_b b'}.$$

It therefore follows that $\prod_b \mathfrak{p}^{\omega_b b'}$ is always a principal $\mathfrak{o}$-ideal. Note that the special hypothesis (4.11) is not needed for this conclusion: for, if $\mathfrak{p}'$ is a Galois conjugate of $\mathfrak{p}$, $\mathfrak{p}' = \mathfrak{p}^{\omega_c}$ say, then $\prod_b \mathfrak{p}^{\omega_c \omega_b b'} = \left(\prod_b \mathfrak{p}^{\omega_b b'}\right)^{\omega_c}$ is principal.

For fixed $k$, the only hypothesis on $p$ was that $p \equiv 1 \bmod (k)$. We shall now give a second result of this type. Let $a \in \mathbb{Z}$, $0 < a < k$, $(a, k) = 1$. Again by (4.2.*b*)

$$\tau(\theta)^a \tau(\theta^a)^{-1} \in \mathbb{Q}[k].$$

Let $\{c\}$ denote the remainder of an integer $c$ on division by $k$. We have

$$\theta^a(x) \equiv x^{m\{ab'\}} \bmod \mathfrak{P}^{\omega_b}.$$

Hence

$$\mathfrak{o}_{K'}\tau(\theta^a) = p \prod_b \mathfrak{P}^{-\omega_b\{ab'\}}$$

while

$$\mathfrak{o}_{K'}\tau(\theta)^a = p^a \prod_b \mathfrak{P}^{-\omega_b ab'}.$$

Therefore

$$\mathfrak{o}_{K'}\tau(\theta)^a\tau(\theta^a)^{-1} = p^{a-1}\prod_b \mathfrak{P}^{-\omega_b(ab'-\{ab'\})}.$$

Of course $(ab' - \{ab'\}) = k[ab']$, where $[ab']$ denotes the integer part of $ab'/k$. Hence

$$\mathfrak{o}_{K'}\tau(\theta)^a\tau(\theta^a)^{-1} = p^{a-1}\prod_b \mathfrak{p}^{-\omega_b[ab']}.$$

Restating our earlier result together with what we have just proved, we have

**Theorem 54.** *Let $p \equiv 1 \bmod (k)$ and let $\mathfrak{p}$ denote a prime ideal in $\mathbb{Q}[k]$ above $p$. Then the $\mathfrak{o}_{\mathbb{Q}[k]}$ ideals*

$$\prod_b \mathfrak{p}^{\omega_b b'}, \qquad \prod_b \mathfrak{p}^{\omega_b[ab']}$$

*are both principal for any a coprime to k.*

The importance of such results lies in the fact, which we cannot prove here, that each ideal class $c$ of $\mathbb{Q}[k]$ contains an ideal $\mathfrak{p}$ as above; this then shows $\prod_b c^{\omega_b b'}$ is always the principal class.

## §5 Elliptic curves

For any field $L$ we may view $\mu_L$, the group of roots of unity in $L$, as the subgroup of points of finite order of the multiplicative group $L^*$. In this section we shall briefly consider certain other abelian group structures, namely elliptic curves, whose points are of considerable arithmetic interest. By adjoining the coordinates of a group $G$ of points of finite order of such a curve, it is possible to obtain a theory which is in many ways analogous to cyclotomic theory. The key point here is that the Galois group of such an extension can frequently be nicely described as a subgroup of $\mathrm{Aut}(G)$.

This section does not belong to the mainstream of the book; it is meant as a brief introduction to encourage the reader to learn more about this important topic.

We begin by describing these group structures in a fairly loose, intuitive way; this is then followed by a rigorous and detailed account.

Throughout this section, we let $L$ denote a subfield of the complex numbers $\mathbb{C}$ and denote by

$$f(X) = X^3 + aX^2 + bX + c$$

a separable polynomial in $L[X]$; we then define $g(X,Y) = Y^2 - f(X)$

in $L[X,Y]$. By writing $g(X,Y)$ in projective form, we shall see that, in addition to the "finite" points of $g$ in $L$:

$$\{(x,y) \in L \times L \mid g(x,y) = 0\}, \tag{5.1}$$

we should also consider an infinite point, which will be denoted by $I$. We write $E(L)$ for the union of the finite points and the infinite point.

Given two distinct points $P, Q \in E(L)$, we may try to define a third point $R \in E(L)$ as follows: let $l$ denote the line connecting $P$ and $Q$; since $g$ has degree three, $l$ will meet $E(L)$ in a further point $R$. Very roughly, the main result of this section is to show that we can use this construction to define an abelian group structure $\oplus$ say, on $E(L)$. To be just a little more precise: $I$ will be the identity of the group; collinear points $P, Q, R$ have trivial sum $P \oplus Q \oplus R = I$; and $P \oplus Q = R^*$, where if $R = (r,s)$, then $R^* = (r,-s)$. (Note that it is intuitively clear that $R, R^*, I$ are collinear and so $R \oplus R^* \oplus I = I$.)

Provided we took care to deal with repetitions, where either $P = Q \neq R$ and $l$ is a tangent at $P$, or where $P = Q = R$ and $P$ is a point of inflexion, then a number of the group axioms would follow at once. For example: (a) $I$ is the identity; (b) $R^*$ is the inverse of $R$; moreover, it is immediately clear that the commutativity condition $P \oplus Q = Q \oplus P$ is satisfied. The associativity condition is, however, somewhat less straightforward. One possibility would be to determine formulae for the $x$ and $y$ coordinates of the two points $(P \oplus Q) \oplus T$, $P \oplus (Q \oplus T)$ for three arbitrary points in $E(L)$, and then check that the two results coincide. Unfortunately this is very tedious and provides very little insight into the situation. Instead, we shall demonstrate the group structure on $E(L)$ by a completely different technique, using valuations and classgroups. In addition it will be extremely rewarding to see how many of the techniques which we have developed for arbitrary Dedekind domains, with algebraic number fields in view, actually give geometric results, which, in turn, give important arithmetic information.

We now briefly recall a few elementary results and definitions for projective spaces. We let $L_*^{(n+1)}$ denote the set of vectors in $L^{(n+1)} = \prod_0^n L$, whose coordinates are not all simultaneously zero. Two points $x, y \in L_*^{(n+1)}$ are said to be equivalent iff there exists $\lambda \in L^*$ such that $x\lambda = y$ i.e. $\lambda x_i = y_i$ for all $i = 0, 1, \ldots, n$. This defines an equivalence relation on $L_*^{(n+1)}$; we write $(x_0 : x_1 : \cdots : x_n)$ for the equivalence class represented by $x = (x_0, x_1, \ldots, x_n)$; we refer to an equivalence class as a projective point, and we write $\mathbb{P}^n(L)$ for the set of all such projective points. Here we shall exclusively be concerned with the cases $n = 1, 2$.

Note that we have embeddings $L^{(2)} \xrightarrow{\alpha} \mathbb{P}^2(L)$ by $(x,y) \mapsto (x : y : 1)$, and $\mathbb{P}^1(L)\xrightarrow{\beta}\mathbb{P}^2(L)$ by $(x:y)\mapsto(x:y:0)$. We denote the image of $\alpha$ by $A^2(L)$, and we call this the finite plane; the image of $\beta$ is denoted by $l_\infty$, and is called the line at infinity. It is readily seen that $\mathbb{P}^2(L)$ is the disjoint union of $l_\infty$ and $A^2(L)$. The homogenisation of $g(X,Y)$ is defined to be the polynomial in three variables

$$(5.2) \qquad g^*(X,Y,Z) := Y^2Z - X^3 - aX^2Z - bXZ^2 - cZ^3.$$

Because $g^*$ is homogeneous, we note that if one representative $(x,y,z)$ of a projective point is a solution to $g^*=0$, i.e. $g^*(x,y,z)=0$, then so are all other representatives. We write $E(L)$ for the set of projective points in $\mathbb{P}^2(L)$ which give solutions to $g^*=0$.

The map $\alpha$ induces a bijection between the finite points of (5.1) and those projective points on $g^*$ which lie in the finite plane $A^2(L)$. This then at least explains some of our earlier terminology, and also shows that we have not lost anything in "going projective". We now check and see what we have gained:

**(5.3)** $E(L)\cap l_\infty = (0:1:0)$.

*Proof.* Since $l_\infty$ is defined by the equation $Z=0$, we set $Z=0$ in (5.2), and see that $g(x,y,0)=0$ iff $x=0$.

$\square$

The point $(0:1:0)$ is the mysterious point at infinity $I$, to which we alluded in the Introduction.

This completes our brief review of the projective space results which we shall need. We now recall and reinterpret the results on valuations which we require.

By a function field $N$ of dimension one over $L$ we mean a field extension of transcendence degree one over $L$ which is linearly disjoint with $\mathbb{C}$ over $L$. In the sequel it is to be inderstood that all function fields which we consider have dimension one. The term valuation $v$ of $N$ is to be taken always to mean a valuation which is trivial on $L^*$, i.e. $v(l)=0$ for all $l\in L^*$; we denote the set of such valuations of $N$ by $V(N)$. Since $v$ is trivial on $L^*$, the residue class field $k_v=\mathfrak{o}_v/\mathfrak{p}_v$ is an extension field of $L$: indeed $L\subset\mathfrak{o}_v$ and $L\cap\mathfrak{p}_v=0$. We say that $v$ has *degree one* if $k_v=L$; we denote by $V_1(N)$ the set of degree one valuations of $N$.

*Example 1.* If $X$ is transcendental over $\mathbb{C}$; then $N=L(X)$ is a function field over $L$. In this case we have seen in (II.2.8) that $V(N)$ is in bijection with the monic irreducible polynomials of $L[X]$, together with the so-called infinite valuation $v_{X,\infty}$, where

$$v_{X,\infty}(p(X)) = -\deg(p(X))$$

for $p \in L[X]$. Note for future reference that $v_{X,\infty}$ has degree one, and that, for any $d \in L$, the element $1/(X-d)$ is a uniformising parameter for this valuation. In particular the above work shows that $V_1(N)$ is in bijection with $v_{X,\infty}$ together with the monic linear polynomials in $L[X]$. This can be reformulated as saying that $V_1(L(X))$ is parameterised by $\mathbb{P}^1(L)$: indeed, to a finite point $(p:q)$ we associate the valuation corresponding to $X - \frac{p}{q}$ and to the infinite point $(1:0)$ we associate $v_{X,\infty}$.

We now modify our previous notation. Let $X$ be as above and let $T$ be algebraically independent with $X$ over $\mathbb{C}$. Since $f(X)$ is separable, the polynomial $g(X,T)$ in $T$ is irreducible over $L(X)$. Let $Y$ now denote a root of $g(X,T)$, so that $g(X,Y)=0$. Then $L(X,Y)$ is a quadratic extension of $L(X)$ and is a function field over $L$. We denote this field by $L(E)$.

*Example 2.* We consider the behaviour of $v_{X,\infty}$ in the extension $L(E)$. We begin by recalling from Theorem 20 that $v_{X,\infty}$ is either split, ramified or inert in the quadratic extension $L(E)/L(X)$. However, by the very definition of $Y$, we know that

$$\frac{1}{X}\left(\frac{Y}{X}\right)^2 = 1 + \frac{a}{X} + \frac{b}{X^2} + \frac{c}{X^3}.$$

Thus if $w$ denotes any valuation of $L(E)$ above $v_{X,\infty}$, then $w(X/Y) = 1/2$; hence $v_{X,\infty}$ ramifies; and we let $v_\infty$ denote the unique valuation of $L(E)$ above $v_{X,\infty}$.

We call a valuation $v$ of $L(E)$ infinite/finite according as $v$ does/does not lie above $v_{X,\infty}$. Another way to view this distinction is to define $\mathfrak{o}_E$ to be the integral closure of the principal ideal domain $L[X]$ in $L(E)$. From Theorem 5 we know that $\mathfrak{o}_E$ is a Dedekind domain, and the finite valuations of $L(E)$ are precisely those which are non-negative on $\mathfrak{o}_E$.

In Theorem 7 we saw that the valuations of an algebraic number field are in natural bijection with the prime ideals; in the present geometric situation we obtain an interpretation of valuations in terms of points on curves. The result which we give goes over to arbitrary function fields; however, for simplicity, we only consider $L(E)$.

**(5.4)** *There exist natural, mutually inverse maps*

$$E(L) \rightleftarrows V_1(L(E)).$$

*Proof.* We start by associating $I \leftrightarrow v_\infty$. It therefore suffices to deal with the finite points of $E(L)$ and the finite, degree one valuations of $L(E)$. Given a finite point $P \in E(L)$, corresponding to the point $(p,q)$ on $g(X,Y) = 0$, we associate a valuation $v_P \in V_1(L(E))$ as follows: suppose initially $q \neq 0$. We claim that there is a unique embedding $L(E) \hookrightarrow L((X-p))$ which extends the embedding $L(X-p) \hookrightarrow L((X-p))$ (c.f. Example 1 following II.3.24). Then, for given $h \in L(E)^*$, we write

$$h = (X-p)^n h_1 \tag{5.4.a}$$

with $h_1 \in L[[X-p]]^*$. The desired valuation is then given by the rule $v_P(h) = n$.

First recall that if $u(X) = \sum_{i\geq 1} a_i X^i \in L[[X]]$ with $a_1 \neq 0$, then there exists a unique $t \in L[[X]]$ such that

$$t(u(X)) = X = u(t(X)). \tag{5.4.b}$$

For simplicity we shall denote $t(X)$ by $u^{-1}(X)$. Next we set $X = X_1 + p$, $Y = Y_1 + q$, so that

$$(Y_1+q)^2 = f(X_1+p) = f(p) + j(X_1)$$

where $j(X_1) \in X_1 L[[X_1]]$. Since $q^2 = f(p)$, this gives

$$Y_1^2 + 2qY_1 = j(X_1). \tag{5.4.c}$$

Thus, setting $u(Y_1) = Y_1^2 + 2qY_1$, we have $Y_1 = u^{-1}(j(X_1))$. Hence the embedding $L(X-p) \hookrightarrow L((X-p))$ together with the map $Y_1 \mapsto u^{-1}(j(X_1))$ induces an embedding

$$L(E) = L(X-p) + Y_1 L(X-p) \hookrightarrow L((X-p)).$$

Conversely, we have shown that under any extension of $L(X-p) \hookrightarrow L((X-p))$ to $L(E)$, we must have that $Y_1$ is mapped to $u^{-1} \circ j(X_1)$.

Lastly we consider the case $q = 0$. In exactly the same way as above, we check that $Y_1^2 = f(X_1+p) = f'(p)X_1 + X_1^2 f_2(X_1)$ for some $f_2 \in L[X_1]$. So, because $f$ is separable, $X-p$ can be expanded in $L[[Y]]$. Thus, in this case, we can write $h(X,Y) = Y^h h_1$ with $h_1 \in L[[Y]]^*$, and then define $v_p(h) = n$.

In both cases, if $v(h) \geq 0$, then clearly $v(h-r) > 0$ for some $r$. (More precisely $r$ is the constant occurring in the expansion of $h$.) This then shows that $v$ has degree 1.

It is interesting to cast the above work in function theoretic language: we say that $h$ has a pole/zero of order $n$ at $P$ if $|v_P(h)| = n$ with $v_P(h)$ negative/positive; and we say that $h$ takes the value $r \in L$ at $P$ with multiplicity $n$, if $v_P(h-r) = n > 0$. Thus the above work shows that $h$ always takes some (possibly infinite) value at each point $P$; furthermore $\mathfrak{o}_{v_P}$ is the ring of functions which are finite at $P$ and $\mathfrak{p}_{v_P}$ is the ideal of functions which vanish at $P$.

We now define the inverse map $V_1(L(E)) \to E(L)$. Let $v$ therefore denote a finite degree one valuation, and let $w$ denote the valuation of $L(X)$ below $v$. By Example 1, we know that $w$ corresponds to some $X - p$, with $p \in L$; hence, in particular, $v(X - p) > 0$; and so $v$ is non-negative on $L[X]$. We write $X = X_1 + p$, so that

$$Y^2 = f(X) = f(X_1 + p) = f(p) + j(X_1)$$

with $j(X_1) \in X_1 L[X_1]$. Hence we deduce that

$$v(Y^2 - f(p)) = v(j(X_1)) \geq v(X_1) > 0.$$

This then shows $Y^2 - f(p) \in \mathfrak{p}_v$. Next observe that $Y^2 \in \mathfrak{o}_v$, hence $Y \in \mathfrak{o}_v$ and so, because $v$ has degree one, $Y - q \in \mathfrak{p}_v$ for some $q \in L$. We have therefore shown $q^2 - f(p) \in \mathfrak{p}_v$, and so $q^2 = f(p)$. $P_v = (p, q)$ is the desired finite point in $E(L)$. Note that if $v(X - p)$, $v(X - p')$, $v(Y - q)$, $v(Y - q')$ were all positive; then $p = p'$ and $q = q'$, so that $P_v$ is certainly well-defined.

It is now entirely routine to show that the maps $P \mapsto v_P$, $v \mapsto P_v$ are mutually inverse.

$\square$

Next we define $\mathrm{Div}(L(E))$, the group of divisors of $L(E)$, to be the free abelian group on $V(L(E))$. A typical element of $\mathrm{Div}(L(E))$ takes the form $\sum n_v v$ with $n_v = 0$ p.p.; we call such elements divisors.

Since $I(\mathfrak{o}_E)$, the group of fractional $\mathfrak{o}_E$-ideals, is isomorphic to the free abelian group on the finite valuations of $L(E)$

(5.5) $$\mathrm{Div}(L(E)) \cong I(\mathfrak{o}_E) \oplus \mathbb{Z} v_\infty.$$

Given $h \in L(E)^*$, define

$$\phi(h) = \sum v(h) v$$

where the sum extends over $v \in V(L(E))$. To show that $\phi(h)$ is a divisor of $L(E)$, we need only observe that those finite $v$ with $v(h) \neq 0$ are precisely those which occur in the factorisation of the ideal $h\mathfrak{o}_E$. Therefore $\phi$ defines a group homomorphism $\phi: L(E)^* \to \mathrm{Div}(L(E))$.

We define $d(h)$, the degree of $h$, to be

$$d(h) = \sum_{v | v(h) > 0} f_v v(h)$$

where $f_v = \dim_L(k_v)$. Note that $f_v$ is finite by the work of Example 1, together with the fact that $L(E)/L(X)$ is finite (in fact quadratic!).

**(5.6)** *If $h \notin L$, then*

(*a*) $d(h) = (L(E) : L(h))$.

(b) *For all* $r \in L$ $d(h) = d(1/h) = d(h+r)$.

*Proof.* Clearly (a) implies (b), since $L(h) = L(1/h) = L(h+r)$. So we now prove (a). First note that since $h$ is transcendental over $L$, $L(E)$ is a finite extension of $L(h)$. Let $w$ denote the valuation of $L(h)$ associated with $h$ (see Example 1). Then from Theorem 20

$$(L(E) : L(h)) = \sum_v e(v|w) f(v|w)$$

where the sum runs through all valuations of $L(E)$ which lie above $w$. Since $w$ has degree one, $f_v = f(v|w)$, and of course

$$v(h) = e(v|w)w(h) = e(v|w).$$

□

Since

$$\sum_v f_v v(h) = \sum_{v|v(h)>0} f_v v(h) + \sum_{v|v(h)<0} f_v v(h)$$
$$= d(h) - d(1/h)$$

we have shown

$$\sum_v f_v v(h) = 0. \tag{5.7}$$

We define the divisor classgroup of $L(E)$, $\mathrm{Cl}(E)$, to be the cokernel of $\phi$

$$\mathrm{Cl}(E) = \frac{\mathrm{Div}(L(E))}{\phi(L(E)^*)}.$$

The divisor classgroup is an extension of the classgroup $\mathrm{Cl}(\mathfrak{o}_E)$ of the Dedekind domain $\mathfrak{o}_E$: indeed, mapping $v_\infty \to 0$, we obtain a surjection

$$\mathrm{Cl}(E) \twoheadrightarrow \mathrm{Cl}(\mathfrak{o}_E). \tag{5.8}$$

We now draw together all the above ideas in studying the map of sets

$$i\colon E(L) \to \mathrm{Cl}(E)$$

given by $i(P) = v_P - v_\infty$.

We wish to show

**Theorem 55.** *$i$ is an injective map, and* $\mathrm{im}(i)$ *is closed under addition in* $\mathrm{Cl}(E)$*; this therefore induces a group structure on* $E(L)$.

Note that, in contradistinction to the arithmetic case, this result implies:

**Corollary.** *If* $L = \mathbb{C}$, *then* $\mathrm{Cl}(\mathfrak{o}_E)$ *is uncountable.*

*Proof.* $E(\mathbb{C})$ is uncountable, since for each $p \in \mathbb{C}$ we have one or two

$q \in \mathbb{C}$ such that $q^2 = f(p)$. Hence by the theorem $\mathrm{Cl}(E)$ is uncountable. However, the kernel of the map (5.8) is $\mathbb{Z}v_\infty$.

□

First we show that $i$ is injective. So we suppose for contradiction that $i(P) = i(Q)$ for distinct $P, Q$ in $E(L)$.

There must then exist $z \in L(E)^*$ such that $\phi(z) = v_P - v_Q$. Recalling that, by (5.4), $v_P$, $v_Q$ both have degree one, and applying (5.6), it follows that $L(E) = L(z)$. We can therefore write $X = \alpha(z)/\beta(z)$, $Y = \gamma(z)/\delta(z)$ wih $\alpha, \beta, \gamma, \delta$ in $L[z]$ and $(\alpha, \beta) = 1 = (\gamma, \delta)$. Moreover, noting that $(L(E) : L(X)) = 2$, by (5.6) we know that $d(X) = 2$; thus $\alpha$ and $\beta$ must have degree less than or equal to 2.

Next we factor $f(X) = \prod_1^3 (X - e_i)$ in $\mathbb{C}[X]$, and we recall that the $e_i$ are distinct (since $f$ is separable). It therefore follows that

$$\frac{\beta^4 \gamma^2}{\delta^2} = \beta \prod_{i=1}^{3} (\alpha - e_i \beta)$$

and so $\beta \prod_i (\alpha - e_i \beta)$ must be a square in $\mathbb{C}[z]$. If $g \in \mathbb{C}[z]$ divided any two of the four terms $\beta$, $\alpha - e_i \beta$; then it would follow that $g \mid \alpha$ and $g \mid \beta$; thus we conclude that these four terms are coprime, and hence each of these terms is a square:

$$\beta = h_0^2 \qquad \alpha - e_i \beta = h_i^2.$$

This in turn implies equalities

$$(e_i - e_j) h_0^2 = (e_i - e_j)\beta = h_j^2 - h_i^2 = (h_j + h_i)(h_j - h_i).$$

Since the $h_i$ are mutually coprime, both $h_i \pm h_j$ are squares; however, they both have degree at most one, and are therefore both constants; this implies that all the $h_k$ are constants which is absurd, and so provides the required contradiction.

We now complete the proof of the theorem by showing $\mathrm{im}(i)$ to be closed under addition. By a (finite) line $l$ in $L(E)$ we mean an element of the form $\alpha + \beta X + \gamma Y$ where $\alpha, \beta, \gamma$ lie in $L$, with $\beta$ and $\gamma$ not both zero. Since $v_\infty(X) = -2$ and $v_\infty(Y) = -3$, we deduce from (5.6) that

$$d(l) = \begin{cases} 2 & \text{if } \gamma = 0 \\ 3 & \text{if } \gamma \neq 0. \end{cases}$$

When $\gamma = 0$ we shall call $l$ a vertical line. Note that if $l$ is non-vertical and if $v_i(l) > 0$ for two distinct degree one valuations $v_i$, then by (5.6) the third valuation with $v(l) > 0$ must also have degree one. Similarly if $v_1(l) \geq 2$ with $v_1$ having degree one; then again the remaining valuation with $v(l) > 0$ also has degree one.

Let $v, w$ denote two finite degree one valuations of $L(E)$. Initially suppose $v \neq w$, and let $l_{v,w}$ denote the line connecting $P_v$ to $P_w$; then $l_{v,w}$ will be vertical iff $P_w = P_v^*$, and by the above discussion:

(5.9.$a$) If $P_v^* = P_w$: $\phi(l_{v,w}) = v + w - 2v_\infty$

(5.9.$b$) If $P_v^* \neq P_w$: $\phi(l_{v,w}) = v+w+u-3v_\infty$ for some $u \in V_1(L(E))$.

In fact if $v = w$, then we need only take $l_{v,w}$ to be the tangent at $P_v$; indeed, as in the proof of (5.4), $v(l_{v,v}) \geq 2$, and so (5.9) still holds.

From (5.9.$a$) we conclude that $-\text{im}(i) = \text{im}(i)$; and so from (5.9.$a$,$b$) we conclude that in all cases $i(P_v) + i(P_w) \in -\text{im}(i) = \text{im}(i)$.

At this point the reader may find it helpful to consult Figure (5.10) in order to understand the underlying geometric ideas.

□

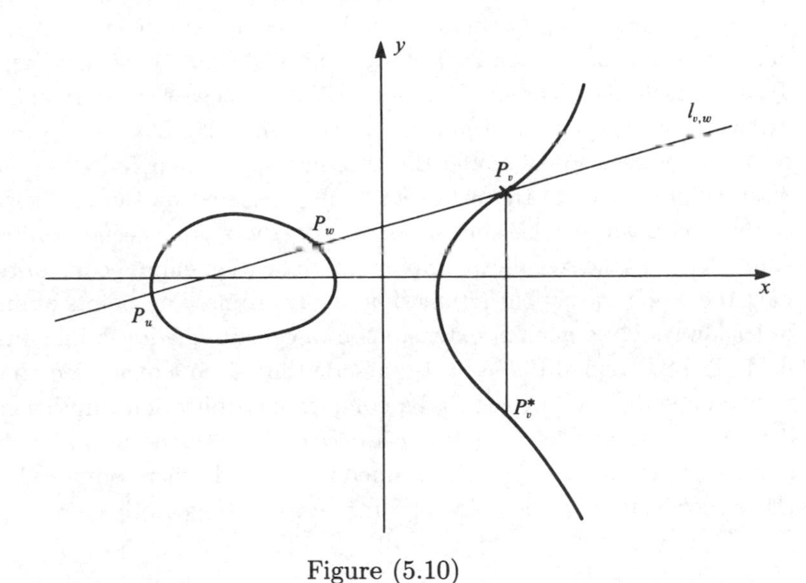


Figure (5.10)

Clearly the group structure of $E(L)$ will be of great interest when one views $g(X,Y) = 0$ as a Diophantine equation over $L$. Following a conjecture of Poincaré, the celebrated theorem of Mordell-Weil shows that $E(L)$ is a finitely generated $\mathbb{Z}$-module, whenever $L$ is a number field. There then remains the problem of obtaining practical criteria for determining whether or not $E(L)$ is finite; this is currently an extremely active area of research.

We conclude this section by briefly returning to our original idea of generalising cyclotomy. The reader who is familiar with the theory of Riemann surfaces will immediately recognise $E(\mathbb{C})$ as a compact Riemann surface of genus one. As such $E(\mathbb{C})$ is topologically isomorphic to a torus. Viewed in this way, the determination of the points of order $n$ on $E(\mathbb{C})$ amounts to finding points of order $n$ on a torus. From the geometric point of view this beautifully extends cyclotomy, which, of course, studies the points of order $n$ of the circle.

In fact, under suitable hypotheses, we have the following analogues:

(a) The $n$th roots of unity are given by evaluating the transcendental function $e^{2\pi ix}$ at $\frac{1}{n}\mathbb{Z}$. In the same way, the coordinates of points of order $n$ of an elliptic curve can be obtained by evaluating two transcendental functions – the Weierstrass $\wp$-function and its derivative associated with a certain lattice $\Lambda$ – at $\frac{1}{n}\Lambda$.

(b) In his famous programme of twenty-three problems for the twentieth century, Hilbert raised the question of using singular values of transcendental functions to generate all the abelian extensions of a given number field. All cyclotomic extensions have abelian Galois group over $\mathbb{Q}$. Conversely the theorem of Kronecker-Weber shows that every abelian extension of $\mathbb{Q}$ is contained in some cyclotomic field. In general the extensions obtained by adjoining the coordinates of the points of order $n$ of an elliptic curve to the field $\mathbb{Q}(a, b, c)$, generated by the coefficients of the curve, will not be abelian. However, there is a special family of elliptic curves (the so-called complex multiplication elliptic curves) that have the property that the adjunction of $x$-coordinates of points of finite order always gives abelian extensions of a certain quadratic imaginary field. The theorem of Hasse-Fueter asserts that given a quadratic imaginary number field $K$, we can find a complex multiplication elliptic curve defined over a fixed abelian extension $H$ of $K$ with the property that any abelian extension of $K$ is contained in the field generated over $H$ by the $x$-coordinate of some point of finite order of the elliptic curve.

# VII

# Diophantine Equations

In the introduction we saw how properties of algebraic numbers could be used to solve certain diophantine equations. In this chapter we shall consider three applications of the techniques, which we have so far developed, to diophantine theory: Fermat's Last Theorem, quadratic forms, and certain cubic equations.

The evolution of algebraic number theory has been very strongly influenced both by attempts to prove Fermat's Last Theorem and by the development of the theory of quadratic forms: indeed, they have arguably been the two most important influences on the subject. However, it is important to understand that here we shall solely be concerned with seeing how our preceding work can be used to obtain information on these topics. For full accounts of these two topics the reader is referred to [R] and [C].

In the third section we shall consider a number of cubic equations. We shall begin by describing a method for the determination of integral solutions to various equations of the form: $Y^2 = X^3 + k$; we then conclude with the theorem of Delaunay and Nagell on integral solutions to $X^3 + dY^3 = 1$. These results are instances of the very deep and beautiful arithmetic theory of elliptic curves.

## §1 Fermat's Last Theorem

We consider the following statement:

**(1.1)** *For an integer $n > 2$, if $x, y, z$ are integer solutions to*

$$X^n + Y^n = Z^n$$

*then at least one of $x, y, z$ is zero.*

In the margin of his copy of Bachet's edition of the works of Diophantus, Fermat made a note which amounts to (1.1); he said that he had a proof, but that space prevented him from providing the details. For this reason (1.1) is usually referred to as Fermat's Last Theorem.

Clearly it would suffice to prove the result for $n = 4$ and for all odd primes $n$. As yet, however, (1.1) is still unproved, although the result has been verified for all primes $n < 125,000$. Evidence for the truth of (1.1) has recently been reinforced in two ways. As a consequence of G. Faltings' proof of the Mordell conjecture, it follows that given any $n > 2$, $X^n + Y^n = Z^n$ has at most a finite number of integral solutions $x, y, z$ with $(x, y, z) = 1$. Work of K. Ribet shows that Fermat's Last Theorem would follow if a conjecture of Weil, on the parameterisation of elliptic curves defined over $\mathbb{Q}$, were true.

In this section we show

**Theorem 56.** *Let $p$ be an odd prime. If the class number of $\mathbb{Q}[p]$ is not divisible by $p$, then there is no integer solution $(x, y, z)$ to $X^p + Y^p = Z^p$ with $p \nmid xyz$.*

In general the non-existence of an integral solution to $X^p + Y^p = Z^p$ with $p \nmid xyz$ is called the *first case* of Fermat's Last Theorem. Treatment of the (usually harder) case where $p \mid xyz$ and $(x, y, z) = 1$ is then referred to as the *second case*.

In fact, when $\mathbb{Q}[p]$ has class number prime to $p$, one can deal with both cases and so show (1.1), when $n$ is such a prime. However, for the sake of brevity, we shall only consider the first case here. It is not known whether there are an infinite number of primes $p$ with the property that $\mathbb{Q}[p]$ has class number coprime to $p$; however, all primes $p < 37$ do possess this property.

*Proof of Theorem 56.* Let $\zeta$ denote a primitive $p$th root of unity. By factoring out the H.C.F. we may, without loss of generality, suppose $(x, y, z) = 1$. Since $\{0, -1, 1\}$ are the only cubes mod(9), we see that the result is clear when $p = 3$; thus from now on we assume $p > 3$.

Suppose for contradiction $x^p + y^p = z^p$ with $p \nmid xyz$; then of course

$$\prod_{j=0}^{p-1} (x + \zeta^j y) = z^p. \tag{1.2}$$

Let $\mathfrak{o}$ denote the ring of integers of $\mathbb{Q}[p]$, so that, by Theorems 45 and 46, $\mathfrak{o} = \mathbb{Z}[\zeta]$, and $\mathfrak{p} = (1-\zeta)\mathfrak{o}$ is the unique prime ideal of $\mathfrak{o}$ above $p$. We show

**(1.3)** *The $\{x + \zeta^j y\}$ are all pairwise coprime in $\mathfrak{o}$.*

Indeed, if a prime ideal $\mathfrak{q}$ of $\mathfrak{o}$ divides $x + \zeta^j y$, $x + \zeta^k y$ for distinct $j, k$ in $[0, p-1]$; then $\mathfrak{q} \mid (\zeta^j - \zeta^k)y\mathfrak{o} = \mathfrak{p}y$, and $\mathfrak{q} \mid (\zeta^k - \zeta^j)x\mathfrak{o} = \mathfrak{p}x$. Since $x, y, z$ are coprime, $x$ and $y$ are coprime, and therefore $\mathfrak{q} = \mathfrak{p}$. Thus $\mathfrak{p} \mid x^p + y^p = z^p$, and so $p \mid z$ which is a contradiction.

□

From (1.2) and (1.3) it follows that for each $j$, $(x + \zeta^j y)\mathfrak{o} = \mathfrak{a}_j^p$, for some $\mathfrak{o}$-ideal $\mathfrak{a}_j$. Therefore, since the classgroup of fractional $\mathfrak{o}$-ideals has order coprime to $p$, it follows that $\mathfrak{a}_j$ is principal, $\mathfrak{a}_j = \alpha_j\mathfrak{o}$ say.

For brevity we write $\alpha$ in place of $\alpha_1$, so that $x + \zeta y = \alpha^p u$ for some $u \in \mathfrak{o}^*$. By (VI.1.19) we can write $u = \rho\eta$, where $\rho$ is totally real and $\eta$ is a $p$th root of unity; hence

(1.4) $$x + \zeta y = \alpha^p \rho\eta.$$

Observe that for some $a \in \mathbb{Z}$, $\alpha \equiv a \bmod \mathfrak{p}$, since $p$ ramifies completely in $\mathbb{Q}[p]$. As $\zeta^j \equiv 1 \bmod \mathfrak{p}$ for all $j$, it follows that $\zeta^j\alpha - a \in \mathfrak{p}$ for all $j$, and so

$$\prod_{j=0}^{p-1}(a - \zeta^j\alpha) = a^p - \alpha^p \in p\mathfrak{p}.$$

Writing $b = a^p$, we have

(1.5) $$\alpha^p \equiv b \bmod p\mathfrak{p} \qquad (b \in \mathbb{Z}).$$

Piecing together (1.5) and (1.4) gives

(1.6) $$x + \zeta y \equiv b\rho\eta \bmod p\mathfrak{p}.$$

(1.7) If $\eta = 1$, then, on applying complex conjugation, we have

$$x + \zeta y \equiv b\rho \equiv x + \zeta^{-1}y \bmod p\mathfrak{p}.$$

Subtracting gives $(\zeta - \zeta^{-1})y \in p\mathfrak{p}$, and so $p \mid y$ which is a contradiction.
(1.8) Similarly, if $\eta \neq 1$ but $\zeta\eta^{-1} = 1$, then we get

$$x\eta^{-1} + y \equiv b\rho \equiv x\eta + y \bmod p\mathfrak{p}$$

which implies $p \mid x$ – also a contradiction.

Next note that since $\mathfrak{o}$ has $\mathbb{Z}$-basis $\{\zeta^i \mid i = 1, 2, \ldots, p-1\}$, then the same set is a $\mathbb{Z}/p\mathbb{Z}$ basis of $\mathfrak{o}/p\mathfrak{o}$: indeed, it is trivially a spanning set; on the other hand, if $\sum_1^{p-1} a_i\zeta^i \in p\mathfrak{o}$ with $a_i \in \mathbb{Z}$ and not all $a_i \equiv 0 \bmod (p)$, then $\frac{1}{p}\sum_1^{p-1} a_i\zeta^i$ can be written in the form $\sum_1^{p-1} b_i\zeta^i$, with $b_i \in \mathbb{Z}$, which gives a non-trivial relation $\sum_1^{p-1}(a_i - pb_i)\zeta^i = 0$, which is absurd. Thus, for $a_i \in \mathbb{Z}$,

$$(1.9) \qquad \sum_1^{p-1} a_i\zeta^i \in p\mathfrak{o} \qquad \text{iff all } a_i \in p\mathbb{Z}.$$

From (1.6) and its complex conjugate, we deduce that

$$(1.10) \qquad x\eta^{-1} + \zeta\eta^{-1}y - x\eta - \zeta^{-1}\eta y \equiv 0 \bmod p\mathfrak{p}.$$

By (1.7), (1.8) we can exclude the cases $\eta = 1$ and $\zeta = \eta$ and by definition $\zeta \neq 1$, so that by (1.9) we can deduce $p \mid xy$ whenever at least three of, and hence all four, of the roots of unity $\{\eta, \zeta\eta^{-1}, \eta^{-1}, \zeta^{-1}\eta\}$ are distinct. It therefore follows that $\eta = \zeta\eta^{-1}$, and we now show that this also leads to a contradiction: indeed, by (1.10),

$$x(\eta^{-1} - \eta) \equiv y(\eta^{-1} - \eta) \bmod p\mathfrak{p},$$

so that finally, as $(\eta^{-1} - \eta)\mathfrak{o} = \mathfrak{p}$, we have shown that $p \nmid xy$ implies $p \mid x - y$. But $x^p + y^p = z^p$ implies $(-x)^p = y^p + (-z)^p$; therefore, interchanging the roles of $x$ and $z$, we conclude that $y \equiv -z \bmod (p)$. Hence $y^p + y^p \equiv -y^p \bmod (p)$, and so, because $p > 3$, $p \mid y$ – which is again a contradiction.

□

## §2 Quadratic forms

In the introduction we considered the problem of which integers were represented by the quadratic form $X^2 + Y^2$, and furthermore we noticed that the corresponding question for the quadratic form $X^2 + 6Y^2$ seemed distinctly harder because, unlike the Gaussian integers $\mathbb{Z}[i]$, the ring of integers $\mathbb{Z}[\sqrt{-6}]$ is not a principal ideal domain. The theory of quadratic forms has played quite a fundamental role in the development of number theory, and this section is devoted to the diophantine problem of determining which numbers are represented by certain quadratic forms.

We begin our study of quadratic diophantine equations by describing the integral solutions $(x, y)$ of

$$(2.1) \qquad X^2 - dY^2 = 1$$

for a given square-free integer $d > 1$. The equation (2.1) is usually referred to as Pell's equation; this is because Euler had believed work, done by Wallis on the equation, to be due to the mathematician John Pell.

First we consider (2.1) when $d \not\equiv 1 \bmod (4)$. We set $K = \mathbb{Q}(\sqrt{d})$; we henceforth fix an embedding $K \hookrightarrow \mathbb{R}$, and view $\sqrt{d}$ as positive. Since, in this case, $\mathfrak{o}_K = \mathbb{Z}[\sqrt{d}]$, clearly $(x, y)$ is an integral solution to (2.1) iff the algebraic integer $x + y\sqrt{d}$ has norm 1. From (IV.4.2) we know that this is equivalent to $x + y\sqrt{d} \in (\pm 1)U_K^+$, where as in V§1 $U_K^+$ is the group of totally positive units. Furthermore, by Dirichlet's Unit Theorem, there exists a fundamental unit $v = a + b\sqrt{d}$ such that

$$(\pm 1)U_K^+ = (\pm 1)v^{\mathbb{Z}}.$$

It is now entirely straightforward to list recursively all solutions once $v$ is known. Indeed, writing $v^n = a_n + b_n\sqrt{d}$, then

$$\begin{aligned} a_{n+1} + b_{n+1}\sqrt{d} = v^{n+1} &= (a + b\sqrt{d})(a_n + b_n\sqrt{d}) \\ &= (a_n a + d b_n b) + (a_n b + b_n a)\sqrt{d} \end{aligned}$$

and so in this case we obtain the recursion formulae

$$\begin{aligned} a_{n+1} &= a_n a + d b_n b \\ b_{n+1} &= a_n b + b_n a. \end{aligned} \tag{2.2}$$

It therefore now remains to consider the case $d \equiv 1 \bmod (4)$. Clearly an integral solution $x + y\sqrt{d}$ will again correspond to a unit in the ring $\mathbb{Z}[\sqrt{d}]$ with norm 1; however, the matter is now a little more complicated since $\mathfrak{o}_K = \mathbb{Z}[\frac{1+\sqrt{d}}{2}]$, and so we wish to study

$$V_K = (\pm 1)U_K^+ \cap \mathbb{Z}[\sqrt{d}].$$

We assert that this is the same as the group of norm 1 units of the ring $\mathbb{Z}[\sqrt{d}]$: trivially the norm 1 units of $\mathbb{Z}[\sqrt{d}]$ lie in $V_K$; conversely, if $a + b\sqrt{d} \in V_K$, then its inverse $a - b\sqrt{d} \in \mathbb{Z}[\sqrt{d}]$. As previously, we let $v = a + b\sqrt{d}$ denote a generator of $U_K^+$. Recall from (III.2.29) that for $d \equiv 1 \bmod (4)$, 2 is split (resp. is inert) in $K$ if $d \equiv 1 \bmod (8)$ (resp. $d \equiv 5 \bmod (8)$). Clearly

$$\frac{\mathfrak{o}_K}{2\mathfrak{o}_K} \cong \begin{cases} \mathbb{F}_2 \times \mathbb{F}_2 & \text{if 2 split} \\ \mathbb{F}_4 & \text{if 2 inert} \end{cases} \tag{2.3}$$

and so

$$|(\mathfrak{o}_K/2\mathfrak{o}_K)^*| = \begin{cases} 1 & \text{if 2 split} \\ 3 & \text{if 2 inert.} \end{cases} \tag{2.4}$$

The reduction map $\mathfrak{o}_K \to \mathfrak{o}_K \bmod 2\mathfrak{o}_K$ induces a group homomorphism

$$\rho\colon (\pm 1)U_K^+ \to (\mathfrak{o}_K/2\mathfrak{o}_K)^*.$$

Since $\mathfrak{o}_K = \mathbb{Z}[\frac{1+\sqrt{d}}{2}]$, it follows that

$$\mathbb{Z}[\sqrt{d}] = \mathbb{Z} + 2\mathfrak{o}_K. \tag{2.5}$$

We shall now show

$$V_K = \ker(\rho). \tag{2.6}$$

Clearly if $w \in V_K$, then, by (2.5), $\rho(w) \equiv 1 \bmod 2\mathfrak{o}_K$, and so $V_K \subset \ker \rho$;

conversely, if $w \in \ker \rho$, then of course $w \in 1 + 2\mathfrak{o}_K$, and so by (2.5) $w$ lies in $V_K$.

In conclusion, by (2.4) and (2.6), we have shown that if $d \equiv 1 \bmod (8)$, then

$$(\pm 1)U_K^+ = V_K;$$

while if $d \equiv 5 \bmod (8)$, then

$$((\pm 1)U_K^+ : V_K) = 1 \text{ or } 3$$

depending on whether or not the generator of $U_K^+$ lies in $\mathbb{Z}[\sqrt{d}]$ or not.

We summarise the above results in:

**Theorem 57.** *Let $v$ denote a generator of $U_K^+$; then $(x, y)$ is an integral solution to (1.1) iff*

$$x + y\sqrt{d} \in \begin{cases} (\pm 1)U_K^+ & \text{if } v \in \mathbb{Z}[\sqrt{d}]; \\ (\pm 1)(U_K^+)^3 & \text{if } v \notin \mathbb{Z}[\sqrt{d}]. \end{cases}$$

*Furthermore the latter case can only arise when $d \equiv 5 \bmod (8)$.*

It should be noted that both cases occur for discriminants congruent to 5 mod (8); for instance $V_K = (\pm 1)U_K^+$ when $d = 37$; but $(\pm 1)U_K^+ \neq V_K$ when $d = 5$.

In the above we have seen how a question concerning the set of solutions of a particular quadratic equation can be completely determined by working in the quadratic number field $\mathbb{Q}(\sqrt{d})$. More generally, we now wish to use the arithmetic of quadratic number fields to determine the solubility, or otherwise, of equations of the form $Q(X, Y) = n$ for a wide variety of quadratic forms $Q$. With this in mind, we briefly establish some elementary facts for binary quadratic forms.

Let $Q(X, Y) = aX^2 + bXY + cY^2$ denote a binary quadratic form with rational coefficients $a, b, c$, and suppose that for all integers $m, n$, $Q(m, n)$ is always an integer. Since

$$a = Q(1, 0), \quad c = Q(0, 1), \quad a + b + c = Q(1, 1)$$

we see immediately that in fact $a, b, c$ must all lie in $\mathbb{Z}$. In the sequel, for brevity, we shall often refer to such binary integral quadratic forms as quadratic forms. Of course we have already encountered some quadratic forms in (V,§1). There we considered the problem of representing integers by binary forms with rational arguments (see for instance (V.1.27)). Here we shall mainly be concerned with the problem of representing integers by quadratic forms with integral arguments. We shall say that $Q$ *represents* $r$ if $Q(m, n) = r$ for some integers $m, n$. It is frequently useful to only consider representations where $m$ and $n$ are coprime; such a representation is called *proper*.

It is often advantageous to view $Q$ in terms of its associated bilinear form and write

$$(2.7)\qquad Q(X,Y) = \underline{X}^T B_Q \underline{X}$$

where

$$B_Q = \begin{pmatrix} a & b/2 \\ b/2 & c \end{pmatrix}, \qquad \underline{X} = \begin{pmatrix} X \\ Y \end{pmatrix}.$$

(Note that of course $B_Q$ will not in general be an integral matrix). We define the discriminant of $Q$, which we denote by $d_Q$, by

$$(2.8)\qquad d_Q = b^2 - 4ac = -4\det(B_Q).$$

The quadratic form $Q$ is said to be *non-degenerate* iff the associated bilinear form $B_Q$ is non-degenerate, i.e. iff $d_Q \neq 0$. Note also that $Q$ represents 0 non-trivially (i.e. $Q(x,y) = 0$ for integers $x, y$ not both of which are zero) iff $d_Q$ is a square in $\mathbb{Z}$. In order to avoid trivialities, we shall henceforth always suppose $d_Q$ to be a non-square.

If $d_Q > 0$, then $Q$ certainly takes on both positive and negative values. On the other hand, if $d_Q < 0$, then $Q$ is positive definite if $a$ is positive, and is negative definite if $a$ is negative. In the sequel when $d_Q$ is negative we shall *always* assume $a$ to be chosen to be positive.

Let $\alpha$,$\beta$,$\gamma$,$\delta$ all lie in $\mathbb{Z}$ and let $A = \begin{pmatrix} \alpha & \beta \\ \gamma & \delta \end{pmatrix}$. If we fully understand the behaviour of the quadratic form $Q(X,Y)$, then, by substitution, the same will go for the form $Q'(X,Y)$ given by

$$(2.9)\qquad \begin{aligned} Q'(X,Y) &= Q(\alpha X + \beta Y, \gamma X + \delta Y) \\ &= \underline{X}^T A^T B_Q A \underline{X}. \end{aligned}$$

Hence we note

$$(2.10)\qquad B_{Q'} = A^T B_Q A; \qquad d_{Q'} = d_Q \det(A)^2.$$

For this process to be reversible, we need to know that the inverse matrix of $A$ has coefficients in $\mathbb{Z}$, i.e. we need $\det(A) = \pm 1$; indeed, in due course we shall find it important to consider only substitutions with positive determinant, and so we only consider $A$ with $\det(A) = +1$. The set of matrices with integral coefficients and with determinant 1 forms a group under matrix multiplication; this group is usually denoted by $\mathrm{SL}_2(\mathbb{Z})$.

The above discussion now motivates the following definition of equivalence for quadratic forms: two quadratic forms $Q(X,Y)$, $Q'(X,Y)$ are said to be equivalent iff

$$Q'(X,Y) = Q(\alpha X + \beta Y, \gamma X + \delta Y)$$

for some matrix $A \in \mathrm{SL}_2(\mathbb{Z})$. It follows immediately from (2.10) that this notion of equivalence is indeed an equivalence relation. In fact equivalence of quadratic forms preserves all the invariants that we shall be concerned with. First we note that if $Q$ represents an integer $r$, then so does an equivalent form $Q'$: indeed with the above notation, if $Q(m,n) = r$, then

$$r = \underline{m}^T B_Q \underline{m} = \underline{m}'^T A^T B_Q A \underline{m}' = \underline{m}'^T B_{Q'} \underline{m}'$$

where
$$\underline{m} = \begin{pmatrix} m \\ n \end{pmatrix}, \underline{m}' = A^{-1}\underline{m}$$

Secondly we note from (2.10) that two equivalent quadratic forms necessarily have the same discriminant. This observation naturally leads us to ask whether two quadratic forms with the same discriminant are necessarily equivalent; in general the answer is no; however, there are only a finite number of equivalence classes of quadratic forms with a given discriminant.

We next consider how we might use quadratic number fields to construct quadratic forms. To this end, let $d$ denote a square-free integer with $d \neq 1$; let $K = \mathbb{Q}(\sqrt{d})$ and let $d_K$ denote the discriminant of $K/\mathbb{Q}$. As in (V.1.24) we define the *norm form* $q_K$ by

(2.11)
$$\begin{cases} q_K(X,Y) = N_{K/\mathbb{Q}}(X + Y\sqrt{d}) = X^2 - dY^2 & \text{if } d \not\equiv 1 \bmod (4) \\ q_K(X,Y) = N_{K/\mathbb{Q}}(X + Y\left(\frac{1+\sqrt{d}}{2}\right)) & \\ \qquad = X^2 + XY + \left(\frac{1-d}{4}\right)Y^2 & \text{if } d \equiv 1 \bmod (4). \end{cases}$$

In both cases note that $q_K$ has discriminant

$$d_{q_K} = d_K.$$

More generally it is interesting to note

**(2.12)** *If $Q$ is a binary integral quadratic form, as above, then $d_Q = d_K f^2$ for some integer $f$, where $K = \mathbb{Q}(\sqrt{d_Q})$. Furthermore, if $d_Q = d_K$, then $Q$ is primitive, i.e. $(a,b,c) = 1$.*

*Proof.* Since $d_Q$ is a non-square, $ac \neq 0$ and so we can write $Q(X,Y) = a(X + \mu Y)(X + \mu' Y)$ for $\mu = (b - \sqrt{d_Q})/2a \in K = \mathbb{Q}(\sqrt{d_Q})$, where $\mu'$ is the Galois conjugate of $\mu$. Writing $\{1, \omega\}$ for the basis of $\mathfrak{o}_K$ used in (2.11), we can write

$$\begin{pmatrix} 1 \\ \mu \end{pmatrix} = A \begin{pmatrix} 1 \\ \omega \end{pmatrix}$$

for a matrix $A$ with rational coefficients. This then shows $d_Q = d_K(a^2 \det(A)^2)$. As $d_Q$ is an integer and as the only possible square factor in $d_K$ is 4, we need only show that if $4 \mid d_K$, then $4 \mid d_Q$. Since

$d_Q = b^2 - 4ac \equiv 0, 1 \bmod (4)$, it follows that if $4 \nmid d_Q$, then $4 \nmid d_K$. Suppose now that $Q = aX^2 + bXY + cY^2$ has discriminant $d_K$. We need to show $(a, b, c) = 1$. On the one hand, if $d_K \equiv 1 \bmod (4)$, then the result is clear since $d_K$ is square-free and since $d_K = b^2 - 4ac$. On the other hand, if $d_K \not\equiv 1 \bmod (4)$, then $d_K = 4d$ with $d$ square-free; hence, as in the previous case, we deduce that $(a, b, c)$ divides 2. Suppose for contradiction that $(a, b, c) = 2$; then

$$\text{(2.12.a)} \qquad d = \frac{d_K}{4} = \left(\frac{b}{2}\right)^2 - 4\left(\frac{a}{2}\right)\left(\frac{c}{2}\right) \equiv \left(\frac{b}{2}\right)^2 \bmod (4).$$

However, by hypothesis $d \not\equiv 1 \bmod (4)$, and $d \not\equiv 0 \bmod (4)$ since $d$ is square-free; therefore $d \equiv 2$ or $3 \bmod (4)$, and this is incompatible with the congruence (2.12.a).

□

Henceforth we shall always restrict attention to the case where $d_Q = d_K$.

Because the set of values represented by $q_K$ coincides with the set $N_{K/\mathbb{Q}}(\mathfrak{o}_K)$, it follows that

**(2.13)** *The set of values represented by $q_K$ is closed under multiplication.*

Presently we shall see that equivalence classes other than the norm class, do not possess this property.

Because of (2.13), we shall frequently restrict attention to determining whether or not a given prime number $p$ is represented by $q_K$. Here we note

**(2.14)** *$p$ is represented by $q_K$ iff $p$ is not inert in $K/\mathbb{Q}$, and some – and hence all – prime ideals $\mathfrak{p}$ of $\mathfrak{o}_K$ above $p$ have trivial narrow class in $C_K^+$.*

*Proof.* If, $p$ is represented by $q_K$, i.e. $p = N_{K/\mathbb{Q}}(\alpha)$ for some $\alpha \in \mathfrak{o}_K$, then $p\mathbb{Z} = \mathcal{N}_{K/\mathbb{Q}}(\alpha\mathfrak{o}_K)$. So $\alpha\mathfrak{o}_K \neq \mathfrak{o}_K$ and, if $\mathfrak{p}$ is a prime ideal factor of $\alpha\mathfrak{o}_K$, then we must have $p\mathbb{Z} = \mathcal{N}_{K/\mathbb{Q}}\mathfrak{p}$ and hence $\alpha\mathfrak{o}_K = \mathfrak{p}$. As $N_{K/\mathbb{Q}}(\alpha) > 0$, we may take $\alpha$ totally positive. Therefore $c^+(\mathfrak{p}) = 1$. Conversely, if $c^+(\mathfrak{p}) = 1$ for some $\mathfrak{p}$ above $p$, then $\mathfrak{p} = \alpha\mathfrak{o}_K$ for some totally positive $\alpha \in \mathfrak{o}_K$, and hence $N_{K/\mathbb{Q}}(\alpha) = p$.

□

At this point it is instructive to reconsider very briefly the case $d = -6$; then $d_K = -24$ and $q_K = X^2 + 6Y^2$; however, the form $Q_2 = 2X^2 + 3Y^2$

also clearly has discriminant $-24$. The two forms cannot be equivalent since $Q_2$ represents 5 by $(x, y) = (1, 1)$, while $q_K$ cannot represent 5, for otherwise

$$5 = x^2 + 6y^2$$

which is absurd.

In summary, we conclude that there will, in general, be equivalence classes of discriminant $d_K$ other than the norm class. However, we can easily generalise the construction of $q_K$ to obtain all equivalence classes with discriminant $d_K$.

We keep the previous notation and we let $\sigma$ denote the non-trivial automorphism of $K$. Let $\mathfrak{a}$ denote a fractional $\mathfrak{o}_K$-ideal and let $a_1, a_2$ denote a $\mathbb{Z}$-basis of $\mathfrak{a}$. From (II.1.39) we know that

$$\begin{vmatrix} a_1 & a_2 \\ a_1^\sigma & a_2^\sigma \end{vmatrix}^2 = d_K \mathbf{N}\mathfrak{a}^2.$$

We fix an embedding $K \subset \mathbb{C}$. In the sequel $\sqrt{d_K}$ is always to be understood as meaning the positive square root if $d_K > 0$, and as meaning the square root with positive imaginary part if $d_K < 0$. We shall call the ordered basis $(a_1, a_2)$ of $\mathfrak{a}$ a *normalised* basis when

(2.15).
$$\begin{vmatrix} a_1 & a_2 \\ a_1^\sigma & a_2^\sigma \end{vmatrix} = \sqrt{d_K} \mathbf{N}\mathfrak{a}$$

Clearly, given a basis $\{a_1, a_2\}$ of $\mathfrak{a}$, then exactly one of the ordered pairs $(a_1, a_2)$, $(a_2, a_1)$ will be normalised. Given a normalised basis $(a_1, a_2)$ of $\mathfrak{a}$, we define the quadratic form

(2.16)
$$Q_{a_1,a_2}(X, Y) = \mathbf{N}\mathfrak{a}^{-1} N_{K/\mathbb{Q}}(a_1 X + a_2 Y).$$

Observe that for any $x, y \in \mathbb{Z}$, $a_1 x + a_2 y \in \mathfrak{a}$, and $\mathbf{N}\mathfrak{a} \mid N_{K/\mathbb{Q}} b$ for any $b \in \mathfrak{a}$; we therefore know that $Q_{a_1,a_2}(x, y)$ is always integral.

**(2.17)** *A positive integer $n$ is represented by $Q_{a_1,a_2}$ iff $n$ is a norm of some integral ideal with the same narrow class as $\mathfrak{a}$.*

*Proof.* If $n = \mathbf{N}\mathfrak{b}$, with $\mathfrak{b} = \alpha\mathfrak{a}^{-1}$ integral and $\alpha$ totally positive, then $n = N_{K/\mathbb{Q}}(\alpha)(\mathbf{N}\mathfrak{a})^{-1}$; thus if $\alpha = a_1 x + a_2 y$, then $Q_{a_1,a_2}(x, y) = n$. Conversely, if $n = Q_{a_1,a_2}(x, y)$, then $n = N_{K/\mathbb{Q}}(\alpha)(\mathbf{N}\mathfrak{a})^{-1} = \mathbf{N}\alpha\mathfrak{a}^{-1}$ (where $\alpha$ is in $\mathfrak{a}$). Since $N_{K/\mathbb{Q}}(\alpha) > 0$, it follows that either $\alpha$ or $-\alpha$ must be totally positive. Finally observe that $\mathbf{N}\mathfrak{b} = \mathbf{N}\mathfrak{b}^\sigma, \langle\sigma\rangle = \mathrm{Gal}(K/\mathbb{Q})$ and $c^+(\mathfrak{b}^\sigma) = c^+(\mathfrak{b}^{-1})$.

□

Writing $Q_{a_1,a_2}(X, Y)$ out in full, we obtain

$$Q_{a_1,a_2}(X, Y) = \mathbf{N}\mathfrak{a}^{-1}(a_1 a_1^\sigma X^2 + (a_1 a_2^\sigma + a_1^\sigma a_2) XY + a_2 a_2^\sigma Y^2)$$

and so $Q_{a_1,a_2}$ has discriminant

$$\begin{aligned} d_{Q_{a_1,a_2}} &= \frac{1}{\mathbf{N}\mathfrak{a}^2}((a_1 a_2^\sigma + a_1^\sigma a_2)^2 - 4a_1^2 a_1^{2\sigma} a_2^2 a_2^{2\sigma}) \\ &= \frac{1}{\mathbf{N}\mathfrak{a}^2}(a_1 a_2^\sigma - a_1^\sigma a_2)^2 \\ &= \mathbf{N}\mathfrak{a}^{-2} \begin{vmatrix} a_1 & a_2 \\ a_1^\sigma & a_2^\sigma \end{vmatrix}^2 = d_K. \end{aligned}$$

In addition, we also note that $X^2$ has coefficient $\mathbf{N}\mathfrak{a}^{-1} N_{K/\mathbb{Q}}(a_1)$; thus, if $d_K < 0$, then $Q_{a_1,a_2}$ is necessarily positive definite, which is in agreement with our previously established convention.

Choosing an alternative normalised basis $(b_1, b_2)$ of $\mathfrak{a}$ corresponds to the substitution $X \mapsto \alpha X + \gamma Y$, $Y \mapsto \beta X + \delta Y$ where $\begin{pmatrix} \alpha & \gamma \\ \beta & \delta \end{pmatrix}$ is the invertible integral matrix such that

$$\begin{aligned} b_1 &= \alpha a_1 + \beta a_2 \\ b_2 &= \gamma a_1 + \delta a_2. \end{aligned}$$

Applying $\sigma$ to the above equalities, we obtain the matrix equality

$$\begin{pmatrix} b_1 & b_2 \\ b_1^\sigma & b_2^\sigma \end{pmatrix} = \begin{pmatrix} a_1 & a_2 \\ a_1^\sigma & a_2^\sigma \end{pmatrix} \begin{pmatrix} \alpha & \gamma \\ \beta & \delta \end{pmatrix}.$$

Since both bases are normalised, we conclude that $\begin{pmatrix} \alpha & \gamma \\ \beta & \delta \end{pmatrix} \in \mathrm{SL}_2(\mathbb{Z})$ and so $Q_{a_1,a_2}$ and $Q_{b_1,b_2}$ are equivalent. Furthermore, we assert that if we replace $(a_1, a_2)$ by a normalised basis for any fractional $\mathfrak{o}_K$-ideal with the same narrow class as $\mathfrak{a}$, then we still get an equivalent form: indeed, by the above, it suffices to show $Q_{a_1,a_2}$ and $Q_{\lambda a_1,\lambda a_2}$ are equivalent whenever $\lambda \in K_+$. First note that since $\lambda$ is totally positive

$$\begin{aligned} \begin{vmatrix} \lambda a_1 & \lambda a_2 \\ \lambda^\sigma a_1^\sigma & \lambda^\sigma a_2^\sigma \end{vmatrix} &= N_{K/\mathbb{Q}}(\lambda)\sqrt{d_K}\mathbf{N}\mathfrak{a} \\ &= \sqrt{d_K}\mathbf{N}\mathfrak{a}\lambda. \end{aligned}$$

Thus $(\lambda a_1, \lambda a_2)$ is normalised, and so $Q_{\lambda a_1,\lambda a_2}$ is indeed well defined. Again, since $N_{K/\mathbb{Q}}(\lambda) = \mathbf{N}(\lambda\mathfrak{o}_K)$,

$$\begin{aligned} Q_{\lambda a_1,\lambda a_2}(X,Y) &= \frac{N_{K/\mathbb{Q}}(\lambda a_1 X + \lambda a_2 Y)}{\mathbf{N}\mathfrak{a}\lambda} \\ &= N_{K/\mathbb{Q}}(a_1 X + a_2 Y)/\mathbf{N}\mathfrak{a} \\ &= Q_{a_1,a_2}(X,Y). \end{aligned}$$

as required.

We now write $\mathcal{Q}_K$ for the set of equivalence classes of binary integral quadratic forms with discriminant $d_K$. Then, in summary, we have constructed a map

$$\kappa: C_K^+ \to \mathcal{Q}_K$$

where, for a given narrow class $c$, $\kappa(c)$ is the equivalence class of $Q_{a_1,a_2}$ for some normalised basis of an ideal $\mathfrak{a}$ with narrow class $c$.

We shall now show

**Theorem 58.** *The map $\kappa: C_K^+ \to \mathcal{Q}_K$ is a bijection.*

*$\kappa$ surjective.* Given $Q = aX^2 + bXY + cY^2$, define $\mathfrak{a}$ to be the fractional $\mathfrak{o}_K$-ideal generated by $a$ and $\frac{b-\sqrt{d_K}}{2}$. We claim that in fact

$$a\left(\frac{d_K + \sqrt{d_K}}{2}\right) \in \mathbb{Z}a + \mathbb{Z}\frac{b - \sqrt{d_K}}{2} \tag{2.18.a}$$

$$\left(\frac{b - \sqrt{d_K}}{2}\right)\left(\frac{d_K + \sqrt{d_K}}{2}\right) \in \mathbb{Z}a + \mathbb{Z}\frac{b - \sqrt{d_K}}{2}. \tag{2.18.b}$$

First observe that

$$a \cdot \frac{d_K + \sqrt{d_K}}{2} = -a \cdot \frac{(b - \sqrt{d_K})}{2} + a \cdot \frac{(b + d_K)}{2}$$

and since $d_K \equiv b \bmod (2)$ this proves (2.18.$a$). Secondly we note that

$$\begin{aligned}\frac{d_K + \sqrt{d_K}}{2} \cdot \frac{(b - \sqrt{d_K})}{2} &= \frac{d_K}{2}\left(\frac{(b - \sqrt{d_K})}{2}\right) + \frac{b\sqrt{d_K} - d_K}{4} \\ &= \frac{d_K}{2}\left(\frac{(b - \sqrt{d_K})}{2}\right) + \frac{b\sqrt{d_K} - b^2 + 4ac}{4} \\ &= \left(\frac{d_K - b}{2}\right)\left(\frac{(b - \sqrt{d_K})}{2}\right) + ac \qquad (2.19)\end{aligned}$$

and this yields (2.18.$b$). Together the two inclusions (2.18.$a$,$b$) show that

$$\{a, (b - \sqrt{d_K})/2\}$$

is a $\mathbb{Z}$-basis of $\mathfrak{a}$.

If $a$ is positive (which, by convention, is always the case when $d_K < 0$), then we set $\lambda = 1$; on the other hand if $a$ is negative, then we set $\lambda = \sqrt{d_K}$. We conclude our proof of the surjectivity of $\kappa$ by considering the fractional ideal $\lambda\mathfrak{a}$, which, by the above, has basis

$$a_1 = \lambda a \qquad a_2 = \lambda \cdot \left(\frac{(b - \sqrt{d_K})}{2}\right).$$

This is a normalised basis since

$$\begin{aligned}\begin{vmatrix} a_1 & a_2 \\ a_1^\sigma & a_2^\sigma \end{vmatrix} &= N_{K/\mathbb{Q}}\lambda \begin{vmatrix} a & \frac{b-\sqrt{d_K}}{2} \\ a & \frac{b+\sqrt{d_K}}{2} \end{vmatrix} \\ &= a\sqrt{d_K} N_{K/\mathbb{Q}}\lambda.\end{aligned}$$

Furthermore we note that this implies

$$\mathbf{N}\mathfrak{a}\lambda = aN_{K/\mathbb{Q}}\lambda. \tag{2.20}$$

Finally we consider the quadratic form

$$Q_{a_1,a_2}(X,Y) = \frac{N_{K/\mathbb{Q}}\lambda}{\mathbf{N}\mathfrak{a}\lambda} N_{K/\mathbb{Q}}(aX + Y\frac{b-\sqrt{d_K}}{2})$$

by (2.20)

$$= \frac{1}{a}\left(a^2X^2 + abXY + Y^2\frac{b^2 - d_K}{4}\right).$$

Since $d_K = b^2 - 4ac$, we conclude that $\kappa(c^+(\mathfrak{a})) = Q(X,Y)$.

□

*$\kappa$ injective.* Suppose we have $\mathfrak{o}_K$ fractional ideals $\mathfrak{a}$, $\mathfrak{b}$ with normalised bases $(a_1, a_2)$, $(b_1, b_2)$ respectively, such that $Q_{a_1,a_2}(X,Y)$ is equivalent to $Q_{b_1,b_2}(X,Y)$. By a change of basis we may then assume that actually

$$Q_{a_1,a_2}(X,Y) = Q_{b_1,b_2}(X,Y) \tag{2.21}$$

We are required to show that $\mathfrak{a} = \lambda\mathfrak{b}$ for some $\lambda \in K$ with $N_{K/\mathbb{Q}}(\lambda) > 0$.

Since $Q_{a_1,a_2}(1,Y)$ has roots $-\frac{a_1}{a_2}$, $-\frac{a_1^\sigma}{a_2^\sigma}$, we deduce from (2.21) that *either*

$$\frac{a_1}{a_2} = \frac{b_1}{b_2} \tag{2.22.a}$$

*or*

$$\frac{a_1}{a_2} = \frac{b_1^\sigma}{b_2^\sigma} \tag{2.22.b}$$

If (2.22.a) holds, then set $\lambda = a_1/b_1 = a_2/b_2$. Thus $a_1 = \lambda b_1$, $a_2 = \lambda b_2$ and so $\mathfrak{a} = \lambda\mathfrak{b}$. Furthermore $N_{K/\mathbb{Q}}(\lambda) > 0$, since

$$\begin{aligned}\sqrt{d_K}\mathbf{N}\mathfrak{a} = \begin{vmatrix} a_1 & a_2 \\ a_1^\sigma & a_2^\sigma \end{vmatrix} &= N_{K/\mathbb{Q}}(\lambda)\begin{vmatrix} b_1 & b_2 \\ b_1^\sigma & b_2^\sigma \end{vmatrix} \\ &= N_{K/\mathbb{Q}}(\lambda)\sqrt{d_K}\mathbf{N}\mathfrak{b}.\end{aligned}$$

So suppose now that (2.22.b) holds, and set $\lambda = a_1/b_1^\sigma = a_2/b_2^\sigma$; then

$$\begin{vmatrix} a_1 & a_2 \\ a_1^\sigma & a_2^\sigma \end{vmatrix} = N_{K/\mathbb{Q}}\lambda \begin{vmatrix} b_1^\sigma & b_2^\sigma \\ b_1 & b_2 \end{vmatrix}$$

and since both bases are normalised we deduce that $N_{K/\mathbb{Q}}(\lambda) < 0$. However, we also know that

$$\mathbf{N}\mathfrak{b}^{-1}(b_1X + b_2Y)(b_1^\sigma X + b_2^\sigma Y) = Q_{b_1,b_2}(X,Y)$$

by (2.21)

$$\begin{aligned}&= Q_{a_1,a_2}(X,Y) \\ &= \mathbf{N}\mathfrak{a}^{-1}(a_1X + a_2Y)(a_1^\sigma X + a_2^\sigma Y) \\ &= \mathbf{N}\mathfrak{a}^{-1}\lambda\lambda^\sigma(b_1^\sigma X + b_2^\sigma Y)(b_1X + b_2Y)\end{aligned}$$

and so we conclude $\mathbf{N}\mathfrak{a} = \mathbf{N}\mathfrak{b} N_{K/\mathbb{Q}}(\lambda)$ which contradicts $N_{K/\mathbb{Q}}(\lambda) < 0$. We have therefore shown that (2.22.$b$) is impossible.

□

As an immediate consequence of Theorem 58 we have

**Corollary 1.** *The number of different equivalence classes of quadratic forms with discriminant $d_K$ is $|C_K^+|$, and so in particular is finite.*

There exists a more general technique, known as the reduction of quadratic forms, which both shows: firstly that there are a finite number of equivalence classes of quadratic forms for a given discriminant $d$ (not just for discriminants of the form $d_K$); secondly, it provides a technique for verifying when two given forms with the same discriminant are equivalent.

Next we note the following elementary, but frequently useful, result:

**(2.23)** *Let $n$ be a given integer and, if $d_K < 0$, suppose also that $n > 0$. Then $n$ is properly represented by an integral quadratic form with discriminant $d_K$ iff there exists an integral solution to the congruence $X^2 \equiv d_K \bmod (4n)$.*

*Proof.* First suppose that $b$ solves the congruence with $b^2 - d_K = 4nc$; then the form $Q = nX^2 + bXY + cY^2$ has discriminant $d_K$ and represents $n$ properly since $Q(1,0) = n$.

Conversely, suppose that $Q = aX^2 + bXY + cY^2$ is an integral quadratic form with discriminant $d_K$ which represents $n$ properly by

$$n = ar^2 + brs + cs^2. \tag{2.24}$$

Hence we see that

$$\begin{aligned} 0 &\equiv 4a^2r^2 + 4bars + 4acs^2 \bmod (4n) \\ &\equiv (2ar + bs)^2 - (b^2 - 4ac)s^2 \bmod (4n). \end{aligned} \tag{2.25}$$

By the Chinese Remainder Theorem, in order to show that we can solve $X^2 \equiv d_K \bmod (4n)$, it suffices to show that we can solve

$$X^2 \equiv d_K \bmod (p^v)$$

for each prime divisor $p$ of $4n$, where $p^v \| 4n$. If $p \nmid s$, then (2.25) leads to a solution of this congruence. If $p | s$, then $p \nmid r$ and so we may use the analogous congruence to (2.25) obtained by interchanging the roles of $a, r$ with those of $c, s$.

□

From Theorem 58 and (2.23) we deduce

**(2.26)** *If $C_K^+ = \{1\}$, then $n$ is properly represented by the norm form $q_K$ iff $X^2 \equiv d_K \bmod (4n)$ is soluble, and provided $n > 0$ if $d_K < 0$.*

Thus we see that we have a highly satisfactory state of affairs when the narrow classgroup of $K$ is trivial. We shall, in fact, obtain a similarly exhaustive result whenever $C_K^+$ has exponent 2. We shall achieve this by use of the so-called Genus Theory of quadratic fields. We begin by briefly recalling a number of results from Chapters V and VI:

**(2.27)** $[C_K^+ : (C_K^+)^2] = 2^{m-1}$ *where $m$ denotes the number of ramified primes in $K/\mathbb{Q}$.*

*Proof.* This follows from Theorem 39 and the fact that $[C_K^+ : (C_K^+)^2] = |C_{K,2}^+|$.

□

We now seek to gain a better understanding of the group $C_K^+/(C_K^+)^2$. The key observation to keep in mind here is that for a given fractional $\mathfrak{o}_K$-ideal $\mathfrak{a}$, the image of the narrow class of $\mathfrak{a}$ in $C_K^+/(C_K^+)^2$ is determined by $\mathbf{N}\mathfrak{a}$: indeed, given a further fractional ideal $\mathfrak{b}$ with $\mathbf{N}\mathfrak{b} = \mathbf{N}\mathfrak{a}$, then $\mathbf{N}\mathfrak{a}\mathfrak{b}^{-1} = 1$ and so $\mathcal{N}_{K/\mathbb{Q}}(\mathfrak{a}\mathfrak{b}^{-1}) = \mathbb{Z}$; therefore by (III.1.22) $\mathfrak{a}\mathfrak{b}^{-1} = \mathfrak{c}^{\sigma-1}$ for some fractional ideal $\mathfrak{c}$; this then implies $\mathfrak{a} = \mathfrak{b}\mathfrak{c}^{1+\sigma}\mathfrak{c}^{-2}$, and so $c^+(\mathfrak{a}) = c^+(\mathfrak{b})c^+(\mathfrak{c})^{-2}$.

In order to carry this analysis of $C_K^+/(C_K^+)^2$ further, we now need to recall a number of results on residue class characters.

From (VI,§3) we know that $K$ is contained in the cyclotomic field $\mathbb{Q}[|d_K|]$. Let $\lambda_K: (\mathbb{Z}/d_K\mathbb{Z})^* \to \{\pm 1\}$ denote the quadratic character associated to $K$, i.e. if $H$ is the subgroup of $(\mathbb{Z}/d_K\mathbb{Z})^*$ corresponding to $K$, then $\ker \lambda_K = H$.

As shown in (III.3.7), $d_K$ can be written uniquely as a product of prime discriminants $d_K = q_1 \cdots q_m$. We let $\lambda_i$ denote the unique quadratic character mod$(q_i)$ such that $\lambda_K \to \prod_1^m \lambda_i$ under the decomposition of (VI.4.1)

$$(\mathbb{Z}/d_K\mathbb{Z})^{*\wedge} \cong \prod_{i=1}^{m} (\mathbb{Z}/q_i)^{*\wedge}.$$

Let $H_i = \ker \lambda_i \subset (\mathbb{Z}/q_i\mathbb{Z})^*$ and define $\tilde{H}$ to be the subgroup of $(\mathbb{Z}/d_K\mathbb{Z})^*$ which corresponds to $\prod_1^m H_i$ under the isomorphism

$$(\mathbb{Z}/d_K\mathbb{Z})^* \cong \prod_1^m (\mathbb{Z}/q_i\mathbb{Z})^*.$$

We therefore conclude that

$$(H : \tilde{H}) = 2^{m-1}. \qquad (2.28)$$

Recall from Ex. 8 to Chapter V that, given any class $c \in C_K^+$, we can always find an $\mathfrak{o}_K$-ideal coprime to $d_K$ which has class $c$. Given an $\mathfrak{o}_K$-ideal $\mathfrak{a}$ which is coprime to $d_K$, we see that $\mathbf{N}\mathfrak{a} \bmod (d_K)$ lies in $H$: on the one hand, if the prime ideal $\mathfrak{p}$ is inert then $\mathbf{N}\mathfrak{p} = p^2$ and so $\mathbf{N}\mathfrak{p} \bmod (d_K)$ is a square; on the other hand, if $\mathfrak{p}$ is split, then the Frobenius of any prime $\mathfrak{P}$ of $\mathbb{Q}[|d_K|]$ above $\mathfrak{p}$ must lie in $\mathrm{Gal}(\mathbb{Q}[|d_K|]/K)$, i.e. $\mathbf{N}\mathfrak{p}$ has image in $H$.

We define $f(\mathfrak{a})$ to be the class of $\mathbf{N}\mathfrak{a} \bmod \tilde{H}$. We note that if $\mathfrak{a} = \alpha\mathfrak{b}^2$ with both $\alpha, \mathfrak{b}$ coprime to $d_K$ and with $\alpha$ a totally positive element of $\mathfrak{o}_K$, then

$$\mathbf{N}\mathfrak{a} = \mathbf{N}\mathfrak{b}^2 N_{K/\mathbb{Q}}\alpha.$$

We claim that $N_{K/\mathbb{Q}}\alpha$, and hence $\mathbf{N}\mathfrak{a}$, lies in $\tilde{H}$. As observed above, $N_{K/\mathbb{Q}}\alpha = \mathbf{N}\alpha\mathfrak{o}_K$ certainly lies in $H$. Since $\lambda_K \to \prod_1^m \lambda_i$, we see that in order to show that $N_{K/\mathbb{Q}}\alpha$ lies in $\tilde{H}$ it suffices to prove $N_{K/\mathbb{Q}}\alpha$ lies in $H_i$ for all but one of the $i$. If $K = \mathbb{Q}(\sqrt{d})$, with $d$ a square free integer, then by (II.1.34) we know that $N_{K/\mathbb{Q}}\alpha$ is a square $\mathrm{mod}(d)$. The result then follows since $d_K$ is either $d$ or $4d$. Thus, in summary, we have shown that $f$ induces a homomorphism

$$\theta\colon C_K^+/(C_K^+)^2 \to H/\tilde{H}. \tag{2.29}$$

By Dirichlet's Theorem on arithmetic progressions (which will be proved in Chapter VIII), we know that, given $n \bmod (d_K)$ in $H$, we can find a prime number $p \equiv n \bmod (d_K)$. Since $p \bmod (d_K)$ lies in $H$, $p$ must split in $K/\mathbb{Q}$; hence $f(\mathfrak{p}) = \mathbf{N}\mathfrak{p} \bmod (d_K) = n \bmod (d_K)$ and so we have shown $\theta$ to be surjective. From (2.27) and (2.28), it therefore follows that $\theta$ is an isomorphism.

We can now deduce a series of important results from the fundamental isomorphism (2.29). Our first such result is originally due to Gauss.

**(2.30)** *Let $p$ denote a split prime of $K$, $p\mathfrak{o}_K = \mathfrak{p}\bar{\mathfrak{p}}$. The equation $pZ^2 = X^2 - dY^2$ has a non-trivial integral solution in $X, Y, Z$ iff the class of $p$ in $H$ lies in $\tilde{H}$, i.e. iff $\theta(c^+(\mathfrak{p})) \in \tilde{H}$.*

It is important to understand the spirit of this classical result: whether the equation has an integral solution or not can be decided entirely by rational residue class characters: such a solution exists iff $\lambda_i(p) = 1$ for all $i$.

*Proof.* In (V.1.27) we saw that for such $p$, the equation is soluble iff $c^+(\mathfrak{p}) \in (C_K^+)^2$. The result follows since $\theta$ is an isomorphism.

□

**Theorem 59.** *Let $Q$ denote an integral binary quadratic form with discriminant $d_K$ (which is positive definite if $d_K < 0$). Then $Q$ represents the prime number $p$ iff the class corresponding to $Q$ under $\kappa$ is of the form $c^+(\mathfrak{p})$, for some prime ideal $\mathfrak{p}$ of $\mathfrak{o}$ with $\mathbf{N}\mathfrak{p} = p$.*

*Proof.* By (2.17) $\mathfrak{p}$ is represented by $Q$, iff $\mathfrak{p} = \mathbf{N}\mathfrak{b}$, $\mathfrak{b}$ integral, where $Q$ corresponds to $c^+(\mathfrak{b})$. But then $\mathfrak{b}$ is a prime ideal with norm $p$. □

We now make use of the isomorphism

$$\mathrm{Gal}(\mathbb{Q}[|d_K|]/\mathbb{Q}) \cong (\mathbb{Z}/d_K\mathbb{Z})^*$$

of Theorem 44, and we view the $\lambda_i$ as characters of this Galois group. Let $\tilde{K}$ denote the subfield of $\mathbb{Q}[|d_K|]$ which corresponds to the subgroup $\tilde{H} \subset (\mathbb{Z}/d_K\mathbb{Z})^*$, i.e. so that the above isomorphism restricts to $\mathrm{Gal}(\mathbb{Q}[|d_K|]/\tilde{K}) \cong \tilde{H}$. Then, by (VI.3.1), we know that

$$\tilde{K} = K(\sqrt{q_1}, \ldots, \sqrt{q_m}).$$

We can thus reformulate (2.29) to say

$$C_K^+/(C_K^+)^2 \cong \mathrm{Gal}(\tilde{K}/K). \tag{2.31}$$

In (III.3.8) we have already seen that $\tilde{K}$ is a non-ramified Galois extension of $K$, whose Galois group has exponent 2. Indeed, it can be shown that the isomorphism (2.31) extends naturally to an isomorphism between $C_K^+$ and the Galois group of the maximal abelian non-ramified extension of $K$ in $\mathbb{Q}^c$; unfortunately though, this deep and beautiful description of $C_K^+$ lies beyond the scope of this book.

We conclude this section on quadratic forms by considering quadratic forms with discriminant $d_K$ in the special case when $C_K^+$ has exponent 2, so that $C_K^+ = C_{K,2}^+ = C_K^+/(C_K^+)^2$. In this case we know from Theorem 39 that the classes of $C_K^+$ are represented exactly twice by the $2^m$ ideals

$$\mathfrak{a} = \prod_1^m \mathfrak{p}_i^{\epsilon_i} \qquad \epsilon_i = 0, 1$$

where $\mathfrak{p}_i$ is the unique prime of $\mathfrak{o}_K$ dividing $q_i$. Thus, by Theorem 58, we can quickly list all forms with discriminant $d_K$. Given $\mathfrak{a}$ as above, we choose a normalised basis and, for brevity, we abuse notation and write $Q_\mathfrak{a}(X, Y)$ for the resulting form.

**(2.32)** *Suppose $C_K^+$ has exponent 2, let $n, r$ denote integers which are coprime to $d_K$, and suppose $n$ is the norm of an $\mathfrak{o}_K$-ideal $\mathfrak{n}$. If $r$ is* *represented by* $Q_{\mathfrak{a}}$*, then* $n$ *is represented by* $Q_{\mathfrak{a}}$ *iff* $\lambda_i(nr) = 1$ *for each* $i = 1, \ldots, m$.

*Proof.* Replacing $\mathfrak{a}$ by another ideal with the same narrow class, we may suppose $\mathfrak{a}$ to be coprime to $d_K$. By (2.17), $n$ will be represented by $Q_{\mathfrak{a}}$ iff there exists an element $\alpha$ of $\mathfrak{a}$ with

$$n\mathbf{N}\mathfrak{a} = N_{K/\mathbb{Q}}\alpha. \tag{2.33}$$

Suppose firstly that $n$ is represented by $Q_{\mathfrak{a}}$. As shown previously when establishing (2.29), $N_{K/\mathbb{Q}}\alpha \in \tilde{H}$, and thus $\lambda_i(n\mathbf{N}\mathfrak{a}) = 1$ for each $i$. Similarly we deduce that $\lambda_i(r\mathbf{N}\mathfrak{a}) = 1$ for all $i$, and so $\lambda_i(nr) = 1$ for all $i$.

Conversely, suppose each $\lambda_i(nr) = 1$. Since $r$ is represented by $Q_{\mathfrak{a}}$, from the above we know $\lambda_i(r\mathbf{N}\mathfrak{a}) = 1$ for each $i$ and so $\lambda_i(n\mathbf{N}\mathfrak{a}) = 1$ for each $i$. It therefore follows that $n\mathbf{N}\mathfrak{a}$ mod $(d_K)$ lies in $\tilde{H}$; hence, because (2.29) is an isomorphism, with $(C_K^+)^2 = \{1\}$, by hypothesis we deduce that $\mathfrak{n}\mathfrak{a} = b\mathfrak{o}_K$ for some totally positive element $b$ of $\mathfrak{a}$. Taking norms we conclude

$$n = \mathbf{N}\mathfrak{a}^{-1}N_{K/\mathbb{Q}}b$$

and so $n$ is represented by $Q_{\mathfrak{a}}$.

□

Note that since the norm form $q_K$ obviously represents 1, (2.32) gives a particularly clean result for that form.

To illustrate the above ideas in practice, we again briefly consider the particular case $d = -6$. Then $K = \mathbb{Q}(\sqrt{-6})$, $d_K = -24$, and, as has already been seen, $C_K^+ = C_K$ has order 2. We write $\mathfrak{p}_2, \mathfrak{p}_3$ for the unique prime ideals above 2, 3 resp; observe that $(-1, \sqrt{-6})$, $(-2, \sqrt{-6})$, $(-3, \sqrt{-6})$, $(-6, \sqrt{-6})$ are normalised bases for $\mathfrak{o}_K$, $\mathfrak{p}_2$, $\mathfrak{p}_3$, $\mathfrak{p}_2\mathfrak{p}_3$ resp. Thus the equivalence classes with discriminant $-24$, are represented by

$$q_K = Q_{\mathfrak{p}_2\mathfrak{p}_3} = X^2 + 6Y^2$$

$$Q_{\mathfrak{p}_2} = Q_{\mathfrak{p}_3} = 2X^2 + 3Y^2.$$

A prime number $p$, coprime to 6, will be an ideal norm iff $p$ is split in $K$. By (III.2.29) this occurs iff $\left(\frac{-6}{p}\right) = 1$, i.e. by the quadratic reciprocity law iff $p \equiv 1, 5, 7, 11 \bmod (24)$. Since

$$\left(\frac{-6}{p}\right) = +1 \iff \left(\frac{-2}{p}\right) = \left(\frac{3}{p}\right) \iff \left(\frac{2}{p}\right) = \left(\frac{p}{3}\right)$$

we deduce that $\lambda_K \to \lambda_3\lambda_8$ with $\lambda_3(p) = \left(\frac{p}{3}\right)$ and $\lambda_8(p) = (-1)^{(p^2-1)/8}$.

Clearly $Q_{\mathfrak{p}_2}$ represents 5 by $(x, y) = (1, 1)$; thus we can conclude that a prime number $p \equiv 1, 5, 7, 11 \bmod (24)$ will be represented by $q_K$ iff

$$\lambda_3(p) = \lambda_8(p) = 1$$

by $Q_{\mathfrak{p}_2}$ iff

$$\lambda_3(5p) = \lambda_8(5p) = 1$$

i.e.

$$\lambda_3(p) = \lambda_8(p) = -1.$$

In terms of congruences, such a prime will therefore be represented by $q_K$ iff $p \equiv 1, 7 \bmod (24)$, and will be represented by $Q_{\mathfrak{p}_2}$ iff $p \equiv 5, 11 \bmod (24)$.

## §3 Cubic equations

In this section we shall consider two types of cubic diophantine equations. In both cases we shall see that the nature of the criteria we obtain, together with the degree of difficulty involved, is quite different from the quadratic results of the previous section. As in the previous section, the results obtained here provide striking illustrations of the power of the techniques that we have now developed.

From the work in (VI,§5) we know that the set of rational solutions to

(3.1) $$Y^2 = X^3 + k$$

together with the point at infinity, form a group for any given non-zero integer $k$. Here we consider the problem of determining *integral* solutions to such an equation. Note that from the formula for the sum of two points (see exercise 5 in Chapter VI), it is clear that the sum of two integral solutions will not in general be an integral solution.

The method which we use goes through in a similar way, without great changes, for many, though by no means all, $k$. As an instance of the difficulties that can arise, the enthusiastic reader is invited to consider the case when $k = 17$, where there is a solution

$$(x, y) = (5234, \pm 378, 661).$$

Provided $k \neq 0$, it follows from a general result of Siegel that (3.1) can only have a finite number of integral solutions; we shall not, however, rely on this result here. We now show

**(3.2)** *The only integral solutions to* $Y^2 = X^3 - 13$ *are* $(x, y) = (17, \pm 70)$.

*Proof.* We put $K = \mathbb{Q}(\sqrt{-13})$; so $\mathfrak{o}_K = \mathbb{Z}[\sqrt{-13}]$, $d_K = -52$; since $K$ is imaginary $C_K^+ = C_K$, and using the Minkowski bound we quickly check that $C_K$ has order 2.

Let $(x, y)$ denote an integral solution to the given equation; this then yields a factorisation of $\mathbb{Z}[\sqrt{-13}]$ ideals

$$(x)^3 = (y + \sqrt{-13})(y - \sqrt{-13}). \tag{3.3}$$

Let $\mathfrak{b} = (y + \sqrt{-13}, y - \sqrt{-13})$, and note $\mathfrak{b} \supset (2\sqrt{-13})$. Thus $\mathfrak{b} = \mathfrak{p}_2^\alpha \mathfrak{p}_{13}^\beta$ where $\mathfrak{p}_2, \mathfrak{p}_{13}$ denote the unique $\mathfrak{o}_K$ prime ideals dividing 2, 13 respectively. Thus we can write $(y + \sqrt{-13}) = \mathfrak{c}\mathfrak{p}_2^\alpha \mathfrak{p}_{13}^\beta$ and $(y - \sqrt{-13}) = \mathfrak{c}'\mathfrak{p}_2^{\alpha'} \mathfrak{p}_{13}^{\beta'}$ with each of $\mathfrak{c}$ and $\mathfrak{c}'$ coprime to $\mathfrak{p}_2\mathfrak{p}_{13}$. Applying $\sigma$, the non-trivial automorphism of $K$, we see that $\mathfrak{c}'\mathfrak{p}_2^{\alpha'} \mathfrak{p}_{13}^{\beta'} = \mathfrak{c}^\sigma \mathfrak{p}_2^\alpha \mathfrak{p}_{13}^\beta$, and we conclude that $\mathfrak{c}' = \mathfrak{c}^\sigma$, $\alpha = \alpha'$, $\beta = \beta'$. By the definition of $\mathfrak{b}$, together with the fact that $\mathfrak{b} \supset (2\sqrt{-13})$, it follows that $\mathfrak{c}$ and $\mathfrak{c}^\sigma$ are coprime; however, by (3.3), $\mathfrak{c}\mathfrak{c}^\sigma \mathfrak{p}_2^{2\alpha} \mathfrak{p}_{13}^{2\beta}$ must be a cube, and so $\mathfrak{c}$ itself must be a cube, $\mathfrak{c} = \mathfrak{a}^3$ say, with $\alpha = 3c$, $\beta = 3d$ for integers $c, d$:

$$\begin{aligned} (y + \sqrt{-13})\mathfrak{o}_K &= \mathfrak{a}^3 \mathfrak{p}_2^{3c} \mathfrak{p}_{13}^{3d} \\ (y - \sqrt{-13})\mathfrak{o}_K &= \mathfrak{a}^{3\sigma} \mathfrak{p}_2^{3c} \mathfrak{p}_{13}^{3d} \end{aligned} \tag{3.4}$$

where $\mathfrak{a}$ is an $\mathfrak{o}_K$-ideal with the property that $\mathfrak{a} + \mathfrak{a}^\sigma = \mathfrak{o}_K$. Since $3 \nmid h_K$, we deduce from (3.4) that $\mathfrak{a}\mathfrak{p}_2^c \mathfrak{p}_{13}^d$ must be principal, $(y + \sqrt{-13})\mathfrak{o}_K = (a + b\sqrt{-13})^3 \mathfrak{o}_K$. Moreover, since $U_K = \{\pm 1\}$, by changing the sign of $a$ and $b$ if necessary, we can put $y + \sqrt{-13} = (a + b\sqrt{-13})^3$. Equating coefficients of 1 and $\sqrt{-13}$, we obtain equalities

$$y = a^3 - 39ab^2 \qquad 1 = b(3a^2 - 13b^2). \tag{3.5}$$

Hence we conclude $b = \pm 1$; we now consider these two possibilities:

if $b = +1$, then $1 = 3a^2 - 13$, which is absurd
if $b = -1$, then $1 = -3a^2 + 13$, and so $a^2 = 4$, i.e. $a = \pm 2$.

Substituting back in (3.5), we obtain $y = \pm 70$, and so $x^3 = 70^2 + 13 = 4913$, as required.

□

The above provides an instance where it is the arithmetic of a quadratic number field which is central to the solution of a cubic equation. In general, however, it is the arithmetic of cubic fields which dominates the solution of such equations. This is, in particular, true of the problem which we now treat: we consider the integral solutions of

$$X^3 + dY^3 = 1 \tag{3.6}$$

for a cube-free integer $d > 0$. Clearly if $d = 1$, then this clearly has no non-trivial integral solutions.

Prior to our considering (3.6), we begin by extending Dirichlet's Unit Theorem to subrings of rings of integers of finite index. Let $\mathfrak{o}_K$ denote, as usual, the ring of integers of a number field $K$; we call a subring $\mathfrak{A}$ of $\mathfrak{o}_K$ an *order in* $K$ if the index $[\mathfrak{o}_K : \mathfrak{A}]$ is finite. (See Ex. 6 to Chapter II). The study of orders is a theme which recurs throughout algebraic number theory. We have already seen, in solving Pell's equation, that the unit group of an order can play a vital role in solving diophantine problems. In the previous section we restricted attention to quadratic forms $Q$ with $d_Q = d_K$; the study of the general case (with $d_Q = d_K f^2$ as in (2.12)) again involves a very substantial use of orders.

Given an order $\mathfrak{A}$, we write $U(\mathfrak{A})$ for the group of units of $\mathfrak{A}$. We note

**(3.7)** $U(\mathfrak{A}) = \mathfrak{A} \cap U_K$.

*Proof.* The inclusion $U(\mathfrak{A}) \subset \mathfrak{A} \cap U_K$ is obvious. Conversely, let $u \in U_K \cap \mathfrak{A}$; we must exhibit an inverse to $u$ in $\mathfrak{A}$. Since $u \in U_K$, it has a characteristic polynomial of the form

$$X^n + a_1X^{n-1} + \cdots + a_{n-1}X \pm 1 \qquad a_i \in \mathbb{Z}$$

and so, since $u$ is a root of its characteristic polynomial, we have

$$u(u^{n-1} + a_1u^{n-2} + \cdots + a_{n-1}) = \mp 1.$$

□

The relationship between the $U_K$ and $U(\mathfrak{A})$ is clarified somewhat by

**(3.8)** *Suppose* $(\mathfrak{o}_K : \mathfrak{A}) = m$ *and* $j = |(\mathfrak{o}_K/\mathfrak{o}_K m)^*|$; *then* $(U_K)^j$ *is a subgroup of* $U(\mathfrak{A})$ *and so, in particular,* $U(\mathfrak{A})$ *has the same* $\mathbb{Z}$*-rank as* $U_K$.

*Proof.* Reduction mod $m\mathfrak{o}_K$ yields a homomorphism

$$\rho: U_K \to (\mathfrak{o}_K/m\mathfrak{o}_K)^*.$$

Thus, for $v \in U_K$, $\rho(v^j) = \rho(v)^j = 1$, and therefore $v^j \in 1 + m\mathfrak{o}_K \subset \mathfrak{A}$.

□

Next we write the cube-free integer $d > 1$ as $d = fg^2$, where $f$ and $g$ are both positive square-free integers; we then write $\overline{d} = f^2g$, and we let $\theta$ denote a solution to $X^3 = d$. We write $\overline{\theta} = fg\theta^{-1}$, so that $\overline{\theta}^3 = \overline{d}$; we set $K = \mathbb{Q}(\theta)$ and note that $\overline{\theta} \in K$. We shall view $K$ as a subfield of $\mathbb{R}$ via the unique real embedding $K \hookrightarrow \mathbb{R}$.

Clearly we have a chain of orders in $K$

$$\mathbb{Z}[\theta] \subset \mathbb{Z}[\theta, \overline{\theta}] \subset \mathfrak{o}_K. \tag{3.9}$$

The relationship between the latter two orders is made clear by the following result of Dedekind:

**(3.10)** *If* $d \not\equiv \pm 1 \bmod (9)$, *then* $\mathfrak{o}_K = \mathbb{Z} + \mathbb{Z}\theta + \mathbb{Z}\overline{\theta}$. *If* $d \equiv \pm 1 \bmod (9)$, *then* $\mathfrak{o}_K = \mathbb{Z} \cdot \frac{1+f\theta+g\overline{\theta}}{3} + \mathbb{Z}\theta + \mathbb{Z}\overline{\theta}$.

*Proof.* From (I.1.5) we know that $\mathrm{Disc}(X^3 - d) = -27d^2$. For any prime number $p$ with $p \nmid 3d$, from (III.2.20) it follows that $\mathfrak{d}(\mathbb{Z}[\theta])_p = \mathbb{Z}_p$, and so by (3.9) $\mathbb{Z}[\theta]_p = \mathfrak{o}_{K,p}$.

Let $\mathfrak{A}$ now denote the order $\mathbb{Z}[\theta, \overline{\theta}]$. If $p \mid f$, then $\theta$ is a uniformising parameter for $K$ at $p$, and so by Theorem 24, $\mathbb{Z}[\theta]_p = \mathfrak{A}_p = \mathfrak{o}_{K,p}$. Similarly if $p \mid g$, then $\mathbb{Z}[\overline{\theta}]_p = \mathfrak{A}_p = \mathfrak{o}_{K,p}$. Thus it now remains to determine under what circumstances $\mathfrak{A}_3 = \mathfrak{o}_{K,3}$ when $3 \nmid fg$. This then implies that $3 \mid f \pm g$ for some choice of sign. If $3 \| f \pm g$, then we note that

$$N_{K/\mathbb{Q}}(\theta \pm \overline{\theta}) = f^2 g \pm g f^2 = fg(f \pm g).$$

It then follows easily that, in this case, $\theta \pm \overline{\theta}$ is a root of an Eisenstein polynomial over $\mathbb{Q}_3$. Hence, by Theorem 24, $\mathfrak{o}_{K,3} = \mathbb{Z}_3[\theta \pm \overline{\theta}]$. So now we may suppose that $3 \nmid d$ and that $f \equiv \pm g \bmod (9)$, i.e. $f^2 \equiv g^2 \bmod (9)$. Since $\mathfrak{d}(\mathfrak{A}) = \mathfrak{d}(\mathfrak{o}_K)[\mathfrak{o}_K : \mathfrak{A}]^2$, and since

$$\mathfrak{d}(\mathfrak{A}) \mid \mathfrak{d}(\mathbb{Z}[\theta]) = 27d^2\mathbb{Z}$$

we see that $[\mathfrak{o}_K : \mathfrak{A}]^2$ divides 27, and so

$$[\mathfrak{o}_K : \mathfrak{A}] = 1 \text{ or } 3.$$

It will therefore suffice to show that under the above hypotheses $\alpha = \frac{1}{3}(1 + f\theta + g\overline{\theta})$ is an algebraic integer. To show this we calculate the characteristic polynomial of $\alpha$. Using the relations $\theta\overline{\theta} = fg$, $\theta^2 = g\overline{\theta}$, $\overline{\theta}^2 = f\theta$, we have

$$(3.11) \qquad \begin{cases} \alpha = \dfrac{1}{3} + \dfrac{f}{3}\theta + \dfrac{g}{3}\overline{\theta} \\[2ex] \alpha\theta = \dfrac{fg^2}{3} + \dfrac{1}{3}\theta + \dfrac{fg}{3}\overline{\theta} \\[2ex] \alpha\overline{\theta} = \dfrac{f^2 g}{3} + \dfrac{fg}{3}\theta + \dfrac{1}{3}\overline{\theta} \end{cases}$$

and so
(3.12)

$$c_{\alpha,K/\mathbb{Q}}(X) = X^3 - X^2 + \frac{1}{3}(1 - f^2g^2)X - \frac{1}{27}(1 + g^4 f^2 + f^4 g^2 - 3f^2 g^2).$$

We write $f^2 = 1 + 3\lambda$, $g^2 = 1 + 3\mu$, so that by hypothesis $\lambda \equiv \mu \bmod (3)$. It is then clear from (3.12) that $\alpha$ is an algebraic integer since we have a congruence mod (27)

$$-27c_{\alpha,K/\mathbb{Q}}(0) \equiv 9(\lambda^2 + \mu\lambda + \mu^2) \equiv 0.$$

This shows that $\mathfrak{o}_K$ differs from $\mathfrak{A}$ iff $3 \nmid fg$ and $f^2 \equiv g^2 \bmod (9)$. Since $d = fg^2$, this occurs iff $d \equiv \pm 1 \bmod (9)$.

To complete the proof we note that since $\theta\overline{\theta} \in \mathbb{Z}$, $\{1, \theta, \overline{\theta}\}$ is a $\mathbb{Z}$-basis of $\mathbb{Z}[\theta, \overline{\theta}]$. In addition if $d \equiv \pm 1 \bmod (9)$, then

$$1 = 3 \cdot \left(\frac{1 + f\theta + g\overline{\theta}}{3}\right) - f\theta - g\overline{\theta}$$

and so

$$\mathfrak{o}_K = \mathbb{Z} \cdot \left(\frac{1 + f\theta + g\overline{\theta}}{3}\right) + \mathbb{Z}\theta + \mathbb{Z}\overline{\theta}.$$

□

After the above preliminaries, we now return to the equation (3.6).

**Theorem 60 (Delaunay-Nagell).** *The equation $X^3 + dY^3 = 1$ has at most one non-trivial solution (i.e. apart from $(x, y) = (1, 0)$).*

Before embarking on the proof, we first make a number of comments: firstly, for ease of presentation we shall only prove the theorem for $d > 10$ (See [C2] for a local proof which does not require $d > 10$). Secondly, for the reader's interest we mention that if $(x, y)$ is a non-trivial solution, then $x + y\theta$ is a fundamental unit of the order $\mathbb{Z}[\theta]$; however, we shall not prove this result here. Note though, as Nagell observed, this unit will not in general be fundamental in $\mathbb{Z}[\theta, \overline{\theta}]$, as is shown by the equality

$$-19 + 7\sqrt[3]{20} = (1 + \sqrt[3]{20} - \sqrt[3]{50})^2.$$

Lastly, we note the striking difference between the above result and our result for Pell's equation where we showed that there are always an infinite number of solutions; the crucial difference here is that we are not just concerned with a unit group of an order, but rather with its intersection with the 2-dimensional subspace of $K$ given by $\mathbb{Q} + \mathbb{Q}\theta$.

As a first step in the proof of the theorem we establish the following result, which is of some interest in its own right:

**(3.14)** *With the above notation and hypotheses (and so in particular assuming $d > 10$), suppose $(a\theta + c)^n = A\theta + C$ for integers $a, c, A, C$ with $(ad, c) = 1$, and with $n \geq 1$; then $n$ must necessarily be 1.*

*Remark.* The condition $d > 10$ is crucial here since $(\sqrt[3]{10} - 1)^5 = 99 - 45\sqrt[3]{10}$.

*Proof.* Suppose for contradiction $n > 1$. Using the basis $\{1, \theta, \theta^2\}$ of $K$ and equating coefficients of $\theta^2$ in the expansion of $(a\theta + c)^n = A\theta + C$, we see that

$$\binom{n}{2} c^{n-2} a^2 + \binom{n}{5} c^{n-5} a^5 d + \cdots = 0. \tag{3.15}$$

So, on dividing by $\binom{n}{2} a^2$,

$$-c^{n-2} = \sum_{k \geq 1} \binom{n-2}{3k} \frac{2c^{n-3k-2} a^{3k} d^k}{(3k+1)(3k+2)}. \tag{3.16}$$

If 3 divided $d$, then 3 would divide all terms in the right-hand side of (3.16), and so $3 \mid c$ which would contradict $(ad, c) = 1$. So from now on we may suppose $3 \nmid d$.

Next choose a prime $q$ dividing $d$, and suppose $q^v \| d$. Since $d$ is cube-free, it follows that $v = 1$ or 2. We observe that if $q^v > 2$, then for $k \geq 2$

$$q^{vk} \geq 3^k > 3k + 2. \tag{3.17}$$

Moreover, provided $q^v$ is not 2, 4, nor 5, then the inequality $q^{vk} > 3k + 2$ also holds for $k = 1$. Thus, provided $q^v$ takes some value distinct from these three, then, because the denominator terms $3k + 1$, $3k + 2$ are coprime, we can deduce that each term on the right of (3.16) is divisible by $q$; hence $q \mid c$ and this contradicts $(ad, c) = 1$. In fact we can reason in the same way when $q^v = 4$: for, by (3.17), all terms with $k \geq 2$ are even, while when $k = 1$, we have denominator 20 and the numerator is divisible by 8. Lastly note that since $d > 10$, we can always find some $q$ such that $q^v$ is distinct from both 2 and 5.

□

We now prove Theorem 60. Let $u$ denote the fundamental unit of the order $\mathfrak{A} = \mathbb{Z}[\theta, \bar{\theta}]$ with $0 < u < 1$. Let $v = x + y\theta$ where $(x, y)$ is a non-trivial solution to (3.6). Clearly $v$ is an algebraic integer which in fact lies in $\mathfrak{A}$, and furthermore $v \in U_K$ since

$$N_{K/\mathbb{Q}} v = \prod_{i=0}^{2} (x + \omega^i \theta y) = x^3 + dy^3 = 1 \tag{3.18}$$

where $\omega^3 = 1$, $\omega \neq 1$. Thus, by (3.7), $v \in U(\mathfrak{A})$.

Since $x^3$ and $dy^3$ are non-zero integers with sum 1, they must have different signs; hence so must their real cube roots, and so, using the inequality $|\,|a| - |b|\,| < |a - b|$ for complex numbers $a, b$ with $\{o, a, b\}$ not collinear,

$$|x + \omega\theta y| > |x + \theta y|.$$

Here, of course, $|.|$ denotes the usual absolute value of $\mathbb{C}$. Thus, by (3.18), $0 < v < 1$. We can therefore find a non-negative integer $m$ such that $v = u^m$. We now choose the integral solution to (3.6), $(x, y)$, which yields the minimal such $m$, and we suppose for contradiction that we have a further non-trivial integral solution $(\lambda, \mu)$ with $\lambda + \mu\theta = u^M$. By the minimality condition $M > m$, and so we can write $M = mn + r$ with $0 \leq r < m$. Since $(x, yd) = 1 = (\lambda, \mu d)$, it follows immediately from (3.14) that $r \neq 0$.

We now set

$$\epsilon = u^r = \alpha + \beta\theta + \gamma\overline{\theta}$$

and so deduce the equality

$$(\lambda + \mu\theta) = (x + y\theta)^n(\alpha + \beta\theta + \gamma\overline{\theta}).$$

Using the basis $\{1, \theta, \overline{\theta}\}$ of $K$, equating coefficients for $\overline{\theta}$, and using $\theta^3 = fg^2$, $\theta\overline{\theta} = fg$, we obtain

$$\begin{aligned} 0 = \alpha g \left[\binom{n}{2} y^2 x^{n-2} + \cdots\right] + \beta g \left[\binom{n}{1} x^{n-1} y + \cdots\right] \\ + \gamma\left[x^n + \binom{n}{3} x^{n-3} y^3 d + \cdots\right]. \end{aligned}$$

We therefore conclude that $\gamma x^n \equiv 0 \bmod (gy)$; however, $x^3 + dy^3 = 1$ implies $x^3 \equiv 1 \bmod (gy)$, and so we have shown

$$\gamma \equiv 0 \bmod (gy). \tag{3.19}$$

As previously, we continue to view $K$ as a subfield of $\mathbb{R}$. We let $\sigma\colon K \hookrightarrow \mathbb{C}$ be the imaginary embedding with $\theta^\sigma = \omega\theta$, where $\omega^3 = 1$, $\omega \neq 1$; and we write $\tau\colon K \hookrightarrow \mathbb{C}$ for the imaginary embedding obtained by composing $\sigma$ with complex conjugation, so that $\theta^\tau = \omega^2\theta$. We note that since $|x| = |v - y\theta| \leq |v| + |y|\theta \leq 1 + |y|\theta$, then

$$\begin{aligned} |v^\sigma|^2 &= |x^2 - xy\theta + y^2\theta^2| \\ &\leq 1 + 2|y|\theta + y^2\theta^2 + |y|\theta + y^2\theta^2 + y^2\theta^2 \\ &\leq 3(1 + |y|\theta)^2 \end{aligned}$$

and so

$$|v^\sigma| \leq \sqrt{3}(1 + |y|\theta). \tag{3.20}$$

Furthermore since $\epsilon = u^r$, $v = u^m$ with $r < m$, we note that

$$\begin{cases} v < \epsilon < 1 \\ 1 < |\epsilon^\sigma| = |\epsilon^\tau| < |v^\sigma| = |v^\tau|. \end{cases} \tag{3.21}$$

Re-choosing $\sigma$ and $\tau$ if necessary, we can write

$$\begin{aligned} \epsilon^\sigma &= \alpha + \omega\beta\theta + \omega^2\gamma\overline{\theta} \\ \epsilon^\tau &= \alpha + \omega^2\beta\theta + \omega\gamma\overline{\theta} \end{aligned}$$

hence

$$3\gamma\bar{\theta} = \epsilon + \epsilon^{\sigma}\omega + \epsilon^{\tau}\omega^2. \tag{3.22}$$

By the minimality condition on $m$, $\gamma$ must be non-zero; therefore, from (3.19), it follows that

$$|3gy\bar{\theta}| \leq |\epsilon| + |\epsilon^{\sigma}| + |\epsilon^{\tau}|$$

by (3.21)

$$\leq 1 + 2|v^{\sigma}|$$

by (3.20)

$$\leq 1 + 2\sqrt{3}(1 + |y|\theta). \tag{3.23}$$

Next note that if $\xi$, $\eta$ are real numbers with $\xi \geq 1$, $\eta \geq 2$; then $6 - 2\sqrt{3} > 5/2$ implies $3\eta^2 - 2\sqrt{3}\eta > 5$ and so

$$3\xi\eta^2 - 2\sqrt{3}\xi\eta > 5. \tag{3.24}$$

Noting that $g\bar{\theta} = \theta^2$, (3.23) gives

$$3|y|\theta^2 < 1 + 2\sqrt{3}(1 + |y|\theta).$$

This then contradicts (3.24) if we set $\eta = \theta > 2$ (since $d > 10$) and $\xi = |y|$. We therefore conclude that (3.6) has at most one non-trivial solution.

□

# VIII

## L- functions

In this chapter we shall develop and apply a range of analytic methods in order to obtain some very powerful arithmetic results. As has already been mentioned in the Introduction, we shall obtain a limit formula (in §2) which provides us with vital information on class groups and units; this theorem is central to this chapter and underpins the majority of results in the remaining sections. We then introduce the notion of a Dirichlet L-function and we use it to prove the celebrated Dirichlet theorem on primes in arithmetic progressions. Next we combine L-function methods with the limit formula of §2 to obtain a number of beautiful results for cyclotomic and quadratic number fields. These sharpen considerably our earlier work on fields of low degree.

### §1 Dirichlet Series

Throughout the whole of this chapter $x$ will always denote a real variable. Given a positive real number $r$, in the usual way $r^x$ denotes the function $\exp(x \log(r))$.

A *Dirichlet series* is a series $\sum_{n=1}^{\infty} a_n n^{-x}$ where $\{a_n\}_{n=1}^{\infty}$ is a sequence of complex numbers. In one way or another virtually all the results of this chapter depend on the notion of a Dirichlet series; their basic properties needed subsequently will be derived in the present section. Throughout, we shall always make free use of standard theorems from analysis.

**(1.1)** *If $\sum_1^\infty a_n n^{-x}$ converges absolutely for $x = x_0$, then the series converges absolutely for all $x \geq x_0$ ; moreover the series converges uniformly on $[x_0, \infty)$ and so defines a continuous function thereon.*

*Proof.* Assume $x > x_0$ and consider $a_n \neq 0$; then

$$\frac{|a_n n^{-x}|}{|a_n n^{-x_0}|} = n^{x_0 - x} \to 0 \text{ as } n \to \infty$$

and so $\sum a_n n^{-x}$ converges by the Comparison Test. Since

$$0 \leq |a_n n^{-x}| \leq |a_n n^{-x_0}|$$

we can apply the Weierstrass $M$-test and deduce that the series converges uniformly on $[x, \infty)$. The continuity of the function defined by the series on this interval then follows from the fact that a uniform limit of continuous functions is continuous.

□

*Example.* The series $\sum_1^\infty n^{-x}$ is known to converge for $x > 1$, but to diverge at $x = 1$. We denote the function so defined on $(1, \infty)$ by $\zeta(x)$; this is the so-called *Riemann zeta-function.*

Suppose now that the two Dirichlet series $\sum a_n n^{-x}$, $\sum b_n n^{-x}$ converge absolutely for $x \geq x_0$, and denote the functions that they define on $[x_0, \infty)$ by $A(x)$, $B(x)$ respectively. Define

$$c_n = \sum_{d|n} a_d b_{n/d}$$

where the sum extends over all positive divisors $d$ of $n$. Then by the Cauchy Theorem for products of series we know that

**(1.2)** $\sum c_n n^{-x}$ *converges absolutely for $x \geq x_0$ ; moreover the function $C(x)$ defined by this series equals $A(x)B(x)$.*

Next we give an extremely useful and practical criterion for establishing the conditional convergence of series.

**(1.3) (Abel Summation)** *Let $\{a_n\}$,$\{b_n\}$ be two sequences of complex numbers and let $A_n = \sum_{i=m+1}^n a_i$ ; then*

$$\sum_{i=m+1}^{n} a_i b_i = A_n(b_{n+1}) - \sum_{m+1}^{n} A_i(b_{i+1} - b_i) - A_m b_{m+1}.$$

*If furthermore, $\{b_n\}$ is real and monotonic decreasing with $lim_{n\to\infty}b_n = 0$ and if $|A_n|$ is bounded, then the series $\sum a_i b_i$ is convergent.*

*Proof.* We have the obvious identity

$$\sum_{m+1}^{n} a_i b_i = \sum_{m+1}^{n} (A_i - A_{i-1}) b_i$$

where we take $A_0 = 0$ when $m = 0$ (the empty sum). Hence

$$\begin{aligned}\sum_{m+1}^{n} a_i b_i &= \sum_{m+1}^{n} A_i b_i - \sum_{m+1}^{n} A_{i-1} b_i \\ &= \sum_{m+1}^{n} A_i b_i - \sum_{m+1}^{n} A_i b_{i+1} + A_n b_{n+1} - A_m b_{m+1}\end{aligned}$$

and this gives the first part.

Given $\varepsilon > 0$, we show that the partial sum satisfies

$$\left|\sum_{m+1}^{n} a_i b_i\right| < \varepsilon \quad \text{for } n > m > N(\varepsilon).$$

By hypothesis we can find $K > |A_n|$ for all $n$, and we choose $N$ sufficiently large to guarantee that $b_n < \varepsilon/4K$ for all $n \geq N$. Then by the first part

$$\begin{aligned}|\sum_{m+1}^{n} a_i b_i| &\leq |A_n| b_{n+1} + |A_m| b_{m+1} + \sum_{m+1}^{n} |A_r|(b_r - b_{r+1}) \\ &\leq K\frac{\varepsilon}{4K} + K\frac{\varepsilon}{4K} + K\sum_{m+1}^{n}(b_r - b_{r+1}) \\ &\leq \frac{\varepsilon}{2} + K(b_{m+1} - b_{n+1}) \leq \frac{3\varepsilon}{4} < \varepsilon.\end{aligned}$$

□

We now use this result to obtain a standard result on power series:

**(1.4) (Abel's Theorem)** *If $\sum a_n$ converges to $a$, then $\sum a_n x^n$ converges uniformly on $[0, 1]$; in particular, if $f(x)$ denotes the function so defined, then*

$$\lim_{x\to 1} f(x) = a.$$

*Proof.* For given $m$, set $A_n = \sum_{m+1}^{n} a_r$. Because $\sum a_i$ converges, we know that given $\varepsilon > 1$, we can find $N$ so that $|A_n| < \varepsilon/2$ whenever $n > m \geq N$. Now choose $x \in [0, 1]$ and set $b_r = x^r$ in (1.3); this then gives

$$\left|\sum_{m+1}^{n} a_r x^r\right| \leq |A_n b_{n+1}| + |A_m b_{m+1}| + \sum_{m+1}^{n} |A_i|\,|b_{i+1} - b_i|$$
$$\leq \frac{\varepsilon}{2}\left(x^{n+1} + x^{m+1} + x^{m+1} - x^{n+1}\right)$$
$$\leq \frac{\varepsilon}{2} \cdot 2x^{m+1} \leq \varepsilon.$$

Hence the series $\sum a_r x^r$ converges uniformly in $[0,1]$, and therefore the function $f(x)$ is continuous on $[0,1]$.

□

Next we apply (1.3) to Dirichlet series:

**(1.5)** *If the partial sums* $|\sum_1^n a_r|$ *are all bounded, then* $\sum a_r r^{-x}$ *converges for all* $x > 0$*; furthermore, for any* $y > 0$ *the convergence is uniform on* $[y, \infty)$*; hence the function defined by the series is continuous on* $(0, \infty)$.

*Proof.* The first part follows directly from (1.3) on setting $b_r = r^{-x}$. To show that the convergence is uniform on $[y, \infty)$, for $x \geq 0$, we put $a'_r = a_r r^{-y}$, $b'_r = r^{-x}$, $A'_n = \sum_{m+1}^n a'_r$. Then, by (1.3), for $n > m$

$$\sum_{m+1}^{n} a_r r^{-(x+y)} = A'_n b'_{n+1} - A'_m b'_{m+1} + \sum_{m+1}^{n} A'_i (b'_{i+1} - b'_i).$$

Since $\sum a_r r^{-y}$ is convergent, it follows that, given $\varepsilon > 0$, we can find $N$ sufficiently large to guarantee that $|A'_n| < \varepsilon/2$ for all $n > m \geq N$. Since the $b'_i$ are monotonic decreasing and are bounded above by 1:

$$\left|\sum_{m+1}^{n} a_r r^{-(x+y)}\right| \leq \frac{\varepsilon}{2}(b'_{n+1} + b'_{m+1} + b'_{m+1} - b'_{n+1}) \leq 2\frac{\varepsilon}{2} b'_{m+1} \leq \varepsilon$$

and so the convergence is indeed uniform in $x$.

□

As an instance of the kind of application that we have in mind for the above work, observe that (1.5) immediately shows that $\sum (-1)^n n^{-x}$ converges for all $x > 0$. On the other hand, we have seen previously that $\zeta(x) = \sum n^{-x}$ defines a continuous function on $(1, \infty)$, but that $\zeta(x)$ becomes unbounded as $x \to 1$.

More precisely we shall now show

**(1.6)** *Let* $\phi(x) = \zeta(x) - 1/(x-1)$*. Then* $\phi(x)$ *is continuous on* $(1, \infty)$

*and* $0 \le \phi(x) \le 1$. *In particular*

$$\lim_{x \to 1+} (x-1)\zeta(x) = 1.$$

Here $\lim_{x \to 1+}$ denotes the limit as $x$ tends to 1 from above.

*Proof.* The continuity of $\phi$ follows from that of $\zeta$. Now let $n \ge 1$, $x > 1$ and let $n \le y \le n+1$, so that

$$(n+1)^{-x} \le y^{-x} \le n^{-x}$$

hence

$$(n+1)^{-x} \le \int_n^{n+1} y^{-x}\, dy \le n^{-x}$$

and so, summing over all $n \ge 1$,

$$\zeta(x) - 1 \le \int_1^{\infty} y^{-x}\, dy \le \zeta(x).$$

whence

$$\zeta(x) - 1 \le \frac{1}{(x-1)} \le \zeta(x)$$

and this shows that $0 \le \phi(x) \le 1$. Therefore

$$\lim_{x \to 1+} (x-1)\phi(x) = 0$$

which gives the final part of the result.

□

*Euler products.* We begin this subsection by recalling a number of standard definitions and results on the convergence of products. A product $\prod_1^\infty (1+b_n)$, for a sequence of complex numbers $\{b_n\}$ with each $b_n \ne -1$, is said to *converge* if the partial products $\Pi_m = \prod_1^m (1+b_n)$ converge to a non-zero value. The product is said to *converge absolutely* if $\prod_1^\infty (1+|b_n|)$ converges. It is a standard fact that $\prod_1^\infty (1+b_n)$ converges if it converges absolutely; furthermore, an absolutely convergent product, when re-ordered, still converges absolutely to the same value.

The following is a very useful criterion for absolute convergence of products:

**(1.7)** $\prod_1^\infty (1+b_n)$ *converges absolutely iff the series* $\sum_1^\infty b_n$ *converges absolutely.*

*Proof.* Without loss of generality we may suppose that all $b_n \ge 0$. Put $\Pi_n = \prod_1^n (1+b_r)$, $\Sigma_n = \sum_1^n b_r$ and note that both $\{\Pi_n\}$ and $\{\Sigma_n\}$ are now both monotonic increasing sequences. Since $\Pi_n \geq 1+\Sigma_n$, it is clear that $\Sigma_n$ converges if $\Pi_n$ does.

For $b \geq 0$, we know from the series expansion of the exponential function that $e^b \geq 1+b$ and so $e^{\Sigma_n} \geq \Pi_n$; therefore, if $\Sigma_n$ converges, $\Pi_n$ is bounded above and hence is convergent.

□

A function $n \mapsto a_n$ will be called *multiplicative* if $a_{nm} = a_n a_m$ whenever the integers $n$, $m$ are coprime.

**(1.8)** *Let $\{a_n\}$ be a multiplicative sequence. The following two conditions are equivalent:*

(*a*) $\sum_{r=1}^{\infty} a_{p^r} p^{-rx}$ *converges absolutely for each prime number $p$ and the product $\prod_p \left(1+\sum_{r=1}^{\infty} |a_{p^r}| p^{-rx}\right)$ running over all primes $p$ converges.*

(*b*) $\sum_{n\geq 2} a_n n^{-x}$ *converges absolutely.*
*Furthermore, either of these two equivalent conditions implies*

(*c*) $\prod_p \left(1+\sum_{r\geq 1} a_{p^r} p^{-rx}\right) = 1+\sum_{n\geq 2} a_n n^{-x}$.

*Proof.* (*b*)⇒(*a*): Since $\sum a_{p^r} p^{-rx}$ is a subseries of $\sum a_n n^{-x}$, it converges absolutely. For any finite set $S$ of prime numbers

$$\prod_{p\in S} \left(1+\sum |a_{p^r}| p^{-rx}\right) = 1+\sum_{S} |a_n| n^{-x} \tag{1.8.d}$$

where $\sum_S$ ranges over all $n \geq 2$ which are divisible only by primes in $S$. The product in (*a*) is therefore convergent since the right-hand side of (1.8.*d*) is bounded above by $1+\sum_1^{\infty} |a_n| n^{-x}$.

(*a*)⇒(*b*). Let $C$ denote the value of the product in (*a*). Given $m$, let $S_m$ denote the set of primes which are less than or equal to $m$; then

$$1+\sum_{n=2}^{m} |a_n| n^{-x} \leq 1+\sum_{S_m} |a_n| n^{-x}$$

and by (1.8.*d*), the right-hand term is bounded above by $C$; hence the left-hand terms converge.

Suppose now that conditions (*a*) and (*b*) are satisfied. Arguing as in (1.8.*d*), we have

$$\prod_{p\in S_m} \left(1+\sum a_{p^r} p^{-rx}\right) = 1+\sum_{S_m} a_n n^{-x}. \tag{1.8.e}$$

Now let $m \to \infty$. Then both sequences in (1.8.*e*) converge, the left-hand sequence by (*a*), and the right-hand sequence by (*b*). This therefore gives (*c*).

□

The product on the left hand side of (1.8.c) is called the *Euler product* of the function defined by the right-hand side.

## §2 The Dedekind zeta function

Throughout this section $x$ will denote a real variable in the interval $(1, \infty)$, and $K$ will denote a number field of degree $n$ over $\mathbb{Q}$. The Dedekind zeta function of $K$ is the Dirichlet series

$$\zeta_K(x) = \sum \mathbf{N}\mathfrak{a}^{-x} \tag{2.1}$$

where the sum runs over all non-zero $\mathfrak{o}_K$-ideals $\mathfrak{a}$. This function contains a considerable amount of arithmetic information some of which will be displayed in this section. We begin by first considering when this series converges.

**(2.2)** *$\zeta_K(x)$ converges for all $x > 1$ and, moreover, for such $x$*

$$\zeta_K(x) = \prod_{\mathfrak{p}} (1 - \mathbf{N}\mathfrak{p}^{-x})^{-1}$$

*where the product extends over all prime ideals of $\mathfrak{o}_K$.*

*Proof* Consider first the right-hand side product in (2.2) and note that

$$1 + \mathbf{N}\mathfrak{p}^{-x} < (1 - \mathbf{N}\mathfrak{p}^{-x})^{-1} \leq (1 + 2\mathbf{N}\mathfrak{p}^{-x}).$$

If $\mathfrak{p}$ lies above the rational prime number $p$, then $\mathbf{N}\mathfrak{p} = p^f$ for some integer $f$, and moreover for each prime number $p$ there are at most $(K : \mathbb{Q})$ prime ideals $\mathfrak{p}$ above $p$ (see Theorem 20). Thus, on writing $(1 - \mathbf{N}\mathfrak{p}^{-x})^{-1} = 1 + b_{\mathfrak{p}}$, we see that

$$\sum_{\mathfrak{p}} b_{\mathfrak{p}} \leq 2(K : \mathbb{Q}) \sum_{p} p^{-x}$$

where the left-hand sum runs over all prime ideals of $\mathfrak{o}_K$ and the right-hand sum runs over all prime numbers $p$. The right-hand series converges for $x > 1$ by comparison with the series $2(K : \mathbb{Q}) \sum n^{-x}$. From (1.7) we conclude that the product in (2.2) is absolutely convergent. Next observe that

$$(1 - \mathbf{N}\mathfrak{p}^{-x})^{-1} = 1 + \sum_{r=1}^{\infty} \mathbf{N}\mathfrak{p}^{-rx},$$

so, by the implication (a)⇒(b) in (1.8), together with the unique factorisation of ideals, we see that the series $\sum \mathbf{N}\mathfrak{a}^{-x}$ converges absolutely. Finally, the equality in (2.2) follows from (1.8.c).

□

In §1 we have seen that $\lim_{x\to 1+}(x-1)\zeta_{\mathbf{Q}}(x) = 1$. The main purpose of this section is to obtain the corresponding result for $\zeta_K(x)$:

**Theorem 61**

$$\lim_{x\to 1+}(x-1)\zeta_K(x) = \frac{2^{s+t}\pi^t R_K h_K}{W_K|d_K|^{1/2}}.$$

With the notation of earlier chapters: $s$ (resp. $2t$) is the number of real (resp. imaginary) embeddings of $K$; $h_K$ is the class number of $K$; $R_K$ is the regulator of $K$; $W_K$ is the number of roots of unity in $K$; and $d_K$ is the discriminant of $K$.

At the same time as proving Theorem 61, we shall also be able to study the function $M_K(x)$, which is defined as the number of $\mathfrak{o}_K$-ideals $\mathfrak{a}$ with absolute norm $\mathbf{N}\mathfrak{a} \le x$. (Recall that from (IV.1.4) we know that $M_K(x)$ is always finite). A subsidiary aim of this section will be to show

**Theorem 62**

$$\lim_{x\to\infty}\frac{M_K(x)}{x} = \frac{2^{s+t}\pi^t R_K h_K}{W_K|d_K|^{1/2}}.$$

It is interesting to note that this result tells us that, in the limit, $M_K(x)$ grows linearly with $x$.

Although in fact we prove Theorem 62 on the way towards proving Theorem 61, we could deduce Theorem 62 from Theorem 61 straight away by applying a standard Tauberian theorem from analysis.

Let $c$ denote a given class in the class group $C_K$, and choose an $\mathfrak{o}_K$-ideal $\mathfrak{b}$ in $c^{-1}$. Then the map $\mathfrak{a} \mapsto \mathfrak{a}\mathfrak{b}$ gives a bijection between $\mathfrak{o}_K$-ideals with class $c$ and principal $\mathfrak{o}_K$-ideals which are divisible by $\mathfrak{b}$. We define the partial zeta function by

$$\zeta_c(x) = \sum N\mathfrak{a}^{-x}$$

with the sum running over all $\mathfrak{o}_K$-ideals $\mathfrak{a}$ with class $c$. By comparison with $\zeta_K(x)$, we see that $\zeta_c(x)$ converges absolutely for $x > 1$; furthermore, because $\zeta_K(x)$ converges absolutely

(2.3) $$\zeta_K(x) = \sum_{c\in C_K}\zeta_c(x).$$

By the above remarks we can rewrite $\zeta_c(x)$ as

$$\zeta_c(x) = \mathbf{N}\mathfrak{b}^x \sum_{\alpha\mathfrak{o}_K\subset\mathfrak{b}} \mathbf{N}(\alpha\mathfrak{o}_K)^{-x}$$

(2.3.a) $$= \mathbf{N}\mathfrak{b}^x {\sum_{\alpha\in\mathfrak{b}}}' |N_{K/\mathbf{Q}}(\alpha)|^{-x}$$

where $\sum'$ denotes the sum over a system of associates of the non-zero elements of $\mathfrak{b}$. Thus our immediate task is to try to describe a fundamental domain which will naturally parameterise such a system of associates. With this in mind we recall from Chapter IV that we have maps:

$$(2.4)\qquad \begin{array}{ccc} K^* & \xrightarrow{\phi} & (\mathbb{R}^{(s)} \times \mathbb{C}^{(t)})^* \\ & \searrow^{\psi} & \downarrow L \\ & & \mathbb{R}^{(s+t)} \end{array}$$

where for $a \in K^*$

(2.4.*a*) $\phi(a) = (a^{\sigma_1}, \ldots, a^{\sigma_{s+t}})$

(2.4.*b*) $\psi(a) = (\log|a^{\sigma_1}|, \ldots, \log|a^{\sigma_s}|, 2\log|a^{\sigma_{s+1}}|, \ldots, 2\log|a^{\sigma_{s+t}}|)$

and where for $\mathbf{y} = (y_1, \ldots, y_{s+t}) \in (\mathbb{R}^{(s)} \times \mathbb{C}^{(t)})^*$

(2.4.*c*) $L(\mathbf{y}) = (\log|y_1|, \ldots, \log|y_s|, 2\log|y_{s+1}|, \ldots, 2\log|y_{s+t}|)\,.$

We define two vectors in $\mathbb{R}^{(s+t)}$

$$\mathbf{l}_0 = (1, \ldots, 1) \qquad \mathbf{l}_{s+t} = (\underbrace{1, \ldots, 1}_{s}, \underbrace{2, \ldots, 2}_{t})$$

thus for $a \in K^*$

$$\begin{aligned} \psi(a) \cdot \mathbf{l}_0 &= \sum_{i=1}^{s} \log|a^{\sigma_i}| + \sum_{i=s+1}^{s+t} 2\log|a^{\sigma_i}| \\ &= \sum_{i=1}^{n} \log|a^{\sigma_i}| \\ &= \log|N_{K/\mathbb{Q}}(a)|. \end{aligned} \qquad (2.5)$$

We write $\Lambda$ for the group $\psi(U_K)$; from (IV.4.6) we know that $\Lambda$ is a lattice in the hyperplane $H \subset \mathbb{R}^{(s+t)}$ which is perpendicular to $\mathbf{l}_0$:

$$(2.6)\qquad H = \{\mathbf{h} \in \mathbb{R}^{(s+t)} \mid \mathbf{h} \cdot \mathbf{l}_0 = 0\}.$$

If $\{u_i \mid 1 \leq i \leq s+t-1\}$ denotes a fundamental system of units of $U_K$, then we define $\mathbf{l}_i = \psi(u_i)$; so that, by the above, $\{\mathbf{l}_i \mid 1 \leq i \leq s+t\}$ is an $\mathbb{R}$-basis of $\mathbb{R}^{(s+t)}$.

Given any $\mathbf{y} \in (\mathbb{R}^{(s)} \times \mathbb{C}^{(t)})^*$, we expand $L(\mathbf{y})$ as

$$L(\mathbf{y}) = \sum_{i=1}^{s+t} \xi_i \mathbf{l}_i$$

and we refer to the $\xi_i = \xi_i(y)$ as the logarithmic coordinates of $\mathbf{y}$. By abuse of language, for $a \in K^*$ we call the logarithmic coordinates of $\phi(a)$, the logarithmic coordinates of $a$.

We now consider how multiplying $a \in K^*$ by a unit alters the logarithmic coordinates of $a$. Clearly, by multiplying $a$ by a suitable unit $u$, we can guarantee that for all $j < s+t$, $\xi_i(au) \in [0,1)$; in addition, since

$\ker \psi = \mu_K$ (see (IV.4.5.$a$)), we can ensure that either $a^{\sigma_1} > 0$ if $s > 0$, or that $\arg(a^{\sigma_1}) \in [0, 2\pi/W_K)$ if $s = 0$. Thus any $a \in K^*$ possesses an associate $a'$ such that:

**(2.7)**

($a$) *The logarithmic coordinates* $\xi_i(a')$ *all lie in* $[0,1)$ *for* $1 \le i \le s + t - 1$.

($b$) $\arg(a'^{\sigma_1}) \in [0, 2\pi/W_K)$.

(Note that, if $\sigma_1$ is a real embedding, then $W_K = 2$ and so the condition ($b$) does indeed mean that $a'^{\sigma_1}$ is positive.)

Conversely, any element of $K^*$ can have only one associate satisfying the conditions (2.7): indeed, if $a$ and $au$ both satisfy (2.7) for some $u \in U_K$, then on the one hand, for $i < s+t$ the logarithmic coordinates $\xi_i(u)$ of $u$ all lie in $(-1, 1)$; however, they are all integers and so they are zero; moreover, by (2.6)

$$\xi_{s+t}(u) = \psi(u) \cdot \mathbf{l}_0 = 0$$

which shows that $\psi(u) = 0$, and so $u$ must be a root of unity. By the second condition $\arg(u^{\sigma_1}) \in (-2\pi/W_K, 2\pi/W_K)$, and therefore $u = 1$, as required. This therefore suggests that we consider the fundamental domain $\mathfrak{F}$ cut out by those $z \in (\mathbb{R}^s \times \mathbb{C}^t)^*$ such that

(2.8.$a$) $$\xi_i(z) \in [0,1) \qquad 1 \le i \le s+t-1$$

(2.8.$b$) $$\arg(z_1) \in [0, 2\pi/W_K).$$

To fix ideas we consider $\mathfrak{F}$ in detail for the two quadratic cases when $(s,t) = (2,0)$ or $(0,1)$. If $(s,t) = (0,1)$, then there is no restriction on the size of $\xi_1(z)$, but $\arg(z_1) \in [0, 2\pi/W_K)$; therefore $\mathfrak{F}$ coincides with the shaded region of the $z_1$ complex plane in figure (2.9.$a$): here broken lines denote the fact that the line is not in $\mathfrak{F}$; likewise $*$ at the origin indicates that the origin is not in $\mathfrak{F}$.

Suppose now that $(s,t) = (2,0)$. We view $K$ as embedded in $\mathbb{R}$ and let $\sigma_1 = id$; we let $u$ then denote the unique fundamental unit of $K$ such that $0 < u < 1$. Let $z = (z_1, z_2) \in \mathbb{R}_{>0} \times \mathbb{R}^*$. The logarithmic coordinates of $z$ are obtained by solving the equation $(\log|z_1|, \log|z_2|) = \xi_1 \mathbf{l}_1 + \xi_2 \mathbf{l}_2$. Since $\mathbf{l}_1 = (\log u, \log|u^{\sigma_2}|) = (\log u, -\log u)$, it follows that

$$\log z_1 = \xi_1 \log u + \xi_2$$
$$\log|z_2| = -\xi_1 \log u + \xi_2.$$

Thus $z_1|z_2| = e^{2\xi_2}$ and $z_1/|z_2| = u^{2\xi_1}$. Since $\xi_1 \in [0,1)$, the region $\mathfrak{F}$ is given by those $(z_1, z_2)$ with $z_1 > 0$, $|z_2| \neq 0$ such that

$$z_1 \le |z_2| < u^{-2} z_1.$$

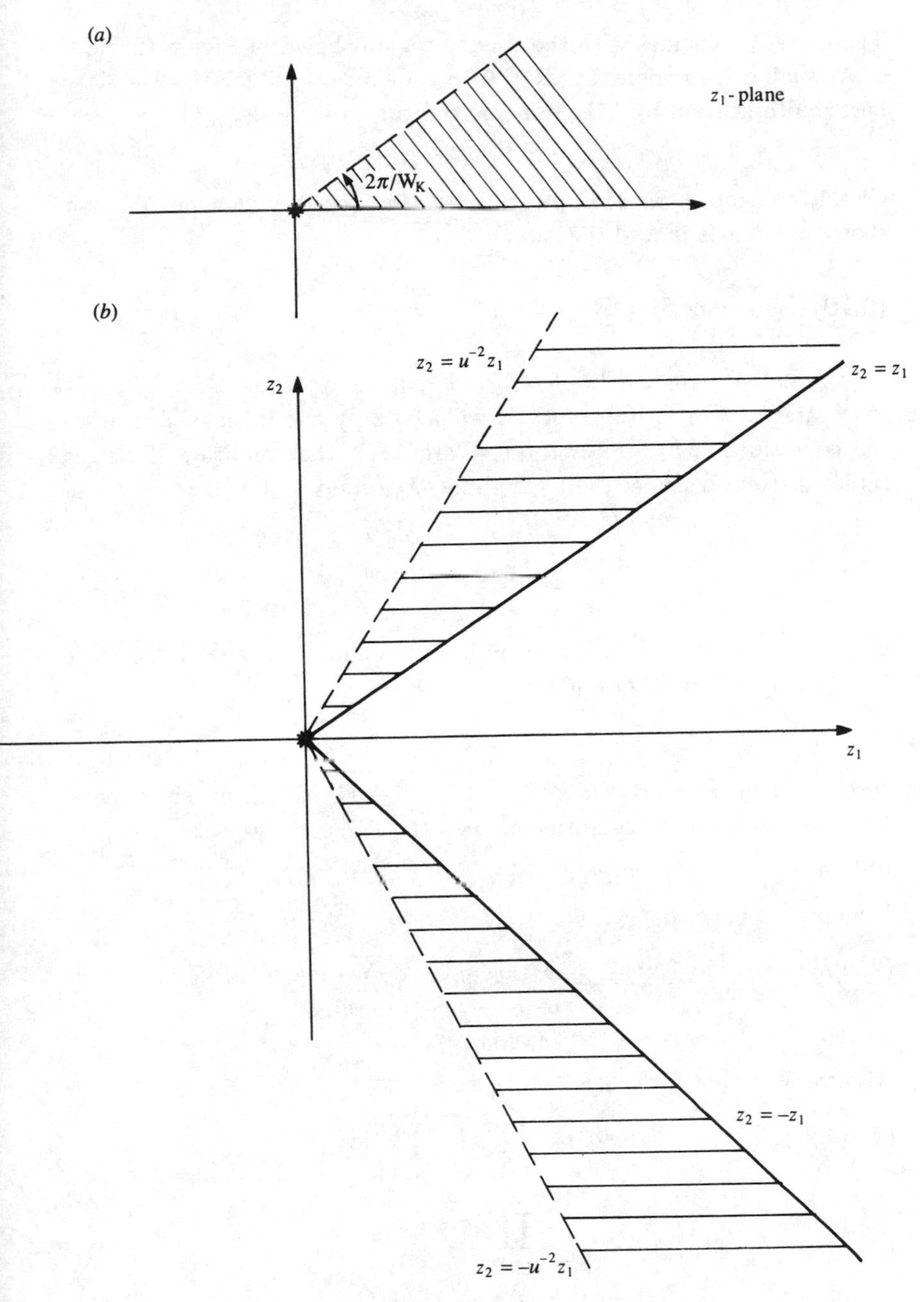


Figure (2.9)

Therefore $\mathfrak{F}$ coincides with the shaded region shown in Figure (2.9.*b*).

We shall call a region $\mathfrak{R} \subset \mathbb{R}^s \times \mathbb{C}^t$ a *cone* if $\mathfrak{R} = \lambda\mathfrak{R}$ for any $\lambda \in \mathbb{R}_{>0}$; here multiplication by $\lambda$ denotes multiplication by scalars, i.e.

$$\lambda(r_1, \ldots, r_{s+t}) = (\lambda r_1, \ldots, \lambda r_{s+t}).$$

Clearly the above two examples suggest that $\mathfrak{F}$ will be a cone. We now show that this is indeed the case:

**(2.10)** *$\mathfrak{F}$ is a cone in $\mathbb{R}^s \times \mathbb{C}^t$.*

*Proof.* Let $\lambda > 0$ and let $f \in \mathfrak{F}$. It suffices to show that $\lambda f \in \mathfrak{F}$. Since $\lambda(\mathbb{R}^s \times \mathbb{C}^t)^* \subset (\mathbb{R}^s \times \mathbb{C}^t)^*$, we need only check that $\lambda f$ satisfies the conditions (2.8). As $\arg(f_1) = \arg(\lambda f_1)$, the condition (2.8.*b*) is satisfied. Next consider the logarithmic coordinates of $\lambda f$:

$$\begin{aligned} L(\lambda f) &= L((\lambda, \ldots, \lambda) \cdot f) \\ &= L(\lambda, \ldots, \lambda) + L(f) \\ &= \log(\lambda)\mathbf{l}_{s+t} + L(f). \end{aligned}$$

Therefore only the final logarithmic coordinate is changed, i.e. for $i < s+t$, $\xi_i(\lambda f) = \xi_i(f) \in [0,1)$.

□

Next we define a map $N : (\mathbb{R}^{(s)} \times \mathbb{C}^{(t)})^* \to \mathbb{R}^*$ which we shall see is intimately related to the norm: for $\mathbf{y} = (y_1, \ldots, y_{s+t})$ we set

(2.11.*a*) $$N(\mathbf{y}) = |y_1 \ldots y_s| . y_{s+1}\bar{y}_{s+1} \cdots y_{s+t}\bar{y}_{s+t}.$$

It follows at once that

(2.11.*b*) $$\begin{aligned} \log(N(\mathbf{y})) &= \sum_{i=1}^{s} \log|y_i| + 2\sum_{i=s+1}^{s+t} \log|y_i| \\ &= L(\mathbf{y}).\mathbf{l}_0. \end{aligned}$$

Also we note that for $a \in K^*$

(2.11.*c*) $$\begin{aligned} N(\phi(a)) &= \prod_{i=1}^{s} |a^{\sigma_i}| \prod_{i=s+1}^{s+t} |a^{\sigma_i}|^2 \\ &= \prod_{i=1}^{n} |a^{\sigma_i}| \\ &= |N_{K/\mathbb{Q}}(a)|. \end{aligned}$$

Presently we shall see that for our arithmetic purposes we need to study the sub-region $\mathfrak{G}$ of $\mathfrak{F}$ given by

$$\mathfrak{G} = \{f \in \mathfrak{F} \mid 0 < N(f) \leq 1\}.$$

Thus, of course, in the two quadratic cases $\mathfrak{G}$ corresponds to the shaded regions:

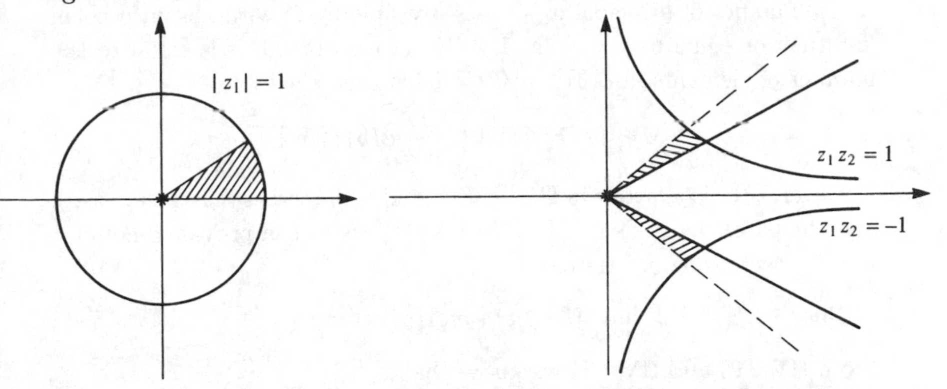


**(2.12)** *For any $\eta > 0$, $a \in K^*$ satisfies $|N_{K/\mathbb{Q}}(a)| \leq \eta$ iff the unique element $b \in K^*$ such that $a = bu$ with $\phi(b) \in \mathfrak{F}$ and $u \in U_K$, has the property that $\phi(b) \subset \eta^{1/n}\mathfrak{G}$.*

*Proof.* Write $\phi(b) = \lambda^{1/n}h$ with $\lambda \in (0,1]$, $h \in \mathfrak{h}$ where

$$\mathfrak{h} = \{f \in \mathfrak{F} \mid N(f) = 1\}$$

so that by (2.11.$b$) $L(h).\mathbf{l}_0 = 0$ for $h \in \mathfrak{h}$.

However, by (2.11.$b$,$c$)

$$\begin{aligned} \log(|N_{K/\mathbb{Q}}(a)|) &= L(\phi(a)).\mathbf{l}_0 = \psi(a).\mathbf{l}_0 \\ &= (\psi(b) + \psi(u)).\mathbf{l}_0 \\ \text{by (2.6)} \quad &= \psi(b).\mathbf{l}_0 \\ &= L(h).\mathbf{l}_0 + \log(\lambda) \\ \text{by the above} \quad &= \log(\lambda) \end{aligned}$$

and this shows that $\lambda = |N_{K/\mathbb{Q}}(a)|$. Therefore we calculate that $|N_{K/\mathbb{Q}}(a)| \leq \eta$ iff $\lambda \leq \eta$; since $\phi(b) = \lambda^{1/n}h$ with $h \in \mathfrak{h}$, it follows that this occurs iff $\phi(b) = \eta^{1/n}g$ with $g \in \mathfrak{G}$.

□

In due course we shall show that $\mathfrak{G}$ is always bounded and Jordan measurable; we denote its volume by $v_K$; at the end of this section we shall determine the exact value of $v_K$. Our first instance of the importance of the region $\mathfrak{G}$ is in showing

(2.13) $$\lim_{x\to\infty} \frac{M_K(x)}{x} = \frac{2^t h_K v_K}{|d_K|^{1/2}}.$$

*Proof.* We choose a class $c \in C_K$. Let $\mathfrak{b}$ denote an $\mathfrak{o}_K$-ideal with class

$c^{-1}$. Let $M_{K,c}(x)$ denote the number of $\mathfrak{o}_K$-ideals $\mathfrak{c}$ with class $c$ and with $\mathbf{N}\mathfrak{c} \leq x$. Reasoning as prior to (2.3), this is seen to be the same as the number of principal $\mathfrak{o}_K$-ideals divisible by $\mathfrak{b}$, with absolute norm less than or equal to $y = x\mathbf{N}\mathfrak{b}$. By the above work, this is equal to the number of elements in $\phi(\mathfrak{b}) \cap y^{1/n}\mathfrak{G}$. This then shows

$$M_{K,c}(x) = \mathrm{Card}\left(\frac{1}{y^{1/n}}\phi(\mathfrak{b}) \cap \mathfrak{G}\right).$$

However, $\phi(\mathfrak{b})$ is a lattice in $\mathbb{R}^{(s)} \times \mathbb{C}^{(t)}$ and so $y$ increases as $x$ increases, and the parallelepipeds in $\frac{1}{y^{1/n}}\phi(\mathfrak{b})$ which intersect $\mathfrak{G}$ provide $\varepsilon$-nets for $\mathfrak{G}$ with decreasing $\varepsilon$. Therefore

$$\lim_{x\to\infty} M_{K,c}(x)\frac{1}{y}.\mathrm{vol}\,(\phi(\mathfrak{b})) = v_K \tag{2.13.a}$$

From (IV.2.9) and (IV.3.3) we know that

$$\mathrm{vol}\,(\phi(\mathfrak{b})) = 2^{-t}|d_K|^{1/2}\mathbf{N}\mathfrak{b} \tag{2.13.b}$$

and so

$$\lim_{x\to\infty} \frac{M_{K,c}(x)}{x} = \frac{2^t v_K}{|d_K|^{1/2}}. \tag{2.13.c}$$

Summing over the different classes in $C_K$, then gives the result.

□

For brevity denote the lattice $\phi(\mathfrak{b})$ in $\mathbb{R}^{(s)} \times \mathbb{C}^{(t)}$ by $B$ and put $b_K = \mathrm{vol}\,(B)$.

**Theorem 63.**

$$\lim_{x\to 1+} (x-1)\zeta_c(x) = \frac{v_K}{b_K}\mathbf{N}\mathfrak{b}.$$

*Proof.* For any real $r > 0$, we put $M(r) = \mathrm{Card}\,(B \cap r^{1/n}\mathfrak{G})$. Since $b_K = 2^{-t}|d_K|^{1/2}\mathbf{N}\mathfrak{b}$ (by (2.13.*b*)), it follows from (2.13.*a*) that

$$\lim_{r\to\infty} \frac{M(r)}{r} = \lim_{r\to\infty} \frac{M_{K,c}(r \cdot \mathbf{N}\mathfrak{b}^{-1})}{r} = \frac{v_K}{b_K}. \tag{2.14.a}$$

We shall prove the theorem by considering $M(r)$ in an alternative way. Because $M(r)$ is always finite, we can order the points $f_i \in B \cap \mathfrak{F}$ for $i \in \mathbb{N}$ in such a way that $N(f_i) \succeq N(f_{i+1})$ for all $i$. We then define $r_i = N(f_i)$; note for any $i$ and for any $\varepsilon > 0$

$$M(r_i - \varepsilon) < i \leq M(r_i)$$

so dividing by $r_i = N(f_i)$, we have

$$\frac{M(r_i - \varepsilon)}{(r_i - \varepsilon)}\left(\frac{r_i - \varepsilon}{r_i}\right) < \frac{i}{N(f_i)} \leq \frac{M(r_i)}{r_i}.$$

Therefore, on letting $i \to \infty$ and using (2.14.$a$), we have shown

$$\text{(2.14.}b\text{)} \qquad \lim_{i\to\infty} \frac{i}{N(f_i)} = \frac{v_K}{b_K}.$$

Because $B \cap \mathfrak{F}$ contains the $\phi$ image of exactly one representative for each equivalence class of associates in $\mathfrak{b}\backslash 0$, it follows from (2.3.$a$) that

$$\text{(2.14.}c\text{)} \qquad \zeta_c(x) = \mathbf{N}\mathfrak{b}^x \sum N(f_i)^{-x}.$$

Now by (2.14.$b$) we know that given $\varepsilon > 0$, there exists $k$ such that for all $i \geq k$

$$\text{(2.14.}d\text{)} \qquad \left(\frac{v_K}{b_K} - \varepsilon\right)^x \leq \frac{i^x}{N(f_i)^x} \leq \left(\frac{v_K}{b_K} + \varepsilon\right)^x.$$

Since $N(f_i)^{-x} \leq (v_K/b_K + \varepsilon)i^{-x}$ when $i \geq k$, it follows by comparison with the series $\sum i^{-x}$, that the series $\sum N(f_i)^{-x}$ converges for all $x > 1$. On dividing by $i^x$ and summing over all $i \geq k$ in (2.14.$d$), we obtain an inequality of the form

$$(c-\varepsilon)^x h(x) \leq \mathbf{N}\mathfrak{b}^{-x}\zeta_c(x) - g(x) \leq (c+\varepsilon)^x h(x)$$

where

$$g(x) = \sum_{i=1}^{k-1} N(f_i)^{-x}, \quad h(x) = \sum_{i\geq k} i^{-x}, \quad c = v_K/b_K\,.$$

So, multiplying through by $x - 1$, we have

$$\text{(2.14.}e\text{)} \quad (g(x) + (c-\varepsilon)^x h(x))(x-1) \leq \mathbf{N}\mathfrak{b}^{-x}\zeta_c(x)(x-1) \leq (g(x) + (c+\varepsilon)^x h(x))(x-1)$$

Now let $x \to 1+$. Then $g(x)(x-1) \to 0$ and $(c \pm \varepsilon)^x \to c \pm \varepsilon$; also, since $(x-1)\zeta(x) \to 1$ (from (1.6)) and since $\zeta(x) - h(x) = \sum_{i<k} i^{-x}$ is continuous on $(0,\infty)$, it follows that $h(x)(x-1) \to 1$. This then shows that given $\varepsilon > 0$ and $\varepsilon_1 > 0$, there exists $\delta = \delta(\varepsilon, \varepsilon_1) > 0$ such that for all $x : 0 < x - 1 < \delta$

$$c - \varepsilon - \varepsilon_1 \leq (x-1)\mathbf{N}\mathfrak{b}^{-x}\zeta_c(x) \leq c + \varepsilon + \varepsilon_1.$$

We have therefore shown that $\lim_{x\to 1+} (x-1)\mathbf{N}\mathfrak{b}^{-x}\zeta_c(x)$ exists and equals $v_K/b_K$.

□

We conclude this section by showing

**(2.15)** *$\mathfrak{G}$ is bounded, Jordan measurable and*

$$\operatorname{vol}(\mathfrak{G}) = \frac{2^s \pi^t R_K}{W_K}.$$

Before embarking on the proof of this result we note that by combining it with Theorem 63 and the fact that $b_K$ (= vol $(B)$) $= 2^{-t}|d_K|^{1/2}\mathbf{N}\mathfrak{b}$, we get

$$\lim_{x\to 1+}(x-1)\zeta_c(x) = \frac{2^{s+t}\pi^t R_K}{W_K|d_K|^{1/2}}.$$

Summing over the classes $c \in C_K$, then gives Theorem 61. Similarly we may combine (2.15) with (2.13) to deduce

$$\lim_{x\to\infty}\frac{M_K(x)}{x} = \frac{2^t h_K \cdot v_K}{|d_K|^{1/2}} = \frac{2^{s+t}\pi^t R_K h_K}{|d_K|^{1/2}W_K}$$

which establishes Theorem 62.

We start to prove (2.15) by first showing that $\mathfrak{G}$ is bounded. Since each element $g \in \mathfrak{G}$ can be written as $g = \lambda h$ with $\lambda \in (0,1]$, $h \in \mathfrak{h}$, it will suffice to prove that $\mathfrak{h}$ is a bounded region; it will therefore be enough to show that $L(\mathfrak{h})$ is bounded. For any $h \in \mathfrak{h}$ we write $L(h) = \sum_1^{s+t}\xi_i \mathbf{l}_i$. Since $\mathfrak{h}$ is a subset of $\mathfrak{F}$, by (2.8.$a$) we know that $\xi_i \in [0,1)$ for $i < s+t$; moreover, $\xi_{s+t}$ must be zero, since by (2.11.$b$)

$$0 = \log(N(h)) = L(h)\cdot \mathbf{l}_0 = n\xi_{s+t}.$$

**(2.16)** *For $u \in U_K$, the $\mathbb{R}$-linear map $\theta_u : \mathbb{R}^{(s)} \times \mathbb{C}^{(t)} \to \mathbb{R}^{(s)} \times \mathbb{C}^{(t)}$ given by componentwise multiplication by $\phi(u)$, i.e.*

$$\theta_u(z_1,\ldots,z_{s+t}) = (u^{\sigma_1}z_1,\ldots,u^{\sigma_{s+t}}z_{s+t})$$

*is volume preserving.*

*Proof.* We have a commutative diagram of $\mathbb{R}$-vector space isomorphisms

$$\begin{array}{ccc} K\otimes\mathbb{R} & \xrightarrow{\phi} & \mathbb{R}^{(s)}\times\mathbb{C}^{(t)} \\ \downarrow{l_u} & & \downarrow{\theta_u} \\ K\otimes\mathbb{R} & \xrightarrow{\phi} & \mathbb{R}^{(s)}\times\mathbb{C}^{(t)} \end{array}$$

where $l_u(k\otimes r) = ku\otimes r$. Thus $\det(\theta_u) = \det(l_u) = N_{k/\mathbb{Q}}(u) = \pm 1$. □

Next observe that if $\{\zeta_k\}$ denote the roots of unity in $K$, then by (2.8.$b$) the $\theta_{\zeta_k}(\mathfrak{G})$ are all disjoint and, by the above, they all have the same volume. Let $\tilde{\mathfrak{G}} = \cup_k \theta_{\zeta_k}(\mathfrak{G})$; then $\tilde{\mathfrak{G}}$ consists of $\mathbf{z} \in (\mathbb{R}^s \times \mathbb{C}^t)^*$ such that

(2.17.$a$) $$0 < N(\mathbf{z}) \le 1$$

(2.17.$b$) $$L(\mathbf{z}) = \sum_1^{s+t}\xi_k \mathbf{l}_k$$

with $\xi_k \in [0,1)$ for all $k < s+t$.

We write $\bar{\mathfrak{G}}$ for the set of elements in $\tilde{\mathfrak{G}}$ whose first $s$ coordinates are all positive, so that

$$\text{(2.18)} \qquad \mathrm{vol}(\tilde{\mathfrak{G}}) = 2^s \mathrm{vol}(\bar{\mathfrak{G}}) \qquad W_K \mathrm{vol}(\mathfrak{G}) = \mathrm{vol}(\tilde{\mathfrak{G}}).$$

We now carry out a series of transformations which enable us to show that $\bar{\mathfrak{G}}$ is Jordan measurable and to determine its volume.

In the usual way, for $\mathbf{z} \in \mathfrak{G}$ we write $L(\mathbf{z}) = \sum_1^{s+t} \xi_k \mathbf{l}_k$ so that by (2.6) and (2.11.$b$)

$$\log(N(\mathbf{z})) = L(\mathbf{z}) \cdot \mathbf{l}_0 = n\xi_{s+t}$$

and hence

$$\text{(2.19)} \qquad L(\mathbf{z}) = \sum_1^{s+t-1} \xi_k \mathbf{l}_k + \frac{1}{n} \log |N(\mathbf{z})| \mathbf{l}_{s+t}.$$

We now make our first change of variables and put

$$r_k = z_k \qquad 1 \le k \le s$$

and for $s < j$, with $z_j = x_j + iy_j$, $x_j, y_j \in \mathbb{R}$ we put $z_j = r_j e^{i\phi_j}$, $r_j \in \mathbb{R}$, $\phi_j \in [0, 2\pi)$ where $i = \sqrt{-1}$. Then

$$\prod_{l=1}^{s} dz_l \prod_{l>s}^{s+t} dx_l dy_l = \prod_{l>s}^{s+t} r_l \prod_{l=1}^{s+t} dr_l \prod_{l>s}^{s+t} d\phi_l$$

since the Jacobian matrix consists of diagonal blocks of the form (1) and $\begin{pmatrix} \cos\phi & \sin\phi \\ -r\sin\phi & r\cos\phi \end{pmatrix}$. By (2.17) this transforms $\bar{\mathfrak{G}}$ into the region $\mathfrak{R}$ of $(r_1 \ldots r_{s+t}, \phi_{s+1}, \ldots, \phi_{s+t})$ space:

(2.20)

($a$) $\quad r_k > 0 \quad 1 \le k \le s+t$

($b$) $\quad \phi_k \in [0, 2\pi) \quad s < k \le s+t$

($c$) $\quad \prod_{k=1}^{s+t} r_k^{e_k} \le 1$ where $e_k = \begin{cases} 1 & k \le s, \\ 2 & k > s. \end{cases}$

($d$) $\quad \log r_k = \frac{1}{n} \log(\prod_1^{s+t} r_h^{e_h}) + \sum_{l=1}^{s+t-1} \xi_l \log(|u_l^{\sigma_k}|)$, where $\xi_l \in [0,1)$ for all $l < s+t$.

($a$) and ($b$) are clear; condition ($c$) comes from (2.17.$a$); condition ($d$) comes from (2.19), on reading off the $k$-th coordinates of $L(\mathbf{z})$. Note that $\mathfrak{R}$ is bounded because $\bar{\mathfrak{G}}$ is bounded and $\mathfrak{R}$ is obtained by a partial change to polar coordinates. It now suffices to show that the resulting $\mathfrak{R}$-integral exists and to determine its value. To do this we transform $\mathfrak{R}$ to a new (but obvious!) region, by changing variables to the

$(\xi_1, \ldots, \xi_{s+t}, \phi_{t+1}, \ldots, \phi_{s+t})$-space: here $\xi_1, \ldots, \xi_{s+t-1}$ are the values in $(d)$, while $\xi_{s+t} = \prod_1^{s+t} r_k^{e_k}$. Thus $\mathfrak{R}$ transforms to the hypercube $\mathfrak{S}$:

(2.21)

$(a)$ $\quad \xi_l \in [0,1) \quad l < s+t$
$(b)$ $\quad \xi_{s+t} \in (0,1]$
$(c)$ $\quad \phi_j \in [0,2\pi) \quad s < j \le s+t.$

From (2.20.$d$)

$$\log(r_j) = \frac{1}{n}\log(\xi_{s+t}) + \sum_{l=1}^{s+t-1} \xi_l \log|u_l^{\sigma_j}|.$$

Differentiating with respect to $\xi_{s+t}$ gives

$$\frac{\partial r_j}{\partial \xi_{s+t}} = \frac{1}{n}\frac{r_j}{\xi_{s+t}}$$

while for $l < s+t$ we get

$$\frac{\partial r_j}{\partial \xi_l} = r_j \log|u_l^{\sigma_j}|.$$

Thus the Jacobian $J$ of this transformation is $\dfrac{1}{n\xi_{s+t}}\displaystyle\prod_1^{s+t} r_k$ times the absolute value of the determinant:

$$\begin{vmatrix} 1 & \log|u_1^{\sigma_1}| & \cdots & \log|u_{s+t-1}^{\sigma_1}| \\ \vdots & \vdots & & \vdots \\ 1 & \log|u_1^{\sigma_{s+t}}| & \cdots & \log|u_{s+t-1}^{\sigma_{s+t}}| \end{vmatrix}$$

From (IV, §4) we know that, by adding all rows to the last row, the value of this determinant is $2^{-t}nR_K$, and so

$$J = \frac{R_K}{2^t\xi_{s+t}}\prod_1^{s+t} r_K.$$

Using $\xi_{s+t} = \prod_1^{s+t} r_k^{e_k}$, this gives

$$J\prod_{j>s}^{s+t} r_j = 2^{-t}R_K.$$

Clearly this constant function is Riemann integrable on the hypercube $\mathfrak{S}$, and the value of this integral is $\pi^t R_K$. Thus by (2.18)

$$\mathrm{vol}(\mathfrak{G}) = \frac{2^s}{W_K}\mathrm{vol}(\bar{\mathfrak{G}}) = \frac{2^s\pi^t R_K}{W_K}.$$

□

## §3 Dirichlet $L$-functions

We shall now introduce some further Dirichlet series, associating with each (extended) primitive residue class character $\chi$ a function $L(x, \chi)$. These functions will play a fundamental role in obtaining explicit formulae for the class numbers of cyclotomic fields and in establishing the Dirichlet theorem on primes in arithmetic progressions. Throughout this section $x$ will always denote a real variable in $(0, \infty)$.

Let $\chi$ now denote an extended primitive residue class character. We define the Dirichlet $L$-function of $\chi$ to be

$$L(x, \chi) = \sum_{n=1}^{\infty} \chi(n) n^{-x} \tag{3.1}$$

whenever the right-hand series converges.

**Theorem 64.** *Suppose that $\chi \neq \epsilon$ (the trivial character, with $\epsilon(n) = 1$ for all $n \in \mathbb{N}$). Then the Dirichlet series $\sum_{n=1}^{\infty} \chi(n) n^{-x}$ converges for $x > 0$, defining a continuous function $L(x, \chi)$ on $(0, \infty)$. It converges absolutely for $x > 1$ and, in that range, satisfies the equation*

$$L(x, \chi) = \prod_{p} (1 - \chi(p) p^{-x})^{-1} \tag{3.2}$$

*where the product extends over all primes $p$.*

*Remark 1.* Clearly $L(x, \epsilon)$ is the Riemann $\zeta$-function $\zeta(x)$ of §1.

*Remark 2.* For $x \in (0, 1]$, the convergence is only conditional; it is therefore to be understood that the terms occur in increasing order of $n$. Indeed, we shall see in §4 that the "partial series" $\sum \chi(p) p^{-1}$ diverges to $\infty$ where $p$ runs over the primes in a given prime residue class modulo the conductor $f(\chi)$.

*Remark 3.* It can be shown that if one views $L(x, \chi)$ as a function of a complex variable, then, for $\chi \neq \epsilon$, it becomes an analytic function which can be extended to the whole complex plane.

*Proof of Theorem 64.* The absolute convergence for $x > 1$ follows from §1; indeed, $|\chi(n) n^{-x}| = n^{-x}$ or 0, and so the result follows by comparison with the series $\zeta(x)$.

Next observe that

$$(1 - \chi(p) p^{-x})^{-1} = \sum_{r=0}^{\infty} \chi(p)^r p^{-rx} \qquad (x > 1)$$

and (3.2) is then an immediate consequence of (1.7).

Let $f = f(\chi)$ be the conductor of $\chi$. From the orthogonality relations for abelian characters (see (A10.$a$) in Appendix A), we know that because $\chi \neq \epsilon$

$$\sum_{j=m+1}^{m+f} \chi(j) = 0.$$

Therefore, writing $N = fq + r$, $0 \leq r < f$, we get

$$\left|\sum_{n=1}^{N} \chi(n)\right| \leq \left|\sum_{fq+1}^{N} \chi(n)\right| \leq r < f.$$

We can thus apply (1.5) to deduce that $\sum a_n n^{-x}$ converges for all $x \geq 0$, and furthermore we know that the convergence is uniform, which shows that $L(x, \chi)$ is continuous on $(0, \infty)$.

□

Let now $\psi \in \Psi_{\mathbb{Q}[f]}$ denote the Galois character corresponding to $\chi$ (see diagram (VI.2.7)). Then, for a prime number $p$ not dividing $f$, we have $\chi(p) = \psi(r_f^{-1}(p))$, where $r_f$ is the isomorphism of Theorem 44, i.e. $r_f^{-1}(p)$ is the Frobenius of $p$ in $\mathrm{Gal}(\mathbb{Q}[f]/\mathbb{Q})$. In these terms, and using (3.2), the $L$-function can now be rewritten for $x > 1$ as

$$L(x, \chi) = \prod_{p \nmid f} (1 - \psi(r_f^{-1}(p))p^{-x})^{-1}. \tag{3.3}$$

In this form the notion of an $L$-function is seen to reflect the decomposition of prime ideals; the right-hand term of (3.3) can be readily generalised to any representation of the Galois group of an arbitrary Galois extension of number fields; we shall return to this point of view again in §7; in particular, there we shall see that we can write the Dedekind zeta function as a product of such generalised $L$-functions.

We next come to a product formula connecting Dirichlet $L$-functions with Dedekind zeta functions of cyclotomic fields. This, in conjunction with the limit formula of §2, then leads to a new classnumber formula (as was explained in the Introduction).

**Theorem 65.** *Let $K$ denote a cyclotomic field. Then*

(*a*)
$$\prod_{\chi \in \tilde{\Theta}_K} L(x, \chi) = \zeta_K(x) \qquad (x > 1)$$

(*b*)
$$\prod_{\substack{\chi \in \tilde{\Theta}_K \\ \chi \neq \epsilon}} L(1, \chi) = \frac{2^{s+t}\pi^t R_K h_K}{W_K |d_K|^{1/2}}.$$

Here the notation is that of Theorem 61 of §2.

*Proof.* In Theorem 47 multiply through by $T^{-fge}$ and then, for a given prime number $p$, we set $T = p^{-x}$; this gives the equation

$$\prod_{\chi \in \tilde{\Theta}_K} (1 - \chi(p)p^{-x}) = \prod_{\mathfrak{p}|p} (1 - \mathbf{N}\mathfrak{p}^{-x}) \tag{3.4}$$

where the product on the right runs over the prime ideals of $\mathfrak{o}_K$ above $p$. Taking inverses, from (2.2) and (3.2) we see that the factors belonging to $p$ in the product expansions of $\prod L(x, \chi)$ and of $\zeta_K(x)$ coincide. Equation $(a)$ then follows since both products converge absolutely for $x > 1$.

To obtain $(b)$ we rewrite $(a)$ for $x > 1$ in the form

$$\prod_{\substack{\chi \in \tilde{\Theta}_K \\ \chi \neq \epsilon}} L(x, \chi) = [\zeta_K(x)/(x-1)][(x-1)/\zeta(x)].$$

We then take limits as $x \to 1$ from above: by Theorem 64 the limit on the left is

$$\prod_{\substack{\chi \in \tilde{\Theta}_K \\ \chi \neq \epsilon}} L(1, \chi)$$

while Theorem 61 and (1.6) give the limit on the right. □

## §4 Primes in an arithmetic progression

Here we give the promised proof of Dirichlet's theorem. It turns out that the class number formula of Theorem 65 plays a crucial role, which exhibits the interconnection between two seemingly unrelated aspects of number theory.

**Theorem 66.** *Let $m$ denote a positive integer and let $a$ be an integer which is coprime to $m$. Then, as $x \to 1$ from above,*

$$\lim_{x \to 1+} ((\sum_{p \equiv a} p^{-x})/(\sum_{p} p^{-x})) = \lim_{x \to 1+} ((\sum_{p \equiv a} p^{-x})/(\log\left(\frac{1}{(x-1)}\right)) \\ = 1/\phi(m).$$

*where $\phi$ denotes the Euler function. Here $\sum_p$ is the sum over all primes, and $\sum_{p \equiv a}$ is the sum over those primes congruent to $a$* mod $(m)$.

**(4.1) Corollary.** *There exist infinitely many primes $p \equiv a$* mod $(m)$.

**(4.2)** *If* $(a, m) = 1 = (b, m)$, *then*

$$\lim_{x \to 1+} ((\sum_{p \equiv a} p^{-x})/(\sum_{p \equiv b} p^{-x})) = 1.$$

(4.2) can be interpreted as saying that, in some sense, the primes are equidistributed over the invertible residue classes $\bmod(m)$.

In the proof of this theorem we shall always use the letter $H$, adorned in various ways, to denote functions of a real variable $x$ which are continuous on $[1, \infty)$; in particular such functions are of course bounded as $x \to 1+$. We shall show that for any extended primitive residue class character $\chi$ and for $x > 1$

$$\log L(x, \chi) = \sum_p \chi(p)p^{-x} + H(x, \chi). \tag{4.3}$$

Here $\sum_p$ is as before and, as always, log denotes the principal branch of the logarithm function on $\mathbb{C}$ (see the résumé in §5). Indeed, for every prime number $p$, the sum $\sum_{n=2}^{\infty} \chi(p)^n/np^{xn}$ converges absolutely for $x \in [1, \infty)$ and satisfies

$$\left| \sum_{n=2}^{\infty} \chi(p)^n/np^{xn} \right| \leq \sum_{n=2}^{\infty} p^{-xn} = \frac{p^{-2x}}{1 - p^{-x}} \leq 2p^{-2x}.$$

Therefore, by (1.1), $H(x, \chi) = \sum_p \sum_{n=2}^{\infty} \chi(p)^n/np^{xn}$ converges uniformly for $x \geq 1$, and we have

$$|H(x, \chi)| \leq 2 \sum_p p^{-2x} \leq 2 \sum_{n=1}^{\infty} n^{-2}.$$

Thus, indeed, $H(x, \chi)$ is continuous and bounded on $[1, \infty)$.

On the other hand, by (3.2), we know that for $x > 1$

$$\begin{aligned} \log L(x, \chi) &= \sum_p \log((1 - \chi(p)p^{-x})^{-1}) \\ &= \sum_p \chi(p)p^{-x} + H(x, \chi) \end{aligned}$$

where the series $\sum_p \chi(p)p^{-x}$ converges absolutely for such $x$.

Now let $\chi$ run through the extended primitive residue class characters whose conductor divides $m$. Then by (4.3)

$$\sum_\chi \chi(a)^{-1} \log L(x, \chi) = \sum_\chi \sum_p \frac{\chi(a)^{-1}\chi(p)}{p^x} + \sum_\chi \chi(a)^{-1} H(x, \chi). \tag{4.4}$$

By Theorem 64 $L(x, \chi)$ is continuous on $(0, \infty)$ when $\chi \neq \epsilon$; moreover – and this is crucial – by Theorem 65(*b*) $\prod_{\chi \neq \epsilon} L(1, \chi) \neq 0$ and so obviously none of the $L(1, \chi)$ equal 0; therefore we can find some interval $U = [1, u)$

such that $\log L(x,\chi)$ is continuous on $U$ for all such $\chi \neq \epsilon$. So for $x \in U$, the left-hand side of (4.4) is given by

$$\sum_{\chi} \chi(a)^{-1} \log L(x,\chi) = \log \zeta(x) + H_1(x). \tag{4.5}$$

Next if $p \equiv a \bmod (m)$, then by orthogonality (see (A7) in the Appendix)

$$\sum_{\chi} \chi(a)^{-1} \chi(p) = \phi(m).$$

On the other hand, if $(p,m) = 1$ and if $p \not\equiv a \bmod (m)$, then

$$\sum_{\chi} \chi(a)^{-1} \chi(p) = 0.$$

Finally, the remaining case $(p,m) > 1$ only contributes finitely many primes. Therefore we conclude that the right-hand side of (4.4) is given by

$$\sum_{\chi} \sum_{p} \frac{\chi(a)^{-1}\chi(p)}{p^x} + \sum_{\chi} \chi(a^{-1}) H(x,\chi) = \phi(m) \sum_{p \equiv a} p^{-x} + H_2(x). \tag{4.6}$$

Now divide both sides in (4.4) by $\log \zeta(x)$. From (4.5), (4.6) we then conclude

$$1 = \left[\phi(m) \left(\sum_{p \equiv a} p^{-x}\right) / \log \zeta(x)\right] + h(x) \tag{4.7}$$

where $h(x)$ is continuous on $[1,\infty)$ with $h(1) = 0$.

To obtain the first result in the theorem, we use (4.3) once more, now with $\chi = \epsilon$; this tells us that

$$\lim_{x \to 1+} \frac{\log(\zeta(x))}{\sum_p p^{-x}} = 1.$$

For the second part of the theorem, we use (1.6) which implies that

$$\lim_{x \to 1+} \log(\zeta(x)) + \log(x-1) = 0.$$

□

## §5 Evaluation of $L(1,\chi)$ and explicit class number formulae for cyclotomic fields

The power of the product formula of Theorem 65 lies in the fact that the values $L(1,\chi)$ which occur can be evaluated in terms of known functions. This will be done here and, as an illustration of the method we derive, for a prime number $p$, an explicit class number formula for $K = \mathbb{Q}[p]$ and also for its maximal real subfield. Further applications of the main result to quadratic fields will follow in §6.

We shall need some basic facts about the complex logarithm $\log(1+z)$.

This is a multivalued function; if $1+z=re^{i\alpha}$, $r>0$ and $-\pi<\alpha<\pi$ then $\log(1+z)=\log(r)+i\alpha+2\pi in$, where $\log(r)$ is the real logarithm. Then the choice of $n\in\mathbb{Z}$ indicates the branch. The *principal branch* is that for which $n=0$ for the above range of $\alpha$; in other words, the principal branch, or "principal value", is characterised by its analytic properties and the fact that for real $z>-1$, $\log(1+z)$ is real. For $|z|<1$ this principal value is given by the power series

$$\log(1+z)=\sum_{n=1}^{\infty}(-1)^{n-1}\frac{z^n}{n}$$

and this extends to values on $|z|=1$ for which the power series converges (see (1.4)).

We shall also need to recall the definition and certain results on Gauss sums $\tau(\chi)$ (always normalised as in (VI.4.2)). For brevity write $e(z)=e^{2\pi iz}$. Then we have

**Theorem 67.**

(*a*) *If $\chi$ is even, i.e. $\chi(-1)=1$, and if $\chi\neq\epsilon$, then*

$$\begin{aligned}L(1,\chi)&=-\frac{\tau(\chi)}{f(\chi)}\sum_{0<a<f(\chi)}\overline{\chi}(a)\log|1-e(-a/f)|\\&=-\frac{\tau(\chi)}{f(\chi)}\sum_{0<a<f(\chi)}\overline{\chi}(a)\log(\sin(\pi a/f))\end{aligned}$$

*where $\overline{\chi}$ is the complex conjugate of the character $\chi$, i.e. $\overline{\chi}(a)=\chi(a)^{-1}$, when $\chi(a)\neq 0$.*

(*b*) *If $\chi$ is odd, i.e. $\chi(-1)=-1$, then*

$$L(1,\chi)=\frac{\pi i\tau(\chi)}{f(\chi)^2}\sum_{0<a<f(\chi)}\overline{\chi}(a)a.$$

(Recall that $\overline{\chi}(a)=0$ whenever $(a,f(\chi))>1$.)

*Proof.* Throughout $\chi$ denotes an extended primitive residue class character which is different from $\epsilon$; we write $f$ for the conductor $f(\chi)$ of $\chi$. The symbol $\sum'$ will stand for summation over a complete set of representatives of the prime residue classes mod$(f)$.

The first step is to prove that for $x>1$

$$\sum_{n=1}^{\infty}\chi(n)n^{-x}=f^{-1}\tau(\chi){\sum_b}'[\overline{\chi}(b)\sum_{n=1}^{\infty}e\left(\frac{-nb}{f}\right)n^{-x}]. \tag{5.1}$$

Here, by definition, all the $b$ are coprime to $f$. Note that the infinite series appearing here are all absolutely convergent for the given range of $x$; this will therefore justify any rearrangement of terms. We first of all rewrite the left-hand side of (5.1) in the form

$$\sum_a{}' \chi(a) \sum_{n\equiv a \bmod (f)} n^{-x}.$$

Given $n$, we know by (A.10) that

$$\sum_{b=0}^{f-1} e((a-n)b/f) = \begin{cases} f & \text{if } n \equiv a \bmod (f) \\ 0 & \text{otherwise.} \end{cases}$$

Therefore we now get

$$\begin{aligned}\sum_{n=1}^{\infty} \chi(n) n^{-x} &= f^{-1} \sum_a{}' [\chi(a) \sum_{n=1}^{\infty} [n^{-x} \sum_{b=0}^{f-1} e((a-n)b/f)]] \\ &= f^{-1} \sum_{b=0}^{f-1} \left\{ \sum_a{}' \chi(a) e(ab/f) \sum_{n=1}^{\infty} e(-nb/f) n^{-x} \right\}.\end{aligned}$$

Now by (VI.4.2.$b$,$b'$)

$$\sum_a{}' \chi(a) e\left(\frac{ab}{f}\right) = \begin{cases} \chi(b)^{-1}\tau(x) & \text{if } (b,f)=1, \\ 0 & \text{otherwise.} \end{cases}$$

Substituting this into the preceding expression gives (5.1). From (1.5) we know that if $(b,f)=1$ then $\sum_{n=1}^{\infty} e(-nb/f)n^{-x}$ on the right-hand side of (5.1) converges and is continuous for $x > 0$; the same is true for the left-hand side by Theorem 64. Therefore we may set $x = 1$ in (5.1). Moreover,

$$\sum_{n=1}^{\infty} e(-nb/f) n^{-1} = \sum_{n=1}^{\infty} e(-b/f)^n n^{-1}$$

is the logarithmic series for $-\log(1-z)$ at $z = e\left(\frac{-b}{f}\right)$. Hence finally we have shown

$$L(1,\chi) = -f^{-1}\tau(\chi) \left[ \sum_b{}' \overline{\chi}(b) \log(1 - e\left(\frac{1}{f}\right)^{-b}) \right] \tag{5.2}$$

with log denoting the principal value. Observe that of course $e\left(\frac{1}{f}\right) = e^{2\pi i/f}$ is the normalised primitive root of unity. Replacing $b$ by $-b$ we get

$$L(1,\chi) = -f^{-1}\tau(\chi)\chi(-1) \left[ \sum_b{}' \overline{\chi}(b) \log(1 - e\left(\frac{1}{f}\right)^{b}) \right]. \tag{5.3}$$

From this point on we are obliged to deal with even and odd characters separately.

First suppose $\chi$ to be even. Adding (5.2) and (5.3) we get

$$2L(1,\chi) = -f^{-1}\tau(\chi)\sum_b{}' \overline{\chi}(b) \log\left\{(1-e\left(\frac{1}{f}\right)^b)(1-e\left(\frac{1}{f}\right)^{-b})\right\}$$

$$= -f^{-1}\tau(\chi)\sum_b{}' \overline{\chi}(b) \log|1-e\left(\frac{1}{f}\right)^{-b}|^2$$

which immediately gives us the first equation in the theorem. Next, for real $\alpha$, we have

$$1-e^{-i\alpha} = e^{-i\alpha/2}2i\sin(\alpha/2) = 2\sin(\alpha/2)e^{i(\pi-\alpha)/2}.$$

Thus, whenever $0<\alpha<2\pi$, we have

(5.4) $$\log(1-e^{-i\alpha}) = \log(2\sin(\alpha/2)) + i(\pi-\alpha)/2$$

for the principal branch of the logarithm. Taking the logarithms of absolute values we have

$$\log|1-e^{-i\alpha}| = \log 2 + \log\sin(\alpha/2).$$

Putting $\alpha = 2\pi a/f$, we deduce from the first equation in Theorem 67 that

$$L(1,\chi) = -\frac{\tau(\chi)}{f}\left[\sum_a \overline{\chi}(a)\log\sin\left(\frac{a\pi}{f}\right) + \sum_a \overline{\chi}(a)\log 2\right].$$

From (A.10.$a$) it follows that $\sum_a \overline{\chi}(a) = 0$, and so we have established the second equation in the theorem for even $\chi$.

Now consider the case of an odd character $\chi$. In conjunction with (5.4) we also have

$$\log(1-e^{i\alpha}) = \log(2\sin(\alpha/2)) + i(\alpha-\pi)/2$$

for $0<\alpha<2\pi$; and so

(5.5) $$\log\left(\frac{1-e^{-i\alpha}}{1-e^{i\alpha}}\right) = i(\pi-\alpha) \qquad \text{for } 0<\alpha<2\pi.$$

Note that of course $\alpha = 2\pi b/f$ satisfies this equality for $0<b<f$. Therefore, adding (5.2) and (5.3), we then get

$$2L(1,\chi) = f^{-1}\tau(\chi)\sum_{0<b<f}\overline{\chi}(b)i\left(\frac{2\pi b}{f}-\pi\right).$$

Again by (A.10.$a$) $\sum\overline{\chi}(b) = 0$, and so we obtain the stated equation for the final part of the theorem.

□

We shall transform the formula for $L(1,\chi)$ when $\chi$ is even. For now write $\eta = e\left(\frac{1}{f}\right)$. We know from Theorem 45 that $(1-\eta^{-1})/(1-\eta^{-a})$ is a unit of $\mathbb{Q}[f]$ provided that $(a,f)=1$. However, we now wish to obtain a unit of the maximal real subfield $\mathbb{Q}[f]_+$ of $\mathbb{Q}[f]$. To this end write $\eta^{1/2} = e(1/2f)$. Then

$$\omega_a = \omega_{a,f} = (\eta^{1/2} - \eta^{-1/2})/(\eta^{a/2} - \eta^{-a/2}) \qquad (a \neq \pm 1) \tag{5.6}$$

is a unit of $\mathbb{Q}[2f]$, as it differs from $(1-\eta^{-1})/(1-\eta^{-a})$ by a root of unity factor $\eta^{1/2} \cdot \eta^{-a/2}$. If $\mathbb{Q}[2f] = \mathbb{Q}[f]$, then $\omega_a \in \mathbb{Q}[f]$. However, this is also true if $\mathbb{Q}[2f]$ is different from $\mathbb{Q}[f]$: for then $\mathrm{Gal}(\mathbb{Q}[2f]/\mathbb{Q}[f])$ is generated by the automorphism $\eta^{1/2} \mapsto -\eta^{1/2}$ which clearly fixes $\omega_a$; moreover, on taking complex conjugates we still have $\overline{\omega}_a = \omega_a$. Thus indeed $\omega_a \in \mathbb{Q}[f]_+$. Since $\chi$ is even $\overline{\chi}(-a) = \overline{\chi}(a)$, and since $|1-\eta^{-a}| = |\eta^{a/2} - \eta^{-a/2}| = |\eta^{-a/2} - \eta^{a/2}|$, we can rewrite the first equation of Theorem 67 as

$$L(1,\chi) = -\frac{2\tau(\chi)}{f}\sum_a^{+} \overline{\chi}(a) \log|\eta^{a/2} - \eta^{-a/2}|. \tag{5.7}$$

Both here and in the sequel $\sum_a^+$ is the sum over a half-system of prime residues mod$(f)$, i.e. a set $S$ so that $a \neq \pm a'$ for $a, a' \in S$ and so that, if $-S = \{-a \mid a \in S\}$, then $S \cup -S$ is a complete set of representatives of prime residues mod$(f)$.

In fact we can go one step further. We assert that

$$0 = \frac{2\tau(\chi)}{f}\sum_a^{+} \overline{\chi}(a) \log|\eta^{1/2} - \eta^{-1/2}|.$$

Indeed, the set of even characters is precisely the set of characters of the quotient group $(\mathbb{Z}/f\mathbb{Z})^*$ modulo the subgroup generated by the residue class of $-1$; while the sum $\sum_a^+$ runs through a set of representatives of this group. So we can use (A.10.a) again.

Adding the last equation to (5.7), we finally obtain

$$L(1,\chi) = 2\frac{\tau(\chi)}{f}\sum_a^{+} \overline{\chi}(a) \log|\omega_{a,f}|. \tag{5.8}$$

This gives rise to the following beautiful class number formula for $\mathbb{Q}[p]_+$, for any odd prime number $p$:

**Theorem 68.** *The group $\Omega_p$ of units generated by the*

$$\omega_{a,p} = \left(e\left(\frac{1}{2p}\right) - e\left(\frac{-1}{2p}\right)\right) \Big/ \left(e\left(\frac{a}{2p}\right) - e\left(\frac{-a}{2p}\right)\right)$$
$$= \sin(\pi/p)/\sin(\pi a/p) \qquad (a,p) = 1$$

*is of finite index in the full group $Y_p$ of units of $\mathbb{Q}[p]_+$ and moreover*

$$[Y_p : \Omega_p] = h_{\mathbb{Q}[p]_+}.$$

*Proof.* (5.7) together with Theorem 65 gives

$$\prod_{\substack{\chi \text{ even} \\ \chi \neq \epsilon}} \left( \frac{2\tau(\chi)}{p} \sum_a^+ \overline{\chi}(a) \log |\eta^{a/2} - \eta^{-a/2}|^{-1} \right) = \frac{2^{(p-1)/2} R_{\mathbb{Q}[p]_+} h_{\mathbb{Q}[p]_+}}{W_{\mathbb{Q}[p]_+} |d_{\mathbb{Q}[p]_+}|^{1/2}}.$$

As $W_{\mathbb{Q}[p]_+} = 2$, the powers of 2 cancel out. Next, as $(\mathbb{Q}[p]_+ : \mathbb{Q})$ is coprime to $p$, the ramification of $p$ is tame (see III, §3); more precisely $e_p = \frac{p-1}{2}$. Hence in this case we know by Theorem 26 that the different $\mathcal{D}$ of $\mathbb{Q}[p]_+/\mathbb{Q}$ is $\mathfrak{p}^{(p-3)/2}$, where $\mathfrak{p}$ denotes the unique prime ideal of $\mathfrak{o}_{\mathbb{Q}[p]_+}$ above $p$. Taking norms gives

$$d_{\mathbb{Q}[p]_+} = p^{(p-3)/2} \qquad |d_{\mathbb{Q}[p]_+}|^{1/2} = \sqrt{p}^{(p-3)/2}.$$

As $\chi$ is even, by (VI.4.4.$a$), $\tau(\chi)\tau(\overline{\chi}) = f$. Thus the contribution arising from every pair of complex conjugate characters $\chi, \overline{\chi}$ is $f = p$. If $p \equiv 1 \bmod (4)$, there are exactly $\frac{p-1}{2} - 2 = \frac{p-5}{2}$ such characters, i.e. $((p-5)/4)$ such pairs. In addition, the quadratic character $\chi \bmod (p)$ is then even, and, by Theorem 50, $f/\tau(\chi) = \sqrt{p}$. This shows that the total contribution of factors $f(\chi)/\tau(\chi)$ is

$$\sqrt{p}^{\frac{p-5}{2}+\frac{1}{2}} = \sqrt{p}^{\frac{p-3}{2}}.$$

If $p \equiv -1 \bmod (4)$, then the quadratic character is odd, so we have $((p-3)/4)$ pairs of complex conjugate characters and the contribution is again $\sqrt{p}^{(p-3)/2}$. Taking all of this into account, we now have

$$\prod_\chi \left( \sum_a^+ \overline{\chi}(a) \log |\eta^{a/2} - \eta^{-a/2}|^{-1} \right) = R_{\mathbb{Q}[p]_+} h_{\mathbb{Q}[p]_+}. \tag{5.9}$$

Again we view $\sum^+$ as running through the quotient group $G_+$ of $(\mathbb{Z}/p\mathbb{Z})^*$ modulo the image of $< -1 >$, and we may view the product $\prod_\chi$ as running over the characters of $G_+$ which are different from $\epsilon$, the trivial character. We can thus apply the formula (A.15) – which we have established precisely for this purpose! The left-hand side of (5.9) is then exactly

$$\det \left( \log \left( \frac{|\eta^{b/2} - \eta^{-b/2}|}{|\eta^{a^{-1}b/2} - \eta^{-a^{-1}b/2}|} \right) \right) = \det(\log |\omega_{a^{-1}}^{\sigma_b}|)$$

where $\sigma_b$ is the automorphism of $\mathbb{Q}[p]_+/\mathbb{Q}$ induced by $\eta \mapsto \eta^b$. Here $a \neq \pm 1 \neq b$. On the other hand we may index a system of fundamental units of $\mathbb{Q}[p]_+$ by $a \in G_+$, $a \neq 1$: $\{u_{a^{-1}}\}$ say. Then

$$R_{\mathbb{Q}[p]_+} = |\det(\log |u_{a^{-1}}^{\sigma_b}|)|.$$

Hence

$$R_{\mathbb{Q}[p]_+} |\det(A)| = |\det(\log |\omega_{a^{-1}}^{\sigma_b}|)|$$

where $A$ denotes the matrix which expresses the $\omega_{a^{-1}}$ in terms of the basis $\{u_{a^{-1}}\}$ of $Y_p$. Thus we end up with the equation

$$|\det(A)| = |h_{\mathbb{Q}[p]_+}| = h_{\mathbb{Q}[p]_+}.$$

Therefore $\det(A) \neq 0$, and hence $\Omega_p$ is of finite index in $Y_p$, with this index being precisely $|\det(A)|$, i.e. $h_{\mathbb{Q}[p]_+}$.

□

Next we come to the class number formula for $\mathbb{Q}[p]$, with $p$ again an odd prime

**Theorem 69.**

$$h_{\mathbb{Q}[p]} = h^* h_{\mathbb{Q}[p]_+}$$

*where $h^*$ is an integer and satisfies*

$$h^* = \frac{(-1)^{(p-1)/2}}{(2p)^{(p-3)/2}} \prod_{\chi \text{ odd}} \left( \sum_{0<a<p} \chi(a)a \right).$$

*Proof.* We know from (VI.1.20) that the class group of $\mathbb{Q}[p]_+$ embeds into that of $\mathbb{Q}[p]$. This immediately establishes the assertion that $h^*$ is an integer.

For the sake of brevity we write $h = h_{\mathbb{Q}[p]}$, $h_0 = h_{\mathbb{Q}[p]_+}$ and

$$R = R_{\mathbb{Q}[p]} \qquad R_0 = R_{\mathbb{Q}[p]_+}$$

for the regulators. A basis for the group of units of $\mathbb{Q}[p]_+ \bmod (\pm 1)$ is, by (VI.1.19), also a basis for the group of units of $\mathbb{Q}[p]$ modulo the roots of unity. Defining both regulators with respect to such a basis, we see that each row of $R$ is twice that of $R_0$. Thus

(5.10) $$R = 2^{(p-3)/2} R_0.$$

Write $d = |d_{\mathbb{Q}[p]}|$, $d_0 = d_{\mathbb{Q}[p]_+}$ for the absolute values of the discriminants. We have already proved that $d_0 = p^{(p-3)/2}$. In the same way one proves that $d = p^{p-2}$. Thus

(5.11) $$p\sqrt{d}/\sqrt{d_0} = p^{(p+3)/4}$$

where it is understood that all the square roots are positive.

From Theorem 65 we have

(5.12.a) $$\prod_{\chi \neq \epsilon} L(1,\chi) = \frac{\pi^{(p-1)/2} 2^{(p-1)/2} hR}{2p\sqrt{d}}$$

where the product runs over the primitive characters mod $(p)$. Similarly

(5.12.b) $$\prod_{\substack{\chi \neq \epsilon \\ \chi(-1)=1}} L(1,\chi) = \frac{2^{(p-1)/2} h_0 R_0}{2\sqrt{d_0}}.$$

Dividing (5.12.$a$) by (5.12.$b$) and using (5.10), (5.11), we get

$$\prod_{\chi(-1)=-1} L(1,\chi) = \frac{h^* 2^{(p-3)/2}\pi^{(p-1)/2}}{p^{(p+3)/4}}. \tag{5.13}$$

From Theorem 67 we can write the left-hand side of (5.13) as

$$\prod_{\chi(-1)=-1} L(1,\chi) = \frac{\pi^{(p-1)/2} i^{(p-1)/2}}{p^{p-1}} \prod_{\chi(-1)=-1} \tau(\chi) \sum_{a=1}^{p-1} \chi(a)a. \tag{5.14}$$

We shall show that

$$\prod_{\chi(-1)=-1} \tau(\chi) = i^{\left(\frac{p-1}{2}\right)} p^{(p-1)/4}. \tag{5.15}$$

The theorem then follows by comparing (5.13) and (5.14), using (5.15).

To establish (5.15) assume first that $p \equiv 1 \bmod (4)$. Then the odd characters $\chi$ occur in complex conjugate pairs, and there are $(p-1)/4$ such pairs. For each pair we know that by (VI.4.4.$a$)

$$\tau(\chi)\tau(\overline{\chi}) = -p.$$

The product in (5.15) is then $(-1)^{(p-1)/4}p^{(p-1)/4}$, as asserted. Next if $p \equiv -1 \bmod (4)$, then the quadratic character is odd, and by Theorem 50 it contributes $ip^{1/2}$. We are left with $(p-3)/4$ complex conjugate pairs of characters. Thus the total contribution is $((-1)^{(p-3)/4}i)p^{(1/2)+(p-3)/4} = i^{(p-1)/2}p^{(p-1)/4}$, as required.

□

## §6 Quadratic fields, yet again

We return again to quadratic fields now that we have at our disposal new sophisticated analytic techniques. Here we obtain results on class numbers in terms of some quite elementary invariants; these results lead to fundamental theorems concerning the distribution of quadratic residues. Throughout this section we always treat real and imaginary quadratic fields separately.

First let $K$ be an imaginary quadratic field with discriminant $d_K$ and associated primitive residue class character $\lambda_K$ (see (VI,§3)). By (VI.2.12), $\lambda_K(-1) = -1$. From Theorems 65 and 67, we know that, provided $K$ is neither $\mathbb{Q}[4]$ nor $\mathbb{Q}[3]$, i.e. $d_K < -4$, then

$$\frac{h_K d_K^2}{i\tau(\lambda_K)|d_K|^{1/2}} = \sum_{\substack{(a,d_K)=1 \\ 0<a<|d_K|}} \lambda_K(a)a. \tag{6.1.$a$}$$

By Theorem 50, and (VI.3.5) for $p = 2$, we have

$$\tau(\lambda_K) = i|d_K|^{1/2} \tag{6.1.$b$}$$

whenever $|d_K|$ is a prime power, and so

$$h_K d_K = \sum \lambda_K(a)a \qquad (d_K < -4). \tag{6.2}$$

Equations (6.1.$b$) and (6.2) actually also hold for composite $d_K$ – see Exercise 8 to Chapter VI. The two excluded cases, when $W_K = 4$ or 6 are trivial. Here we pause to note that (6.2) tells us that the elementary expression $\sum \lambda_K(a)a$ is always negative; this easily stated assertion depends both on the analytic formula and on the determination of the sign of the Gauss sum!

For an odd prime discriminant $-p$, the formula (6.2) can be transformed to give an even more direct determination of $h_K$ and an even clearer statement on the distribution of quadratic residues. Let here $R_p$ and $N_p$ respectively denote the number of quadratic residues and quadratic non-residues mod$(p)$ in the interval $1 \le a \le (p-1)/2$. Then we have

**Theorem 70.**

$$h_{\mathbb{Q}(\sqrt{-p})} = \begin{cases} R_p - N_p & \text{if } p \equiv -1 \bmod (8) \\ \frac{1}{3}(R_p - N_p) & \text{if } p \equiv 3 \bmod (8),\ p \neq 3. \end{cases}$$

*Proof.* Here $\lambda_K(a) = \left(\frac{a}{p}\right)$. Let $\sum$ and $\sum'$ denote sums over $a$ coprime to $p$, $\sum$ in the interval $1 \le a \le p-1$, $\sum'$ in the interval $1 \le a \le (p-1)/2$. Then by (6.2)

$$ph_K = -\sum \left(\frac{a}{p}\right) a = -\sum{}' \left(\frac{a}{p}\right) (a - (p-a))$$

since $\left(\frac{p-a}{p}\right) = -\left(\frac{a}{p}\right)$. As $\sum' \left(\frac{a}{p}\right) = R_p - N_p$, we obtain

$$ph_K = (-2\sum{}' \left(\frac{a}{p}\right) a) + p(R_p - N_p). \tag{6.3.$a$}$$

Next denote by $\sum''$ the sum over even $a$ in the range $1 \le a \le p-1$. Then $p-a$ will run over the odd integers in the given range, and so

$$\begin{aligned} ph_K &= -\sum{}'' \left(\frac{a}{p}\right) (a - (p-a)) \\ &= -2\sum{}'' \left(\frac{a}{p}\right) a + p\sum{}'' \left(\frac{a}{p}\right). \end{aligned}$$

Replace $a$ by $2a$ with $a$ running now from 1 to $(p-1)/2$. Then we obtain

$$ph_K = (-4\left(\frac{2}{p}\right)\sum{}' \left(\frac{a}{p}\right) a) + p\left(\frac{2}{p}\right)(R_p - N_p). \tag{6.3.$b$}$$

Now multiply (6.3.$a$) by $2\left(\frac{2}{p}\right)$ and then subtract this from (6.3.$b$); this then gives

$$ph_K(1-2\left(\frac{2}{p}\right)) = -p(R_p - N_p)\left(\frac{2}{p}\right).$$

Thus indeed

$$h_K = \begin{cases} R_p - N_p & \text{if } \left(\frac{2}{p}\right) = 1 \\ \frac{1}{3}(R_p - N_p) & \text{if } \left(\frac{2}{p}\right) = -1. \end{cases}$$

□

An immediate corollary of this result is

**(6.4)** *For $p \equiv -1 \bmod (4)$, $p > 3$, $R_p > N_p$, i.e. there are more quadratic residues than quadratic non-residues in the interval* $[1, \frac{(p-1)}{2}]$.

It is a striking fact that this result, whose statement requires only the definition of a quadratic residue, has no known elementary proof.

Next observe that

$$R_p - N_p \equiv R_p + N_p \equiv \frac{p-1}{2} \bmod (2).$$

Thus, as a further consequence of the above theorem, we get a new proof that $h_{\mathbb{Q}(\sqrt{-p})}$ is odd for $p \equiv -1 \bmod (4)$.

*Example.* $h_{\mathbb{Q}(\sqrt{-31})} = 3$. By repeatedly applying the Quadratic Reciprocity Law (see Theorem 40), one readily verifies that 1, 2 ,4, 5, 7, 8, 9, 10, 14 are quadratic residues, while 3, 6, 11, 12, 13, 15 are quadratic non-residues. Hence $R_{31} = 9$, $N_{31} = 6$, and so by the theorem $h_{\mathbb{Q}(\sqrt{-31})} = 3$.

We now consider the case when $K$ is a real quadratic number field. By (VI.2.12) $\lambda_K(-1) = +1$. We let $\eta_K$ denote that fundamental unit of $K$ which is greater than 1. Then from Theorems 65 and 67, we deduce that

$$2h_K \log \eta_K = -\frac{\tau(\lambda_K)}{|d_K|^{1/2}} \sum_{\substack{(a,d_K)=1 \\ 1 \le a < d_K}} \lambda_K(a) \log(\sin\left(\frac{\pi a}{d_K}\right)). \tag{6.5}$$

By Theorem 50, and by (VI.3.5) for $p = 2$, we get $\tau(\lambda_K) = |d_K|^{1/2}$ when $K = \mathbb{Q}(\sqrt{p})$, for $p \equiv 1 \bmod (4)$ or $p = 2$; hence

$$2h_K \log \eta_K = -\sum \lambda_K(a) \log \sin\left(\frac{\pi a}{d_K}\right). \tag{6.6}$$

In fact this result holds in general – see Exercise 8 to Chapter VI. Moreover, as $\lambda_K(d_K - a) = \lambda_K(a)$ we can rewrite this as

$$h_K \log \eta_K = -\sum_{\substack{0 < a < d_K/2 \\ (a,d_K)=1}} \lambda_K(a) \log \sin\left(\frac{\pi a}{d_K}\right) \tag{6.7}$$

where all the logarithms are just the usual real logarithm. Put

$$\prod\nolimits_K^* = \prod \sin\left(\frac{\pi a}{d_K}\right)^{-\lambda_K(a)} \tag{6.8}$$

where the product runs over $a$ coprime to $d_K$, in the interval $[1, d_K/2]$. Applying the exponential function to (6.7) we get

**Theorem 71.**

$$\prod\nolimits_K^* = \eta_K^{h_K}.$$

For $K = \mathbb{Q}(\sqrt{p})$, $p$ a prime number $\equiv 1 \bmod (4)$, $\prod_K^*$ is the same as the value $\prod_p^*$ defined in (VI,§3). The above theorem clearly implies that $\prod_K^* > 1$; and so, in particular, we have now completed the proof of Theorem 51.

The last theorem also tells us something about on the distribution of quadratic residues for a prime $p \equiv 1 \bmod (4)$: indeed, we now know that

$$\prod_{a^-} \sin\left(\frac{\pi a^-}{p}\right) > \prod_{a^+} \sin\left(\frac{\pi a^+}{p}\right) \tag{6.9}$$

where $a^+$ runs through the quadratic residues and $a^-$ runs through the quadratic non-residues in $1 \le a \le \frac{p-1}{2}$. Finally, recall that

$$N_{K/\mathbb{Q}}\left(\prod\nolimits_K^*\right) = -1 \qquad (K = \mathbb{Q}(\sqrt{p}))$$

cf. (VI.3.13). This then yields another proof that $h_{\mathbb{Q}(\sqrt{p})}$ is odd and that $\eta_{\mathbb{Q}(\sqrt{p})}$ has norm $-1$.

## §7 Brauer relations

The underlying aim of this section is to derive various non-trivial identities between Dedekind zeta functions of different number fields; then, by considering their behaviour as $x \to 1+$ and using Theorem 61, we shall be able to obtain some powerful identities between the class numbers and regulators of the fields involved. In particular this technique will give us new results for various biquadratic, cubic and sextic number fields.

As an illustration of the ideas involved, consider a biquadratic number field $K$ with quadratic subfields $k_i$, and let $\lambda_i$ denote the primitive residue class character associated with $k_i$. Then, applying Theorem 65(a) four times, we see that for $x > 1$

$$\zeta_K(x) = \zeta(x) \prod_{i=1}^{3} L(x, \lambda_i)$$

$$\zeta_{k_i}(x) = \zeta(x) L(x, \lambda_i) \qquad (i = 1, 2, 3),$$

and therefore it follows that

$$\zeta_K(x)\zeta^2(x) = \prod_{i=1}^{3} \zeta_{k_i}(x).$$

On multiplying by $(x-1)^3$ and letting $x \to 1+$, the limit formula in Theorem 61, together with (1.6), yields an expression for

$$2^{s(K)+t(K)}\pi^{t(K)}R_K h_K / W_K |d_K|^{1/2}$$

in terms of the corresponding invariants for the subfields $k_i$. This may then be used to obtain a relationship between the units and class numbers of the $k_i$. Thus, to some extent, we shall be able to reduce the determination of the biquadratic invariants to calculations in the quadratic subfields. The purpose of the present section is to systematise and generalise this situation.

In what follows we assume the reader to be familiar with elementary group representation theory over the field of complex numbers $\mathbb{C}$. The notion of determinant, for elements in the group algebra, will play a basic role throughout: if $\Gamma$ is a finite (not necessarily abelian) group, if $V$ is a $\mathbb{C}\Gamma$ module (always with a finite dimension over $\mathbb{C}$ and with $\Gamma$ usually acting on the right), and if $x \in \mathbb{C}\Gamma$, then we write $\det_V(x)$ for the determinant of the $\mathbb{C}$-linear endomorphism of $V$ given by the action of $x$ on $V$; since this value depends only on the isomorphism class of $V$, it only depends on the character, $\chi$ say, of $V$; therefore we shall often denote such a determinant by $\det_\chi(x)$. If $\chi$, $\phi$ are both characters of $\Gamma$, then, directly from the definition, we see that

$$\det_{\chi+\phi}(x) = \det_\chi(x)\det_\phi(x). \tag{7.1}$$

For a subgroup $\Delta$ of $\Gamma$ and for a $\mathbb{C}\Gamma$-module $V$, we define

$$V^\Delta = \{v \in V \mid v\delta = v \quad \forall \delta \in \Delta\}$$

that is to say $V^\Delta$ is the vector space of $\Delta$-fixed points in $V$. Note also that if in addition $\Delta$ is normal in $\Gamma$, then $V^\Delta$ is a $\mathbb{C}\Gamma$-module, and can be viewed as a representation for the quotient group $\Gamma/\Delta$ by the rule $v\gamma\Delta = v\gamma$ for $\gamma \in \Gamma$ and for $v \in V^\Delta$. If $x \in \mathbb{C}\Gamma$ has image $\bar{x} \in \mathbb{C}\Gamma/\Delta$, then from the definition it follows that

$$\det_{V^\Delta}(x) = \det_{V^\Delta}(\bar{x}).$$

**(7.2)** *Let $\Lambda$ denote a subgroup of $\Gamma$ and denote by $\Sigma_\Lambda$ the sum $\sum_{\lambda\in\Lambda}\lambda$ over the elements of $\Lambda$. Then $V^\Lambda = V\Sigma_\Lambda$.*

*Proof.* Since $\Sigma_\Lambda\lambda = \Sigma_\Lambda$, it follows that $V\Sigma_\Lambda \subset V^\Lambda$. Conversely, if

$v \in V^{\Lambda}$, then $v\Sigma_{\Lambda} = v|\Lambda|$ and so

$$v = v\frac{1}{|\Lambda|}\Sigma_{\Lambda} \in V\Sigma_{\Lambda}.$$

□

We write $\mathbb{C}(\Lambda\backslash\Gamma)$ for the $\mathbb{C}$-vector space on the cosets $\Lambda\backslash\Gamma$; this is then a $\mathbb{C}\Gamma$-module in the natural way via the (right) $\Gamma$-action on $\Lambda\backslash\Gamma$, that is to say

$$(\sum \xi_{\Lambda\omega}\Lambda\omega)\gamma = \sum \xi_{\Lambda\omega}\Lambda\omega\gamma$$

for $\xi_{\Lambda\omega} \in \mathbb{C}$ and with the sums running over the distinct cosets $\Lambda\omega$ of $\Lambda\backslash\Gamma$.

The first major result of this section will be to obtain a new way of writing the Dedekind zeta function $\zeta_K(x)$ of a number field $K$. We shall see that this new point of view naturally leads us to introduce a new type of $L$-function.

**Theorem 72 (Artin).** *Let $N/\mathbb{Q}$ be a finite Galois extension with Galois group $\Gamma$, let $\Lambda \subset \Gamma$ and let $K = N^{\Lambda}$ be the subfield of $N$ fixed by $\Lambda$. For a given prime number $p$, let $\mathfrak{p}_1, \ldots, \mathfrak{p}_g$ be the prime ideals of $\mathfrak{o}_K$ above $p$ and let $\mathfrak{P}$ be a chosen prime ideal of $\mathfrak{o}_N$ above $p$. Denote the decomposition group (resp. the inertia group) of $\mathfrak{P}$ in $N/\mathbb{Q}$ by $\Delta_p$ (resp. $I_p$) and let $\sigma_p \in \Delta_p/I_p$ denote the Frobenius automorphism of $\mathfrak{P}$, i.e. $\sigma_p$ is an element of $\Delta_p$ whose restriction to the inertia field $L_{\mathfrak{P}}^{I_p}$ is the Frobenius automorphism. Then*

$$(7.3.a) \qquad \prod_{i=1}^{g}(1 - \mathbf{N}\mathfrak{p}_i^{-x}) = \det\nolimits_{\mathbb{C}(\Lambda\backslash\Gamma)^{I_p}}(1 - \sigma_p p^{-x})$$

*and thus on taking the product over all $p$,*

$$(7.3.b) \qquad \zeta_K(x) = \prod_p \det\nolimits_{\mathbb{C}(\Lambda\backslash\Gamma)^{I_p}}(1 - \sigma_p p^{-x})^{-1} \qquad (x > 1).$$

Before proving this result, we first investigate some of its implications. For any character $\psi$ of $\Gamma$, afforded by a $\mathbb{C}\Gamma$-module $V$, the above decomposition suggests that we define

$$(7.4) \qquad L(x,\psi) = \prod_p \det\nolimits_{V^{I_p}}(1 - \sigma_p p^{-x})^{-1} \qquad (x > 1).$$

$L(x,\psi)$ is called the *Artin L-function of* $\psi$. If $n$ denotes the degree of $\psi$ (i.e. $n = \dim_{\mathbb{C}}(V)$), if $m_p = \dim_{\mathbb{C}}(V^{I_p})$ and if $\{\eta_i\}$ denote the eigenvalues of $\sigma_p$ on $V^{I_p}$, then

$$\begin{aligned}\left|\det\nolimits_{V^{I_p}}(1 - \sigma_p p^{-x})\right|^{-1} &= \prod_i \left|1 - \eta_i p^{-x}\right|^{-1} \\ &\le (1 + 2p^{-x})^{m_p} \le (1 + 2p^{-x})^n.\end{aligned}$$

By the convergence of the series $2n \sum p^{-x}$ for $x > 1$, we know from (1.7) that the product in (7.4) converges absolutely for $x > 1$.

If $\phi$ denotes a further character of $\Gamma$, then by (7.1) and (7.4) we see that

$$L(x, \phi + \psi) = L(x, \phi) L(x, \psi). \tag{7.5}$$

The case when we take $N$ to be a cyclotomic field then becomes particularly interesting, since on the one hand by Theorem 72 $\zeta_K(x) = L(x, \rho)$ with $\rho$ the character of $\mathbb{C}\Gamma$ (i.e. the regular character of $\Gamma$). On the other hand $\rho = \sum \psi$ where $\psi$ ranges over all elements of $\hat{\Gamma}$, so that by (7.5)

$$\zeta_K(x) = L(x, \rho) = \prod_{\psi \in \hat{\Gamma}} L(x, \psi). \tag{7.6}$$

With the terminology of (VI, §2), let $\chi$ denote the primitive Dirichlet character associated with a given (Galois) character $\psi \in \hat{\Gamma}$, and write $f = f(\chi)$ for the conductor; then by definition

$$L(x, \chi) = \prod_{p} (1 - \chi(p) p^{-x})^{-1} = \prod_{p \nmid f} (1 - \chi(p) p^{-x})^{-1}.$$

However, from (VI.2.12), we know that $p \nmid f$ iff $\psi$ is trivial on $I_p$; moreover, if $p \nmid f$ then $\chi(p) = \psi(\sigma_p)$. This then shows that each $p$-term in $L(x, \chi)$ is equal to the corresponding $p$-term in $L(x, \psi)$, and so

$$L(x, \psi) = L(x, \chi), \tag{7.7}$$

a result which we have essentially already seen in (3.3). This apparently innocent equality is of fundamental importance, since it provides two completely different ways of looking at the same object; moreover, it suggests various further $L$-function relationships, which are of outstanding arithmetic interest.

It is also interesting to note that, for a cyclotomic field $N$, (7.6) and (7.7) immediately give a proof of the product formula of Theorem 65:

$$\zeta_K(x) = \prod_{\chi \in \tilde{\Theta}_N} L(x, \chi).$$

Another way of viewing matters is to say that, thanks to (7.7), (7.6) provides a generalisation of Theorem 65.

Prior to proving Theorem 72, we first need to establish a few elementary algebraic results. As previously, $\Gamma$ denotes a finite group, and we now let $S$ denote a finite $\Gamma$-set, with $\Gamma$ acting on the right. We know that $S$ can be written as a disjoint union of $\Gamma$-orbits $S_i$ and that if $s_i \in S_i$ has stabilizer $\Lambda_i$, then $S_i$ is isomorphic as a $\Gamma$-set to the right coset space $\Lambda_i \backslash \Gamma$. Thus we have an isomorphism of $\Gamma$-sets $S \cong \bigcup_i \Lambda_i \backslash \Gamma$ (disjoint union).

The $\mathbb{C}$-vector space on $S$, which we write as $\mathbb{C}(S)$, has a natural $\mathbb{C}\Gamma$-module structure via the rule

$$(\sum \lambda_s s)\gamma = \sum \lambda_s(s\gamma).$$

Given a further $\Gamma$-set $T$ it can happen that $\mathbb{C}(S)$ and $\mathbb{C}(T)$ are isomorphic $\mathbb{C}\Gamma$-modules, but that $S$ and $T$ are *not* isomorphic as $\Gamma$-sets. Presently we shall see that it is precisely such phenomena that provide the algebraic basis for the main results of this section.

Note that, with the above notation, there is an isomorphism of $\mathbb{C}\Gamma$-modules

(7.8) $$\mathbb{C}(S) \cong \bigoplus_i \mathbb{C}(\Lambda_i\backslash\Gamma).$$

Recall that for $\Lambda \subset \Gamma$, $\Sigma_\Lambda = \sum_{\lambda\in\Lambda}\lambda$. The map $\Lambda\gamma \mapsto \Sigma_\Lambda\gamma$ induces an isomorphism of $\Gamma$-sets from $\Lambda\backslash\Gamma$ to the set of elements $\Sigma_\Lambda\gamma$ with $\gamma$ running through a complete set of right representatives of $\Lambda$ in $\Gamma$. Then $\{\Sigma_\Lambda\gamma\}$ is a basis for the right $\mathbb{C}\Gamma$-ideal $\Sigma_\Lambda\mathbb{C}\Gamma$; we therefore have an isomorphism of $\mathbb{C}\Gamma$-modules

(7.9) $$\mathbb{C}(\Lambda\backslash\Gamma) \cong \Sigma_\Lambda\mathbb{C}\Gamma.$$

Let $\Delta \subset \Gamma$ denote a further subgroup of $\Gamma$, and view the cosets $\Lambda\backslash\Gamma$ as a $\Delta$-set, by restricting action from $\Gamma$ to $\Delta$. From the very definition of a double coset, there is an isomorphism of $\Delta$-sets

(7.10) $$\Lambda\backslash\Gamma \cong \bigcup_i \Lambda\backslash\Lambda\gamma_i\Delta$$

where $\{\gamma_i\}$ denotes a system of double coset representatives. Here $\Lambda\backslash\Lambda\gamma\Delta$ is the quotient set of the double coset $\Lambda\gamma\Delta$ modulo left action by $\Lambda$; its elements are the right cosets $\Lambda\gamma\delta$ as $\delta$ varies over $\Delta$. Writing $\Lambda^\gamma = \gamma^{-1}\Lambda\gamma$, we claim that for $\gamma \in \Gamma$, the map $\phi(\lambda\backslash\Lambda\gamma\delta) = (\Lambda^\gamma \cap \Delta)\delta$ induces an isomorphism of $\Delta$-sets

(7.11) $$\phi\colon \Lambda\backslash\Lambda\gamma\Delta \cong (\Lambda^\gamma \cap \Delta)\backslash\Delta.$$

*Proof.* To see that $\phi$ is well defined and injective note that for $\delta$, $\delta' \in \Delta$, $\Lambda\gamma\delta = \Lambda\gamma\delta'$ iff $\delta'\delta^{-1} \in \Delta \cap \Lambda^\gamma$. It is then clear that $\phi$ is onto and that it respects $\Delta$-action.

□

Piecing together (7.8–11), we have shown

**(7.12)** *Let $\Lambda$, $\Delta$ denote subgroups of $\Gamma$; then there is an isomorphism of $\mathbb{C}\Delta$-modules*

$$\mathbb{C}(\Lambda\backslash\Gamma) \cong \bigoplus_i \mathbb{C}((\Lambda^{\gamma_i} \cap \Delta)\backslash\Delta)$$

*with $\{\gamma_i\}$ as in (7.10).*

The last of our preliminary results on permutation modules concerns a description of the fixed points of a coset space $\mathbb{C}(\Lambda\backslash\Gamma)$ by a normal sub-group.

**(7.13)** *Let $\Omega$ be a subgroup of the finite group $\Delta$ and $I$ a normal sub-group of $\Delta$; then there is an isomorphism of $\mathbb{C}\Delta$-modules*

$$\mathbb{C}(\Omega\backslash\Delta)^I \cong \mathbb{C}(\Omega I\backslash\Delta).$$

*Proof.* From (7.9) $\mathbb{C}(\Omega\backslash\Delta) \cong \Sigma_\Omega\mathbb{C}\Delta$, so by (7.2) $\mathbb{C}(\Omega\backslash\Delta)^I \cong \Sigma_\Omega\mathbb{C}\Delta\Sigma_I$. Since $I$ is normal in $\Delta$, $\mathbb{C}\Delta\Sigma_I = \Sigma_I\mathbb{C}\Delta$ and therefore

$$\mathbb{C}(\Omega\backslash\Delta)^I \cong \Sigma_\Omega\Sigma_I\mathbb{C}\Delta.$$

Using (7.9) again, together with the fact that $\Sigma_I\Sigma_\Omega = |I \cap \Omega|\Sigma_{I\Omega}$, we deduce that

$$\Sigma_\Omega\Sigma_I\mathbb{C}\Delta = \Sigma_{I\Omega}\mathbb{C}\Delta \cong \mathbb{C}(\Omega I\backslash\Delta).$$

□

We are now in a position to prove both Theorem 72 and the other main theoretical result of this section.

*Proof of Theorem 72.* With the notation of the statement of the theorem we write $I$, $\Delta$ for the inertia and decomposition groups of $\mathfrak{P}$ in $\Gamma$. Applying (7.12) gives an isomorphism of $\mathbb{C}\Delta$-modules

$$\mathbb{C}(\Lambda\backslash\Gamma) \cong \bigoplus_i \mathbb{C}((\Lambda^{\gamma_i} \cap \Delta)\backslash\Delta)$$

and so by (7.13)

$$\mathbb{C}(\Lambda\backslash\Gamma)^I \cong \bigoplus_i \mathbb{C}((\Lambda^{\gamma_i} \cap \Delta)I\backslash\Delta). \tag{7.14}$$

Since $\Delta/I$ is a cyclic group, the same is clearly true of $\Delta/(\Lambda^{\gamma_i} \cap \Delta)I$, and we denote its order by $k_i$. Because the image of $\sigma_p$ in this group is necessarily a generator, we have

$$\det_{\mathbb{C}((\Lambda^{\gamma_i}\cap\Delta)I\backslash\Delta)}(1 - \sigma_p p^{-x}) = \prod(1 - \eta p^{-x}) = 1 - p^{-xk_i} \tag{7.15}$$

where the middle product extends over all $k_i$th roots of unity.

Hence, by (7.14) and (7.1), it follows that

$$\det_{\mathbb{C}(\Lambda\backslash\Gamma)^I}(1 - \sigma_p p^{-x}) = \prod_i (1 - p^{-xk_i}). \tag{7.16}$$

Next, merely as a notational convenience, we also endow the field $N$

with left $\Gamma$-action, via the rule ${}^{\gamma}n = n^{\gamma^{-1}}$ for $n \in N$, $\gamma \in \Gamma$. With this convention recall from (III.1.20), that there is a bijection between the double cosets $\Lambda\backslash\Gamma/\Delta$ and the prime ideals of ${}^{\Lambda}N$ above $p$ given by the map $\Lambda\gamma\Delta \mapsto {}^{\gamma}\mathfrak{P} \cap {}^{\Lambda}N$; furthermore, recall that $(\Delta \colon I(\Lambda^{\gamma_i} \cap \Delta)) = f_i$, the residue class extension degree of ${}^{\gamma_i}\mathfrak{P} \cap {}^{\Lambda}N$. Since

$$\prod_i (1 - \mathbf{N}\mathfrak{p}_i{}^{-x}) = \prod_i (1 - p^{-xf_i})$$

the theorem follows from (7.16) on noting that

$$f_i = (\Delta \colon I(\Lambda^{\gamma_i} \cap \Delta)) = k_i.$$

□

Suppose now that two $\Gamma$-sets $S$ and $T$ have the property that $\mathbb{C}(S)$ and $\mathbb{C}(T)$ are isomorphic $\mathbb{C}\Gamma$-modules. We shall show how such a relationship automatically gives an identity in Dedekind zeta functions. Suppose that $S$ and $T$ have orbit decompositions:

$$S \cong \dot{\bigcup_a} \Lambda_a\backslash\Gamma \qquad T = \dot{\bigcup_b} \Upsilon_b\backslash\Gamma. \tag{7.17}$$

Since $\mathbb{C}(S)$ and $\mathbb{C}(T)$ are isomorphic, we know that for each prime number $p$

$$\det_{\mathbb{C}(S)^{I_p}}(1 - \sigma_p p^{-x}) = \det_{\mathbb{C}(T)^{I_p}}(1 - \sigma_p p^{-x})$$

and so by (7.17)

$$\prod_a \det_{\mathbb{C}(\Lambda_a\backslash\Gamma)^{I_p}}(1 - \sigma_p p^{-x}) = \prod_b \det_{\mathbb{C}(\Upsilon_b\backslash\Gamma)^{I_p}}(1 - \sigma_p p^{-x})$$

Therefore, by Theorem 72, we have shown the first part of

**Theorem 73** (Brauer). *Let $N_a$ (resp. $N_b$) denote the subfield of $N$ fixed by $\Lambda_a$ (resp. $\Upsilon_b$). Then*

*(a)*
$$\prod_a \zeta_{N_a}(x) = \prod_b \zeta_{N_b}(x)$$

*(b)*
$$\prod_a \frac{h_{N_a} R_{N_a}}{W_{N_a}|d_{N_a}|^{1/2}} = \prod_b \frac{h_{N_b} R_{N_b}}{W_{N_b}|d_{N_b}|^{1/2}}.$$

To prove part (*b*) we consider the behaviour of both sides in part (*a*) as $x \to 1+$. From Theorem 61 we know that $\prod_a \zeta_{N_a}(x)$ is of the form $(x-1)^{-n}\phi(x)$ with $\phi(x)$ continuous on $[1, \infty)$, where $n$ denotes the number of of orbits in $S$. From (*a*) we conclude straightaway that $S$ and $T$ must have the same number of $\Gamma$-orbits. Therefore on multiplying by $(x-1)^n$ on both sides of (*a*), and considering the limit as $x \to 1+$,

Theorem 61 gives

$$\prod_a \frac{2^{s(N_a)+t(N_a)}\pi^{t(N_a)}h_{N_a}R_{N_a}}{W_{N_a}|d_{N_a}|^{1/2}} = \prod_b \frac{2^{s(N_b)+t(N_b)}\pi^{t(N_b)}h_{N_b}R_{N_b}}{W_{N_b}|d_{N_b}|^{1/2}}.$$

Thus to prove part (*b*) it suffices to prove

$$(7.18.a) \qquad \sum_a s(N_a) = \sum_b s(N_b)$$

$$(7.18.b) \qquad \sum_a t(N_a) = \sum_b t(N_b).$$

To prove (7.18) first note that

$$\begin{aligned}\sum_a s(N_a) + 2t(N_a) &= \sum_a (\Gamma\colon \Lambda_a) = \dim_{\mathbb{C}}(\mathbb{C}(S)) \\ &= \dim_{\mathbb{C}}(\mathbb{C}(T)) = \sum_b (\Gamma\colon \Upsilon_b) \\ &= \sum_b s(N_b) + 2t(N_b).\end{aligned}$$

It will therefore suffice to prove

$$(7.18.c) \qquad \sum_a s(N_a) + t(N_a) = \sum_b s(N_b) + t(N_b).$$

Let $\Sigma$ denote the stabiliser of any Archimedean absolute value of $N$, so that $\Sigma$ has order 1 (resp. 2) if $N$ is totally real (resp. totally imaginary). From (III.1.20) we know that $s(N_a)+t(N_a)$, the number of Archimedean absolute values of $N_a$, is the number of elements in the double coset $\Lambda_a\backslash\Gamma/\Sigma$, which we denote by $|\Lambda_a\backslash\Gamma/\Sigma|$. This then shows that

$$\begin{aligned}\sum_a s(N_a) + t(N_a) &= \sum_a |\Lambda_a\backslash\Gamma/\Sigma| \\ &= \dim_{\mathbb{C}}(\mathbb{C}(S)^{\Sigma}).\end{aligned}$$

Since $\mathbb{C}(S)^{\Sigma} \cong \mathbb{C}(T)^{\Sigma}$, running the argument backwards, with $T$ in place of $S$, gives (7.18.*c*).

□

We remark that it can also be shown that

$$\prod_a |d_{N_a}| = \prod_b |d_{N_b}|$$

which gives a further improvement to part (*b*) of the theorem; however, we do not prove this identity here.

In the sequel, for brevity we shall often write

$$\xi(N_a) = \frac{h_{N_a}R_{N_a}}{W_{N_a}|d_{N_a}|^{1/2}} \qquad \xi_0(N_a) = \frac{h_{N_a}R_{N_a}}{W_{N_a}}.$$

Thus, with this notation, part (*b*) of the theorem becomes

$$\prod_a \xi(N_a) = \prod_b \xi(N_b). \tag{7.19}$$

The remainder of this section is devoted to applications of the above theorem. As our first example of two Γ-sets which provide an interesting Brauer relation, we consider the case where Γ is an elementary group of order $l^2$, with $l$ a prime number: thus Γ can be written as the direct product of two cyclic groups of order $l$. Let $S$ and $T$ denote Γ-sets with orbit decompositions:

$$S = (\Gamma\backslash\Gamma)^{(l)} \dot{\bigcup} \Gamma$$

$$T = \dot{\bigcup_{\Delta}} \Delta\backslash\Gamma.$$

Here $\Gamma\backslash\Gamma$ is a singleton with a trivial Γ-action, $(\Gamma\backslash\Gamma)^{(l)}$ denotes the disjoint union of $l$-copies of this Γ-set, and the union in the definition of $T$ extends over the $l+1$ distinct subgroups of $\Delta$ of order $l$ in Γ.

The trace of $\gamma \in \Gamma$ on $\mathbb{C}(S)$ is given by the number of $\gamma$-fixed elements in $S$, denoted $|S^{<\gamma>}|$. Thus to show $\mathbb{C}(S) \cong \mathbb{C}(T)$, it suffices to prove

**(7.20)** *For each* $\gamma \in \Gamma$, $|S^{<\gamma>}| = |T^{<\gamma>}|$.

*Proof.* The result is clear when $\gamma = 1$, since $|S| = l^2 + l = |T|$. So let $\gamma \in \Gamma$, $\gamma \neq 1$. We note that

$$(\Gamma\backslash\Gamma)^{<\gamma>} = \Gamma\backslash\Gamma \text{ and } \Gamma^{<\gamma>} = \emptyset$$

$$(\Delta\backslash\Gamma)^{<\gamma>} = \begin{cases} \Delta\backslash\Gamma & \text{if } \gamma \in \Delta; \\ \emptyset & \text{if } \gamma \notin \Delta. \end{cases}$$

Therefore $|S^{<\gamma>}| = l = |T^{<\gamma>}|$. □

(7.20) in conjunction with Theorem 73 yields the following relation

**(7.21)** *Let* $K/\mathbb{Q}$ *have Galois group* Γ *(as above). Then* $\xi(\mathbb{Q})^l \xi(K) = \prod_\Delta \xi(K^\Delta)$.

This equality contains a considerable amount of information on the relationship between the class numbers and units of an elementary extension $K/\mathbb{Q}$ of degree $l^2$, and those of its subfields. We now pursue the detailed analysis of this relation in the particular case $l = 2$; this is the case mentioned at the start of this section.

## Biquadratic fields

We adopt the notation of §2 of Chapter V: so, in particular, we let $K$ be a biquadratic field and we let $k_1$, $k_2$, $k_3$, denote the quadratic subfields of $K$, with $k_1$ real if $K$ is imaginary. From (7.21) we know that

$$\xi(\mathbb{Q})^2\xi(K) = \xi(k_1)\xi(k_2)\xi(k_3). \tag{7.22}$$

Next we consider the problem of the discriminantal contribution.

**(7.23)** *If 2 is not totally ramified in a biquadratic number field $K$, then*

$$|d_K| = \prod_i |d_{k_i}| \tag{7.23.a}$$

*and so*

$$\xi_0(\mathbb{Q})^2\xi_0(K) = \xi_0(k_1)\xi_0(k_2)\xi_0(k_3). \tag{7.23.b}$$

[As remarked previously, in fact the results of (7.23) hold without this condition on 2.]

*Proof.* Let $p$ denote a prime: we show (7.23.$a$) by checking that the $p$-valuations of the integers on each side agree. We begin by observing that, in all cases, the inertia group $I$ of $p$ in $K/\mathbb{Q}$ is cyclic: for, if $p = 2$, then by hypothesis $|I| \leq 2$; while if $p \neq 2$, then $p$ is at most tamely ramified, and so $I$ is cyclic by Theorem 28.

If $I = (1)$, then all discriminants in question are coprime to $p$ and the result is immediate. If $I \neq (1)$, then $|I| = 2$ and we suppose $k_i = K^I$. In this case $p$ ramifies in each $k_j/\mathbb{Q}$ for $j \neq i$, but is non-ramified in $k_i/\mathbb{Q}$; therefore, by the formula for discriminants in (III.2.15) applied to the tower of number fields $K \supset k_j \supset \mathbb{Q}$, we have

$$d_{K,p} = d_{k_j,p}^2 \text{ for each } j \neq i$$

where $d_{K,p}$ denotes the $p$-part of $d_K$, etc.

We have therefore shown that

$$d_{K,p} = d_{k_1,p}d_{k_2,p}d_{k_3,p}$$

which establishes (7.23.$a$). Lastly we note that (7.23.$b$) is a direct consequence of (7.22) and (7.23.$a$).

□

We write $V_K = U_{k_1}U_{k_2}U_{k_3}$. Our aim in the sequel is to use (7.23.$b$) to show how the knowledge of all the $h_{k_i}$ together with the unit index $(U_K{:}\,V_K)$, determines $h_K$. The index $(U_K{:}\,V_K)$ was studied in some detail in (V, §2) in the case when $K$ is imaginary. At this stage, we remark that it has not been shown that, in general, $V_K$ has finite index in $U_K$. Although one could prove this fairly easily by algebraic means, it will follow as an immediate consequence of our Brauer relation.

**(7.24)** *Let $\nu_K = \mu_K \cap V_K$, then*

$$4W_K = (\mu_K : \nu_K) \prod_{i=1}^{3} W_{k_i}.$$

*Proof.* We prove this result by considering the various possible cases for $L = \mathbb{Q}(\mu_K)$. We let $i = \sqrt{-1}$ and we write $\omega$ for a non-trivial cube root of unity. By (V.2.1) we need only consider the three following cases.

(1) $W_K = 8$ or $12$; then $K = L$ and the result is easily verified, since the $k_i$ are explicitly determined.

(2) $W_K = 4$ or $6$; then $L$ is a quadratic subfield of $K$ and the only roots of unity in the two remaining quadratic subfields are $< \pm 1 >$.

(3) $W_K = 2$. Thus $W_{k_i} = 2$ for all $i$ and the result is immediate.

□

Recall that $\psi: U_K \to \mathbb{R}^{(s+t)}$ denotes the logarithmic homomorphism associated with $K$, defined in (IV, §4). Applying $\psi$ to $V_K$, we obtain a commutative diagram with exact rows

$$\begin{array}{ccccccccc} 1 & \longrightarrow & \mu_K & \longrightarrow & U_K & \longrightarrow & \psi(U_K) & \longrightarrow & 0 \\ & & \uparrow & & \uparrow & & \uparrow & & \\ 1 & \longrightarrow & \nu_K & \longrightarrow & V_K & \longrightarrow & \psi(V_K) & \longrightarrow & 0 \end{array}$$

So, by the Snake Lemma, we see that if $V_K$ has finite index in $U_K$, then

$$(U_K : V_K) = (\mu_K : \nu_K)(\psi(U_K) : \psi(V_K)). \quad (7.25)$$

Let $r = s(K) + t(K) - 1$ denote the Dirichlet rank of $U_K$, and let $\{u_i \mid 1 \le i \le r\}$ denote a system of fundamental units of $U_K$. As previously we put $n = (K : \mathbb{Q})$, and, for the moment, we write $nR(V_K)$ for the absolute value of the determinant of the matrix

$$\begin{vmatrix} \psi(v_1) \\ \vdots \\ \psi(v_r) \\ \mathbf{1}_{s+t} \end{vmatrix}$$

where $\{v_i \mid 1 \le i \le r\}$ denotes any set of units with the property that they span $V_K/\nu_K$.

We write $\psi(v_i) = \sum_j a_{ij}\psi(u_j)$, $a_{ij} \in \mathbb{Z}$, so that $R(V_K) = |\det(a_{ij})| R_K$

and $\psi(V_K)$ has finite index in $\psi(U_K)$ iff $\det(a_{ij}) \neq 0$; moreover when this index is finite it is $|\det(a_{ij})|$, i.e. in that case $R(V_K) \neq 0$ and

$$\frac{R(V_K)}{R_K} = (\psi(U_K) \colon \psi(V_K)). \tag{7.26}$$

Next we show

**(7.27)**

$$\prod_{i=1}^{3} R_{k_i} = R(V_K) \times \begin{cases} 1/4 & \textit{if } K \textit{ is real} \\ 1/2 & \textit{if } K \textit{ is imaginary.} \end{cases}$$

*Proof.* We consider separately the case where $K$ is imaginary or real. If $K$ is imaginary, then by definition $1 = R_{k_2} = R_{k_3}$, while

$$R_{k_1} = |\log|u|| \text{ and } R(V_K) = \left|\log(|u|^2)\right| = 2\,|\log|u||$$

where as usual $u$ denotes a fundamental unit of $k_1$.

If $K$ is real, then by definition $R(V_K) = \left|\log|u_i^{\sigma_j}|\right|$, where $i, j =$ 1, 2, 3. The result then follows on noting that

$$(a) \qquad \log\left|u_i^{\sigma_j}\right| = \begin{cases} \log|u_i| & \text{if } i = j \\ -\log|u_i| & \text{if } i \neq j \end{cases}$$

$$(b) \qquad \det\begin{pmatrix} a & -a & -a \\ -b & b & -b \\ -c & -c & c \end{pmatrix} = -4abc.$$

□

It is now relatively straightforward to combine (7.23.$b$) with above results to obtain

**Theorem 74.** *$R(V_K) \neq 0$ and, if 2 is not totally ramified in $K/\mathbb{Q}$, then*

$$h_K = 2^{-\frac{1}{2}(s(K)+t(K))}(U_K \colon V_K)\prod_{i=1}^{3} h_{k_i}.$$

*Proof.* By (7.23.$b$) together with the definition of $\xi_0$, we know that

$$\xi_0(\mathbb{Q})^2 h_K = \prod h_{k_i}\left(\frac{\prod R_{k_i}}{R_K}\right)\left(\frac{W_K}{\prod W_{k_i}}\right).$$

By (7.27) we immediately conclude that $R(V_K) \neq 0$. Using (7.24) and the fact that $\xi_0(\mathbb{Q}) = \frac{1}{2}$, we deduce that

$$h_K = \left(\prod h_{k_i}\right)\left(\frac{\prod R_{k_i}}{R_K}\right)(\mu_K \colon \nu_K).$$

Thus by (7.25,26,27)

$$h_K = \left(\prod h_{k_i}\right)(U_K\colon V_K) \times \begin{cases} 1/4 & \text{if } K \text{ is real;} \\ 1/2 & \text{if } K \text{ is complex.} \end{cases}$$

□

**Corollary 1.** *If $K$ is imaginary, with $k_1 = \mathbb{Q}(\sqrt{p})$, for some prime number $p \equiv 1 \bmod (4)$; then*

$$h_K = \frac{1}{2} h_{k_1} h_{k_2} h_{k_3}.$$

*Proof.* Since 2 is non-ramified in $k_1/\mathbb{Q}$, we know by the above theorem that

$$h_K = \frac{1}{2} h_{k_1} h_{k_2} h_{k_3} (U_K\colon V_K). \tag{7.28}$$

Now, in (V, 2.4) we showed that $\psi(U_K) = \psi(V_K)$; while by (V, 2.1) we know that $\mu_K = \nu_K$, unless $K$ is either the field of eighth, or twelfth, roots of unity. Since neither of these two fields has its real subfield of our given form, we conclude that $\mu_K = \nu_K$, and so by (7.25) $U_K = V_K$.

□

**Corollary 2.** *If $K = \mathbb{Q}(\sqrt{p}, \sqrt{-q})$ with $p$, $q$ primes, $p \equiv 1 \bmod (4)$, $q \equiv -1 \bmod (4)$; then, writing $k_2 = \mathbb{Q}(\sqrt{-q})$, $h_{k_1}$ and $h_{k_2}$ are both odd; moreover $h_{k_3} \equiv 2 \bmod (4)$ iff $h_K$ is odd.*

*Proof.* From (V, §1) we know that $h_{k_1}$ and $h_{k_2}$ are both odd. The remainder of the result then follows from Corollary 1.

□

**Corollary 3.** *If $K = \mathbb{Q}(\sqrt{-p_1}, \sqrt{-p_2})$ where $p_1$, $p_2$ are distinct prime numbers with $p_1 \equiv -1 \equiv p_2 \bmod (4)$; then*

$$h_K = h_{k_1} h_{k_2} h_{k_3}.$$

*Proof.* Since 2 is non-ramified in $K/\mathbb{Q}$, we can again apply the theorem to deduce (7.28) in this case. As in Corollary 1 we note that $\mu_K = \nu_K$; while by (V, 2.5) $(\psi(U_K)\colon \psi(V_K)) = 2$. By (7.25) we conclude that $(U_K\colon V_K) = 2$, and the result follows.

□

We conclude this sub-section on biquadratic fields by giving three worked examples.

*Example 1.* $K = \mathbb{Q}(\sqrt{5}, \sqrt{-3})$; then $k_1 = \mathbb{Q}(\sqrt{5})$. Using the Minkowski method of Chapter IV it is readily verified that $\mathbb{Q}[\sqrt{5}]$ and $\mathbb{Q}[\omega]$ ($\omega^3 = 1 \neq \omega$) have class number one, while $\mathbb{Q}[\frac{1+\sqrt{-15}}{2}]$ has class number two. We can then apply Corollary 1 to deduce that $h_K = 1$.

This example provides an instance of a number field with non-trivial class group which can be embedded in a number field with class number one. In general this cannot always be achieved.

*Example 2.* $K = \mathbb{Q}(\sqrt{-3}, \sqrt{-7})$; then $k_1 = \mathbb{Q}(\sqrt{21})$. By the Minkowski method one easily checks that all three quadratic subfields have class number one; then by Corollary 3 we conclude that $h_K = 1$.

*Example 3.* Let $\zeta$ denote a primitive eighth root of unity and let $K = \mathbb{Q}(\zeta)$. In this case 2 is totally ramified in $K/\mathbb{Q}$, and so we cannot immediately apply Theorem 74. However, by Theorem 46 and applying (I.1.5) to $X^4 + 1$, we know $|d_K| = 2^8$; on the other hand $d_{\mathbb{Q}(\sqrt{2})} = 8$, $d_{\mathbb{Q}(\sqrt{-2})} = -8$, $d_{\mathbb{Q}(\sqrt{-1})} = -4$; so that by direct verification

$$|d_K| = \prod_{j=1}^{3} |d_{k_j}|$$

and we may therefore apply the equation of Theorem 74. Since $k_1 = \mathbb{Q}(\sqrt{2})$, we take $u_1 = 1 + \sqrt{2}$ which has norm $-1$. We therefore note that $\mathbb{Q}(\sqrt{u_1})/\mathbb{Q}$ is non-abelian, and so, as in the proof of (V.2.4), we see that $U_K = <\zeta> \times u_1^{\mathbb{Z}}$, $V_K = <i> \times u_1^{\mathbb{Z}}$. It is very easily verified that all three $k_j$ have class number one: indeed their rings of integers are all Euclidean domains. It therefore follows that $h_K = \frac{1}{2}(U_K\colon V_K) = 1$.

## Frobenius groups

For our second application of Theorem 73 we consider the family of finite groups known as Frobenius groups. We shall call a finite group $\Gamma$ a *Frobenius group* if

(7.29.*a*) $\Gamma$ possesses a proper sub-group $T$ with the property that $T^\gamma \cap T = (1)$ whenever $\gamma \notin T$.

(7.29.*b*) $\Gamma$ possesses a normal sub-group $\Sigma$ such that $T \cap \Sigma = (1)$, $\Gamma = \Sigma T$.

In fact a theorem of Frobenius shows that if $\Gamma$ satisfies condition (7.29.*a*), then condition (7.29.*b*) is automatically satisfied; however, we shall not prove this fact here.

The first aim of this sub-section is to show

**Theorem 75.** *Let $K$ be a finite Galois extension of $\mathbb{Q}$ with $\Gamma = \mathrm{Gal}(K/\mathbb{Q})$ a Frobenius group, as above; then*

$$\xi(\mathbb{Q})^{|T|}\xi(K) = \xi(K^T)^{|T|}\xi(K^\Sigma).$$

By Theorem 73 it suffices to consider the two $\Gamma$-sets

$$S = (\Gamma\backslash\Gamma)^{|T|} \dot{\bigcup} \Gamma \qquad R = (T\backslash\Gamma)^{|T|} \dot{\bigcup} \Sigma\backslash\Gamma$$

and show that for any cyclic group $\Xi$ of $\Gamma$

$$|S^\Xi| = |R^\Xi|. \tag{7.30}$$

Prior to calculating the $\Gamma$-action on the various orbits, we prove

**(7.31)** *Given $\xi \in \Gamma$*

(7.31.$a$) $\Sigma\gamma = \Sigma\gamma\xi$ iff $\xi \in \Sigma$;

(7.31.$b$) *for $\delta \in \Gamma - \Sigma$ there exists $\gamma \in \Gamma$ such that $\delta^\gamma = \gamma^{-1}\delta\gamma \in T$. Furthermore if $\delta^\tau \in T$ then $\tau\gamma^{-1} \in T$.*

*Proof.* (7.31.$a$) is clear since $\Sigma$ is normal in $\Gamma$. To prove (7.31.$b$) we consider the union $\mathcal{S}$ of all elements in the groups $T^\gamma$ different from the identity

$$\mathcal{S} = \bigcup_{\gamma\in\Gamma}(T^\gamma - 1).$$

Since $\Sigma$ is normal in $\Gamma$, with $\Sigma \cap T = (1)$, it follows that $\mathcal{S} \subset \Gamma - \Sigma$. We wish to prove $S = \Gamma - \Sigma$; we do this by verifying $|S| = |\Gamma| - |\Sigma|$. By property (7.29.$b$), $T^\gamma \cap T^\delta = (1)$ unless $\gamma\delta^{-1} \in T$. Thus conjugation by $\Gamma$ on $T$ yields $(\Gamma : T)$ different sub-groups whose only pairwise common element is (1). Thus $|\mathcal{S}| = (|T|-1)(\Gamma : T)$, which now establishes the first part of (7.31.$b$). Next if $\delta^\gamma$ and $\delta^\tau$ both lie in $T$, then $T^{\gamma^{-1}} \cap T^{\tau^{-1}} \neq (1)$, whence $T \cap T^{\tau^{-1}\gamma} \neq (1)$; and so $\tau^{-1}\gamma \in T$.

□

We now construct the table for the number of fixed points

| | $\xi = 1$ | $\xi \in \Sigma - 1$ | $\xi \in \Gamma - \Sigma$ |
|---|---|---|---|
| $\Gamma\backslash\Gamma$ | 1 | 1 | 1 |
| $\Gamma$ | $\lvert\Gamma\rvert$ | 0 | 0 |
| $T\backslash\Gamma$ | $\lvert\Sigma\rvert$ | 0 | 1 |
| $\Sigma\backslash\Gamma$ | $\lvert T\rvert$ | $\lvert T\rvert$ | 0 |

The entries for $\Gamma$ and $\Gamma\backslash\Gamma$ are immediate. $\xi$ has fixed point $\Sigma\gamma$ in $\Sigma\backslash\Gamma$ iff $\Sigma\gamma = \Sigma\gamma\xi$ iff $\xi \in \Sigma$ by (7.31.$a$). So next consider a fixed point for $\xi$ in $T\backslash\Gamma$: $T\gamma$ is fixed iff $T\gamma\xi = T\gamma$ iff $\xi^{\gamma^{-1}} \in T$. If $\xi = 1$, clearly all cosets are fixed; if $\xi \in \Sigma - 1$, then $\xi^{\gamma^{-1}} \in \Sigma - 1$ since $\Sigma$ is normal, and so $\xi$

has no fixed points, because $\Sigma \cap T = (1)$. Lastly the result for $\xi \in \Gamma - \Sigma$ follows by (7.31.*b*).

From the above table we see immediately that (7.30) holds in all cases. This then completes our proof of Theorem 75.

## Cubic and sextic fields

We now return to the situation first considered in (V, §3): namely we consider a cubic extension $K$, of negative discriminant $d_K$ and with normal closure $N = K.L$, where $L = \mathbb{Q}(\sqrt{d_K})$. We adopt the notation of that section, so that $\Gamma = \mathrm{Gal}(N/\mathbb{Q})$, $\Sigma = <\sigma> = \mathrm{Gal}(N/L)$, $T = <\tau> = \mathrm{Gal}(N/K)$. Since $\gamma^{-1}\tau\gamma = \tau$ iff $\gamma \in <\tau>$; $<\sigma> \cap <\tau> = 1$; $<\sigma> \lhd \Gamma$; $\Gamma = \Sigma.T$: we see that $\Gamma$ is a Frobenius group and so we can apply Theorem 75 to deduce that

$$\xi(\mathbb{Q})^2\xi(N) = \xi(K)^2\xi(L). \tag{7.32}$$

It is our intention to use this equality to show

**Theorem 76.** *$V_N$ has finite index in $U_N$ and*

$$\frac{3h_N}{|d_N|^{1/2}} = \frac{{h_K}^2}{|d_K|}\frac{h_L}{|d_L|^{1/2}}(\mathfrak{U}_N : \mathfrak{V}_N).$$

(Recall that $V_N = U_K U_{K_2} U_{K_3}$, where $K_2$, $K_3$ denote the conjugate fields of $K$ in $N$, and that we view $K$ as contained in $\mathbb{R}$; also $\mathfrak{U}_N = U_N/\mu_N$ and $\mathfrak{V}_N = V_N\mu_N/\mu_N$.)

Define

$$A = \frac{W_{\mathbb{Q}}^2 W_N}{W_K^2 W_L} \qquad B = \frac{R_K^2 R_L}{R_N}.$$

In order to be able to use (7.32) to deduce Theorem 76, it will suffice to show

$$A = 1 \tag{7.33.a}$$

$$B = \frac{1}{3}(\mathfrak{U}_N : \mathfrak{V}_N). \tag{7.33.b}$$

*Proof.* By (V.3.9) we know that $\mu_N = \mu_L$; while $\mu_{\mathbb{Q}} = \mu_K = \pm 1$, since $K$ possesses a real embedding. This proves (*a*).

In order to prove (*b*), we first note that of course $R_L = 1$, since $L$ is quadratic imaginary. Writing $R(V_N)$ for the $N$-regulator of $V_N$, we recall that, as previously, $V_N$ has finite index in $U_N$ iff $R(V_N) \neq 0$ and in that case

$$\frac{R(V_N)}{R_N} = (\mathfrak{U}_N : \mathfrak{V}_N).$$

We choose $u > 1$ to be a fundamental unit of $K$; then by the definition of the regulator

$$R_K = \log|u|$$

while $R(V_N)$ is the absolute value of the determinant

$$\begin{vmatrix} 2\log|u| & 2\log|u_2| \\ 2\log|u_2| & 2\log|u_3| \end{vmatrix}$$

Since $u\,u_2\,u_3 = 1$ and $|u_2| = |u_3|$, we see that $\log|u| = -2\log|u_2|$ and so

$$R(V_N) = 12\log^2|u_2| = 3\log^2|u|.$$

□

As an illustration of Theorem 76 in action we now consider the Ishida polynomials $f(X) = X^3 + lX - 1$, where $l$ is an even, positive integer such that $4l^3 + 27$ is square-free; let $v$ denote the unique real root of $f$, set $K = \mathbb{Q}(v)$, and let $N$ denote the normal closure of $K/\mathbb{Q}$. We shall now show that

$$3h_N = h_K^2 h_L \tag{7.34}$$

*Proof.* First we consider the unit index in Theorem 76: by (V.3.10) we know that $\mathfrak{U}_N = \mathfrak{V}_N$.

We next consider the discriminantal factors. Since $f$ has discriminant $d = -(4l^3 + 27)$ which is square-free, we deduce that $\mathfrak{o}_K = \mathbb{Z}[v]$ and $d_K = d$; furthermore, since $d \equiv 1 \bmod (4)$, we note that also $d_L = d$. We claim that $N/L$ is non-ramified at all primes of $L$: so that by the tower formula for discriminants we shall have $d_N\mathbb{Z} = d^3\mathbb{Z}$, and so

$$|d_N| = |d^3| = |d_K^2|\,|d_L|.$$

To prove that $N/L$ is non-ramified, we first observe that, since $d = -(4l^3 + 27)$ is square-free, $3 \nmid l$; hence 3 is non-ramified in $K/\mathbb{Q}$. Next we let $p$ denote a ramified prime of $K/\mathbb{Q}$. If $p$ were totally ramified in $K/\mathbb{Q}$, then $K_p/\mathbb{Q}_p$ would be tamely ramified, and so by (III.2.14) and Theorem 26, $p^2 \mid d_K = d$ which is impossible. We therefore conclude that no prime $p$ has inertia group containing $\Sigma$. Alternatively proceed as for (III.4.15).

□

*Example.* Let $K = \mathbb{Q}(\alpha)$, where $\alpha$ is the real root of $X^3 + 2X - 1$. Then $d_K = -59$ and so by (V.3.4) $U_N = V_N$. The calculation of the class number of $L = \mathbb{Q}(\sqrt{-59})$ is entirely straight-forward, and one finds

$h_L = 3$. To calculate $h_K$, we use the Minkowski bound to conclude that it suffices to consider the class of integral $\mathfrak{o}_K$-ideals $\mathfrak{a}$ with norm

$$\mathbf{N}\mathfrak{a} \leq \frac{8}{9\pi}.\sqrt{59} < 3.$$

However, $X^3 + 2X - 1 \equiv (X+1)(X^2 + X + 1) \bmod (2)$, and so $\mathfrak{p} = (\alpha - 1, 2)$ is the unique non-trivial $\mathfrak{o}_K$-ideal with norm $< 3$. Substituting $\alpha = \phi + 1$ in the minimal equation for $\alpha$, we see that $\phi^3 + 3\phi^2 + 5\phi + 2 = 0$, and so $\mathfrak{p} = \phi\mathfrak{o}_K$. We therefore have shown that $K$ has class number one, and so, by (7.34), $N$ also has class number one.

# Appendix A

## Characters of Finite Abelian Groups

This appendix is included for the reader who is unfamiliar with the basics of the theory of abelian characters. We shall use the results of (I.§1) to derive all the character theoretic results which we use in the text (with the exception of (VIII,§7)).

Throughout this section $K$ denotes a fixed algebraically closed field of characteristic zero, and $\Gamma$ will always denote a finite abelian group. The group of characters of $\Gamma$ is $\mathrm{Hom}(\Gamma, K^*)$, i.e. the group of all homomorphisms $\chi: \Gamma \to K^*$; we denote this group by $\hat{\Gamma}$. Note that, as $K$ is fixed, we do not include the dependence of $\hat{\Gamma}$ on $K$ in our notation.

The theory of abelian group characters provides a powerful and elegant application of the theory of finite commutative algebras; moreover group characters play an important role at a number of points in our treatment of algebraic number theory, in particular in Chapters VI and VIII. As well as studying the various properties of $\hat{\Gamma}$ as a group, we can also use $\hat{\Gamma}$ to describe $K$-vector spaces on which $\Gamma$ acts.

We shall make heavy use of the notion of a direct product

(A1) $$\Gamma = \Gamma_1 \times \Gamma_2 \times \cdots \times \Gamma_n$$

of abelian groups, and so we quickly run over the basic definitions. The elements of the product (A1) are vectors $(\gamma_1, \ldots, \gamma_n)$ with $\gamma_i \in \Gamma_i$. The group operation, written multiplicatively, is defined componentwise, i.e.

$$(\gamma_1, \ldots, \gamma_n)(\delta_1, \ldots, \delta_n) = (\gamma_1\delta_1, \ldots, \gamma_n\delta_n).$$

We have projections

$$\pi_i: \Gamma \to \Gamma_i \qquad \pi_i(\gamma_1, \ldots, \gamma_n) = \gamma_i$$

with the property that, given homomorphisms of groups

$$f_i: \Sigma \to \Gamma_i,$$

there is a unique homomorphism

$$f: \Sigma \to \Gamma$$

so that $f_i = \pi_i \circ f$ for all $i$: indeed, for $\sigma \in \Sigma$, we take $f(\sigma) = (f_1(\sigma), ..., f_n(\sigma))$. This property characterises $\{\Gamma, \pi_i\}_i$ uniquely to within isomorphism.

We shall also view $\Gamma$ as a "direct sum": by this we mean that there are injections $\kappa_i: \Gamma_i \to \Gamma$ for all $i$, given by

$$\kappa_i(\gamma_i) = (1_1, \ldots, \gamma_i, \ldots, 1_n)$$

where $1_j$ denotes the identity of $\Gamma_j$. These injections have the property that, given homomorphisms $g_i: \Gamma_i \to \Omega$ of abelian groups, then there is a unique homomorphism $g: \Gamma \to \Omega$ with $g \circ \kappa_i = g_i$ for all $i$: indeed, we take

$$g(\gamma_1, \ldots, \gamma_n) = g_1(\gamma_1) \cdots g_n(\gamma_n).$$

We claim that

(A2) $$\hat{\Gamma} = \hat{\Gamma}_1 \times \hat{\Gamma}_2 \times \cdots \times \hat{\Gamma}_n.$$

For, given $\chi_i: \Gamma_i \to K^*$, then, by the above, there is a unique homomorphism $\chi: \Gamma \to K^*$ with $\chi_i = \chi \circ \kappa_i$. The map

$$\hat{\Gamma}_1 \times \cdots \times \hat{\Gamma}_n \to \hat{\Gamma} \qquad (\chi_1, \ldots, \chi_n) \mapsto \chi$$

is clearly a homomorphism of groups, and, by the defining property $\chi_i = \chi \circ \kappa_i$, we see that the map is injective. In fact it is also surjective, since $\chi \in \hat{\Gamma}$ is the image of

$$(\chi \circ \kappa_1, \ldots, \chi \circ \kappa_n).$$

Next we consider in detail the special case when $\Gamma$ is cyclic of order $n$, on a chosen generator $\gamma$. If $\chi \in \hat{\Gamma}$, then $\chi(\gamma^i) = \chi(\gamma)^i$; thus $\chi(\gamma)$ must be an $n$th root of unity in $K$; conversely, if $\zeta$ denotes an $n$th root of unity in $K$, then $\chi(\gamma^i) = \zeta^i$ defines a character of $\Gamma$. Thus we have shown that $\chi \mapsto \chi(\gamma)$ induces an isomorphism between $\hat{\Gamma}$ and $\mu_n$, the group of $n$th roots of unity of $K$. Since $K$ has characteristic zero, $\mu_n$ is a cyclic group of order $n$. Thus we have shown that if $\Gamma$ is cyclic, then $\Gamma \cong \hat{\Gamma}$.

By the structure theory of finite abelian groups, every such group is the product (A1) of cyclic groups $\Gamma_i$. Therefore we conclude that

**(A3)** $\Gamma \cong \hat{\Gamma}$ *for any finite abelian group* $\Gamma$.

We shall now introduce the group ring $K\Gamma$ of the finite abelian group $\Gamma$

over the field $K$. This enables us to apply the theory of finite commutative algebras developed in (I,§1), in order to gain a deeper understanding of group characters. As a $K$-vector space $K\Gamma$ has a basis consisting of the group elements $\{\gamma \mid \gamma \in \Gamma\}$. Thus addition in $K\Gamma$ is componentwise, i.e. for $a_\gamma, b_\gamma \in K$ we have

$$\sum_{\gamma\in\Gamma} a_\gamma\gamma + \sum_{\gamma\in\Gamma} b_\gamma\gamma = \sum_{\gamma\in\Gamma}(a_\gamma + b_\gamma)\gamma.$$

Multiplication is bilinear and is given by multiplication in $\Gamma$; thus

$$\begin{aligned}(\sum_{\gamma\in\Gamma} a_\gamma\gamma)(\sum_{\delta\in\Gamma} b_\delta\delta) &= \sum_{\gamma,\delta} a_\gamma b_\delta \gamma\delta \\ &= \sum_{\sigma\in\Gamma}(\sum_{\gamma\in\Gamma} a_\gamma b_{\gamma^{-1}\sigma})\sigma.\end{aligned}$$

It is clear that $K\Gamma$ is a finite dimensional commutative $K$-algebra. We shall now find its idempotents. This will determine the structure of $K\Gamma$ and will prove to be a fundamental tool in establishing the basic properties of characters. For $\chi \in \hat{\Gamma}$ define

(A4) $$e_\chi = |\Gamma|^{-1}\sum_{\gamma\in\Gamma}\gamma\chi(\gamma^{-1}).$$

**Theorem A.**

(*a*) *The map $\chi \mapsto e_\chi$ is a bijection from $\hat{\Gamma}$ onto the set of primitive idempotents of $K\Gamma$.*

(*b*) *For each $\chi$, $c \mapsto ce_\chi$ is an isomorphism $K \cong K\Gamma e_\chi$ of $K$-algebras.*

*Proof.* Immediately from (A4) we have

(A5) $$\begin{aligned}e_\chi\gamma &= |\Gamma|^{-1}\sum_\delta \delta\gamma\chi(\delta)^{-1} \\ &= |\Gamma|^{-1}(\sum_\delta \delta\gamma\chi(\delta\gamma)^{-1})\chi(\gamma) \\ &= e_\chi\chi(\gamma)\end{aligned}$$

for all $\gamma \in \Gamma$. Therefore

$$e_\chi^2 = |\Gamma|^{-1}\sum_\gamma e_\chi\chi(\gamma)^{-1}\gamma = |\Gamma|^{-1}\sum_\delta e_\chi = e_\chi$$

and so $e_\chi$ is a non-zero idempotent. By (A5), $(\sum_\gamma a_\gamma\gamma) \mapsto e_\chi(\sum_\gamma a_\gamma\gamma) = \sum_\gamma e_\chi\chi(\gamma)a_\gamma$ and so $e_\chi K\Gamma = e_\chi K$ which gives (*b*). By (I.1.34) this implies that $e_\chi$ is a primitive idempotent. If $\chi, \phi \in \hat{\Gamma}$, $\chi \neq \phi$, then by definition $e_\chi \neq e_\phi$. Thus the map $\chi \mapsto e_\chi$ is an injective map from $\hat{\Gamma}$ into the set of primitive idempotents of $K\Gamma$. By (I.1.36) the $e_\chi$ are mutually orthogonal, so by (I.1.32) the sum $\sum_\chi K\Gamma e_\chi$ of subspaces of $K\Gamma$ is direct. Hence, using $|\Gamma| = |\hat{\Gamma}|$ (from (A3)),

$$\dim_K(\sum_\chi K\Gamma e_\chi) = |\hat{\Gamma}| = |\Gamma| = \dim_K(K\Gamma).$$

Thus we have shown

$$K\Gamma = \sum_\chi K\Gamma e_\chi \cong \prod_\chi (K\Gamma e_\chi).$$

Therefore, by (I.1.35), the map $\chi \mapsto e_\chi$ surjects onto the set of primitive idempotents.

□

We now derive a number of consequences of this result.

**(A6)** *$K\Gamma$ is separable over $K$, i.e.* $\mathrm{Rad}\,(K\Gamma) = (0)$.

*Proof.* The above shows that $K\Gamma$ is a product of copies of $K$.

□

Since $\sum_\chi e_\chi = 1$, we have that for $\sigma \in \Gamma$

$$\begin{aligned}\sigma = \sum_\chi e_\chi \sigma &= |\Gamma|^{-1} \sum_{\chi,\gamma} \sigma\gamma\chi(\gamma)^{-1} \\ &= |\Gamma|^{-1}(\sum_{\delta\in\Gamma} \delta \sum_\chi \chi(\sigma)\chi(\delta)^{-1}).\end{aligned}$$

Comparing coefficients of the basis elements $\{\delta \mid \delta \in \Gamma\}$, we have shown

**(A7) (1st Orthogonality relation)**

$$\sum_\chi \chi(\sigma)\chi(\delta)^{-1} = \begin{cases} |\Gamma| & \delta = \sigma \\ 0 & \textit{otherwise.} \end{cases}$$

*In particular, when $\delta = 1$:*

(A7.$a$)
$$\sum_\chi \chi(\sigma) = \begin{cases} |\Gamma| & \text{if } \sigma = 1 \\ 0 & \text{otherwise.} \end{cases}$$

It is an immediate consequence of (A7) that

**(A8)** *If $\delta \neq \sigma$, then there exists $\chi \in \hat{\Gamma}$ with $\chi(\delta) \neq \chi(\sigma)$.*

Next observe that the map $t_\gamma : \hat{\Gamma} \to K^*$, given by $t_\gamma(\chi) = \chi(\gamma)$, is a homomorphism. Thus $\gamma \mapsto t_\gamma$ is a homomorphism $\Gamma \to \hat{\hat{\Gamma}}$. By (A8) it is injective; however, by (A3) the two groups have the same order. Therefore

**(A9)** *$\gamma \mapsto t_\gamma$ is an isomorphism $\Gamma \cong \hat{\hat{\Gamma}}$.*

Next observe that

$$e_\chi = \sum_\phi e_\chi e_\phi = |\Gamma|^{-1} \sum_{\phi,\gamma} e_\phi \chi(\gamma)\gamma^{-1}$$
$$= |\Gamma|^{-1} \sum_{\phi,\gamma} e_\phi \phi(\gamma)^{-1}\chi(\gamma).$$

Comparing coefficients of the basis $\{e_\phi \mid \phi \in \hat{\Gamma}\}$, we have

**(A10) (2nd Orthogonality relation)**

$$\sum_\gamma \chi(\gamma)\phi(\gamma)^{-1} = \begin{cases} |\Gamma| & \textit{if } \chi = \phi \\ 0 & \textit{otherwise.} \end{cases}$$

*In particular, letting $\phi = \epsilon$, the identity character, (with $\epsilon(\gamma) = 1$ for all $\gamma \in \Gamma$), we have:*

(A10.$a$)
$$\sum_\gamma \chi(\gamma) = \begin{cases} |\Gamma| & \textit{if } \chi = \epsilon \\ 0 & \textit{otherwise.} \end{cases}$$

Alternatively, observe that we can deduce (A10) from (A7) on replacing $\Gamma$ by $\hat{\Gamma}$, $\hat{\Gamma}$ by $\hat{\hat{\Gamma}}$ and by identifying $\Gamma$ with $\hat{\hat{\Gamma}}$ via (A9).

Now let $f: \Gamma \to \Omega$ denote a homomorphism of finite abelian groups. If $\phi \in \hat{\Omega}$ then $\phi \circ f: \Gamma \to K^*$ is a character of $\Gamma$. Thus we have a map $\hat{f}: \hat{\Omega} \to \hat{\Gamma}$ given by

$$\hat{f}(\phi) = \phi \circ f.$$

This is clearly a homomorphism. If, moreover, we have a further homomorphism $g: \Delta \to \Gamma$, then

$$\widehat{f \circ g} = \hat{g} \circ \hat{f}.$$

Furthermore, if $f$ is an isomorphism, then clearly $\hat{f}$ is also an isomorphism.

Recall that a sequence

$$1 \to A \xrightarrow{\alpha} B \xrightarrow{\beta} C \to 1$$

of abelian groups is said to be *exact* if $\alpha$ is injective, $\beta$ is surjective, and $\operatorname{Im} \alpha = \ker \beta$.

**(A11)** *If the sequence*

$$1 \to \Delta \xrightarrow{g} \Gamma \xrightarrow{f} \Omega \to 1$$

of finite abelian groups is exact, then so is the sequence

$$1 \to \hat{\Omega} \xrightarrow{\hat{f}} \hat{\Gamma} \xrightarrow{\hat{g}} \hat{\Delta} \to 1.$$

*Proof.* Suppose that $\hat{f}(\phi) = \epsilon_\Gamma$ for $\phi \in \hat{\Omega}$, where $\epsilon_\Gamma$ denotes the identity character of $\Gamma$; then $\phi(f(\gamma)) = 1$ for all $\gamma \in \Gamma$. As $f$ is surjective, $\phi(\omega) = 1$ for all $\omega \in \Omega$, i.e. $\phi = \epsilon_\Omega$. Thus indeed $\ker \hat{f} = \{\epsilon_\Omega\}$, and so $\hat{f}$ is injective.

Next assume $\chi = \hat{f}(\phi)$, $\phi \in \hat{\Omega}$, so that $\chi(\gamma) = \phi(f(\gamma))$. Then for $\delta \in \Delta$, $(\hat{g}\chi)(\delta) = \chi(g(\delta)) = \phi(f \circ g(\delta)) = \phi(1_\Omega) = 1$; hence $\hat{g}(\chi) = \epsilon_\Delta$, and so $\operatorname{Im} \hat{f} \subset \ker \hat{g}$. Conversely, if $\chi \in \ker \hat{g}$, then $\chi$ is trivial on $\operatorname{Im} g$; hence $\chi(\gamma)$ depends only on $\gamma \bmod \operatorname{Im} g$, i.e. on $f(\gamma)$. Define $\phi$ on $\Omega$ by $\phi(f(\gamma)) = \chi(\gamma)$. Then indeed $\phi \in \hat{\Omega}$ and $\chi = \hat{f}(\phi)$.

Finally we show that $\hat{g}$ is surjective: this is the deepest part of (A11) and is a consequence of Theorem A above. Without loss of generality we may suppose that $\Delta$ is actually a subgroup of $\Gamma$, and we view $K\Delta$ as contained in $K\Gamma$. Let $\phi \in \hat{\Delta}$, and let $e_{\Delta,\phi}$ denote the corresponding primitive idempotent of $K\Delta$. It is a non-zero idempotent of $K\Gamma$, and hence can be written as a sum

$$e_{\Delta,\phi} = \sum{}' e_\chi$$

of primitive idempotents of $K\Gamma$. Suppose that $e_{\chi_1}$ occurs in this sum. Then, for all $\delta \in \Delta$,

$$\begin{aligned} \chi_1(\delta) e_{\chi_1} &= \delta e_{\chi_1} = \delta e_{\chi_1} e_{\Delta,\phi} = e_{\chi_1} e_{\Delta,\phi} \phi(\delta) \\ &= e_{\chi_1} \phi(\delta) \end{aligned}$$

and so $\phi(\delta) = \chi_1(\delta)$ for all $\delta \in \Delta$; thus $\phi = \hat{g}(\chi_1)$.

□

The group ring $K\Gamma$ can be used as a tool for analysing representations of $\Gamma$, i.e. homomorphisms into the group of non-singular, linear transformations of a $K$-vector space $V$; we denote this latter group by $\mathrm{GL}(V)$. Let $t$ be a homomorphism $t: \Gamma \to \mathrm{GL}(V)$, so that in particular for each $\gamma \in \Gamma$ we have a linear transformation $t(\gamma) : V \to V$. We then define an action of $K\Gamma$ on $V$ by the rule that for $\alpha = \sum a_\gamma \gamma$

(A12.$a$) $$v \cdot \alpha = \sum_\gamma v \cdot t(\gamma) a_\gamma.$$

In this way $V$ becomes a $K\Gamma$-module with $1_\Gamma$ acting as the identity. Conversely, if $V$ is a $K\Gamma$ module with $1_\Gamma$ acting as the identity map, then define $t(\gamma) \in \mathrm{GL}(V)$ to be the linear transformation such that

(A12.$b$) $$v \cdot t(\gamma) = v\gamma \qquad \text{for all } v \in V.$$

Restricting the action of $K\Gamma$ to $K$, $V$ becomes a $K$-vector space and $t: \Gamma \to \mathrm{GL}(V)$ is a group homomorphism.

We conclude this appendix by using Theorem A to establish certain determinantal formulae which have important applications.

**(A13)** *Consider a map $f: \Gamma \to K$; then*

$$\det((f(\delta\gamma^{-1}))_{\delta,\gamma}) = \prod_{\chi \in \hat{\Gamma}} (\sum_{\gamma} f(\gamma)\chi(\gamma)).$$

*Proof.* We view multiplication of elements of the vector space $K\Gamma$ by the element $\sum_{\gamma \in \Gamma} f(\gamma)\gamma$ as a linear transformation of $K\Gamma$, and evaluate the determinant $d$ of this transformation with respect to the two bases $\{e_\chi \mid \chi \in \hat{\Gamma}\}$ and $\{\gamma \mid \gamma \in \Gamma\}$.

Firstly $\sum_\gamma f(\gamma)\gamma e_\chi = (\sum_\gamma f(\gamma)\chi(\gamma))e_\chi$, and so

$$d = \prod_{\chi} (\sum_{\gamma} f(\gamma)\chi(\gamma)).$$

Next

$$\sum_{\gamma} f(\gamma)\gamma\sigma = \sum_{\delta} f(\delta\sigma^{-1})\delta.$$

Hence $d = \det(f(\delta\gamma^{-1})_{\delta,\gamma})$.

□

*Application.* Let $E$ be a finite Galois extension of $F$, $\Gamma = \mathrm{Gal}(E/F)$. Let $a \in E^*$ and put $f(\gamma) = a^\gamma$ for $\gamma \in \Gamma$. Then by (A13)

(A14) $$\det(a^{\delta\gamma^{-1}}) = \prod_{\chi} (\sum_{\gamma} a^\gamma \chi(\gamma)).$$

We shall also need a variant of (A13)

**(A15)**

$$\det((f(\delta\gamma^{-1}) - f(\delta))_{\delta,\gamma}) = \prod_{\substack{\chi \in \hat{\Gamma} \\ \chi \neq \epsilon_\Gamma}} (\sum_{\gamma} f(\gamma)\chi(\gamma))$$

*where on the left $\delta \neq 1$ and $\gamma \neq 1$.*

*Proof.* We now consider the kernel $I_\Gamma \subset K\Gamma$ of the linear map

$$\lambda \mapsto \lambda e_{\epsilon_\Gamma} \qquad \text{i.e. } \sum \lambda_\gamma \gamma \mapsto \sum \lambda_\gamma.$$

This is a $K\Gamma$-ideal, and again we evaluate the determinant $d'$, of the linear transformation given by multiplication by $\sum f(\gamma)\gamma$, in two different ways.

One basis of $I_\Gamma$ is $\{e_\chi \mid \chi \neq \epsilon_\Gamma\}$. As before, with respect to this basis we get

$$d' = \prod_{\chi \neq \epsilon_\Gamma} \left(\sum_\gamma f(\gamma)\chi(\gamma)\right).$$

Another basis is $\{\gamma - 1 \mid \gamma \in \Gamma,\ \gamma \neq 1\}$: indeed, these are linearly independent elements of $I_\Gamma$; moreover, if $\sum_\gamma a_\gamma \gamma \in I_\Gamma$, i.e. $\sum a_\gamma = 0$, then $\sum a_\gamma \gamma = \sum_{\gamma \neq 1} a_\gamma(\gamma - 1)$. To evaluate the determinant $d'$, we observe that for $\gamma \neq 1$ we have

$$\sum_{\sigma \in \Gamma} f(\sigma)\sigma(\gamma - 1) = \sum_\sigma f(\sigma)(\sigma\gamma - 1) - \sum_\sigma f(\sigma)(\sigma - 1).$$

Put $\delta = \sigma\gamma$ in the first sum on the right-hand side; then $\delta$ runs over $\Gamma$ and we get

$$\begin{aligned}\sum_{\sigma \in \Gamma} f(\sigma)\sigma(\gamma - 1) &= \sum_{\delta \in \Gamma} f(\delta\gamma^{-1})(\delta - 1) - \sum_{\delta \in \Gamma} f(\delta)(\delta - 1) \\ &= \sum_{\substack{\delta \in \Gamma \\ \delta \neq 1}} [f(\delta\gamma^{-1}) - f(\delta)](\delta - 1).\end{aligned}$$

This then gives the required alternative equation for $d'$.

□

# Exercises

## Chapter I

1. Let $N$ denote a finite extension field of a field $L$, and let $\sigma_i$ $(i = 1, \ldots, n)$ denote distinct embeddings of $N$, over $L$, into an algebraic closure of $L$. Show that the $\sigma_i$ are linearly independent over $L$, i.e. if $\sum l_i m^{\sigma_i} = 0$ with $l_i \in L$ given, and for all $m \in N$; then the $l_i$ are all zero.
   If now $N/L$ is finite and separable, deduce that $t_{N/K}$ maps onto $L$.
2. Let $N$ denote a finite Galois extension of $F$, with $\Gamma = \mathrm{Gal}(N/F)$. Let $E, L$ be subextensions of $N$ which are Galois over $F$, let $\Delta = \mathrm{Gal}(N/E)$, $\Sigma = \mathrm{Gal}(N/L)$, and let $M$ denote the compositum of $E$ and $L$ in $N$. Show that the map $e \otimes l \to (el, \ldots, e^{\gamma}l, \cdots)$ induces an isomorphism
$$E \otimes_F L \cong \prod_{\gamma} M$$
   where the $\gamma$ run over a set of representatives of $\Gamma/\Delta\Sigma$.
3. Let $N/K$ be a finite Galois extension of degree $n$ with Galois group $\Gamma = \{\gamma_i \mid i = 1, \ldots, n\}$ and suppose $K$ is infinite.
   (*a*) If $f \in N[X_1, \ldots, X_n]$ has the property that $f(a^{\gamma_1}, \ldots, a^{\gamma_n}) = 0$ for all $a \in N$; show that $f = 0$.
   [Hint: For a basis $\{b_i\}$ of $N/K$ set
$$g(Y_1, \ldots, Y_n) = f(\sum Y_i b_i^{\gamma_1}, \ldots, \sum Y_i b_i^{\gamma_n});$$
   deduce that $g = 0$; and then use the invertibility of the matrix $(b_i^{\gamma_j})$ to show $f = 0$.]

(*b*) Write $X_i = X(\gamma_i)$, and set $f(X_1, \ldots, X_n) = \det(X(\gamma_i\gamma_j))$. Show $f(1, 0, \ldots, 0) \neq 0$ and use part (*a*) to show that there exists $c \in N$ such that $\det(c^{\gamma_i\gamma_j}) \neq 0$. Show that $c$ has the property that $\{c^{\gamma_i} \mid i = 1, \ldots, n\}$ is a $K$-basis of $N$. [Hence $N$ is a free $K\Gamma$-module on $c$; such a basis $\{c^\gamma\}$ is called a normal basis of $N/K$.]

4. Prove that a finite field $k$ of $q$ elements has an extension $k_m$ of degree $m$ for every positive integer $m$. Show that for given $m$, $k_m$ is unique to within isomorphism over $k$. Show that $k_m/k$ is a Galois extension with cyclic Galois group generated by the Frobenius automorphism $x \to x^q$.

5. Let $k$ be a field with $q$ elements, and let $l$ denote an extension of $k$ of degree $m$. Show that for $x \in l$
(*a*) $t_{l/k}(x) = \sum_{i=0}^{m-1} x^{q^i}$
(*b*) $N_{l/k}(x) = x^{q^m - 1/q-1}$.
Hence deduce that both $t_{l/k}$ and $N_{l/k}$ map onto $k$.

6. Show that any unique factorisation domain is integrally closed.

7. Let $m$ be an integer, which is not a square, and with the property that $m \equiv 1 \bmod (4)$. Show that $\mathbb{Z}[\sqrt{m}]$ is not a principal ideal domain.
[Hint: A principal ideal domain is integrally closed.]

8. If $\mathfrak{o}$ is a Noetherian ring, show that the formal power series ring $\mathfrak{o}[[x]]$ is also a Noetherian ring.

9. Show that the following algebraic numbers are all algebraic integers
(*a*) $\sqrt[3]{15}(\sqrt[39]{7} + \sqrt[7]{39})$
(*b*) $(1+i)/\sqrt{2}$
(*c*) $\frac{1}{3}(1 + \sqrt[3]{10} + \sqrt[3]{100})$.

10. (Burnside) Let $\zeta^n = 1$ and assume that $\frac{1}{m}(\sum_{i=1}^m \zeta^{k_i})$ is an algebraic integer. Show that either $\sum_{i=1}^m \zeta^{k_i} = 0$ or $\zeta^{k_1} = \zeta^{k_2} = \cdots = \zeta^{k_m}$.

11. Let $\mathfrak{o}$ be an integrally closed integral domain, and let $f$ and $g$ be monic polynomials in $\mathfrak{o}[x]$. Prove that Disc($f$).Disc($g$) divides Disc($f.g$).

## Chapter II

1. Let $K$ be a perfect field and $L$ denote an extension of $K$ of degree 3. Let $d(L/K)$ denote the field discriminant of $L/K$, $d(L/K) =$

$\det(\mathrm{Tr}_{L/K}(x_i x_j))$ for a $K$-basis $\{x_i\}$ of $L$. By considering the action of the embeddings over $K$ on a square root of $d(L/K)$, show that $L/K$ is Galois iff $d(L/K)$ is a square in $K^*$.

2. Determine the ring of integers of $\mathbb{Q}(\sqrt[3]{2})$ and hence calculate $d_{\mathbb{Q}(\sqrt[3]{2})}$.

3. Let $m$ be a negative, square-free integer which has at least two distinct prime factors. Show that $\mathbb{Z}[\sqrt{m}]$ is not a principal ideal domain.

4. Let $K$ be the subfield of $\mathbb{R}$ obtained by adjoining to $\mathbb{Q}$ the positive numbers $\alpha_n$ where $\alpha_n^{2^n} = 3$ $(n = 1, 2, 3, \ldots)$. Show that the ring $\mathfrak{o}$ of algebraic integers in $K$ is integrally closed and that every prime ideal of $\mathfrak{o}$ is maximal. By considering the ideal of $\mathfrak{o}$ generated by all the $\alpha_n$, or otherwise, prove that $\mathfrak{o}$ is not Noetherian.

5. Let $\mathbb{Z}[X]$ denote the ring of polynomials in an indeterminate $X$ over $\mathbb{Z}$. Show that $\mathbb{Z}[X]$ is Noetherian, is integrally closed in its field of fractions, but is not a Dedekind domain.

6. Let $R$ be a subring of the ring of algebraic integers $\mathfrak{o}$ of an algebraic number field $K$. Establish the equivalence of the following conditions:
   (*a*) As a subgroup of the additive group $\mathfrak{o}$, $R$ is of finite index, $[\mathfrak{o} : R] = f$, say
   (*b*) $R$ contains a basis of $K$ over $\mathbb{Q}$.
   (*c*) The field of fractions of $R$ is $K$.
   Now assume these conditions to hold; prove
   (*i*) $R$ is Noetherian
   (*ii*) Every prime ideal of $R$ is maximal
   (*iii*) If $R \neq \mathfrak{o}$, then $R$ is not integrally closed in $K$.
   (*iv*) If $R \neq \mathfrak{o}$, then $R$ has a non-zero ideal which is not invertible. Every ideal of $R$ which is also an ideal of $\mathfrak{o}$ has this property.
   Rings satisfying conditions (*a*)-(*c*) are called *orders* in $K$. If $K$ is a quadratic field show that the orders in $K$ are in bijection with the natural numbers via $f \to R_f$, where $R_f = \mathbb{Z} + f\mathfrak{o} = \{x \in K \mid x = y + fz,\ y \in \mathbb{Z},\ z \in \mathfrak{o}\}$.

7. Show that $\mathbb{Q}_p$ has no continuous automorphisms, apart from the identity map.

8. Show that $\alpha \in \mathbb{Q}_p$ is a unit iff $X^n = \alpha$ is soluble in $\mathbb{Q}_p$ for infinitely many integers $n$. Deduce that any automorphism of $\mathbb{Q}_p$ must take units to units. Hence show that the identity is the only field automorphism of $\mathbb{Q}_p$.

9. Show that $\mathbb{Q}_p$ is uncountable.

10. Describe the group $\mathbb{Q}_p^*/\mathbb{Q}_p^{*2}$ when (*a*) $p \neq 2$ (*b*) $p = 2$.

11. If $p$ and $q$ are distinct primes, then show that $\mathbb{Q}_q$ and $\mathbb{Q}_p$ are not isomorphic.

12. Show that the congruence $X^8 \equiv 16 \bmod (p)$ is soluble for each prime number $p$.

13. Show that $k((1/T))$, the ring of finitely tailed Laurent series in $1/T$ with coefficients in $k$, is the completion of $k(T)$ with respect to a discrete absolute value associated with the valuation $v_\infty$ of $k(T)$.

14. If $(K, u)$ is a valued field with completion $(\overline{K}, \overline{u})$, show that $u$ is an ultrametric iff $\overline{u}$ is an ultrametric.

15. If $\mathfrak{o}$ is a unique factorisation domain and if $\pi$ is an irreducible element of $\mathfrak{o}$, show that the rule $v(x) = n$, where $n$ is the highest power of $\pi$ dividing $x$, induces a valuation on the field of fractions $K$.
    Now let $F$ be a field, let $\mathfrak{o} = F[X, Y]$, for algebraically independent indeterminates $X, Y$ and let $v$ be the valuation associated with $X$, as above. Show that with the notation of §2, $\mathfrak{o}/\mathfrak{p}_v \cong F[Y]$, $\mathfrak{o}_v/\mathfrak{p}_v \cong F(Y)$.

16. Let $\mathbb{Q}_p^c$ be an algebraic closure of $\mathbb{Q}_p$ and let $\mathbb{Z}_p^c$ be the integral closure of $\mathbb{Z}_p$ in $\mathbb{Q}_p^c$. Show that $\mathbb{Z}_p^c$ is integrally closed and has exactly one (non-zero) prime ideal, which is therefore maximal; show $\mathbb{Z}_p^c$ is not Noetherian, by proving that the maximal ideal is not finitely generated.
    Let $a \in \mathbb{Q}_p^{c*}$, so that $a$ belongs to some finite extension, $K$ say, of $\mathbb{Q}_p$. Define $|a| = p^{-v/e}$, where $v = v_{\mathfrak{p}_K}(a)$ for $v_{\mathfrak{p}_K}$ the valuation of $K$, and where $e = e(K/\mathbb{Q}_p)$ is the ramification index. Prove that this definition is independent of the choice of $K$ (within the stated conditions). Show that $|.|$ is both an ultrametric and a non-discrete absolute value on $\mathbb{Q}_p^c$, and that
    $$\mathbb{Z}_p^c = \{x \in \mathbb{Q}_p^c \mid |x| \leq 1\}.$$
    For each non-negative real $\rho$, define
    $$I_\rho = \{x \in \mathbb{Q}_p^c \mid |x| < \rho\}.$$
    Prove that the map $\rho \to I_\rho$ is a bijection from the non-negative reals to the set of non-finitely generated ideals of $\mathbb{Z}_p^c$. For each non-negative rational $r$, define
    $$P_r = \{x \in \mathbb{Q}_p^c \mid |x| \leq r\}.$$
    Prove that $r \to P_r$ is a bijection from the set of non-negative ratio-

nals to the set of finitely generated ideals of $\mathbb{Z}_p^c$, and show that in fact these are all principal.

17. Show that the series

$$\log(1+x) = \sum_{n=1}^{\infty} \frac{(-1)^{n+1}x^n}{n}$$

converges on $p\mathbb{Z}_p$ with respect to $|.|_p$. Show that for any positive integer $n$, $v_p(n!) < n/p - 1$ and hence show that

$$\exp(x) = \sum_{n=0}^{\infty} \frac{x^n}{n!}$$

converges on $p\mathbb{Z}_p$ if $p > 2$ (resp. on $4\mathbb{Z}_2$ if $p = 2$). Hence show that $(1 + p\mathbb{Z}_p)^\times \cong (p\mathbb{Z}_p)^+$ if $p > 2$, and that $(1 + 4\mathbb{Z}_2)^\times \cong (4\mathbb{Z}_2)^+$ if $p = 2$.

18. Let $L$ denote a finite extension of the $p$-adic field $\mathbb{Q}_p$. Let $v$ denote the valuation associated with $L$, let $n$ denote a positive integer and let $t = v(n)$. Show that for $i > t$, raising to the $n$th power induces an isomorphism $U^{(i)} \cong U^{(i+t)}$. Let $V$ denote an open subgroup of the units of $\mathfrak{o}_L$. Show that $\{u^n \mid u \in V\}$ is open. Show also that $N_{L/\mathbb{Q}_p}$ is continuous on $L^*$, and that it is an open map.

19. Let $\mathfrak{o}$ be a ring and let $M$ be a finitely generated $\mathfrak{o}$ module. Show that the following conditions are equivalent:
    (*a*) $M$ is isomorphic to a direct summand of a finitely generated free $\mathfrak{o}$-module.
    (*b*) For every surjective map of $\mathfrak{o}$-modules $\pi\colon P \to M$, there exists an $\mathfrak{o}$-module homomorphism $i\colon M \to P$ such that $\pi \circ i = \mathrm{id}_M$.
    (*c*) Given homomorphisms of $\mathfrak{o}$-modules $M \xrightarrow{f} T$, $S \xrightarrow{g} T$ with $g$ surjective; then there exists an $\mathfrak{o}$-module homomorphism $M \xrightarrow{h} S$ such that $g \circ h = f$.

20. (Harley Flanders). Let $\mathfrak{o}$ be a Dedekind domain with field of fractions $K$ and let $F = K(X)$ be the rational function field over $K$ in an indeterminate $X$. Extend the definition of the content ideal $\mathfrak{a}_f$ to rational functions $f$ by

$$\mathfrak{a}_{g/h} = \mathfrak{a}_g \cdot \mathfrak{a}_h^{-1}$$

if $g, h \in K[X]$, $h \neq 0$, and $\mathfrak{a}_0 = (0)$. Prove that $\mathfrak{a}_{g/h}$ only depends on $g/h$. If $\mathfrak{b}$ is a fractional $\mathfrak{o}$ ideal define

$$\overline{\mathfrak{b}} = \{f \in F \mid \mathfrak{a}_f \subset \mathfrak{b}\}.$$

Prove that $\overline{\mathfrak{o}}$ is a principal ideal domain with field of fractions $F$ and that the map $\mathfrak{b} \to \overline{\mathfrak{b}}$ defines an isomorphism $I_\mathfrak{o} \cong I_{\overline{\mathfrak{o}}}$, with inverse

$\bar{\mathfrak{b}} \to \bar{\mathfrak{b}} \cap K$. Show that $\bar{\mathfrak{p}}$ is a prime ideal of $\bar{\mathfrak{o}}$ iff $\mathfrak{p}$ is a prime ideal of $\mathfrak{o}$.

21. Show $x^4 + 1$ is reducible in $\mathbb{Q}_p$ for all primes $p > 2$.

## Chapter III

1. Find the ring of integers of $\mathbb{Q}(\theta)$ when
   (*a*) $\theta^3 + \theta + 1 = 0$;
   (*b*) $\theta^3 - 2\theta + 2 = 0$;
   (*c*) $\theta^3 + \theta^2 - 2\theta + 8 = 0$.

2. (*a*) Calculate the ring of integers of $\mathbb{Q}(\zeta)$, where $\zeta$ denotes a primitive $p$th root of unity.
   (*b*) By considering ramification, show that $\mathbb{Q}(\sqrt{\left(\frac{-1}{p}\right)p})$ is the unique quadratic subfield of $\mathbb{Q}(\zeta)$.

3. Let $N/K$ denote an extension of number fields. Show that $\mathfrak{o}_K$ is a $\mathfrak{o}_K$-direct summand of $\mathfrak{o}_N$.
   Suppose now that $K$ is a quadratic imaginary number field, and let $N/K$ denote a non-ramified quadratic extension of $K$. If $\{\pm 1\}$ are the only roots of unity in $N$, show that $\mathfrak{o}_N$ is not free over $\mathfrak{o}_K$.

4. Show how ideals generated by $2, 3, 5, 7$ and $11$, each factorise in $\mathbb{Q}(\sqrt[3]{6})$ and $\mathbb{Q}(\sqrt[3]{10})$.

5. Show that no prime number $p$ stays prime in $\mathbb{Q}(\omega, \sqrt[3]{2})$, where $\omega^3 = 1$, $\omega \neq 1$:
   (*a*) by considering the decomposition group;
   (*b*) by factoring $x^3 - 2 \bmod (p)$.

6. Let $L/K$ denote an extension of algebraic number fields. If $\mathfrak{o}_L$ is a free $\mathfrak{o}_K$-module, show that the discriminant $\mathfrak{d}(L/K)$ is $\mathfrak{o}_K$-principal. Conversely, if $K$ has odd class number, show that $\mathfrak{o}_L$ is $\mathfrak{o}_K$-free if $\mathfrak{d}(L/K)$ is $\mathfrak{o}_K$-principal.

7. Let $\mathfrak{o}$ be a Dedekind domain with field of fractions $K$; let $L/K$ be a finite separable extension and let $\mathfrak{o}_L$ denote the integral closure of $\mathfrak{o}$ in $L$. If $\mathfrak{P}$ is a tamely ramified prime ideal of $\mathfrak{o}_L$ in $L/K$, show $t_{L/K}(\mathfrak{P}) = \mathfrak{P} \cap \mathfrak{o}$. Give an example of a wildly ramified prime ideal which has this property.

8. If $N$ and $L$ are linearly disjoint finite Galois extensions of a number field $K$, and if there is a prime ideal $\mathfrak{P}$ of $\mathfrak{o}_{NL}$ which ramifies in both

$NL/L$ and $NL/N$; show that $\mathfrak{o}_N \otimes_{\mathfrak{o}_K} \mathfrak{o}_L$ identifies with a proper subring of $\mathfrak{o}_{NL}$.

9. Let $K$ denote a finite extension of the rational $p$-adic field and let $L/K$ denote a finite field extension. Let $\mathfrak{P}$ resp. $\mathfrak{p}$ denote the unique maximal ideal of $\mathfrak{o}_L$ resp. $\mathfrak{o}_K$; let $U_L^{(i)} = 1 + \mathfrak{P}^i$, $U_K^{(i)} = 1 + \mathfrak{p}^i$ for $i > 0$ and put $U_L^{(0)} = U_L$, $U_K^{(0)} = U_K$.
   (*a*) If $L/K$ is non-ramified, show that, for all $i \geq 0$, $t_{L/K}(\mathfrak{P}^i) = \mathfrak{p}^i$ and $N_{L/K}(U_L^{(i)}) = U_K^{(i)}$.
   (*b*) If $L/K$ is at most tamely ramified, with ramification index $e$, show that $N_{L/K}(U_L^{(ei)}) = U_K^{(i)}$ for all $i \geq 1$.

## Chapter IV

1. Show that $\mathbb{Z}[\sqrt{2}]$, $\mathbb{Z}[\sqrt{3}]$ are both Euclidean domains. (Hint: embed $\mathbb{Z}[\sqrt{2}] \hookrightarrow \mathbb{R}^2$).

2. Let $K$ be an algebraic number field. Show that $\mathfrak{o}_K$ is a principal ideal domain iff for every $\alpha \in K$, with $\alpha \notin \mathfrak{o}_K$, there exists $\beta, \gamma \in \mathfrak{o}_K$ such that $0 < |N_{K/\mathbb{Q}}(\alpha\beta - \gamma)| < 1$.

3. Using the Minkowski bound show that: $\mathbb{Q}[\sqrt{-23}]$ has class number 3; $\mathbb{Q}[\sqrt{-47}]$ has class number 5; $\mathbb{Q}[\sqrt{-14}]$ has class number 4; $\mathbb{Q}[\sqrt{-41}]$ has class number 8.

4. Show that $\mathbb{Q}(\sqrt[3]{2})$ has class number 1.

5. Let $\theta$ denote a root of $X^3 - X - 4$ and let $K = \mathbb{Q}(\theta)$. Show that $\mathfrak{o}_K$ has $\mathbb{Z}$-basis 1, $\theta$, $\frac{\theta+\theta^2}{2}$ and that
$$\mathfrak{o}_K = \mathbb{Z}[\frac{\theta+\theta^2}{2}];$$
hence show that $K$ has class number 1.
[Hint: Show that $(\theta + \theta^2)/2$ is an algebraic integer and consider the square factors of the discriminant of $X^3 - X - 4$.]

6. Let $\theta$ denote a root of $X^3 - X + 2$ and let $K = \mathbb{Q}(\theta)$. Show that $\mathfrak{o}_K = \mathbb{Z}[\theta]$ and that $K$ has class number 1. [Hint: Consider the square factors of the discriminant of $X^3 - X + 2$ and show that $\frac{1}{2}(a + b\theta + c\theta^2)$ is an algebraic integer iff $a$, $b$ and $c$ are all even.]

7. Suppose $\theta^3 - 7\theta^2 + 14\theta - 7 = 0$, and put $K = \mathbb{Q}(\theta)$. Show that $K/\mathbb{Q}$ is Galois, that $\mathfrak{o}_K = \mathbb{Z}[\theta]$ and that $K$ has class number 1. Show that $e_7(K/\mathbb{Q}) = 3$, and that every element of $\mathfrak{o}_K$, which is coprime to 7, has norm congruent to $\pm 1$ mod (7). Deduce that a prime number $p$ different from 7 stays prime in $K$ unless $p \equiv \pm 1$ mod (7).

8. (Hermite). Let $K$ denote a number field of degree $n$ over $\mathbb{Q}$. Given $1 \leq k \leq s+t$ show that
   (*a*) If $k \leq s$, then there exists non-zero $x \in \mathfrak{o}_K$ with
   $$|x^{\sigma_j}| \leq \frac{1}{2} \qquad 1 \leq j \leq s+t,\ j \neq k$$
   $$|x^{\sigma_k}| \leq 2^{n-t}|d_K|.$$
   (*b*) If $k > s$, then there exists non-zero $x \in \mathfrak{o}_K$ with
   $$|x^{\sigma_j}| \leq \frac{1}{2} \qquad 1 \leq j \leq s+t,\ j \neq k$$
   $$|\mathrm{Re}\, x^{\sigma_k}| \leq \frac{1}{2} \qquad |\mathrm{Im}\, x^{\sigma_k}| \leq 2^{n-t}|d_K|.$$
   Show that $K = \mathbb{Q}(x)$, and hence deduce that for any given real number $N$ there are only finitely many algebraic number fields $K$ of degree $n$, whose discriminant $d_K$ satisfies $|d_K| \leq N$. Now use the proof of Theorem 36 to show that there are only finitely many algebraic number fields whose discriminant has absolute value less than or equal to $N$.

9. If $S$ is a closed, bounded, convex, symmetric set in $\mathbb{R}^n$ with $\mathrm{vol}(S) \geq m2^n$, then show that $S$ contains at least $m$ pairs of points in $\mathbb{Z}^n$ (other than the origin).

10. Let $B$ denote a closed, bounded, convex region in $\mathbb{R}^n$.
    (*a*) If $B$ has no interior points, show that $B$ is Jordan measurable with content zero.
    (*b*) Suppose $B$ has an interior point $\mathbf{0}$. Show that the projection map from a large sphere $S$ centre $\mathbf{0}$ into the boundary $\partial B$ defines a homeomorphism between $S$ and $\partial B$. Use this map to show that $\partial B$ has content zero, so that $B$ is Jordan measurable.

11. Let $r > 1$ and let $q$ denote a prime number. Show that
    $$\frac{T^{q^r} - 1}{T^{q^{r-1}} - 1} = (T^{q^{r-1}} - 1)^{q-1} + \sum_{n=1}^{q-1} \binom{q}{n} (T^{q^{r-1}} - 1)^{n-1}.$$
    By putting $T = (a^{q^r} - 1)/(a^{q^{r-1}} - 1)$ and considering divisors of this number, show that for a given integer $a > 1$, there exists a prime $p$ such that $a$ has order $q^r$ mod $(p)$.

12. If $p = 4n - 1 > 7$, show that $K = \mathbb{Q}(\sqrt{-p})$ has class number one iff $m^2 + m + n$ is prime for all $m\colon 0 \leq m \leq n-2$.
    [Since $\mathbb{Q}(\sqrt{-163})$ has class number one, we obtain the remarkable fact, observed by Euler, that $X^2 + X + 41$ takes on prime values for $X = 1, 2, \ldots, 39$.]

## Chapter V

1. Find the ring of integers, the class number and a fundamental unit of $\mathbb{Q}(\theta)$ when (*a*) $\theta^3+\theta+1=0$ (*b*) $\theta^3-2\theta+2=0$ (*c*) $\theta^3+\theta^2-2\theta+8=0$.

2. Let $a$ and $b$ be positive integers which are not squares. Show that every unit of $\mathbb{Z}[\sqrt{a}, \sqrt{-b}]$ is a unit of $\mathbb{Z}[\sqrt{a}]$.

3. Show that $2-\sqrt[3]{7}$ is a fundamental unit in $\mathbb{Q}(\sqrt[3]{7})$.

4. By first observing that $\sqrt[3]{5}-2$ has norm $-3$ or otherwise, show that $41+24\sqrt[3]{5}+14\sqrt[3]{25}$ is a fundamental unit of $\mathbb{Q}(\sqrt[3]{5})$.

5. By first observing that $2-\sqrt[3]{14}$ has norm $-6$ or otherwise, show that $29+14\sqrt[3]{14}+5\sqrt[3]{196}$ is a fundamental unit of $\mathbb{Q}(\sqrt[3]{14})$.

6. In each of the following cases show that the algebraic number given is a fundamental unit of the given biquadratic field:
   (*a*) $\frac{(1+\sqrt{-1})(1+\sqrt{3})}{2}$ in $\mathbb{Q}(\sqrt{-1}, \sqrt{3})$,
   (*b*) $\frac{10\sqrt{-3}-4\sqrt{-19}}{2}$ in $\mathbb{Q}(\sqrt{-3}, \sqrt{-19})$,
   (*c*) $\frac{\sqrt{-11}+\sqrt{-7}}{2}$ in $\mathbb{Q}(\sqrt{-7}, \sqrt{-11})$.

7. Determine the fundamental unit of:
   (*a*) $\mathbb{Q}(\sqrt{-1}, \sqrt{-7})$ (*b*) $\mathbb{Q}(\sqrt{-1}, \sqrt{-19})$ (*c*) $\mathbb{Q}(\sqrt{-3}, \sqrt{-23})$.

8. Given any $c \in C_K^+$ and an ideal $\mathfrak{f}$ of $\mathfrak{o}_K$, show that there is an ideal $\mathfrak{a}$ of $\mathfrak{o}_K$ with class $c$ which is coprime to $\mathfrak{f}$.

## Chapter VI

1. Let $l, p$ be odd primes, with $l \equiv 1 \bmod (3)$. By considering the factorisation of $p$ in the cubic subfield $L$ of $\mathbb{Q}[l]$, show that $l$ splits completely in $\mathbb{Q}(\sqrt[3]{p})$ iff $p$ splits completely in $L$.

2. Let $p$ be a prime number and let $\zeta$ denote a primitive $p^n$th root of unity for $n \geq 1$; let $N_n$ and $t_n$ denote the norm and trace from $\mathbb{Q}(\zeta)$ to $\mathbb{Q}(\zeta^p)$. Show that
   (*a*) $t_n(\mathbb{Z}[\zeta]) = \begin{cases} p\mathbb{Z}[\zeta^p] & \text{if } n > 1 \\ \mathbb{Z} & \text{if } n = 1 \end{cases}$
   (*b*) for $x, y \in \mathbb{Z}[\zeta]$, and for $n > 1$,
   $$N_n(x+y) \equiv N_n(x) + N_n(y) \bmod p\mathbb{Z}[\zeta^p].$$
   Deduce that $N_n(x) \equiv x^p \bmod p\mathbb{Z}[\zeta]$.

3. (Liang) Let $\zeta$ denote a root of unity and let $K = \mathbb{Q}(\zeta+\zeta^{-1})$. Show that $\mathfrak{o}_K = \mathbb{Z}[\zeta+\zeta^{-1}]$.

4. Let $p$ be a fixed prime. By considering the behaviour of prime

divisors of numbers of the form $N^{p-1} + N^{p-2} + \ldots + N + 1$ in $\mathbb{Q}[p]$, show that there are an infinite number of primes $q$ such that $q \equiv 1 \bmod (p)$.

5. Let $P = (x, y)$, $P' = (x', y')$ denote two finite points on an elliptic curve $Y^2 = X^3 + aX^2 + bX + c$, with $a, b, c \in K$. If $x \neq x'$, show that $P + P'$ has $x$ and $y$-coordinates

$$x(P+P') = \left(\frac{y-y'}{x-x'}\right)^2 - a_2 - x - x'$$

$$y(P+P') = -\left(\frac{y-y'}{x-x'}\right) x(P+P') - \left(\frac{yx'-y'x}{x'-x}\right).$$

6. For an elliptic curve $E$, as in (5), show that the group of points of order less than or equal to two in $E(\mathbb{C})$, is isomorphic to $C_2 \times C_2$.

7. Let $N/K$ denote an extension of number fields, with abelian Galois group $\Gamma = \mathrm{Gal}(N/K)$. Suppose that there exists $a \in \mathfrak{o}_N$ such that $(a^\gamma)_{\gamma \in \Gamma}$ is an $\mathfrak{o}_K$-basis of $\mathfrak{o}_N$. Use the Frobenius determinant formula (A14) to show that

$$\delta(N/K) = \prod_{\chi \in \hat{\Gamma}} \left(\sum a^\gamma \chi(\gamma^{-1})\right)^2 \mathfrak{o}_K.$$

Use this decomposition to determine the discriminant of $\mathbb{Q}[p]$.

8. For a primitive quadratic residue class character $\lambda$, define the normalised Gauss sum by

$$\tau(\lambda) = \sum \lambda(x) \exp\left(\frac{2\pi i x}{f}\right)$$

$f = f(\lambda)$ its conductor and the sum extending over a complete system of prime residues $x \bmod (f)$. Assuming Theorem 50, prove that

$$\tau(\lambda) = \begin{cases} +\sqrt{f} & \text{if } \lambda(-1) = 1 \\ i\sqrt{f} & \text{if } \lambda(-1) = -1 \end{cases}$$

where $+\sqrt{f}$ is the positive square root. (Hint: Proceed by induction on the number of distinct primes dividing $f$, using the expression for $\tau(\lambda_1\lambda_2)$ in terms of $\tau(\lambda_1)$, $\tau(\lambda_2)$, where $\lambda_1, \lambda_2$ have coprime conductors.)

9. Let $l$ denote an odd prime and let $\mathfrak{p}$ denote a prime ideal in $\mathbb{Q}[l]$ not dividing $l$. Given $a \in \mathbb{Z}[l]$, the ring of integers in $\mathbb{Q}[l]$, with $a \not\equiv 0 \bmod \mathfrak{p}$, prove that there exists an $l$th root of unity $\left(\frac{a}{\mathfrak{p}}\right)_l \in \mu_l$,

so that

$$a^{(N\mathfrak{p}-1)/l} \equiv \left(\frac{a}{\mathfrak{p}}\right)_l \bmod \mathfrak{p}.$$

Show that this determines $\left(\frac{a}{\mathfrak{p}}\right)_l$ uniquely. Show also that

(*a*) $\left(\frac{a}{\mathfrak{p}}\right)_l \left(\frac{b}{\mathfrak{p}}\right)_l = \left(\frac{ab}{\mathfrak{p}}\right)_l$ for $\mathfrak{p} \nmid a, b \in \mathbb{Z}[l]$.

(*b*) $\left(\frac{a}{\mathfrak{p}}\right)_l = 1 \iff$ there exists $b$ so that $a \equiv b^l \bmod \mathfrak{p}$.

(*c*) If $a \in \mu_l$, then $\left(\frac{a}{\mathfrak{p}}\right)_l = a^{(N\mathfrak{p}-1)/l}$.

Now suppose that $p$ is a prime number, $p \equiv 1 \bmod (l)$. Establish a bijection from the set of prime ideal divisors of $p$ in $\mathbb{Q}[l]$ onto the set of primitive $l$th power residue class characters $\chi$ of $\mathbb{F}_p^*$ so that for all $a \in \mathbb{Z}$, $l \nmid a$, we have

$$\chi(a) = \left(\frac{a}{\mathfrak{p}}\right)_l$$

if $\mathfrak{p} \leftrightarrow \chi$.

10. Let $m$ be an integer, $m > 2$, and let $p$ be a prime number with $p^r \equiv 1 \bmod (m)$ for some odd integer $r$. Let $\mathfrak{p}$ be a prime ideal dividing $p$, in the maximal real subfield $K$ of $\mathbb{Q}[p^n m]$ for given $n \geq 1$. Prove that $K_\mathfrak{p} \cong \mathbb{Q}_p[p^n m]$.

11. Let $p_1, p_2$ be distinct odd prime numbers, and let $d \mid (p_1 - 1, p_2 - 1)$. Show that $\mathbb{Q}[p_1 p_2]$ has a subfield $L$ so that
    (*a*) $(\mathbb{Q}[p_1 p_2] : L) = d$.
    (*b*) $\mathbb{Q}[p_1 p_2]/L$ is non-ramified, i.e. for all prime ideals $\mathfrak{p}$ of $\mathfrak{o}_L$, $e_\mathfrak{p}(\mathbb{Q}[p_1 p_2]/L) = 1$.

12. Suppose $m \geq 2$ and put $E = Q[4]$. Let $K = \mathbb{Q}[2^m]$, let $L = \mathbb{Q}[2^m]_+$ denote the maximal real subfield of $K$ and let $\rho \in \mathrm{Gal}(K/\mathbb{Q})$ denote complex conjugation. Show that if $u \in U_K$ does not lie in $\mu_K U_L$, then $u^{1-\rho} = \zeta$ say, is a primitive $2^m$th root of unity. Prove that $U_K = \mu_K U_L$ by showing that $N_{K/E}\zeta$ is a primitive fourth root of unity, while $N_{K/E}(u)^{1-\rho}$ must be $\pm 1$.

## Chapter VII

1. Let $k$ be a finite field, and let $d \in k$, $c \in k$. Show that there exist $x, y$ in $k$ such that $x^2 + dy^2 = c$.

2. What integers can be expressed in the form (*a*) $X^2 - 5Y^2$, (*b*) $X^2 - 6Y^2$, for integral $X, Y$?

3. Find all integral solutions to $3X^2 - 4Y^2 = 11$.

4. Solve the following equations for integral $X, Y$
   (*a*) $X^3 = Y^2 + 2$
   (*b*) $X^5 = Y^2 + 19$
   (*c*) $X^3 = Y^2 + 54$
   (*d*) $X^3 = Y^2 + 200$.

5. Solve $4X^3 = Y^2 + p$ for integral $X, Y$ when $p \equiv 3 \bmod (4)$, $p \neq 3$, and when 3 does not divide the class number of $\mathbb{Q}(\sqrt{-p})$.

6. What integers can be expressed in the form (*a*) $x^2+5y^2$; (*b*) $x^2+15y^2$; (*c*) $x^2 - 23y^2$; (*d*) $x^2 + 23y^2$ for integral $x, y$?

7. Let $m$ be an even, square-free positive integer, and suppose that $\mathbb{Q}(\sqrt{-m})$ has class number which is not divisible by 3. Show that $X^3 = Y^2 + m$ has at most one solution in natural numbers.

8. Show that $\mathbb{Z}[\sqrt[3]{6}]$ has class number 1; hence show that $X^3 + 6Y^3 = 10Z^3$ has no non-trivial integer solutions.

## Chapter VIII

1. Using the results of Ex.6 of Ch.V show that
$$\mathbb{Q}(\sqrt{-1}, \sqrt{3})$$
$$\mathbb{Q}(\sqrt{-3}, \sqrt{-19})$$
$$\mathbb{Q}(\sqrt{-7}, \sqrt{-11})$$
all have class number 1.

2. Show that $\mathbb{Q}(\sqrt{5}, \sqrt{13})$ has class number 2, and that $\mathbb{Q}(\sqrt{-3}, \sqrt{-23})$ has class number 3.

3. Let $\chi$ be the primitive residue class character with conductor $f(\chi) = 4$. Prove that $L(x, \chi) > 0$ for all $x > 0$.

4. Let $K$ be a cyclotomic field of odd prime degree $l$ over $\mathbb{Q}$. Let $f = \prod p_i$ denote the product of primes $p_i$ which ramify in $K$, with all $p_i \equiv 1 \bmod (l)$. For the units $\omega_{f,a}$ defined in (5.6), write $\omega_{K,a} = N_{\mathbb{Q}[f]+/K}(\omega_{f,a})$, and let $\Omega_K$ denote the subgroup of $K^*$ generated by the $\omega_{K,a}$. Prove that $\Omega_K$ is a subgroup of finite index in the group $U_K$ of units of $K$ and that
$$[U_K : \Omega_K] = h_K.$$

5. For a non-zero real number $x$ define $s(x)$ by $x/|x| = (-1)^{s(x)}$, viewing $s(x)$ as an element of the field $\mathbb{F}_2$ of two elements. Call a sub-

group $V$ of the group of units $U_K$ of a totally real algebraic number field $K$ *full* if $[V : V^2] = 2^{(K:\mathbb{Q})}$. Suppose $V$ is full. For a set of elements $\{v_j\}$ of $V$ whose classes mod $V^2$ form a basis of $V/V^2$ over $\mathbb{F}_2$, consider the matrix $A_V = (s(v_j^{\sigma_k}))$, where $\sigma_k$ ranges over the embeddings $K \to \mathbb{R}$. Prove that the rank $r(A_V)$ of $A_V$ depends only on $V$, and show that, with $h_K^+$ as defined in (V,§1),

(*a*) $$2^{r(A_V)}h_K^+ \mid h_K 2^{(K:\mathbb{Q})}$$

with equality for $V = U_K$.

Now let $K$ be a cyclotomic field of odd prime degree $l$, and let $V$ be the group generated by $\Omega_K$ and $-1$. Prove that if $r(A_V) = (K : \mathbb{Q})$, then $h_K^+$ is odd. (Hasse) State and prove analogues for $K = \mathbb{Q}[p]_+$, $p$ an odd prime.

6. A prime ideal $\mathfrak{p}$ of $\mathfrak{o}_K$ is said to have degree 1 if $\mathbf{N}\mathfrak{p}$ is a prime number. By considering the behaviour of $\zeta_K(x)$ as $x \to 1+$, show that $\mathfrak{o}_K$ always has an infinite number of degree 1 prime ideals.

7. Suppose that $N/\mathbb{Q}$ is Galois with Galois group $A_4$, the alternating group on 4 elements. Express $\zeta_N(x)$ in terms of Dedekind zeta functions of proper subfields of $N$.

8. (Brauer) With the notation of Theorem 73, let $W_{N_a}^{(p)}$ denote the exact power of $p$ dividing $W_{N_a}$ (the number of roots of unity in $N_a$). Show that if $p > 2$ then

$$\prod_a W_{N_a}^{(p)} = \prod_b W_{N_b}^{(p)}$$

but that the above equality fails to hold in general when $p = 2$. [Hint: Let $W_N^{(p)} = p^m$. Note that for $p > 2$, $\mathrm{Gal}\,(\mathbb{Q}[p^n]/\mathbb{Q}) = \Gamma_n$ is always cyclic. For $n \leq m$, let $\chi_n$ denote a faithful abelian character of $\Gamma_n$, which can then also be viewed as a character of $\Gamma$, via the surjection $\Gamma \to \Gamma_n$. Show that $W_{N_a}^{(p)} = p^t$ where

$$t = (\rho_a, \sum_{n=1}^{m} \chi_n)$$

where $\rho_a$ denotes the character of $\mathbb{C}(\Lambda_a\backslash\Gamma)$ and where $(,)$ denotes the standard inner product of character theory.]

## Appendix A

(The notation here is that of Appendix A.)

1. Let $\Gamma$ denote a finite abelian group, let $M = (\chi(\gamma))_{\gamma,\chi}$ for $\gamma \in \Gamma$,

$\chi \in \hat{\Gamma}$, and let $M^* = (\chi^{-1}(\gamma))_{\chi,\gamma}$. Show that

$$MM^* = \operatorname{diag}\begin{pmatrix} |\Gamma| & & \\ & \ddots & \\ & & |\Gamma| \end{pmatrix};$$

hence deduce that $\det(M)^2 = (-1)^N |\Gamma|^{|\Gamma|}$ where $2N$ denotes the number of elements in $\Gamma$ with order greater than 2.

2. Let $V$ be a finitely generated $K\Gamma$-module. Show that $V$ is the direct sum of the $K$ spaces $Ve_\chi$ ($\chi \in \hat{\Gamma}$); furthermore, given another such module $V'$, then $V$ and $V'$ are isomorphic $K\Gamma$-modules iff $\dim_K(Ve_\chi) = \dim_K(V'e_\chi)$ for all $\chi \in \hat{\Gamma}$.

3. Let $E/F$ be a Galois extension of algebraic number fields with abelian Galois group $\Gamma$, viewed as contained in the algebraically closed field $K$. Suppose that $F$ contains all the values $\chi(\gamma)$ ($\chi \in \hat{\Gamma}$, $\gamma \in \Gamma$). Using Ex 3 of Chapter I, prove that $E$ has an $F$-basis $\{a_\chi\}$ ($\chi \in \hat{\Gamma}$) such that

$$a_\chi^\gamma = a_\chi \chi(\gamma) \qquad \text{for all } \gamma \in \Gamma.$$

Show also that $E = F(b_1, \dots, b_r)$ where for all $i$, $b_i^{n_i} \in F^*$ for some $n_i > 0$.

4. (Alternative approach: assume only the definition of $K\Gamma$ and $\hat{\Gamma}$.)
   (*a*) Show that

$$t_{K\Gamma/K}(\gamma) = \begin{cases} |\Gamma| & \text{if } \gamma = 1_\Gamma \\ 0 & \text{otherwise.} \end{cases}$$

   (*b*) By evaluating $\det(t_{K\Gamma/K}(\gamma\delta))$, show that $K\Gamma$ is a commutative separable $K$-algebra, i.e. $K\Gamma = \sum A_i$ (direct sum), where each $A_i$ is a simple ideal; thus, as a $K$-algebra, each $A_i \cong K$, since $K$ is algebraically closed.
   (*c*) For each $i$, $A_i = K\Gamma e_i$ with $e_i$ a primitive idempotent; and show also that $e_i\gamma = e_i\chi(\gamma)$ for some $\chi \in \hat{\Gamma}$. Using this equation and writing $e_i = \sum a_\gamma \gamma$, show that $e_i = e_\chi$, where $e_\chi$ is as defined in (A4).
   (*d*) Deduce that $|\hat{\Gamma}| = |\Gamma|$.
   (*e*) If $\chi \in \hat{\Gamma}$, $\chi \neq \epsilon_\Gamma$; then, by evaluating $\chi(\sigma)\sum_\gamma \chi(\gamma)$, prove that $\sum \chi(\gamma) = 0$.
   (*f*) Deduce (A10).
   (*g*) Use (A10) to prove that the matrix $(\chi(\gamma))_{\chi,\gamma}$ ($\chi \in \hat{\Gamma}$, $\gamma \in \Gamma$) is invertible.
   (*h*) Deduce (A7) and (A7.*a*).

# Suggested Further Reading

*For alternative approaches to some of the material in this book, the reader may find it helpful to consult (GTM are the Graduate Texts in Mathematics):*

[A1] E. Artin, *The Theory of Algebraic Numbers*, notes by G. Wurges, Göttingen, 1956. A short series of elementary lectures in the subject.

[A2] E. Artin, *Algebraic Numbers and Algebraic Functions*, notes by I. Adamson, Princeton–New York University, 1950. This provided the first modern account of a wide range of topics in algebraic number theory.

[H] E. Hecke, *Lectures on the Theory of Algebraic Numbers*, Springer-Verlag GTM77, New York, 1981. A classical account of to algebraic number theory which has played a major role in introducing people to the subject.

*In the course of the book a number of topics have been encountered which point the way to further developments. For a full treatment of such themes we suggest:*

[B] A. Baker, *Transcendental Number Theory*, Cambridge University Press, 1975. In addition to much other material, it describes powerful results in Diophantine theory and in the theory of class numbers of imaginary quadratic number fields.

[C1] J.W.S. Cassels, *Rational Quadratic Forms*, Academic Press, 1979. A full and self-contained account of the theory of rational quadratic forms.

[C2] J.W.S. Cassels, *Local Fields*, LMS Student Texts 3, Cambridge University Press, 1986. A very thorough introduction to the theory of local fields; it includes a wide variety of interesting applications to the theory of Diophantine equations.

[CF] J.W.S. Cassels and A. Fröhlich (editors), *Algebraic Number Theory*, Academic Press, 1967. Provides a brisk introduction to the class field theory of both local and global fields. The final chapter (Tate's thesis) contains an alternative approach to the theory of $L$-functions.

[D1] H. Davenport, *The Higher Arithmetic*, Hutchinson, 1968 and 5th edition Cambridge University Press, 1982. Gives an account of continued fractions and their use for the determination of fundamental units of real quadratic fields; it also provides an alternative approach to the theory of binary quadratic forms.

[D2] H. Davenport, *Multiplicative Number Theory*, revised by H. Montgomery, Springer-Verlag GTM74, 1980. The Riemann zeta function and Dirichlet $L$-functions are used to obtain many results on the distribution of prime numbers.

[I] K. Iwasawa, *Local Class Field Theory*, Oxford University Press, 1986. Gives an elegant account of the theory of abelian extensions of local fields which is based on the systematic use of formal groups. A very readable book.

[IR] K. Ireland and M. Rosen, *A Classical Introduction to Modern Number Theory*, Springer-Verlag GTM84, 1982. A modern approach to the subject; a particularly nice introduction to the arithmetic of elliptic curves. The choices of topics nicely complement those of this book.

[L] S. Lang, *Algebraic Number Theory*, Addison-Wesley, 1970. Reprinted as GTM110, Springer-Verlag, 1986. A very accessible account of class field theory; it also contains some more advanced results in the theory of $L$-functions.

[M] L. Mordell, *Diophantine Equations*, Academic Press, 1969. Contains a wide spread of techniques and examples in the theory of Diophantine equations.

[R] P. Ribenboim, *13 Lectures on Fermat's Last Theorem*, Springer-Verlag, 1979. By providing the history of attempts at solving this problem, the book gives an excellent insight into the wealth of methods available in number theory.

[Se] J-P. Serre, *Corps Locaux*, Hermann, 1968. (English translation: *Local Fields*, Springer-Verlag, GTM67, 1979) Contains a very thorough and elegant treatment of the theory of local fields, and gives a good account of local class field theory by cohomological methods.

[Si] J. Silverman, *The Arithmetic of Elliptic Curves*, Springer-Verlag GTM106, 1985. Starts with an elegant introduction to the analytic, the geometric and the local theory of elliptic curves, and concludes with a study of the group of points of elliptic curves defined over a number field.

[Wa] L. Washington, *Introduction to Cyclotomic Fields*, Springer-Verlag GTM83, New York, 1980. Contains not only a thorough account of classical results in the theory, but also an account of the important $p$-adic methods of Iwasawa theory.

[We] A. Weil, *Basic Number Theory*, Springer-Verlag Grundlehren 144, 1973. Begins with the theory of local fields based on the use of the Haar measure; it then goes on to treat the class field theory, of both local and global fields, from the point of view of division algebras.

# Glossary of Theorems

| | |
|---|---|
| I §1 | Theorem 1 |
| II §1 | Theorems 2, 3, 4, 5, 6 |
| II §2 | Theorems 7, 8, 9 |
| II §3 | Theorems 10, 11, 12 |
| II §4 | Theorems 13, 14, 15 |
| III §1 | Theorems 16, 17, 18, 19, 20, 21 |
| III §2 | Theorems 22, 23 |
| III §3 | Theorems 24, 25, 26 |
| III §4 | Theorems 27, 28, 29, 30 |
| IV §1 | Theorem 31 |
| IV §2 | Theorems 32, 33 |
| IV §3 | Theorems 34, 35, 36 |
| IV §4 | Theorems 37, 38 |
| V §1 | Theorems 39, 40, 41 |
| V §2 | Theorem 42 |
| V §3 | Theorem 43 |
| VI §1 | Theorems 44, 45, 46 |
| VI §2 | Theorem 47 |
| VI §3 | Theorems 48, 49, 50, 51 |
| VI §4 | Theorems 52, 53, 54 |
| VI §5 | Theorem 55 |
| VII §1 | Theorem 56 |
| VII §2 | Theorems 57, 58, 59 |
| VII §3 | Theorem 60 |
| VIII §2 | Theorems 61, 62, 63 |
| VIII §3 | Theorems 64, 65 |
| VIII §4 | Theorem 66 |
| VIII §5 | Theorems 67, 68, 69 |
| VIII §6 | Theorems 70, 71 |
| VIII §7 | Theorems 72, 73, 74, 75, 76 |
| Appendix A | Theorem A |

# Index